# Probability, Statistics, and Data

# Probability, Statistics, and Data

## A Fresh Approach Using R

Darrin Speegle
Bryan Clair

CRC Press is an imprint of the
Taylor & Francis Group, an **informa** business
A CHAPMAN & HALL BOOK

First edition published 2022
by CRC Press
6000 Broken Sound Parkway NW, Suite 300, Boca Raton, FL 33487-2742

and by CRC Press
2 Park Square, Milton Park, Abingdon, Oxon, OX14 4RN



CRC Press is an imprint of Taylor & Francis Group, LLC



ISBN: 978-0-367-43667-4 (hbk)
ISBN: 978-1-032-15441-1 (pbk)
ISBN: 978-1-003-00489-9 (ebk)

DOI: 10.1201/9781003004899

Publisher's note: This book has been prepared from camera-ready copy provided by the authors.

# *Contents*

# *Preface*

This book represents a fundamental rethinking of a calculus based first course in probability and statistics. We offer a **breadth first** approach, where the essentials of probability and statistics can be taught in one semester. The statistical programming language R plays a central role throughout the text through simulations, data wrangling, visualizations, and statistical procedures. Data sets from a variety of sources, including many from recent, open source scientific articles, are used in examples and exercises. Demonstrations of important facts are given through simulations, with some formal mathematical proofs as well.

This book is an excellent choice for students studying data science, statistics, engineering, computer science, mathematics, science, business, or for any student wanting a practical course grounded in simulations.

The book assumes a mathematical background of one semester of calculus along with some infinite series in Chapter 3. Integrals and infinite series are used for notation and exposition in Chapters 3 and 4, but in other chapters the use of calculus is minimal. Since an emphasis is placed on understanding results (and robustness to departures from assumptions) via simulation, most if not all parts of the book can be understood without calculus. Proofs of many results are provided, and justifications via simulations for many more, but this text is not intended to support a proof based course. Readers are encouraged to follow the proofs, but often one wants to understand a proof only after first understanding the result and why it is important.

Our philosophy in this book is to not shy away from messy data sets. The book contains extensive sections and many exercises that require data cleaning and manipulation. This is an essential part of the text.

A one-semester course using this book could reasonably cover most material in Chapters 1-8 in order and then select two or three additional chapters. Sections 2.4, 3.6, 5.6, 8.7 and 8.8 may be omitted or given light coverage. The descriptive statistics in Chapters 6 and 7 are frequently the first part of a statistics course, but we recommend leaving them in the middle as they provide students with a welcome change of pace during the semester. Chapter 9 (Rank Based Tests) is particularly important because it uses simulation techniques developed throughout the text to help students understand power and the effects of assumptions on testing.

There is enough material for a more leisurely and thorough two-semester sequence that would delve deeper into probability theory, spend more time on data wrangling, and cover all of the inference chapters.

Most chapters in the book contain at least one **vignette**. These short sections are not part of the development of the base material. We imagine these vignettes as starting points for further study for some students, or as interesting additions to the main material. Examples include chloropleth maps, data and gender, Stein's paradox, and a treatment of Covid-19 data.

Base R and tidyverse tools are interspersed, depending on which is better for a particular

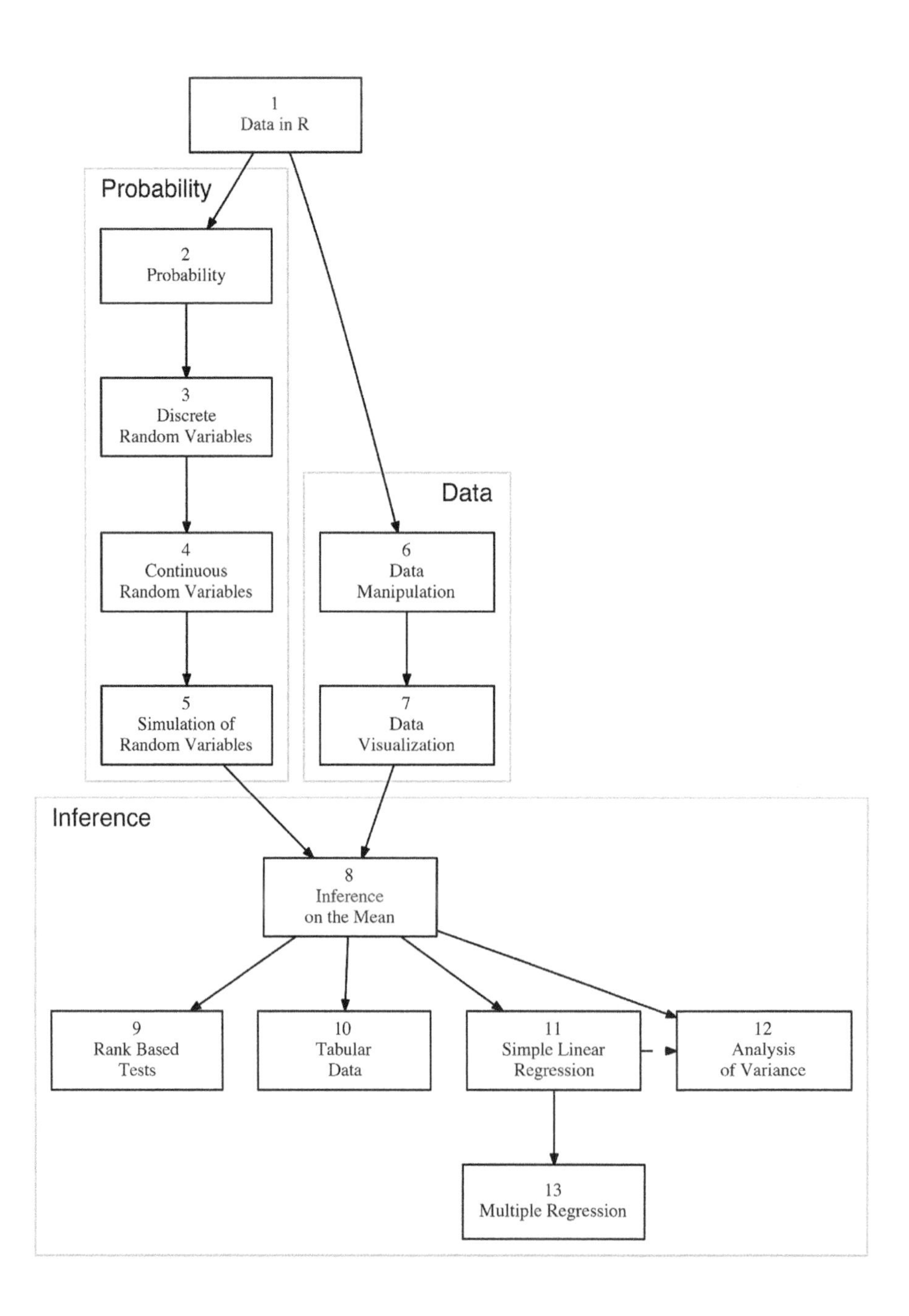


**FIGURE 1** Chapter dependencies.

job, though we don't introduce any tidyverse tools until Chapter 6. We feel that `replicate` and the base R plotting tools are appropriate for doing simulations and creating the types of visualizations that we need in the first chapters. The `dplyr` package is used for most data wrangling beginning in Chapter 6, and `ggplot2` is used for most visualizations beginning in Chapter 7. Other tidyverse tools introduced in this text, in order of emphasis, are `stringr` for string manipulation, `tidyr` for `pivot_longer` and `pivot_wider`, `lubridate` and `janitor::clean_names`.

All data sets for this book are found in freely available R packages. The bulk of the data sets are in the package associated to this book, `fosdata`, and are mostly sourced from open access publications that are linked to in the help pages of the data. We encourage readers to spend time reading the publications that were written using the data in the book. We have taken two approaches to the data from original papers. In some instances, `fosdata` provides essentially all of the data from the published paper. This allows you to explore the data further and think about other visualizations and analyses that would be useful. It also typically requires some wrangling to get the data in a format for the analysis. In other instances, we have simplified the data from the paper quite a bit. In particular, in a few instances we have modified the data by filtering out observations or averaging in order to make it reasonable to assume independence. Please see the links provided in the help pages of `fosdata` for details.

No book like this would be complete without resources for the student who wishes to learn more. Here are some suggestions for further study that the authors have enjoyed:

- *ggplot2*, by Hadley Wickham, gives a nice overview of the capabilities of the `ggplot2` package. Students interested in data visualization would find this book interesting.
- *Advanced R*, by Hadley Wickham, provides much more information on R than what we cover in this book. Computer Science students might enjoy reading this book.
- *The Statistical Sleuth*, by Ramsey and Schafer, will help the student think more like a statistician when dealing with data sets. This book is on a lower level mathematically.
- *Modern Applied Statistics with S*, by Venables and Ripley, is a book that covers more advanced statistical topics without much mathematics.
- *Introductory Statistics with R*, by Peter Dalgaard, is a concise introduction to using R for many types of statistical procedures.
- *Mathematical Statistics with Applications*, by Wackerly, Mendenhall, and Scheaffer, is a more mathematical (but still only requiring multivariate calculus and perhaps basic linear algebra) look at the topics of this book. Students interested in learning how to do the material in this book by hand without access to a computer may enjoy this book.
- *Data Feminism*, by D'Ignazio and Klein, offers a way of thinking about data science and data ethics informed by the ideas of intersectional feminism. About more than just gender, this book investigates the use and abuse of the power of data science.

This book is written in R Markdown using the `bookdown` package, by Yihui Xie. The original idea for a course of this type is due to Michael Lamar. The authors wish to thank Matt Schuelke, Kerith Conron, and Christophe Dervieux for helpful discussions. Thanks to Haijun Gong, Kimberly Druschel, Luis Miguel Anguas, Mustafa Attallah, Xue Li, and Caden Beddingfield for working through early editions. The anonymous reviewers provided useful comments, for which we are also grateful. We had two editors at CRC: John Kimmel, who believed in us from the start, and Lara Spieker, who got us to the finish line. Many thanks to both of you and a happy retirement to John!

## Software Installation

**R** is a programming language, distributed as its own software program.

To install R:

*Mac users*

1. Visit the CRAN archive, at https://cran.r-project.org
2. Find the link that looks like "R-x.x.x.pkg" under the Latest Release heading.
3. Download the "R-x.x.x.pkg" file, double-click it to open, and follow the installation instructions.

*Windows users*

1. Visit the CRAN archive, at https://cran.r-project.org
2. Click on the "Download R for Windows" link at the top of the page.
3. Click on the "base" link.
4. Click the large "Download R x.x.x for Windows" link and save the executable file somewhere on your computer.
5. Run the .exe file and follow the installation instructions.

**RStudio** is a graphical interface to R. R can work without RStudio, but RStudio requires R to work. Though you may choose to use R in its native form, the improvements that come with RStudio are absolutely worth the effort to install it. In fact, once you have RStudio installed, there is little need to ever run the R program itself.

To install RStudio:

1. Go to www.rstudio.com and click on the "Download RStudio" button.
2. Click on "Download RStudio Desktop."
3. Click on the version recommended for your system and install it.

**Libraries** you might choose to install before starting are the `tidyverse` and the book's data set `fosdata`. With RStudio running, find the Console and type:

```
install.packages("tidyverse")
install.packages("remotes")
remotes::install_github(repo = "speegled/fosdata")
```

More details on package installation are in Section 1.8.

# 1

# *Data in R*

The R Statistical Programming Language plays a central role in this book. While there are several other programming languages and software packages that do similar things, we chose R for several reasons:

1. R is widely used among statisticians, especially academic statisticians. If there is a new statistical procedure developed somewhere in academia, chances are that the code for it will be made available in R. This distinguishes R from, say, Python.
2. R is commonly used for statistical analyses in many disciplines. Other software, such as SPSS or SAS is also used and in some disciplines would be the primary choice for some discipline specific courses, but R is popular and its user base is growing.
3. R is free. You can install it and all optional packages on your computer at no cost. This is a big difference between R and SAS, SPSS, MATLAB, and most other statistical software.
4. R has been experiencing a renaissance. With the advent of the tidyverse and RStudio, R is a vibrant and growing community. We also have found the community to be extremely welcoming. The R ecosystem is one of its strengths.

In this chapter, we will begin to see some of the capabilities of R. We point out that R is a fully functional programming language, as well as being a statistical software package. We will only touch on the nuances of R as a programming language in this book.

## 1.1 Arithmetic and variable assignment

We begin by showing how R can be used as a calculator. Here is a table of commonly used arithmetic operators.

TABLE 1.1: Basic arithmetic operators in R.

| Operator | Description | Example |
|---|---|---|
| `+` | addition | `1 + 1` |
| `-` | subtraction | `4 - 3` |
| `*` | multiplication | `3 * 7` |
| `/` | division | `8 / 3` |
| `^` | exponentiation | `2^3` |

The output of the examples in Table 1.1 is given below. Throughout the book, lines that start with `##` indicate output from R commands. These will not show up when you type in

DOI: 10.1201/9781003004899-1

the commands yourself. The [1] in the lines below indicate that there is one piece of output from the command. These **will** show up when you type in the commands.

```
1 + 1
```

```
## [1] 2
```

```
4 - 3
```

```
## [1] 1
```

```
3 * 7
```

```
## [1] 21
```

```
8 / 3
```

```
## [1] 2.666667
```

```
2^3
```

```
## [1] 8
```

A couple of useful constants in R are `pi` and `exp(1)`, which are $\pi \approx 3.141593$ and $e \approx 2.718282$. Here R a couple of examples of how you can use them.

```
pi^2
```

```
## [1] 9.869604
```

```
2 * exp(1)
```

```
## [1] 5.436564
```

R is a functional programming language. If you don't know what that means, that's OK, but as you might guess from the name, functions play a large role in R. We will see many, many functions throughout the book. Every time you see a new function, think about the following four questions:

1. What type of input does the function accept?
2. What does the function do?
3. What does the function return as output?
4. What are some typical examples of how to use the function?

In this section, we focus on functions that do things that you are likely already familiar with from your previous math courses.

We start with `exp`. The function `exp` takes one argument named `x` and returns $e^x$. So, for example, `exp(x = 1)` will compute $e^1 = e$, as we saw above. In R, it is optional as to whether you supply the named version `x = 1` or just `1` as the argument. So, it is equivalent to write `exp(x = 1)` or `exp(1)`. Typically, for functions that are "well-known," the first argument or two will be given without names, then the rest will be provided with their names. Our advice is that if in doubt, include the name.

Next, we discuss the `log` function. The function `log` takes two arguments `x` and `base` and returns $\log_b x$, where $b$ is the `base`. The `x` argument is required. The `base` argument is optional with a default value of $e$. In other words, the default logarithm is the natural logarithm. Here are some examples of using `exp` and `log`.

```
exp(2)
```

```
## [1] 7.389056
```

```
log(8)
```

```
## [1] 2.079442
```

```
log(8, base = 2)
```

```
## [1] 3
```

You can't get very far without storing results of your computations to variables! The way[1] to do so is with the arrow `<-`. Typing `Alt` + `-` is the keyboard shortcut for `<-`.

```
height <- 62 # in inches
height <- height + 2
height <- 3 * height
```

The `# in inches` part of the code above is a *comment*. These are provided to give the reader information about what is going on in the R code, but are not executed and have no impact on the output.

If you want to see what value is stored in a variable, you can

1. type the variable name

```
height
```

```
## [1] 192
```

2. look in the environment box in the upper right-hand corner of RStudio.
3. Use the `str` command. This command gives other useful information about the variable, in addition to its value.

```
str(height)
```

```
##  num 192
```

This says that height contains *num*-eric data, and its current value is 192 (which is 3(62 + 2)). Note that there is a big difference between typing `height + 2` (which computes the value of `height + 2` and displays it on the screen) and typing `height <- height + 2`, which computes the value of `height + 2` *and stores the new value back in* `height`.

It is important to choose your variable names wisely. Variables in R *cannot* start with a number, and for our purposes, they *should* not start with a period. Do not use `T` or `F` as a variable name. Think twice before using c, q, t, C, D, or I as variable names, as they are already defined. It may also be a bad idea (and is one of the most frustrating things to debug on the rare occasions that it causes problems) to use `sum`, `mean`, or other commonly used functions as variable names. `T` and `F` are variables with default values `TRUE` and `FALSE`, which can be changed. We recommend writing out `TRUE` and `FALSE` rather than using the shortcuts `T` and `F` for this reason.

We also misspoke when we said `pi` is a constant. It is actually a variable which is set to 3.141593 when R is started, but can be changed to any value you like.[2] If you find that `pi`

[1] Using `=` for variable assignment is also allowed, as in many other programming languages. The arrow was the original and only assignment operator in R until 2001, and arrow is required by the Google and Tidyverse R style guides. However, some R users prefer to use `=`, and it is one of those things that you just can't reason about. The StackOverflow question, "What are the differences between `=` and `<-` in R?" has over 256K views as of this writing.

[2] Perhaps to 3.2, if you are Edward J. Goodwin trying to enact the "Indiana Pi Bill."

or T or F has been changed from a default, and you want to have them return to the default state, you have a couple of choices. You can restart R by clicking on Session/Restart. This will do more than just reset variables to their default values; it will reset R to its start-up state. Or, you can remove a variable from the R environment by using the function `rm()`. The function `rm` accepts the name of a variable and removes it from memory. As an example, look at the code below:

```
pi
```

```
## [1] 3.141593
```

```
pi <- 3.2
pi
```

```
## [1] 3.2
```

```
rm(pi)
pi
```

```
## [1] 3.141593
```

We end this section with a couple more hints. To remove all of the variables from your working environment, click on the broom icon in the Environment tab in RStudio. You may want to do this from time to time, as R can slow down when it has too many large variables in the current environment.

If you have a longish variable name in your environment, then you can use the `tab` key to auto-complete. Finally, RStudio has an option to restore the data in your working environment when you restart R. We recommend turning this off by going to `RStudio -> Preferences` and unchecking the box "Restore .RData into workspace at startup." This will ensure that each time you start R, all of the commands that you used to create the data in your environment will be run inside that R session.

## 1.2 Help

R comes with built-in help. In RStudio, there is a help tab in the lower right pane. From the console, placing a `?` before an object gives help for that object.

> **Try It Yourself.**
> Type `?log` in the console to see the help page for `log` and `exp`.

Help pages in R have some standard headings. Let's look at some of the main areas in the help page for `log`.

**Description**
: The help page says that `log` computes logarithms, by default natural logarithms.

**Usage**
: `log(x, base = exp(1))` means that `log` takes two arguments, `x` and `base`, and that `base` has a default of `exp(1)`.

**Arguments**
: `x` is a numeric or complex vector, and `base` is a positive or complex number. This might

be confusing for now, because we indicated that `x` was a positive real number above, but the help page indicates that `log` is more flexible than what we have seen so far.

**Examples**

The help page provides code that you can copy and paste into the R console to see what the function does. In this case, it provides among other things, `log(exp(3))`, which is equal to 3.

It can take some time to get used to reading R Documentation. For now, we recommend reading those four headings to see whether there are things you can learn about new functions. Don't worry if there are things in the documentation that you don't yet understand.

## 1.3 Vectors

Data often takes the form of multiple values of the same type. In R, multiple values are stored in a data type called a *vector*. R is designed to work with vectors quickly and easily.

There are many ways to create vectors. Perhaps the easiest is the `c` function:

```
c(2, 3, 5, 7, 11)
```

```
## [1]  2  3  5  7 11
```

The `c` function *combines* the values given to it into a vector. In this case, the vector is the list of the first 5 prime numbers. We can store vectors in variables just like we did with numbers:

```
primes <- c(2, 3, 5, 7, 11)
```

You can also create a vector of numbers in order using the : operator:

```
1:10
```

```
##  [1]  1  2  3  4  5  6  7  8  9 10
```

The `rep` function is a more flexible way of creating vectors. It has a required argument `x`, which is a vector of values that are to be repeated (could be a single value as well). It has optional arguments `times`, `length.out` and `each`. Normally, just one of the additional arguments is specified, and we will not discuss what happens if multiple arguments are specified. The argument `times` specifies the number of times that the value in `x` is repeated. This can either be specified as a single value, in which case the values of `x` are repeated that many times, or as a vector of values the same length as `x`. In this case, the values in `x` are repeated the number of times associated with the vector, but **the ordering is different**. For example:

```
rep(c(2, 3), times = 2)
```

```
## [1] 2 3 2 3
```

```
rep(c(2, 3), times = c(2, 2))
```

```
## [1] 2 2 3 3
```

```
rep(c(2, 3), times = c(2, 3))
```

```
## [1] 2 2 3 3 3
```

Alternatively, you can specify the length of the vector that you are trying to obtain, using `length.out`. R will truncate the last repetition of the vector that you are repeating to force the length to be exactly `length.out`.

```
rep(c(2, 3), length.out = 6)
```

```
## [1] 2 3 2 3 2 3
```

```
rep(c(2, 3), length.out = 3)
```

```
## [1] 2 3 2
```

Finally, setting `each` to a number repeats each value in `x` the same number of times. However, it orders the values as if you had written a vector of values in `times`.

```
rep(c("Bryan", "Darrin"), each = 2)
```

```
## [1] "Bryan"  "Bryan"  "Darrin" "Darrin"
```

```
rep(c("Bryan", "Darrin"), times = 2)
```

```
## [1] "Bryan"  "Darrin" "Bryan"  "Darrin"
```

We see here that the original vector `x` does not have to be numeric!

One last useful function for creating vectors is `seq`. This is a generalization of the : operator described above. We will not go over the entire list of arguments associated with `seq`, but we note that it has arguments `from`, `to`, `by` and `length.out`. We provide a couple of examples that we hope illustrate well enough how to use `seq`.

```
seq(from = 1, to = 11, by = 2)
```

```
## [1]  1  3  5  7  9 11
```

```
seq(from = 1, to = 11, length.out = 21)
```

```
##  [1]  1.0  1.5  2.0  2.5  3.0  3.5  4.0  4.5  5.0  5.5  6.0  6.5  7.0
## [14]  7.5  8.0  8.5  9.0  9.5 10.0 10.5 11.0
```

Once you have created a vector, you may also want to do arithmetic or other operations on it. Most of the operations in R work "as expected" with vectors. Suppose you wanted to see what the square roots of the first 5 primes were. You might guess:

```
primes^(1 / 2)
```

```
## [1] 1.414214 1.732051 2.236068 2.645751 3.316625
```

and you would be right! Returning to the cryptic manual entry in `log`, we recall that it stated that `x` is a numeric *vector*. This is the documentation's way of telling us that `log` is vectorized. If we supply the log function with a vector of values, then `log` will compute the log of each value separately. For example,

```
log(primes)
```

```
## [1] 0.6931472 1.0986123 1.6094379 1.9459101 2.3978953
```

Other commands reduce a vector to a number, for example `sum` adds all elements of the vector, and `max` finds the largest.

```
sum(primes)

## [1] 28

max(primes)

## [1] 11
```

> **Try It Yourself.**
> Guess what would happen if you type `primes + primes`, `primes * primes` and `sum(1/primes)`. Were you right?

The `plot` command creates graphic images from data. This book will only use `plot` for simple graphics, preferring the more powerful `ggplot` command discussed in Chapter 7. In its most basic application, give plot vectors of $x$- and $y$-coordinates, and it will produce a graphic with a point shown for each $(x, y)$ pair.

```
x <- -10:10
y <- x^2
plot(x, y)
```

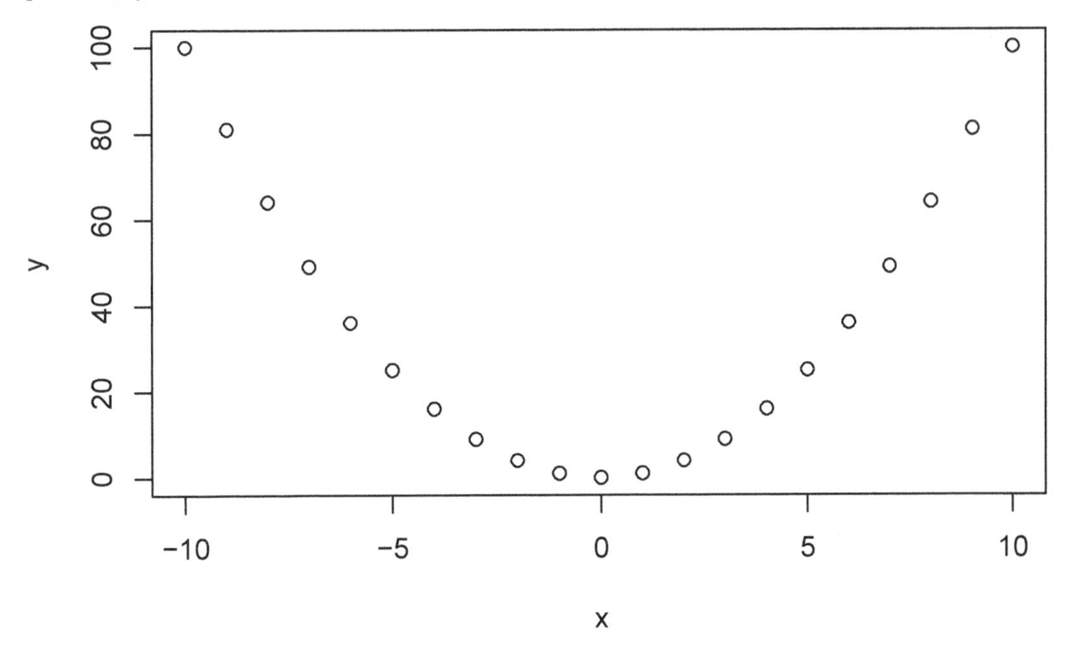


## 1.4 Indexing vectors

To examine or use a single element in a vector, you need to supply its *index*. `primes[1]` is the first element in the vector of primes, `primes[2]` is the second, and so on.

```
primes[1]

## [1] 2
```

```
primes[2]
```

```
## [1] 3
```

You can do many things with indexes. For example, you can provide a *vector* of indices, and R will return a new vector with the values associated with those indices.

```
primes[1:3]
```

```
## [1] 2 3 5
```

You can *remove* a value from a vector by using a - sign.

```
primes[-1]
```

```
## [1]  3  5  7 11
```

You can provide a vector of `TRUE` and `FALSE` values as an index, and R will return the values that are associated with `TRUE`. As a beginner, take care to have the length of the vector of `TRUE` and `FALSE` values be the same length as the original vector.

```
primes[c(TRUE, FALSE, TRUE, FALSE, TRUE)]
```

```
## [1]  2  5 11
```

The construct of providing a Boolean vector (that is, a vector containing `TRUE` and `FALSE`) for indexing is most useful for selecting elements that satisfy some condition. Suppose we wanted to "pull out" the values in `primes` that are bigger than 6. We create an appropriate vector of `TRUE` and `FALSE` values, then index `primes` by it.

```
primes > 6
```

```
## [1] FALSE FALSE FALSE  TRUE  TRUE
```

```
primes[primes > 6]
```

```
## [1]  7 11
```

Observe the use of > for comparison. In R (and most modern programming languages), there are some fundamental comparison operators:

- == equal to
- != not equal to
- > greater than
- < less than
- >= greater than or equal to
- <= less than or equal to

Another important operator is the `%in%`, which is `TRUE` if a value is *in* a vector. For example:

```
4 %in% primes
```

```
## [1] FALSE
```

```
odds <- seq(from = 1, to = 11, by = 2)
primes[primes %in% odds]
```

```
## [1]  3  5  7 11
```

R comes with many built-in data sets. For example, the `rivers` data set is a vector containing the length of major North American rivers. Try typing `?rivers` to see some more information about the data set. Let's see what the data set contains.

```
rivers
```

```
##   [1]  735  320  325  392  524  450 1459  135  465  600  330  336  280
##  [14]  315  870  906  202  329  290 1000  600  505 1450  840 1243  890
##  [27]  350  407  286  280  525  720  390  250  327  230  265  850  210
##  [40]  630  260  230  360  730  600  306  390  420  291  710  340  217
##  [53]  281  352  259  250  470  680  570  350  300  560  900  625  332
##  [66] 2348 1171 3710 2315 2533  780  280  410  460  260  255  431  350
##  [79]  760  618  338  981 1306  500  696  605  250  411 1054  735  233
##  [92]  435  490  310  460  383  375 1270  545  445 1885  380  300  380
## [105]  377  425  276  210  800  420  350  360  538 1100 1205  314  237
## [118]  610  360  540 1038  424  310  300  444  301  268  620  215  652
## [131]  900  525  246  360  529  500  720  270  430  671 1770
```

By typing `?rivers`, we learn that this data set gives the lengths (in miles) of 141 major rivers in North America, as compiled by the US Geological Survey. This data set is explored further in the exercises in this chapter. We will often want to examine only the first few elements when the data set is large. For that, we can use the function `head`, which by shows the first six elements.

```
head(rivers)
```

```
## [1] 735 320 325 392 524 450
```

The `discoveries` data set is a vector containing the number of "great" inventions and scientific discoveries in each year from 1860 to 1959. Try `?discoveries` to see more information about the `discoveries` data set. You might try the examples listed there just to see what they do, but we won't be doing anything like that yet. Let's see what the data set contains.

```
discoveries
```

```
## Time Series:
## Start = 1860
## End = 1959
## Frequency = 1
##   [1]  5  3  0  2  0  3  2  3  6  1  2  1  2  1  3  3  3  5  2  4  4  0
##  [23]  2  3  7 12  3 10  9  2  3  7  7  2  3  3  6  2  4  3  5  2  2  4
##  [45]  0  4  2  5  2  3  3  6  5  8  3  6  6  0  5  2  2  2  6  3  4  4
##  [67]  2  2  4  7  5  3  3  0  2  2  2  1  3  4  2  2  1  1  1  2  1  4
##  [89]  4  3  2  1  4  1  1  1  0  0  2  0
```

If we type `str(discoveries)` we see that the data set is stored as a *time series* rather than as a vector. For our purposes in this section, that will be an unimportant distinction, and we can simply think of the variable as a vector of numeric values.

The first ten elements are:

```
head(discoveries, n = 10)
```

```
##  [1] 5 3 0 2 0 3 2 3 6 1
```

Here are a few more things you can do with a vector:

```
sort(discoveries)

##  [1]  0  0  0  0  0  0  0  0  0  1  1  1  1  1  1  1  1  1  1  1  1  2
## [23]  2  2  2  2  2  2  2  2  2  2  2  2  2  2  2  2  2  2  2  2  2  2
## [45]  2  2  2  3  3  3  3  3  3  3  3  3  3  3  3  3  3  3  3  3  3  3
## [67]  3  4  4  4  4  4  4  4  4  4  4  4  4  5  5  5  5  5  5  5  6  6
## [89]  6  6  6  6  7  7  7  7  8  9 10 12

sort(discoveries, decreasing = TRUE)

##  [1] 12 10  9  8  7  7  7  7  6  6  6  6  6  6  5  5  5  5  5  5  5  4
## [23]  4  4  4  4  4  4  4  4  4  4  4  3  3  3  3  3  3  3  3  3  3  3
## [45]  3  3  3  3  3  3  3  3  3  2  2  2  2  2  2  2  2  2  2  2  2  2
## [67]  2  2  2  2  2  2  2  2  2  2  2  2  2  1  1  1  1  1  1  1  1  1
## [89]  1  1  1  0  0  0  0  0  0  0  0  0

table(discoveries)

## discoveries
##  0  1  2  3  4  5  6  7  8  9 10 12
##  9 12 26 20 12  7  6  4  1  1  1  1

max(discoveries)

## [1] 12

sum(discoveries)

## [1] 310

discoveries[discoveries > 5]

##  [1]  6  7 12 10  9  7  7  6  6  8  6  6  6  7

which(discoveries > 5) + 1859

##  [1] 1868 1884 1885 1887 1888 1891 1892 1896 1911 1913 1915 1916 1922
## [14] 1929
```

When `table` is provided a vector, it returns a table of the number of occurrences of each value in the vector. It will not provide zeros for values that are not there, even if it seems "obvious" to a human that there might have been place for that value. The function `which` accepts a vector of `TRUE` and `FALSE` values, and returns the *indices* in the vector that are `TRUE`. So, in the last line of the code above, adding 1859 to the indices gives the years that had more than 5 great discoveries.

## 1.5 Data types

All data in R has a type. Basic types hold one simple object. Data structures are built on top of the basic types. You can always learn the type of your data with the `str` structure command.

There are six basic types in R, although we won't use `complex` or `raw`. Also, it will not be necessary to make a distinction between `numeric` and `integer` data in this book, since R converts between types automatically.

**numeric**
: Real numbers, stored as some number of decimal places and an exponent. If you type `x <- 2`, then `x` will be stored as `numeric` data.

**integer**
: Integers. If you type `x <- 2L`, then `x` will be stored as an integer.[3] When reading data in from files, R will detect if all elements of a vector are integers and store that data as integer type.

**character**
: A collection of characters, also called a *string*. If you type `x <- "hello"`, then `x` is a `character` variable. Compare `str("hello")` to `str(c(1,2))`. Note that if you want to access the `e` from `hello`, you **cannot** use `x[2]`. Section 6.5 explains string manipulation using the `stringr` package.

**logical**
: Either `TRUE` or `FALSE`. The operators `!`, `&`, and `|` perform Boolean logic NOT, AND, and OR, respectively, on logical data.

**complex**
: A complex number, like `3 + 4i`.

**raw**
: Unstructured information stored as bytes.

There are many different data structures. The most important is the vector, which we have already met. Data frames will be described in Section 1.6. Another important structured type is called a *factor*:

**factor**
: Factor data takes on values in a predefined set. The possible values are called the *levels*. Levels are stored efficiently as numbers, and their names are only used for output. For example, a `rating` variable might take values high, medium, and low. A variable `continent` could be set up to allow only entries of Africa, Antarctica, Asia, Australia, Europe, North America, or South America. Factor type data is common in statistics, and many R functions only work properly when data is in factor form.

Our experience has been that students underestimate the importance of knowing what type of data they are working with. As a first example of the importance of data types, let's return to the `table` function. If we use `table` on a vector of integers, then R simply gives a list of the values that occur together with the number of times that they occur. However, if we use `table` on a `factor`, then R gives a list of all possible levels of the factor together with the number of times that they occur, including zeros. See Exercise 1.11 for an example.

R works really well when the data types are assigned properly. However, some bizarre things can occur when you try to force R to do something with a data type that is different than what you think it is! Whenever you examine a new data set (especially one that you read in from a file!), your first move is to use `str()` on it, followed by `head()`. Make sure that the data is stored the way you want *before* you continue with anything else.

---

[3]L stands for "long," a reference to the number of bits used to store R integers.

### 1.5.1 Missing data

Missing data is a problem that comes up frequently, and R uses the special value `NA` to represent it. `NA` isn't a data type, but a value that can take on *any* data type. It stands for Not Available, and it means that there is no data collected for that value.

As an example, consider the vector `airquality$Ozone`, which is part of base R:

```
airquality$Ozone
```

```
##   [1]  41  36  12  18  NA  28  23  19   8  NA   7  16  11  14  18  14
##  [17]  34   6  30  11   1  11   4  32  NA  NA  NA  23  45 115  37  NA
##  [33]  NA  NA  NA  NA  NA  29  NA  71  39  NA  NA  23  NA  NA  21  37
##  [49]  20  12  13  NA  NA  NA  NA  NA  NA  NA  NA  NA  NA 135  49  32
##  [65]  NA  64  40  77  97  97  85  NA  10  27  NA   7  48  35  61  79
##  [81]  63  16  NA  NA  80 108  20  52  82  50  64  59  39   9  16  78
##  [97]  35  66 122  89 110  NA  NA  44  28  65  NA  22  59  23  31  44
## [113]  21   9  NA  45 168  73  NA  76 118  84  85  96  78  73  91  47
## [129]  32  20  23  21  24  44  21  28   9  13  46  18  13  24  16  13
## [145]  23  36   7  14  30  NA  14  18  20
```

This shows the daily ozone levels (ppb) in New York during the summer of 1973. We would like to find the average ozone level for that summer, using the R function `mean`. However, just applying `mean` to the data produces an `NA`:

```
mean(airquality$Ozone)
```

```
## [1] NA
```

This is because the `Ozone` vector itself contains numerous `NA` values, corresponding to days when the ozone level was not recorded, Most R functions will force you to decide what to do with missing values, rather than make assumptions. To find the mean ozone level for the days with data, we must specify that the `NA` values should be removed with the argument `na.rm = TRUE`:

```
mean(airquality$Ozone, na.rm = TRUE)
```

```
## [1] 42.12931
```

## 1.6 Data frames

Returning to the built-in data set `rivers`, it would be very useful if the `rivers` data set also had the *names* of the rivers also stored. That is, for each river, we would like to know both the name of the river and the length of the river. We might organize the data by having one column, titled `river`, that gave the name of the rivers, and another column, titled `length`, that gave the length of the rivers. This leads us to one of the most common data types in R, the *data frame*. A data frame consists of a number of observations of variables. Some examples would be:

1. The name and length of major rivers.
2. The height, weight, and blood pressure of a sample of healthy adult females.
3. The high and low temperature in St Louis, MO, for each day of 2016.

As a specific example, let's look at the data set `mtcars`, which is a predefined data set in R.

Start with `str(mtcars)`. You can see that `mtcars` consists of 32 observations of 11 variables. The variable names are `mpg, cyl, disp` and so on. You can also type `?mtcars` on the console to see information on the data set. Some data sets have more detailed help pages than others, but it is always a good idea to look at the help page.

You can see that the data is from the 1974 *Motor Trend* magazine. You might wonder why we use such an old data set. In the R community, there are standard data sets that get used as examples when people create new code. The fact that familiar data sets are usually used lets people focus on the new aspect of the code rather than on the data set itself. In this course, we will do a mix of data sets; some will be up-to-date and hopefully interesting. Others will familiarize you with the common data sets that "developeRs" use.

The bracket operator `[ ]` picks out rows, columns, or individual entries from a data frame. It requires two arguments, a row and a column. For example, the weight or `wt` column of `mtcars` is column 6, so to get the third car's weight, use:

```
mtcars[3, 6]
```

```
## [1] 2.32
```

To pick out the third row of the `mtcars` data frame, leave the column entry blank:

```
mtcars[3, ]
```

```
##             mpg cyl disp hp drat   wt  qsec vs am gear carb
## Datsun 710 22.8   4  108 93 3.85 2.32 18.61  1  1    4    1
```

To pick out the first ten cars, we could use `mtcars[1:10,]`. To form a new data frame called `smallmtcars`, that only contains the variables `mpg`, `cyl` and `qsec`, we could use `smallmtcars <- mtcars[,c(1,2,7)]`. Referencing columns by name is also allowed, so `smallmtcars <- mtcars[,c("mpg", "cyl", "qsec")]` works.

Selecting a single column from a data frame is very common, so R provides the '$' operator to make this easier. To produce a vector containing the weights of all cars, for example:

```
mtcars$wt
```

```
##  [1] 2.620 2.875 2.320 3.215 3.440 3.460 3.570 3.190 3.150 3.440 3.440
## [12] 4.070 3.730 3.780 5.250 5.424 5.345 2.200 1.615 1.835 2.465 3.520
## [23] 3.435 3.840 3.845 1.935 2.140 1.513 3.170 2.770 3.570 2.780
```

Both `mtcars[,"wt"]` and `mtcars[,6]` produce the same vector result. Indexing the resulting vector gives the third car's weight:

```
mtcars$wt[3]
```

```
## [1] 2.32
```

As with vectors, providing a Boolean vector will select observations of the data that satisfy certain properties. For example, to pull out all observations that get more than 25 miles per gallon, use `mtcars[mtcars$mpg > 25,]`.

In order to test equality of two values, you use `==`. For example, in order to see which cars have 2 carburetors, we can use `mtcars[mtcars$carb == 2,]`.

Finally, to combine multiple conditions, you can use the vector logical operators `&` for *and*

and `|`, for *or*. As an example, to see which cars either have 2 carburetors or 3 forward gears (or both), we would use `mtcars[mtcars$carb == 2 | mtcars$gear == 3,]`.

Several exercises in this chapter provide practice manipulating data frames. In Chapter 6, we will introduce `dplyr` tools which we will use to do more advanced manipulations, but it is good to be able to do basic things with `[,]` and `$` as well.

**Example 1.1.** The `airquality` data frame is part of base R, and it gives air quality measurements for New York City in the summer of 1973.

```
str(airquality)
```

```
## 'data.frame':    153 obs. of  6 variables:
##  $ Ozone  : int  41 36 12 18 NA 28 23 19 8 NA ...
##  $ Solar.R: int  190 118 149 313 NA NA 299 99 19 194 ...
##  $ Wind   : num  7.4 8 12.6 11.5 14.3 14.9 8.6 13.8 20.1 8.6 ...
##  $ Temp   : int  67 72 74 62 56 66 65 59 61 69 ...
##  $ Month  : int  5 5 5 5 5 5 5 5 5 5 ...
##  $ Day    : int  1 2 3 4 5 6 7 8 9 10 ...
```

From the structure, we see that `airquality` has 153 observations of 6 variables. The `Wind` variable is numeric and the others are integers.

We now find the hottest temperature recorded, the average temperature in June, and the day with the most wind:

```
max(airquality$Temp)
```

```
## [1] 97
```

```
junetemps <- airquality[airquality$Month == 6, "Temp"]
mean(junetemps)
```

```
## [1] 79.1
```

```
mostwind <- which.max(airquality$Wind)
airquality[mostwind, ]
```

```
##    Ozone Solar.R Wind Temp Month Day
## 48    37     284 20.7   72     6  17
```

It got to 97°F at some point, it averaged 79.1°F in June, and June 17 was the windiest day that summer.

The `data.frame` function creates a data frame. It takes `name = value` pairs as arguments, where the `name` will be the column name in the new data frame, and `value` will be the vector of values in that column. Here is a simple example:

```
great_lakes <- data.frame(
  name = c("Huron", "Ontario", "Michigan", "Erie", "Superior"),
  volume = c(3500, 1640, 4900, 480, 12000), # km^3
  max_depth = c(228, 245, 282, 64, 406) # meters
)
str(great_lakes)
```

```
## 'data.frame':    5 obs. of  3 variables:
##  $ name     : chr  "Huron" "Ontario" "Michigan" "Erie" ...
##  $ volume   : num  3500 1640 4900 480 12000
```

```
## $ max_depth: num 228 245 282 64 406
```

## 1.7 Reading data from files

Loading data into R is one of the most important things to be able to do. If you can't get R to load your data, then it doesn't matter what kinds of neat tricks you could have done. It can also be one of the most frustrating things – not just in R, but in general. Your data might be on a web page, in an Excel spreadsheet, or in any one of dozens of other formats each with its own idiosyncrasies. R has powerful packages that can deal with just about any format of data you are likely to encounter, but for now we will focus on just one format, the *CSV* file. Usually, data in a CSV file will have the extension ".csv" at the end of its name. CSV stands for "Comma Separated Values" and means that the data is stored in rows with commas separating the variables. For example, CSV formatted data might look like this:

```
"Gender","Body.Temp","Heart.Rate"
"Male",96.3,70
"Male",96.7,71
"Male",96.9,74
"Female",96.4,69
"Female",96.7,62
```

This would mean that there are three variables: `Gender`, `Body.Temp` and `Heart.Rate`. There are 5 observations: 3 males and 2 females. The first male had a body temperature of 96.3 and a heart rate of 70.

The command to read a CSV file into R is `read.csv`. It takes one argument, a string giving the path to the file on your computer. R always has a *working directory*, which you can find with the `getwd()` command, and you can see with the Files tab in RStudio. If your file is stored in that directory, you can read it with the command `read.csv("mydatafile.csv")`.

More advanced users may want to set up a file structure that has data stored in a separate folder, in which case they must specify the *pathname* to the file they want to load. The easiest way to find the full pathname to a file is with the command `file.choose()`, which will open an interactive dialog where you can find and select the file. Try this from the R console and you will see the full path to the file, which you can then use as the argument to `read.csv`. Using `file.choose()` requires you, the programmer, to provide input. One of the main reasons to use R is that analysis with R is reproducible and can be performed without user intervention. Using interactive functions means your analysis will **not** be reproducible and is better avoided.

If you've tried `read.csv`, you may have noticed that it printed the contents of your CSV file to the console. To actually use the data, you need to store it in a variable as a data frame. Try to choose a name that is descriptive of the actual contents of your data file. For example, to load the file `normtemp.csv`, which contains the gender, body temperature and heart rate data mentioned above, you would type `temp_data <- read.csv("normtemp.csv")`, or provide the full path name.

In other instances, the CSV file that you want to read is hosted on a web page. In this case, it is sometimes easier to read the file directly from the web page by us-

ing `read.csv("http://website/file.csv")`. As an example, there is a csv hosted at http://stat.slu.edu/~speegle/data/normtemp.csv. To load it, use:

```
normtemp <- read.csv("http://stat.slu.edu/~speegle/data/normtemp.csv")
```

**Try It Yourself.**
Download the data set from http://stat.slu.edu/~speegle/data/normtemp.csv onto your own computer and store it some place that you can find it. Use `file.choose()` interactively in the console to get the full path to the file. Use `read.csv()` with the full path to the file inside of the parentheses to load the data set into R.

We can't emphasize enough the importance of looking at your data after you have loaded it. Start by using `str()`, `head()` and `summary()` on your variable after reading it in. As often as not, there will be something you will need to change in the data frame before the data is usable.

To write R data frames to a CSV file, use the `write.csv()` command. If your row names are not meaningful, then often you will want to add `row.names = FALSE`. The command `write.csv(mtcars, "mtcars_file.csv", row.names = FALSE)` writes the variable `mtcars` to the file `mtcars_file.csv`, which is again stored in the directory specified by `getwd()` by default.

## 1.8 Packages

When you first start using R, the commands and data available to you are called "Base R." The R language is extensible, which means that over the years people have added functionality. New functionality comes in the form of a *package*, which may be included in your R distribution or which you may need to install. For example, the `HistData` package contains a few dozen data sets with historical significance.

Happily, installing packages is extremely simple: in RStudio you can click the Packages tab in the lower right panel, and then hit the Install button to install any package you need. Alternatively, you can use the `install.packages` command, like so:

```
install.packages("HistData")
```

Installing packages does require an Internet connection, and frequently when you install one package R will automatically install other packages, called *dependencies*, that the package you want must have to work.

Package installation is not a common operation. Once you have installed a package, you have it forever.[4] However, each time you start R, you need to load any packages you want to use. You do this with the `library` command:

```
library(HistData)
```

Once you have loaded the package, the contents of the package are available to use. `HistData` contains a data set `DrinksWages` with data on drinking and earned wages from 1910. After

[4]Well, at least until you update R to the newest version, which cleans out the packages that you had previously installed.

loading `HistData` you can inspect `DrinksWages` and learn that rivetters were paid well in 1910:

```
head(DrinksWages)
```

```
##   class         trade sober drinks     wage  n
## 1     A papercutter     1      1 24.00000  2
## 2     A      cabmen     1     10 18.41667 11
## 3     A  goldbeater     2      1 21.50000  3
## 4     A   stablemen     1      5 21.16667  6
## 5     A  millworker     2      0 19.00000  2
## 6     A      porter     9      8 20.50000 17
```

```
DrinksWages[which.max(DrinksWages$wage), ]
```

```
##    class    trade sober drinks wage n
## 64     C rivetter     1      0   40 1
```

Some packages are large, and you may only require one small part of them. The `::` double colon operator selects the required object without loading the entire package. For example, `MASS::immer` can access the `immer` data from the `MASS` package without loading the large and messy `MASS` package into your workspace:

```
head(MASS::immer)
```

```
##   Loc Var    Y1    Y2
## 1  UF   M  81.0  80.7
## 2  UF   S 105.4  82.3
## 3  UF   V 119.7  80.4
## 4  UF   T 109.7  87.2
## 5  UF   P  98.3  84.2
## 6   W   M 146.6 100.4
```

Learning R with this book will require you to use a variety of packages. Though you need only install each package one time, you will need to use the `::` operator or load it with `library` each time you start a new R session. One of the more common errors you will encounter is: `Error: object 'so-and-so' not found`, which may mean that `so-and-so` was part of a package you forgot to load.

Much of the data used in this book is distributed in a package called `fosdata`, which you will want to install using the following commands if you have not already done so.

```
install.packages("remotes")
remotes::install_github("speegled/fosdata")
```

**Try It Yourself.**
Look at some of the data sets in the `fosdata` package. Once you have successfully installed `fosdata`, you can see a list of all of the data sets by typing `data(package = "fosdata")`.

Pick a couple of data sets that sound interesting and read about them. For example, if the Bechdel test sounds interesting to you, you can type `?fosdata::bechdel` to read about the data set, and `head(fosdata::bechdel)` to see the first few observations.

One feature of `fosdata` is that many of the data sets are taken directly from recent papers. For example, the `frogs` data set is data relating to the discovery of a new species of frog in

Dhaka, Bangladesh, which is one of the most densely populated cities in the world. The data was used by the authors to show that the morphology of the new frog species differs from those of the same genus. By following the link given in the help page `?fosdata::frogs`, you can find and read the original paper associated with this data.

## 1.9 Errors and warnings

R, like most programming languages, is very picky about the instructions you give it. It pays attention to uppercase and lowercase characters, similar looking symbols like = and == mean very different things, and every bit of punctuation is important.

When you make mistakes (called *bugs*) in your code, a few things may happen: errors, warnings, and incorrect results. Code that runs but that runs incorrectly is usually the hardest problem to fix, since the computer sees nothing wrong with your code and debugging is left entirely to you. The simplest bug is when your code produces an error. Here are a few examples:

```
mean(primse)

## Error in mean(primse): object 'primse' not found
mtcars[, 100]

## Error in `[.data.frame`(mtcars, , 100): undefined columns selected
airquality[airquality$Month = 6,"Temp"]

## Error: <text>:1:29: unexpected '=' ## 1: airquality[airquality$Month =
## ## ^
```

The first is a typical spelling error. In the second, we asked for column 100 of `mtcars`, which has only 11 columns. In the third, we used = instead of ==. You will encounter these sorts of errors all the time and then quickly graduate to much more subtle bugs.

Warnings occur when R detects a potential problem in your code but can continue working. For example, here we try to assign an entire vector to one element of a vector. R cannot do this, so it assigns the first element of the vector and prints a warning message.

```
a <- 1:10
a[5] <- 100:200

## Warning in a[5] <- 100:200: number of items to replace is not a multiple
## of replacement length
a

##  [1]   1   2   3   4 100   6   7   8   9  10
```

Complicated statistical operations such as hypothesis tests and regression analysis frequently produce warnings or messages that the user might not care about. The output of R commands in this book will sometimes omit these messages to save space and focus attention on the important part of the output. If you notice your command producing warnings not shown in the book, either ignore them or dig deeper and learn a little more about R. In your own code, you can use the commands `suppressWarnings` and `suppressMessages` to remove extraneous output for presentation quality work.

Another pitfall that traps many novice R users is the `+` prompt. Working interactively in the console, you hit return and see a `+` instead of the friendly `>` prompt. This means that the command you typed was incomplete; for example, because you opened a parenthesis `(` and failed to close it with `)`. Sometimes this behavior is desirable, allowing a long command to extend over two lines. More often, the `+` is unexpected. You can escape from this situation with the escape key, `ESC`, hence its name.

## 1.10 Useful idioms

Here is a summary list of useful programming idioms that we will use throughout the textbook, for ease of future reference. We assume `vec` is a numeric or integer vector.

**`sum(vec == 3)`**
the number of times that the value 3 occurs in the vector `vec`.

**`mean(vec == 3)`**
the percentage of times that the value 3 occurs in the vector `vec`.

**`table(vec)`**
the number of times that each value occurs in `vec`.

**`max(table(vec))`**
the number of times the most common value in `vec` occurs.

**`length(unique(vec))`**
the number of distinct values that occur in `vec`.

**`vec[vec > 0]`**
a new vector that only includes values in `vec` that are positive.

**`vec[!is.na(vec)]`**
a new vector that only includes the non-missing values in `vec`.

## Vignette: Data science communities

If you are serious about learning statistics, R, data visualization and data wrangling, the best thing you can do is to practice with real data. Finding appropriate data to practice on can be a challenge for beginners, but happily the R world abounds with online communities that share interesting data.

Both beginners and experts post visualizations, example code, and discussions of data from these sources regularly. Look at other developeRs code and decide what you like, and what you don't. Incorporate their ideas into your own work!

**Kaggle** https://kaggle.com

A website that requires no cost registration. The Datasets section of Kaggle allows users to explore, analyze, and share quality data. Most data sets are clearly licensed for use, are available in .csv format, and come with a description that explains the data. Each data set has a discussion page where users can provide commentary and analysis.

Beyond data, Kaggle hosts machine learning competitions for competitors at all levels of expertise. Kaggle also offers R notebooks for cloud based computing and collaboration.

**Tidy Tuesday** Twitter: #TidyTuesday

A project that arose from the R4DS Learning Community. The project posts a new data set each Tuesday. Data sets are suggested by the community and curated by the Tidy Tuesday organizers. Tidy Tuesday data sets are good for learning how to summarize and arrange data in order to make meaningful visualizations with `ggplot2`, `tidyr`, `dplyr`, and other tools in the tidyverse ecosystem.

Data scientists post their visualizations and code on Twitter. Tidy Tuesday data is available through a GitHub repository or with the R package `tidytuesdayR`.

**Data Is Plural** https://www.data-is-plural.com/

A weekly newsletter of useful and curious data sets, maintained by Jeremy Singer-Vine. Data sets are well curated and come with source links. There is a shared spreadsheet with an archive of past data.

**Stack Overflow** https://stackoverflow.com

A community Q&A forum for every computer language and a few other things besides. It has over 300,000 questions tagged `r`. If you ask a search engine a question about R, you will likely be directed to StackOverflow. If you can't find an answer already posted, create a free account and ask the question yourself. It is common to get expert answers within hours.

**R Specific Groups** https://rladies.org/ and https://jumpingrivers.github.io/meetingsR/r-user-groups.html

Both of these groups support R users with educational opportunities. R Ladies is an organization that promotes gender diversity in the R community. They also hold meetups in various locations around the world to get people excited about using R. UseR groups primarily host meetups where they discuss various aspects of R, from beginning to advanced. If you find yourself wanting to get connected to the larger R community, these are good places to start.

## Vignette: An R Markdown primer

RStudio has included a method for making higher quality documents from a combination of R code and some basic markdown formatting, which is called R Markdown. In fact, this book itself was written in an extension of R Markdown called bookdown. The goal of this vignette is to get you started making simple documents in R Markdown.

For students reading this book, writing your homework in R Markdown will allow you to have reproducible results. It is much more efficient than creating plots in R and copy/pasting into Word or some other document. For professionals, R Markdown is an excellent way to produce reports and other finished documents that contain reproducible research.

## R Markdown files

In order to create an R Markdown file inside RStudio, you click on `File`, then `New File` and `R Markdown`. You will be prompted to enter a Title and an Author. For now, leave the Default Output Format as .html, and leave the Document icon highlighted. When you have entered an acceptable title and author, click OK and a new tab in the Source pane should open up. (That is the top left pane in RStudio.) The first thing to do is to click on the `Knit` button right above the source panel. It should knit the document and give you a nice-ish looking document.

The first part of an R Markdown document is the YAML (YAML Ain't Markup Language) header.

```
---
title: "Homework 1"
author: "Riley Student"
date: "1/20/2021"
output: html_document
---
```

The YAML header describes the basic properties of the document. The `output` line says that this document will knit to become an html document.

> **Try It Yourself.**
> Change the output line to say `output: word_document` and knit again to produce a Microsoft Word document. If you have LaTeX installed, try `output: pdf_document`. Finally, try installing the `tufte` package and use `output: tufte::tufte_html`.

One benefit of using the `tufte` package is that in order for the document to look good, it requires you to write more text than you might otherwise. Many beginners tend to have too much R code and output (including plots), and not enough words. If you use the `tufte` package, you will see that the document doesn't look right unless the proportion of words to plots and other elements is better.

## Document content

Everything below the YAML header is R Markdown document content. A simple document consists of markdown formatted text intermixed with R code chunks. The new document you created in RStudio comes with some sample content. You may delete everything below the YAML header and put your own work in its place.

Here is what the content of a homework assignment might look like:

````
1. Consider the `DrinksWages` data set in the `HistData` package.
a. Which trade had the lowest wages?

```{r findwage}
library(HistData)
DrinksWages[which.min(DrinksWages$wage),]
```

We see that **factory worker** had the lowest wages; 12 shillings per week.
If there had been multiple professions with a weekly wage of 12 shillings
````

```
per week, then we would have only found one of them here.

b. Provide a histogram of the wages of the trades

```{r makehist}
hist(DrinksWages$wage)
```

The weekly wages of trades in 1910 was remarkably symmetric. Values ranged
from a low of 12 shillings per week for factory workers to a high of 40
shillings per week for rivetters and glassmakers.
```

Markdown is intended to be easy to read, and is fairly self-explanatory. Nevertheless, here is a breakdown of how we formatted the example content above:

- The `1.` at the beginning of the first line starts a numbered list. The `a.` below starts a lettered sub-list.
- Putting `DrinksWages` inside backquotes causes it to be formatted in a fixed width font, so that it is clear that it is an R object of some sort.
- Putting `**` before and after a word or phrase makes it **bold**. Surround text with single `*` for *italic.*
- The lines beginning with "We see that..." will be formatted as a paragraph. All lines in a paragraph are wrapped together, so you do not need to decide where to break lines. Paragraphs end at a blank line.

For nicely formatted mathematics, R Markdown supports a typesetting language called TeX. TeX is a powerful formula processor that produces high quality mathematics. TeX formulas are placed between `$` characters. For example `$x^2$` produces $x^2$, `$\frac{1}{2}$` produces $\frac{1}{2}$, and `$X_i$` produces $X_i$. If you don't have TeX installed, then you may use the following two R commands to install a tiny version of it:

```
install.packages("tinytex")
tinytex::install_tinytex()
```

### Code chunks

The most powerful feature of R Markdown is the ability to include and run R code within the document. This code is placed inside *chunks* which begin with ` ```{r} ` and end with ` ``` `. When you knit the file, commands in R chunks are run and both the command and its output appear in the knit document. The example above has two chunks: one called `findwage` that finds the lowest wage, and one called `makehist` that creates a histogram of wages. You may name your chunks whatever you please. You can create new R chunks with the keyboard shortcut `Command + Alt + I` instead of typing all the backquotes.

To use a package inside R Markdown, you *must* load the package inside your markdown file. The example does this with `library(HistData)` inside the first R chunk. Once loaded, the package will be available in all subsequent R chunks. Every R chunk should have *some* sort of explanation in words. Just giving the R code is rarely enough for someone to understand what you did.

R chunks have options that control the output they generate. For example, you might want to make your plots smaller to save space. The chunk options `fig.width` and `fig.height` control the size of figures. Some R commands produce messages you may not wish to see in your knit document. You could turn those off with the chunk option message=FALSE. To produce a histogram that is smaller and produces no messages, use:

```
```{r makehist, message=FALSE, fig.width=3, fig.height=3}
hist(DrinksWages$wage)
```
```

A common idiom in R Markdown is to have a setup chunk just after the YAML header. This setup chunk can set up options for the rest of the document and also load libraries you might need. Here is an example of a typical setup chunk:

```
```{r setup, message=FALSE, warning=FALSE}
knitr::opts_chunk$set(echo = TRUE, fig.height = 4)
library(HistData)
```
```

## Onward

R Markdown has many more capabilities than this simple introduction describes (remember, this book was written entirely with R Markdown). The definitive guide to R Markdown is the appropriately titled book by Xie, Allaire and Grolemund.[5]

Some excellent resources to learn more are Grolemund and Wickham,[6] Ismay,[7] or try looking at the tufte template in RStudio by clicking on File -> New File -> R Markdown -> From Template -> Tufte Handout.

## Exercises

Exercises 1.1 – 1.5 require material through Sections 1.1 – 1.3.

**1.1.** Let x <- c(1,2,3) and y <- c(6,5,4). Predict what will happen when the following pieces of code are run. Check your answer.

a. x * 2
b. x * y
c. x[1] * y[2]

**1.2.** Let x <- c(1,2,3) and y <- c(6,5,4). What is the value of x after each of the following commands? (Assume that each part starts with the values of x and y given above.)

a. x + x
b. x <- x + x
c. y <- x + x
d. x <- x + 1

[5]Yihui Xie, J J Allaire, and Garrett Grolemund, *R Markdown: The Definitive Guide* (Chapman & Hall/CRC, 2018).

[6]Garrett Grolemund and Hadley Wickham, *R for Data Science* (O'Reilly Media, 2017), https://r4ds.had.co.nz/r-markdown.html.

[7]Chester Ismay and Patrick C Kennedy, *Getting Used to R, RStudio and R Markdown*, 2019, https://bookdown.org/chesterismay/rbasics/.

**1.3.** Determine the values of the vector `vec` after each of the following commands is run.

a. `vec <- 1:10`
b. `vec <- 1:10 * 2`
c. `vec <- 1:10^2`
d. `vec <- 1:10 + 1`
e. `vec <- 1:(10 * 2)`
f. `vec <- rep(c(1,1,2), times = 2)`
g. `vec <- seq(from = 0, to = 10, length.out = 5)`

**1.4.** In this exercise, you will graph the function $f(p) = p(1-p)$ for $p \in [0, 1]$.

a. Use `seq` to create a vector `p` of numbers from 0 to 1 spaced by 0.2.
b. Use `plot` to plot `p` in the `x` coordinate and `p(1-p)` in the `y` coordinate. Read the help page for `plot` and experiment with the `type` argument to find a good choice for this graph.
c. Repeat, but with creating a vector `p` of numbers from 0 to 1 spaced by 0.01.

**1.5.** Use R to calculate the sum of the squares of all numbers from 1 to 100: $1^2 + 2^2 + \cdots + 99^2 + 100^2$.

---

Exercises 1.6 – 1.7 require material through Section 1.4.

**1.6.** Let `x` be the vector obtained by running the R command `x <- seq(from = 10, to = 30, by = 2)`.

a. What is the length of `x`? (By length, we mean the number of elements in the vector. This can be obtained using the `str` function or the `length` function.)
b. What is `x[2]`?
c. What is `x[1:5]`?
d. What is `x[1:3*2]`?
e. What is `x[1:(3*2)]`?
f. What is `x > 25`?
g. What is `x[x > 25]`?
h. What is `x[-1]`?
i. What is `x[-1:-3]`?

**1.7.** R has a built-in vector `rivers` which contains the lengths of major North American rivers.

a. Use `?rivers` to learn about the data set.
b. Find the mean and standard deviation of the rivers data using the base R functions `mean` and `sd`.
c. Make a histogram (`hist`) of the rivers data.
d. Get the five number summary (`summary`) of rivers data.
e. Find the longest and shortest lengths of rivers in the set.
f. Make a list of all (lengths of) rivers longer than 1000 miles.

---

Exercises 1.8 – 1.11 require material through Sections 1.5 – 1.6.

**1.8.** Consider the built-in data frame `airquality`.

a. How many observations of how many variables are there?
b. What are the names of the variables?

c. What type of data is each variable?
d. Do you agree with the data type that has been given to each variable? What would have been some alternative choices?

**1.9.** There is a built-in data set `state`, which is really seven separate variables with names such as `state.name`, `state.region`, and `state.area`.

a. What are the possible regions a state can be in? How many states are in each region?
b. Which states have area less than 10,000 square miles?
c. Which state's geographic center is furthest south? (Hint: use `which.min`)

**1.10.** Consider the `mtcars` data set.

a. Which cars have 4 forward gears?
b. What subset of `mtcars` does `mtcars[mtcars$disp > 150 & mtcars$mpg > 20,]` describe?
c. Which cars have 4 forward gears and manual transmission? (Note: manual transmission is 1 and automatic is 0.)
d. Which cars have 4 forward gears or manual transmission?
e. Find the mean mpg of the cars with 2 carburetors.

**1.11.** Consider the `mtcars` data set.

a. Convert the `am` variable to a factor with two levels, `auto` and `manual`, by typing the following: `mtcars$am <- factor(mtcars$am, levels = c(0, 1), labels = c("auto", "manual"))`.
b. How many cars of each type of transmission are there?
c. How many cars of each type of transmission have gas mileage estimates greater than 25 mpg?

---

Exercises 1.12 – 1.14 require material through Section 1.8.

**1.12.** This problem uses the data set `hot_dogs` from the package `fosdata`.

a. How many observations of how many variables are there? What types are the variables?
b. What are the three kinds of hot dogs in this data set?
c. What is the highest sodium content of any hot dog in this data set?
d. What is the mean calorie content for Beef hot dogs?

**1.13.** This problem uses the data set `DrinksWages` from the package `HistData`.

a. How many observations of how many variables are there? What types are the variables?
b. The variable `wage` contains the average wage for each profession. Which profession has the lowest wage?
c. The variable `n` contains the number of workers surveyed for each profession. Sum this to find the total number of workers surveyed.
d. Compute the mean wage for all workers surveyed by multiplying wage * n for each profession, summing, and dividing by the total number of workers surveyed.

**1.14.** This problem uses the package `Lahman`, which needs to be installed on your computer. The data set `Batting`, in the `Lahman` package contains batting statistics of all major league baseball players since 1871, broken down by season.

a. How many observations of how many variables are there?
b. Use the command `head(Batting)` to get a look at the first six lines of data.
c. What is the most number of triples (X3B) that have been hit in a single season?

d. What is the playerID(s) of the person(s) who hit the most number of triples in a single season? In what year did it happen?
e. Which player hit the most number of triples in a single season since 1960?

---

Exercise 1.15 requires material through Section 1.10.

**1.15.** Consider the `bechdel` data set in the `fosdata` package.

a. How many movies in the data set pass the Bechdel test?
b. What percentage of movies in the data set pass the Bechdel test?
c. Create a table of number of movies in the data set by year.
d. Which year has the most movies in the data set?
e. How many different values are there in the `clean_test` variable?
f. Create a data frame that contains only those observations that pass the Bechdel test.
g. Create a data frame that contains all of the observations that do **not** have missing values in the `domgross` variable.

# 2

# *Probability*

A primary goal of statistics is to describe the real world based on limited observations. These observations may be influenced by random factors, such as measurement error or environmental conditions. This chapter introduces probability, which is designed to describe random events. Later, we will see that the theory of probability is so powerful that we intentionally introduce randomness into experiments and studies so we can make precise statements from data.

## 2.1 Probability basics

In order to learn about probability, we must first develop a vocabulary that we can use to discuss various aspects of it.

**Definition 2.1.** Terminology for statistical experiments:

- An *experiment* is a process that produces an *observation*.
- An *outcome* is a possible observation.
- The set of all possible outcomes is called the *sample space*.
- An *event* is a subset of the sample space.
- A *trial* is a single running of an experiment.

**Example 2.1.** Roll a die and observe the number of dots on the top face. This is an experiment, with six possible outcomes. The sample space is the set $S = \{1, 2, 3, 4, 5, 6\}$. The event "roll higher than 3" is the set $\{4, 5, 6\}$.

**Example 2.2.** Stop a random person on the street and ask them in which month they were born. This experiment has the twelve months of the year as possible outcomes. An example of an event $E$ might be that they were born in a summer month, $E = \{\text{June}, \text{July}, \text{August}\}$.

**Example 2.3.** Suppose a traffic light stays red for 90 seconds each cycle. While driving you arrive at this light, and observe the amount of time that you are stopped until the light turns green. The sample space is the interval of real numbers $[0, 90]$. The event "you didn't have to stop" is the set $\{0\}$.

Since events are, by their very definition, sets, it will be useful for us to review some basic set theory.

**Definition 2.2.** Let $A$ and $B$ be events in a sample space $S$.

1. $A \cap B$ is the set of outcomes that are in *both* $A$ and $B$.
2. $A \cup B$ is the set of outcomes that are in *either* $A$ or $B$ (or both).

DOI: 10.1201/9781003004899-2

3. $A - B$ is the set of outcomes that are in $A$ and not in $B$.
4. The *complement* of $A$ is $\overline{A} = S - A$. So, $\overline{A}$ is the set of outcomes that are *not* in $A$.
5. The symbol $\emptyset$ is the *empty set*, the set with no outcomes.
6. $A$ and $B$ are *disjoint* if $A \cap B = \emptyset$.
7. $A$ is a *subset* of $B$, written $A \subset B$, if every element of $A$ is also an element of $B$.

**Example 2.4.** Suppose that the sample space $S$ consists of the positive integers. Let $A$ be the set of all positive even numbers, and let $B$ be the set of all prime numbers. Then, $A = \{2, 4, 6, \ldots\}$ and $B = \{2, 3, 5, 7, 11, \ldots\}$. Then,

1. $A \cap B = \{2\}$
2. $A \cup B = \{2, 3, 4, 5, 6, 7, 8, 10, 11, 12, 13, 14, 16, 17, 18, 19, 20, 22, \ldots\}$
3. $B - A$ is the set of *odd* prime numbers.
4. $\overline{A}$ is the set of all positive odd integers.
5. $A$ and $B$ are not disjoint, since both contain the number 2.
6. $A$ is not a subset of $B$ since 4 is in $A$, but 4 is not an element of $B$.

The *probability of an event* is a number between zero and one that describes the proportion of time we expect the event to occur. Since probability lies at the heart of all mathematical statements in this book, we will define it formally in Definition 2.3 and prove its basic properties in Theorem 2.1.

**Definition 2.3.** Let $S$ be a sample space. A valid *probability* satisfies the following **probability axioms**:

1. Probabilities are non-negative real numbers. That is, for all events $E$, $P(E) \geq 0$.
2. The probability of the sample space is 1, $P(S) = 1$.
3. Probabilities are countably additive: If $A_1, A_2, \ldots$ are pairwise disjoint, then

$$P\left(\bigcup_{n=1}^{\infty} A_n\right) = \sum_{n=1}^{\infty} P(A_n)$$

We will not be concerned in this book about carefully describing *all* of the subsets of $S$ which have an associated probability. We will assume that any event of interest will be an event associated with a probability.

Probabilities obey some important rules, which are consequences of the axioms.

**Theorem 2.1.** *Let $A$ and $B$ be events in the sample space $S$.*

1. $P(\emptyset) = 0$.
2. *If $A$ and $B$ are disjoint, then* $P(A \cup B) = P(A) + P(B)$.
3. *If $A \subset B$, then* $P(A) \leq P(B)$.
4. $0 \leq P(A) \leq 1$.
5. $P(A) = 1 - P(\overline{A})$.
6. $P(A - B) = P(A) - P(A \cap B)$.
7. $P(A \cup B) = P(A) + P(B) - P(A \cap B)$.

*Proof.* We sketch the proof of these results. Part 1 follows from countable additivity. Let $A_1, A_2, \ldots$ all be the empty set. Since they are pairwise disjoint, $P(\emptyset) = \sum_{i=1}^{\infty} P(\emptyset)$, which implies $P(\emptyset) = 0$. Part 2 is just a special case of probability Axiom 3 with $A_1 = A$, $A_2 = B$, and $A_3, A_4, \ldots$ all equal to the empty set.

Part 3 follows from letting $A_1 = A$ and $A_2 = B - A$ and applying part 2 and Axiom 1. Part 4 follows from parts 1 and 3, together with Axiom 2.

For part 6, we have that $A = (A \cap B) \cup (A - B)$, where $A \cap B$ and $A - B$ are disjoint. We have $P(A) = P(A \cap B) + P(A - B)$, which gives the result. Part 5 is a special case of part 6.

To prove part 7, we note that $A \cup B = A \cup (B - A)$, where $A$ and $B - A$ are disjoint. Therefore, $P(A \cup B) = P(A) + P(B - A) = P(A) + P(B) - P(A \cap B)$ by parts 2 and 5. ■

One way to assign probabilities to events is *empirically*, by repeating an experiment many times and observing the proportion of times the event occurs. While this can only approximate the true probability, it is sometimes the only approach possible. For example, in the United States, the probability of being born in September is noticeably higher than the probability of being born in January, and these values can only be estimated by observing actual patterns of human births.

Another method is to make an assumption that all outcomes are equally likely, usually because of some physical property of the experiment. When all outcomes are equally likely, we compute the probability of an event $E$ by counting the number of outcomes in $E$ and dividing by the number of outcomes in the sample space. Given an event $E$, we denote the number of outcomes in $E$ by $|E|$.

For example, because (high quality) dice are close to perfect cubes, one believes that all six sides of a die are equally likely to occur.

Using the additivity of disjoint events (axiom 3 in the definition of probability),

$$P(\{1\}) + P(\{2\}) + P(\{3\}) + P(\{4\}) + P(\{5\}) + P(\{6\}) = P(\{1, 2, 3, 4, 5, 6\}) = 1$$

Since all six probabilities are equal and sum to 1, the probability of each face occurring is 1/6. In this case, the probability of an event $E$ can be computed by counting the number of elements in $E$ and dividing by the number of elements in $S$.

**Example 2.5.** Suppose that two six-sided dice are rolled and the numbers appearing on the dice are observed.[1]

The sample space $S$ is given by

[1]All dice in this book will be assumed to be **fair dice** unless otherwise stated. That is, the probability of each die landing on a face is one over the number of faces, and the results of some dice do not affect the results of the other dice.

$$\begin{pmatrix} (1,1),(1,2),(1,3),(1,4),(1,5),(1,6) \\ (2,1),(2,2),(2,3),(2,4),(2,5),(2,6) \\ (3,1),(3,2),(3,3),(3,4),(3,5),(3,6) \\ (4,1),(4,2),(4,3),(4,4),(4,5),(4,6) \\ (5,1),(5,2),(5,3),(5,4),(5,5),(5,6) \\ (6,1),(6,2),(6,3),(6,4),(6,5),(6,6) \end{pmatrix}$$

a. By the symmetry of the dice, we expect all 36 possible outcomes to be equally likely. So the probability of each outcome is $1/36$.
b. The event "The sum of the dice is 6" is represented by

$$E = \{(1,5),(2,4),(3,3),(4,2),(5,1)\}$$

c. The probability that the sum of two dice is 6 is given by

$$P(E) = \frac{|E|}{|S|} = \frac{5}{36},$$

which can be obtained by simply counting the number of elements in each set above.
d. Let $F$ be the event "At least one of the dice is a 2." This event is represented by

$$F = \{(2,1),(2,2),(2,3),(2,4),(2,5),(2,6),(1,2),(3,2),(4,2),(5,2),(6,2)\}$$

and the probability of $F$ is $P(F) = \frac{11}{36}$.
e. $E \cap F = \{(2,4),(4,2)\}$ and $P(E \cap F) = \frac{2}{36}$.
f. $P(E \cup F) = P(E) + P(F) - P(E \cap F) = \frac{5}{36} + \frac{11}{36} - \frac{2}{36} = \frac{14}{36}$.
g. $P(\overline{E}) = 1 - P(E) = \frac{31}{36}$.

**Remark. What is a probability, actually?** Broadly speaking there are two interpretations of probability, known as *frequentist* and *evidential*.

The frequentist interpretation: if the probability of an event $E$ is $p$, then if you repeat the experiment many times, then the proportion of times that the event occurs will eventually be close to $p$.

The evidential interpretation: the probability of an event measures the degree of certainty that a person has about whether the event occurs or not.

Both interpretations are reasonable. To understand the difference, consider the following thought experiment. Suppose that I am about to toss a coin, and I ask you to estimate the probability that the coin will land on heads. Not knowing any reason to think otherwise, you might say that you estimate it to be $p = 0.5$. From the frequentist point of view, that would mean that you believe that if I repeat the experiment infinitely many times, then the proportion of times that it is heads will converge to 0.5. From the certainty of belief point of view, you believe that each outcome (heads/tails) is equally likely.

Now, suppose that I flip the coin and look at it, but don't tell you whether it is heads or tails. At this point there is nothing random that I can repeat, so the frequentist interpretation cannot assign a probability to the result. However, your degree of certainty about the outcome hasn't changed, it is still $p = 0.5$. This example illustrates the importance of the random nature of statistical experiments.

**Remark (Odds).** Sometimes, probabilities are discussed in terms of *odds*. The odds of an event $E$ are $\frac{P(E)}{1-P(E)}$, the ratio of the probability of the event occurring to the probability of the event not occurring. Often these are expressed as a ratio with a colon. For example, the probability of rolling a six on a six-sided die is $1/6$, the probability of not rolling a six is $5/6$, and the odds of rolling a six are given as 1:5, or "one to five."

## 2.2 Simulations

The goal of probability and statistics is to understand the real world. Statistical experiments in the real world are usually slow and often expensive. Instead of running real world experiments, it is easier to model these experiments and then use a computer to imitate the results. This process goes by several names, such as *stochastic simulation*, *Monte Carlo simulation* or *probability simulation*. Since this is the only type of simulation we discuss in this book, we simply call it *simulation*.

The foundational mathematical theorem which justifies using simulations to estimate probabilities as described in this chapter and throughout the book is the Law of Large Numbers, which is described in detail in this context in Section 2.2.2, and in more generality in Section 3.2.

This book places simulation at the center of the study of probability. With a good understanding of how to simulate experiments, you can answer a wide range of questions involving probability. Later in the book, we will use simulation to explore how statistical methods behave under different assumptions about data. Simulation also plays a fundamental role in modern statistical methods such as resampling, bootstrapping, non-parametric statistics, and genetic algorithms.

### 2.2.1 Simulation with `sample`

For an experiment with a finite sample space $S = \{x_1, x_2, \ldots, x_n\}$, the R command `sample()` can simulate one or many trials of the experiment. Essentially, `sample` treats $S$ as a bag of outcomes, reaches into the bag, and picks one.

The syntax of sample is

```
sample(x, size, replace = FALSE, prob = NULL)
```

where the parameters are:

`x` The vector of elements from which you are sampling.

`size`
The number of samples you wish to take.

`replace`
Whether you are sampling with replacement or not. Sampling without replacement means that `sample` will not pick the same value twice, and this is the default behavior. Pass `replace = TRUE` to sample if you wish to sample with replacement.

`prob`
A vector of probabilities or weights associated with `x`. It should be a vector of nonnegative numbers of the same length as `x`. If the sum of `prob` is not 1, it will be normalized. If this value is not provided, then each element of `x` is considered to be equally likely.

The most straightforward use of sample is to choose one element of a vector "at random."

When people say "at random," they usually mean that all outcomes are equally likely to be chosen, and that is how `sample` operates. To get a random number from 1 to 10:

```
sample(x = 1:10, size = 1)
```

```
## [1] 2
```

The `size` argument tells `sample` how many random numbers you want:

```
sample(x = 1:10, size = 8)
```

```
## [1]  9 10  1  5  6  3  7  2
```

Observe that the eight numbers chosen are all different.

Unless you tell it otherwise, `sample` will never choose the same outcome twice.

If you ask `sample` for more than ten different numbers from 1 to 10, you get an error:

```
sample(x = 1:10, size = 30)
```

```
## Error in sample.int(length(x), size, replace, prob): cannot take a
## sample larger than the population when 'replace = FALSE'
```

The `replace` argument of `sample` determines whether sample is allowed to repeat values. The name "replace" comes from the model that sample has a bag of outcomes and is reaching into the bag to draw one. When `replace = FALSE`, the default, once sample chooses a value from the bag of outcomes it won't replace it into the bag. With `replace = TRUE`, sample draws an outcome from the bag, records it, and then puts it back into the bag.

Here we set `replace = TRUE` to get 20 random numbers from 1 to 10:

```
sample(x = 1:10, size = 20, replace = TRUE)
```

```
##  [1] 10  5 10  5  9  6 10  5 10 10  3  1 10  6  8  7  6  2  9 10
```

For `sample` and many other functions in R, you are not required to name the arguments with `x = ...` or `size = ...` as long as these come first and second in the function. For example:

```
sample(1:10, 20, replace = TRUE)
```

```
##  [1] 10  4  8  8  4  4  8  6  5  3  8  3  2  4  8  1  2 10  1  6
```

However, it is often clearer to explicitly name the arguments to complicated functions like `sample`. Use your best judgment, and include the parameter name if there is any doubt.

The `prob` argument of `sample` allows for sampling when outcomes are not all equally likely.

**Example 2.6.** In the United States, human blood comes in four types: O, A, B, and AB. These types occur with the following probability distribution:

| Type | $A$ | $AB$ | $B$ | $O$ |
|---|---|---|---|---|
| Probability | 0.40 | 0.04 | 0.11 | 0.45 |

We can sample thirty blood types from this distribution by defining a vector of blood types and a vector of their probabilities:

```
bloodtypes <- c("O", "A", "B", "AB")
bloodprobs <- c(0.45, 0.40, 0.11, 0.04)
sample(x = bloodtypes, size = 30, prob = bloodprobs, replace = TRUE)
```

```
##  [1] "A"  "O"  "AB" "A"  "O"  "A"  "A"  "O"  "O"  "O"  "O"  "A"  "O"
## [14] "A"  "A"  "A"  "B"  "A"  "A"  "B"  "A"  "O"  "O"  "A"  "A"  "O"
## [27] "O"  "A"  "A"  "O"
```

Observe that a large sample reproduces the original probabilities with reasonable accuracy:

```
sim_data <- sample(
  x = bloodtypes, size = 10000,
  prob = bloodprobs, replace = TRUE
)
table(sim_data)

## sim_data
##    A   AB    B    O
## 3998  425 1076 4501

table(sim_data) / 10000

## sim_data
##      A     AB      B      O
## 0.3998 0.0425 0.1076 0.4501
```

### 2.2.2 Using simulation to estimate probabilities

The goal of simulation is usually to estimate the probability of an event. This is a three-step process:

1. Simulate the experiment many times to produce a vector of outcomes.
2. Test if the outcomes are in the event to produce a vector of TRUE/FALSE.
3. Compute the `mean` of the TRUE/FALSE vector to compute the probability estimate.

Steps 1 and 2 can often be interesting problems, and their solutions can require creativity and expertise. Step 3 relies on the fact that R converts `TRUE` to 1 and `FALSE` to 0 when taking the mean of a vector. If we take the average of a vector of `TRUE/FALSE` values, we get the number of `TRUE` divided by the size of the vector, which is exactly the proportion of times that the event occurred. The theoretical justification for this procedure is given in Theorem 2.2 below.

We illustrate this process by reworking Example 2.5 using simulation.

**Example 2.7.** Suppose that two six-sided dice are rolled and the numbers appearing on the dice are added.

Simulate this experiment by performing 10,000 rolls of each die with `sample` and then adding the two dice:

```
die1 <- sample(x = 1:6, size = 10000, replace = TRUE)
die2 <- sample(x = 1:6, size = 10000, replace = TRUE)
sumDice <- die1 + die2
```

Let's take a look at the simulated data:

```
head(die1)

## [1] 1 4 1 2 5 3
```

```
head(die2)
```

```
## [1] 1 6 1 4 1 3
```

```
head(sumDice)
```

```
## [1]  2 10  2  6  6  6
```

Let $E$ be the event "the sum of the dice is 6," and $F$ be the event "at least one of the dice is a 2." We define these events from our simulated data:

```
eventE <- sumDice == 6
head(eventE)
```

```
## [1] FALSE FALSE FALSE  TRUE  TRUE  TRUE
```

```
eventF <- die1 == 2 | die2 == 2
head(eventF)
```

```
## [1] FALSE FALSE FALSE  TRUE FALSE FALSE
```

Here, $F$ is interpreted as "die 1 is a two or die 2 is a two" and uses R's "or" operator `|`.

From theory, $P(E) = \frac{5}{36} \approx 0.139$, and $P(F) = \frac{11}{36} \approx 0.306$. Using `mean` we find out what percentage of the time our events occurred in the simulation, which estimates the correct probabilities:

```
mean(eventE) # P(E)
```

```
## [1] 0.1409
```

```
mean(eventF) # P(F)
```

```
## [1] 0.2998
```

To estimate $P(E \cap F) = \frac{2}{36} \approx 0.056$ we use R's "and" operator `&`:

```
mean(eventE & eventF)
```

```
## [1] 0.0587
```

It is not necessary to store the TRUE/FALSE vectors in event variables. Here is an estimate of $P(E \cup F) = \frac{14}{36} \approx 0.389$:

```
mean((sumDice == 6) | (die1 == 2 | die2 == 2))
```

```
## [1] 0.382
```

The justification for using simulation to estimate probabilities is given by the following theorem, which in this context is sometimes referred to as *Bernoulli's Theorem*. It is a consequence of the more general *Law of Large Numbers*, Theorem 3.2.

**Theorem 2.2.** *Let $E$ be an event with probability $p$. Let $m_n$ be the number of times that $E$ occurs in $n$ repeated trials, where we assume the outcome of trials do not affect the outcome of other trials. Then*

$$\lim_{n \to \infty} \frac{m_n}{n} = p.$$

While we cannot simulate running infinitely many trials, we can take a large number of trials and expect that the proportion of successes will be approximately the true probability $p$. We also expect that a larger number of trials will, on average, give a better estimate of the true probability than a smaller number of trials. We investigate this further in Example 2.10.

### 2.2.3 Using `replicate` to repeat experiments

The `size` argument to `sample` allowed us to perform many repetitions of an experiment. For more complicated statistical experiments, we use the R function `replicate`, which can take a single R expression and repeat it many times.

The function `replicate` is an example of an *implicit loop* in R. Suppose that `expr` is one or more R commands, the last of which returns a single value. The call

```
replicate(n, expr)
```

repeats the expression stored in `expr` `n` times and stores the resulting values as a vector.

**Example 2.8.** Estimate the probability that the sum of seven dice is larger than 30.

To simulate this event once, we can use sample to roll seven dice, sum to add them, and then test for the event "the sum is larger than 30":

```
dice <- sample(x = 1:6, size = 7, replace = TRUE) # roll seven dice
sum(dice) > 30 # test if the event occurred
```

```
## [1] FALSE
```

The result of this single simulation was `FALSE`. Using `replicate` repeats the experiment many times:

```
replicate(20, {
  dice <- sample(x = 1:6, size = 7, replace = TRUE) # roll seven dice
  sum(dice) > 30 # test if the event occurred
})
```

```
##  [1] FALSE FALSE FALSE FALSE FALSE FALSE FALSE  TRUE FALSE  TRUE FALSE
## [12] FALSE FALSE FALSE FALSE FALSE  TRUE FALSE FALSE FALSE
```

The curly braces `{` and `}` are required to replicate more than one R command. When you put multiple commands inside `{}` you are creating a *code block*. A code block acts like a single statement, and only the result of the last command is saved in the vector via `replicate`.

Using multiple lines for a code block is not required but highly recommended for readability. It is also legal to put the entire code block on a single line and separate each command in the block with a semicolon.

Finally, we want to compute the probability of the event. We replicate 10,000 times for a reasonably accurate estimate:

```
event <- replicate(10000, {
  dice <- sample(x = 1:6, size = 7, replace = TRUE) # roll seven dice
  sum(dice) > 30 # test if the event occurred
})
mean(event)
```

```
## [1] 0.0963
```

When rolling seven dice, there is about a 9.63% probability the sum will be larger than 30. How accurate is our estimate? It is often a good idea to repeat a simulation a couple of times to get an idea about how much variance there is in the results. Running the code a few more times gave answers 0.0947, 0.091, 0.0965, and 0.0867. It seems safe to report the answer as roughly 9%.

The more replications you perform with `replicate()`, the more accurate you can expect your simulation to be. On the other hand, replications can be slow. For events which are not rare, 10,000 trials runs quickly and gives an answer accurate to about two decimal places.

For complicated simulations, we *strongly* recommend that you follow the workflow as presented above; namely,

1. Write code that performs the experiment a single time.
2. Replicate the experiment a small number of times and check the results:

   `replicate(100, {  EXPERIMENT GOES HERE  }))`
3. Replicate the experiment a large number of times and store the result:

   `event <- replicate(10000, {  EXPERIMENT GOES HERE  }))`
4. Compute probability using `mean(event)`.

It is much easier to trouble-shoot your code this way, as you can test each line of your simulation separately.

**Example 2.9.** Three dice are thrown. Estimate the probability that the largest value is a four.

Here is one trial:

```
die_roll <- sample(1:6, 3, TRUE)
max(die_roll) == 4
```

```
## [1] TRUE
```

Here are a few trials, and we observe that sometimes the event occurs and sometimes it does not:

```
replicate(20, {
  die_roll <- sample(1:6, 3, TRUE)
  max(die_roll) == 4
})
```

```
##  [1] FALSE FALSE FALSE FALSE FALSE FALSE FALSE FALSE FALSE FALSE FALSE
## [12] FALSE FALSE FALSE FALSE  TRUE  TRUE FALSE FALSE FALSE
```

Finally, perform many trials and compute the probability of the event:

```
event <- replicate(10000, {
  die_roll <- sample(1:6, 3, TRUE)
  max(die_roll) == 4
})
mean(event)
```

```
## [1] 0.1737
```

When three dice are thrown, the probability that the largest value is four is approximately 17%.

**Example 2.10.** The purpose of this example is to investigate the rate at which the proportion of successes converges to the true probability in a specific setting. To do so, we need to choose an experiment and an event $E$ for which we know $P(E)$.

Suppose two dice are rolled. Let $E$ denote the event that the sum of the dice is six. By counting, we know that $P(E) = 5/36 \approx .138888$. If we use 10 trials to estimate the probability, then the closest we can get is 0.1. We recommend running the code below several times to see that we sometimes get 0.1, but many times also get 0, 0.2 or 0.3.

```
mean(replicate(10, {
  sum(sample(1:6, 2, T)) == 6
}))
```

```
## [1] 0.3
```

On the other hand, when we use 200 trials, we are unlikely to be wrong in the first significant digit.

```
mean(replicate(200, {
  sum(sample(1:6, 2, T)) == 6
}))
```

```
## [1] 0.135
```

If we increase to 10,000 trials, we are even closer on average.

```
mean(replicate(10000, {
  sum(sample(1:6, 2, T)) == 6
}))
```

```
## [1] 0.1413
```

In Figure 2.1, we estimated the probability of $E$ for simulations ranging from 10 trials to 10,000 trials and then plotted the results.

The downside to using more trials is that it takes more time to do the simulation, which can become very important if each trial itself takes a considerable amount of time. As a general rule of thumb, we have found that using 10,000 trials is a good compromise between speed and accuracy.

**Example 2.11.** A fair coin is repeatedly tossed.[2] Estimate the probability that you observe heads for the third time on the 10th toss.

In this example, an outcome of the experiment is ten tosses of the coin. The event "you observe heads for the third time on the 10th toss" is complicated, and most of the work involves testing whether that event occurred.

As before, we build this up in stages. Begin by simulating an outcome, a sample of ten tosses of a coin:

```
coinToss <- sample(c("H", "T"), 10, replace = TRUE)
coinToss
```

```
##  [1] "T" "T" "T" "T" "H" "T" "T" "H" "T" "T"
```

In order for the event to occur, we need for there to be exactly three heads, so we count the number of heads and check whether it is equal to three:

[2]All coins in this book, unless otherwise stated, will be **fair coins** in the sense that the probability of heads is 1/2 and successive trials are independent.

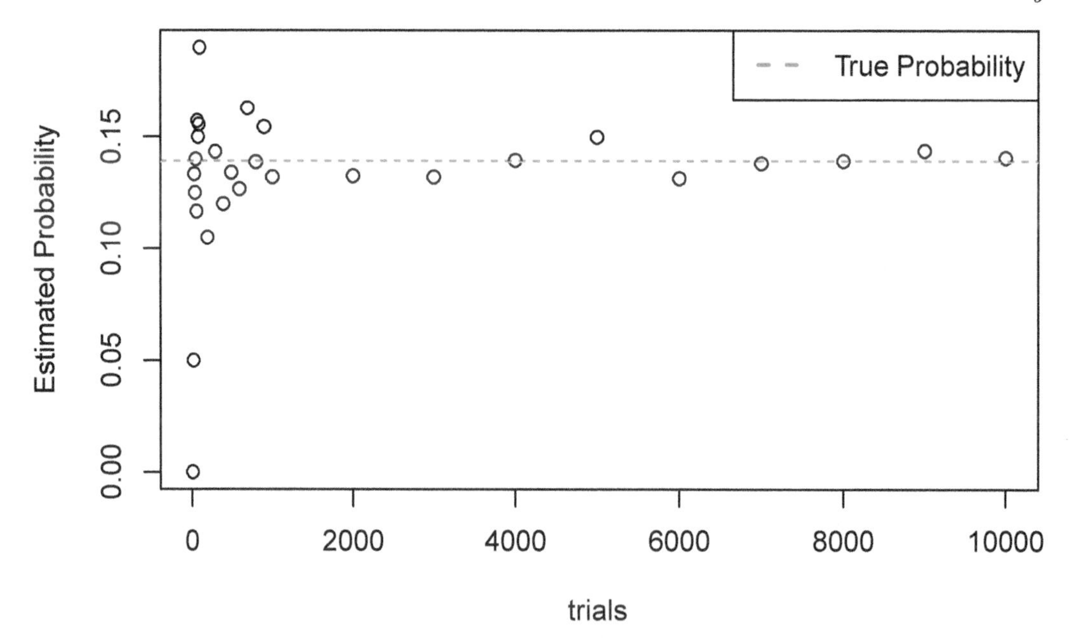


**FIGURE 2.1** Illustration of the Law of Large Numbers. Probability estimates from simulation converge to the true probability as the number of trials increases.

```
sum(coinToss == "H")
```

```
## [1] 2
```

```
sum(coinToss == "H") == 3
```

```
## [1] FALSE
```

Next, we also need to make sure that we had only two heads in the first nine tosses. So, we look only at the first nine tosses:

```
coinToss[1:9]
```

```
## [1] "T" "T" "T" "T" "H" "T" "T" "H" "T"
```

and add up the heads observed in the first nine tosses:

```
sum(coinToss[1:9] == "H") == 2
```

```
## [1] TRUE
```

Note that both of those have to be true in order for the event to occur:

```
sum(coinToss == "H") == 3 & sum(coinToss[1:9] == "H") == 2
```

```
## [1] FALSE
```

We put this inside `replicate` and compute the probability:

```
event <- replicate(10000, {
  coinToss <- sample(c("H", "T"), 10, replace = TRUE)
  (sum(coinToss == "H") == 3) & (sum(coinToss[1:9] == "H") == 2)
})
```

```
mean(event)

## [1] 0.0369
```

The probability of observing heads for the third time on the 10th toss is about 3%.

The test that there were three total heads but only two on the first nine tosses is not the only way to approach this problem. Here are two other tests that give the same result by looking at the tenth coin toss:

```
sum(coinToss == "H") == 3 & coinToss[10] == "H"

## [1] FALSE

sum(coinToss[1:9] == "H") == 2 & coinToss[10] == "H"

## [1] FALSE
```

**Example 2.12 (The Birthday Problem).** Estimate the probability that out of 25 randomly selected people, at least two will have the same birthday. Assume that all birthdays are equally likely, except that none are leap-day babies.

An outcome of this experiment is 25 randomly selected birthdays, and the event $B$ is that at least two birthdays are the same. Simulating the experiment is straightforward, with sample:

```
birthdays <- sample(x = 1:365, size = 25, replace = TRUE)
birthdays

##  [1] 324 167 129 299 270 187 307  85 277 362 330 263 329  79 213  37 105
## [18] 217 165 290 362  89 289 340 326
```

Next, we test for the event $B$. In order to do this, we need to be able to find any duplicates in a vector. R has many, many functions that can be used with vectors. For most things that you want to do, there will be an R function that does it. In this case it is `anyDuplicated()`, which returns the location of the first duplicate if there are any, and zero otherwise. The important thing to learn here isn't necessarily this particular function, but rather the fact that most tasks are possible via some built-in functionality.

```
anyDuplicated(birthdays)

## [1] 21
```

It happens that our sample *did* have two of the same birthday. At location 21 of the birthday vector, the number 362 appears. That same number showed up earlier in location 10. The event $B$ occurred in this example, and we can test for it by checking that the result of `anyDuplicated` is larger than 0. Putting it all together:

```
eventB <- replicate(n = 10000, {
  birthdays <- sample(x = 1:365, size = 25, replace = TRUE)
  anyDuplicated(birthdays) > 0
})
mean(eventB)

## [1] 0.5631
```

The probability of this event is approximately 0.56. Interestingly, we see that it is actually quite likely that a group of 25 people will contain two with the same birthday.

**Try It Yourself.**
Modify the above code to take into account leap years.

**Example 2.13.** Three numbers are picked uniformly at random from the interval $(0,1)$. What is the probability that a triangle can be formed whose side-lengths are the three numbers that you chose?

Solution: We need to be able to simulate picking three numbers at random from the interval $(0,1)$. We will see later in the book that the way to do this is via `runif(3, 0, 1)`, which returns 3 numbers randomly chosen between 0 and 1. We then need to check whether the sum of the two smaller numbers is larger than the largest number. We use the `sort` command to sort the three numbers into increasing order, as follows:

```
event <- replicate(10000, {
  x <- sort(runif(3, 0, 1))
  sum(x[1:2]) > x[3]
})
mean(event)
```

```
## [1] 0.4979
```

**Example 2.14.** According to Rick Wicklin,[3] the proportion of M&M's of various colors produced in the New Jersey M&M factory are as shown in Table 2.1.

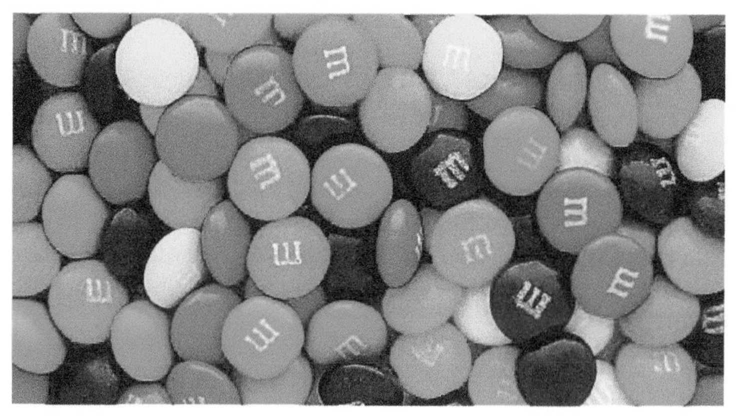

*Image credit: Evan Amos.*

[3] C Purtill, "A Statistician Got Curious about M&M Colors and Went on an Endearingly Geeky Quest for Answers," *Quartz*, March 15, 2017, https://qz.com/918008/the-color-distribution-of-mms-as-determined-by-a-phd-in-statistics/.

**TABLE 2.1** Color distribution of M&M's.

| Color | Percentage |
|---|---:|
| Blue | 25.0 |
| Orange | 25.0 |
| Green | 12.5 |
| Yellow | 12.5 |
| Red | 12.5 |
| Brown | 12.5 |

If you buy a bag from the New Jersey factory that contains 35 M&M's, what is the probability that it will contain exactly 9 Blue and 5 Red M&M's?

To do this, we can use the `prob` argument in the `sample` function, as follows.

```
mm.colors <- c("Blue", "Orange", "Green", "Yellow", "Red", "Brown")
mm.probs <- c(25, 25, 12.5, 12.5, 12.5, 12.5)
bag <- sample(
  x = mm.colors,
  size = 35,
  replace = TRUE,
  prob = mm.probs
)
sum(bag == "Blue") # counts the number of Blue M&M's

## [1] 11

event <- replicate(10000, {
  bag <- sample(
    x = mm.colors,
    size = 35,
    replace = TRUE,
    prob = mm.probs
  )
  sum(bag == "Blue") == 9 & sum(bag == "Red") == 5
})
mean(event)

## [1] 0.0291
```

## 2.3 Conditional probability and independence

Sometimes when considering multiple events, we have information that one of the events has occurred. This new information requires us to reconsider the probability that the other event occurs. For example, suppose that you roll two dice and one of them falls off of the table where you cannot see it, while the other one shows a 4. We would want to update the probabilities associated with the sum of the two dice based on this information. The new probability that the sum of the dice is 2 would be 0, the new probability that the sum of the dice is 5 would be 1/6 because that is just the probability that the die that we cannot see is a "1," and the new probability that the sum of the dice is 7 would also be 1/6 (which is the same as its original probability).

Formally, we have the following definition.

**Definition 2.4.** Let $A$ and $B$ be events in the sample space $S$, with $P(B) \neq 0$. The *conditional probability* of $A$ given $B$ is

$$P(A|B) = \frac{P(A \cap B)}{P(B)}$$

We read $P(A|B)$ as "the probability of $A$ given $B$."

It is important to keep straight in your mind that the fixed idiom $P(A|B)$ means the probability of $A$ given $B$, or the probability that $A$ occurs given that $B$ occurs. $P(A|B)$ does *not* mean the probability of some event called $A|B$.

In mathematics, the vertical bar symbol | is used for conditional probability, and you would write $A \cup B$ for the event "$A$ or $B$." In R, the vertical bar denotes the *or* operator. You do not use the vertical bar in R to work with conditional probability.

The general process of assuming that $B$ occurs and making computations under that assumption is called *conditioning on* $B$. Note that in order to condition on $B$ in the definition of $P(A|B)$, we must assume that $P(B) \neq 0$, since otherwise we would get $\frac{0}{0}$, which is undefined. This also makes some intuitive sense. If we assume that a probability zero event occurs, then probability of further events conditioned on that would need to be undefined.

**Example 2.15.** Two dice are rolled. What is the probability that both dice are 4, given that the sum of two dice is 8?

Solution: Let $A$ be the event "both dice are 4" and $B$ be the event "the sum is 8." Then

$$P(A|B) = P(A \cap B)/P(B) = \frac{1/36}{5/36} = 1/5.$$

Rolling two 4s is the hardest way to get an 8. Check that the probability of rolling one three and one five is 2/5, and also for one two and one six.

With conditional probability, the order of the two events is important. Suppose we reverse the order, and ask: "What is the probability that the sum of two dice is 8, given that both dice are 4?" Now the answer is 1, because if both dice are 4, then the sum is certainly 8. Formally, with the events defined as in Example 2.15:

$$P(B|A) = P(B \cap A)/P(A) = \frac{1/36}{1/36} = 1.$$

**Proposition 2.1.** *Two simple facts about conditional probability are:*

1. $P((A \cap B)|B) = P(A|B)$.
2. $P(A \cup B|B) = 1$.

*Proof.* In words, statement 1 says that the probability of "$A$ and $B$" given $B$ is the probability of $A$ given $B$. Arguing informally, assume that we know that $B$ occurs. Then the probability that both $A$ and $B$ occur is just the probability that $A$ occurs. Using set notation:

$$\begin{aligned} P((A \cap B)|B) &= P((A \cap B) \cap B)/P(B) \\ &= P(A \cap (B \cap B))/P(B) \\ &= P(A \cap B)/P(B) \\ &= P(A|B). \end{aligned}$$

We included parentheses around $(A \cap B)$ above, but we did not need to. Remember, there is no event called "$B|B$," so the only possible interpretation of $P(A \cap B|B)$ is $P((A \cap B)|B)$.

Statement 2 is left as Exercise 2.24.

■

### 2.3.1 Independent events

We have seen examples where the probability of $A$ given $B$ can be larger than $P(A)$, smaller than $P(A)$, or equal to $P(A)$. Of particular interest are pairs of events $A$ and $B$ such that knowledge that one of the events occurs does not impact the probability that the other event occurs.

**Definition 2.5.** Two events are said to be *independent* if knowledge that one event occurs does not give any probabilistic information as to whether the other event occurs. Formally, we say that $A$ and $B$ are independent if $P(A \cap B) = P(A)P(B)$.

Events $A$ and $B$ are said to be *dependent* if they are not independent.

It is not immediately clear why the formal statement in the definition of independence implies the intuitive statement that "the knowledge that one event occurs does not give any probabilistic information as to whether the other event occurs." To see that, we assume $P(B) \neq 0$ and compute:

$$P(A|B) = P(A \cap B)/P(B) = \frac{P(A)P(B)}{P(B)} = P(A)$$

If $P(A) \neq 0$, a similar computation shows that $P(B|A) = P(B)$ and proves the following theorem.

**Theorem 2.3.** *Let $A$ and $B$ be events with non-zero probability in the sample space $S$. The following are equivalent:*

1. *$A$ and $B$ are independent.*
2. *$P(A \cap B) = P(A)P(B)$.*
3. *$P(A|B) = P(A)$.*
4. *$P(B|A) = P(B)$.*

**Remark.** Part 2 of Theorem 2.3 is called the *multiplication rule* for independent events.

Usually independence will be an assumption that we make about events and we will use that assumption to apply the multiplication rule. Here is a simple example.

**Example 2.16.** You and your friend each purchase a bag of M&M's, with colors coming from the probability distribution shown in Example 2.14. What is the probability that both of you get a blue M&M as the first candy out of your bag?

There are two events here: $A_1$ is that your first M&M is blue, and $A_2$ is that your friend's first M&M is blue. It is reasonable to assume these are independent, since what color you draw from your bag should have no effect on the color of M&M your friend draws from their bag. Since both $P(A_1) = 0.25$ and $P(A_2) = 0.25$, the multiplication rule shows that

$$P(\text{you both get blue}) = P(A_1 \cap A_2) = P(A_1)P(A_2) = 0.25 \cdot 0.25 = 0.0625.$$

There is a 6.25% chance you both get a blue M&M first.

It is important to develop an intuition about independence in order to determine whether the assumption of independence is reasonable.

**Example 2.17.** Consider the following scenarios, and determine whether the events indicated are most likely dependent or independent.

1. A day in the last 365 days is selected at random. Event $A$ is that the high temperature in St. Louis, Missouri on that day was greater than 90 degrees. Event $B$ is that the high temperature on that same day in Cape Town, South Africa was greater than 90 degrees.
2. Two coins are flipped, and $A$ is the event that the first coin lands on heads, while $B$ is the event that the second coin lands on heads.
3. Six patients are given a tuberculosis skin test, which requires a professional to estimate the size of the reaction to the tuberculin agent. Two professionals, Alexis and Angel, are randomly chosen. Let $A$ be the event that Alexis estimates the size of the reaction in each of patients 1-5 to be larger than Angel does. Let $B$ be the event that Alexis estimates the size of the reaction in patient 6 to be larger than Angel does.

In scenario 1, the events are dependent. If we know that the high temperature in St. Louis was greater than 90 degrees, then the day was most likely a day in June, July, August, or September, which gives us probabilistic information about whether the high temperature in Cape Town was greater than 90 degrees on that day. (In this case, it means that it was very unlikely, since that is winter in the southern hemisphere.)

In scenario 2, the events are independent, or at least approximately so. One could argue that knowing that one of the coins is heads means the person tossing the coins might be more likely to obtain heads when tossing coins. However, the potential effect here seems so weak based on experience, that it is a reasonable assumption that the events are independent.

In scenario 3, it may be inadvisable to assume that the events are independent. Of course, they may be. It could be that Alexis and Angel are well-trained, and there is no bias in their measurements. However, it is also possible that there is something systematic about how they measure the reactions so that one of them usually measures it as larger than the other one does. Knowing $A$ may be an indication that Alexis does systematically measure reactions as larger than Angel does. (Of course, it would also be interesting to know which one was closer to the true value, but that is not what we are worried about at this point.) Later, we will develop tools that will allow us to make a more quantitative statement about this type of problem.

**Example 2.18.** Two dice are rolled. It is reasonable to assume that events concerning one die are independent of events concerning the other die. For more interesting events involving both dice, we may use the definition of independence to check.

Let $A$ be the event "the sum of the dice is 8," let $B$ be the event "the sum of the dice is 7," and let $C$ be the event "The first die is a 5." Show that:

1. $A$ and $B$ are dependent.
2. $A$ and $C$ are dependent.
3. $B$ and $C$ are independent.

For part 1, $P(A) = 5/36$, and $P(B) = 6/36$ but $P(A \cap B) = 0$ since it's not possible to roll an 8 and a 7 at the same time ($A$ and $B$ are disjoint). Since $P(A)P(B) \neq P(A \cap B)$ the events are dependent.

For part 2, $P(A) = 5/36$ and $P(A|C) = 1/6$ since given a 5 on the first die, there is a $1/6$ chance of rolling a 3 on the second to make a sum of 8. Since $P(A) \neq P(A|C)$, the events $A$ and $C$ are not independent.

Finally, $P(B) = 6/36 = 1/6$ and $P(B|C) = 1/6$. Therefore, $B$ and $C$ are independent. You might expect that knowing one die tells you something about the sum, as it does in part 2. However, a roll of 7 is special. It happens $1/6$ of the time, and if you know the value of one die there is *still* a $1/6$ chance the second die will be the one number needed to make 7.

We now extend the definition of independence to mutual independence of multiple events. The intuition remains the same: mutual independence means that knowing something about some of these events gives no probabilistic information about the others. However, the notion of mutual independence is stronger than just saying that every pair of the events are independent – see Exercise 2.25. We require a multiplication rule for every pair, triple, and so on for every sub-collection of these events.

**Definition 2.6.** A collection of events $A_1, A_2, \ldots, A_n \subset S$ are *mutually independent* if for any sub-collection $A_{i_1}, \ldots, A_{i_k}$ we have

$$P(A_{i_1} \cap A_{i_2} \cap \cdots \cap A_{i_k}) = P(A_{i_1}) \cdot P(A_{i_2}) \cdot \ldots \cdot P(A_{i_k})$$

In this book, we will use Definition 2.6 exclusively as a multiplication rule, where events $A_1, \ldots, A_n$ are assumed to be independent and we apply $P(A_1 \cap \cdots \cap A_n) = P(A_1) \cdot \ldots \cdot P(A_n)$.

**Example 2.19.** Suppose you roll five ordinary dice. What is the probability that all of them show six?

Let event $D_i$ be the event that the $i^{\text{th}}$ die shows six. We don't expect the dice to affect each other, so the events $D_1, \ldots, D_5$ are mutually independent. Then

$$P(\text{all sixes}) = P(D_1 \cap D_2 \cap D_3 \cap D_4 \cap D_5) =$$

$$P(D_1) \cdot P(D_2) \cdot P(D_3) \cdot P(D_4) \cdot P(D_5) = \left(\frac{1}{6}\right)^5 \approx 0.000129.$$

In the next example, we illustrate a method to compute probabilities involving the idea of "at least one" or "any." The trick is to convert logical OR into logical AND using DeMorgan's Law. Example 2.20 also illustrates the use of the `any` command in R, which takes a T/F vector and detects if any entry is TRUE.

**Example 2.20.** Suppose you roll five ordinary dice. What is the probability that at least one six appears?

Let $A$ be the event "at least one six appears." Let $D_i$ be the event that the $i^{\text{th}}$ die is a six. Then $A = D_1 \cup D_2 \cup D_3 \cup D_4 \cup D_5$. Since the $D_i$ are not disjoint, we cannot use the addition rule for disjoint events. However, the $D_i$ events are independent. Observe that "not $A$" is the event "no sixes appear," so that

$$\text{not } A = (\text{not } D_1) \text{ AND } (\text{not } D_2) \text{ AND } \ldots \text{ AND } (\text{not } D_5)$$

Using the multiplication rule for independent events

$$P(\text{not } A) = P(\text{not } D_1) \cdot P(\text{not } D_2) \cdot \ldots \cdot P(\text{not } D_5) = (5/6)^5$$

Finally, $P(A) = 1 - P(\text{not } A) = 1 - (5/6)^5 \approx 0.598$.

To do this with simulation:

```
mean(replicate(10000, {
  roll5 <- sample(1:6, 5, replace = TRUE)
  any(roll5 == 6)
}))
```

```
## [1] 0.5967
```

### 2.3.2 Simulating conditional probability

Simulating conditional probabilities is challenging. In order to estimate $P(A|B)$, we will estimate $P(A \cap B)$ and $P(B)$ and then divide the two answers. This is not the most efficient or best way to estimate $P(A|B)$, but it is easy to do with the tools that we already have developed.

**Example 2.21.** Two dice are rolled. Estimate the conditional probability that the sum of the dice is at least 10, given that at least one of the dice is a 6.

First, we estimate the probability that the sum of the dice is at least 10 *and* at least one of the dice is a 6.

```
eventAB <- replicate(10000, {
  dieRoll <- sample(1:6, 2, replace = TRUE)
  (sum(dieRoll) >= 10) && (6 %in% dieRoll)
})
probAB <- mean(eventAB)
```

Next, we estimate the probability that at least one of the dice is a 6.

```
eventB <- replicate(10000, {
  die_roll <- sample(1:6, 2, replace = TRUE)
  6 %in% die_roll
})
probB <- mean(eventB)
```

Finally, we take the quotient.

```
probAB / probB
```

```
## [1] 0.4560601
```

The correct answer is $P(A \cap B)/P(B) = \frac{5/36}{11/36} = 5/11 \approx 0.4545$.

### 2.3.3 Bayes' Rule and conditioning

The Law of Total Probability allows the computation of the probability of an event by "conditioning" on another event.

**Theorem 2.4 (Law of Total Probability).** *Let $A$ and $B$ be events in the sample space $S$. Then*

$$P(A) = P(A \cap B) + P(A \cap \overline{B}) = P(A|B)P(B) + P(A|\overline{B})P(\overline{B})$$

This formula breaks the probability of $A$ into two pieces, one where $B$ happens and one where $B$ does not happen.

*Proof.* Since the sample space $S = B \cup \overline{B}$, $A = A \cap S = (A \cap B) \cup (A \cap \overline{B})$. Because $A \cap B$ and $A \cap \overline{B}$ are disjoint,

$$P(A) = P\left((A \cap B) \cup (A \cap \overline{B})\right) = P(A \cap B) + P(A \cap \overline{B}).$$

The second equality follows from Definition 2.4, the definition of conditional probability. ∎

**Example 2.22.** The name "Mary" was given to 7065 girls in 1880, and to 11475 girls in 1980. There were 97583 girls born in 1880, and 177907 girls born in 1980. Suppose that a randomly selected girl born in 1880 or 1980 is chosen. What is the probability that the girl's name is "Mary?"

To solve this, let's let $A$ be the event that the randomly selected girl's name is Mary. If we knew what year the girl was born in, then we would have a good idea what to do. We don't, so we condition on the birth year. Let $B$ be the event that the randomly selected girl was born in 1880.

Applying the Law of Total Probability,

$$\begin{aligned} P(A) &= P(A|B)P(B) + P(A|\overline{B})P(\overline{B}) \\ &= \frac{7065}{97583}\frac{97583}{97583+177907} + \frac{11475}{177907}\frac{177907}{97583+177907} \\ &= 0.0676 \end{aligned}$$

The probability that the randomly selected girl's name is Mary is 0.0676.

Bayes' Rule is a simple statement about conditional probabilities that allows the computation of $P(A|B)$ from $P(A)$.

**Theorem 2.5 (Bayes' Rule).** *Let $A$ and $B$ be events in the sample space $S$.*

$$P(A|B) = \frac{P(B|A)P(A)}{P(B)} = \frac{P(B|A)P(A)}{P(B|A)P(A) + P(B|\overline{A})P(\overline{A})}$$

*Proof.* From the definition of conditional probability (Definition 2.4), $P(A \cap B) = P(A|B)P(B)$. Switching $A$ and $B$ gives $P(B \cap A) = P(B|A)P(A)$. Since $A \cap B = B \cap A$, we have $P(A|B)P(B) = P(B|A)P(A)$. Dividing both sides by $P(B)$ proves the first equality. The second equality is simply the Law of Total Probability applied to $P(B)$ in the denominator. ■

Using the evidential interpretation of probability, Bayes' Rule forms the foundation of Bayesian statistics. Suppose we have some prior evidence about the event $A$, in the form of $P(A)$. Then $P(A|B)$ is the knowledge we have about $A$ after accounting for the information that $B$ is true. With this interpretation, the rule is a way to update our evidence for $A$ given new information.

**Example 2.23.** In a certain hotel near the US/Canada border, 70% of hotel guests are American and 30% are Canadian. It is known that 40% of Americans wear white socks, while 20% of Canadians wear white socks. Suppose you randomly select a person and observe that they are wearing white socks. What is the probability that the person is Canadian?

Let $A$ be the event that a randomly person selected is Canadian. We are given that $P(A) = 0.3$ as prior knowledge. Let $B$ denote the event that a randomly selected person is wearing white socks. We are asked to find $P(A|B)$, the probability that a randomly selected person is Canadian, given that they are wearing white socks. Since relatively few Canadians wear white socks, $P(A|B)$ should be lower than 0.3. Bayes' Rule computes the probability exactly:

$$\begin{aligned} P(A|B) &= \frac{P(B|A)P(A)}{P(B|A)P(A) + P(B|\overline{A})P(\overline{A})} \\ &= \frac{0.2 \times 0.3}{0.2 \times 0.3 + 0.4 \times 0.7} = 0.176 \end{aligned}$$

Now that we know the person is wearing white socks, there is only a 0.176 probability they are Canadian.

There is a more general version of the Law of Total Probability and Bayes' Rule.

**Definition 2.7.** We say that $A_1, \ldots, A_k$ is a *partition* of the sample space $S$ if $\cup_{i=1}^k A_i = S$ and $A_i \cap A_j = \emptyset$ whenever $i \neq j$.

**Theorem 2.6 (Law of Total Probability and Bayes Rule).** *Let $A_1, \ldots, A_k$ be a partition of the sample space $S$ and let $B$ be an event. Then,*

$$P(B) = \sum_{i=1}^k P(B \cap A_i) = \sum_{i=1}^k P(B|A_i)P(A_i)$$

*and*

$$P(A_j|B) = \frac{P(B|A_j)P(A_j)}{\sum_{i=1}^k P(B|A_i)P(A_i)}$$

Exercise 2.32 requires this more general form.

## 2.4 Counting arguments

Given a sample space $S$ consisting of equally likely simple events, and an event $E$, recall that $P(E) = \frac{|E|}{|S|}$. For this reason, it can be useful to be able to carefully enumerate the elements in a set. While an interesting topic, this is not a point of emphasis of this book, as (1) we assume that students have seen some basic counting arguments in the past and (2) we emphasize simulation techniques.

This text will only work with two counting rules:

**Proposition 2.2 (Rule of product).** *If there are $m$ ways to do something, and for each of those $m$ ways there are $n$ ways to do another thing, then there are $m \times n$ ways to do both things.*

**Proposition 2.3 (Combinations).** *The number of ways of choosing $k$ distinct objects from a set of $n$ is given by*

$$\binom{n}{k} = \frac{n!}{k!(n-k)!}$$

The R command for computing $\binom{n}{k}$ is `choose(n,k)`.

**Example 2.24.** A coin is tossed 10 times. Some possible outcomes are HHHHHHHHHH, HTHTHTHTHT, and HHTHTTHTTT. Since each toss has two possibilities, the rule of product says that there are $2 \cdot 2 \cdot 2 \cdot 2 \cdot 2 \cdot 2 \cdot 2 \cdot 2 \cdot 2 \cdot 2 = 2^{10} = 1024$ possible outcomes for the experiment. We expect each possible outcome to be equally likely, so the probability of any single outcome is $1/1024$.

Let $E$ be the event "we flipped exactly three heads." This might happen as the sequence HHHTTTTTTT, or TTTHTHTTHT, or many other ways. What is $P(E)$? To compute the probability, we need to count the number of possible ways that three heads may appear. Since the three heads may appear in any of the ten slots, the answer is

$$|E| = \binom{10}{3} = \frac{10 \times 9 \times 8}{3 \times 2 \times 1} = 120.$$

Then $P(E) = 120/1024 \approx 0.117$. We can also estimate $P(E)$ with simulation:

```
event <- replicate(10000, {
  flips <- sample(c("H", "T"), 10, replace = TRUE)
  heads <- sum(flips == "H")
  heads == 3
})
mean(event)
```

```
## [1] 0.1211
```

**Example 2.25.** Suppose that in a class of 10 boys and 10 girls, 5 students are randomly chosen to present work at the board.

a. What is the probability that all 5 students are boys?
b. What is the probability that exactly 4 of the students are girls?

Let $E$ be the event "all 5 students are boys." The sample space consists of all ways of choosing 5 students from a class of 20, so `choose(20,5) = 15504`. The event $E$ consists of all ways of choosing 5 boys from a group of 10, so `choose(10, 5) = 252`. Therefore, the probability is `252/15504 = .016`.

Next, let $A$ be the event "exactly 4 of the students are girls." The sample space is still the same. The event $A$ can be broken down into two tasks: choose the 4 girls and choose the 1 boy. By the multiplication principle, there are `choose(10, 4) * choose(10,1) = 2100` ways of doing that. Therefore, the probability is `2100/15504 = .135`.

**Example 2.26.** A deck of 52 cards has four suits[4] and 13 ranks in each suit: 2,3,4,5,6,7,8,9,10,J,Q,K,A. If you are dealt two cards, what is the probability they have the same rank?

The sample space is all possible two card hands. Since there are 52 cards in the deck, there are $\binom{52}{2} = 1326$ possible hands.

The event "both cards have the same rank" can be broken down into two choices. First, choose the rank those cards will have. There are 13 choices. Next choose two of the four cards with that rank for your hand, for which there are $\binom{4}{2} = 6$ choices. Then there are $13 \times 6 = 78$ ways for your hand to have a pair of the same rank.

The probability of getting a pair is $78/1326 \approx 0.059$.

To estimate with simulation, we build a deck of cards by thinking of the ranks as the numbers 2 through 14 and then using 'rep' to produce four copies:

```
deck <- rep(2:14, 4)
pair <- replicate(10000, {
```

[4] The names of the suits depend on the country of origin. French-suited cards are the ones most commonly used, and have Hearts, Diamonds or Tiles, Clovers or Clubs, and Pikes or Spades. German-suited cards consist of Hearts, Bells, Acorns, and Leaves, while Italian-suited cards consist of Swords, Cups, Coins, and Batons.

```
  hand <- sample(deck, 2)
  hand[1] == hand[2]
})
mean(pair)

## [1] 0.059
```

## Vignette: Negative surveys

Suppose you are trying to get information about a relatively sensitive topic on a survey. For example, you might want to determine how much money people owe on their credit cards at a given moment in time. While this information is not terribly sensitive, it is sensitive enough that some people might not feel comfortable telling the truth in a survey.

One way to combat this problem is with a negative survey.[5] Instead of asking the participant to answer the question correctly, you ask the participant to select one of the answers that is **not** true according to some probability distribution. As an example, we could ask the following:

How much money do you **not** owe on your credit cards as of today?

a. Zero dollars.
b. Between 1 and 1000 dollars.
c. More than 1000 dollars.

We instruct the respondent to randomly select one of the answers that is not correct. In practice, we might want to have more possible answers; we are using three answers to illustrate the mathematics behind the scenes.

Suppose you collect 1000 surveys, and your proportion of answers are:

a. 50%
b. 30%
c. 20%

Now you want to figure out the percentages of people who **owe** certain values, rather than what they do not owe. We use 0.5, 0.3, and 0.2 as our estimates for the true proportion of people who will answer a, b, and c, respectively, when given this survey. Let $A$ be the event that a person owes zero dollars, $B$ be the event they owe between 1 and 1000 dollars, and $C$ be the event that they owe more than 1000 dollars. Let $D$ be the event they **select choice a on the survey**, $E$ be the event they select choice b on the survey, and $F$ be the event they select choice c on the survey. By Theorem 2.4, the Law of Total Probability,

$$\begin{pmatrix} P(D) & = & 0 \times P(A) & + & 1/2 \times P(B) & + & 1/2 \times P(C) \\ P(E) & = & 1/2 \times P(A) & + & 0 \times P(B) & + & 1/2 \times P(C) \\ P(F) & = & 1/2 \times P(A) & + & 1/2 \times P(B) & + & 0 \times P(C) \end{pmatrix}$$

Using some matrix algebra,

[5] F Esponda and Víctor M Guerrero, "Surveys with Negative Questions for Sensitive Items," *Statistics & Probability Letters* 79 (2009): 2456–61.

$$\begin{pmatrix} 0.5 \\ 0.3 \\ 0.2 \end{pmatrix} = \begin{pmatrix} 0 & 1/2 & 1/2 \\ 1/2 & 0 & 1/2 \\ 1/2 & 1/2 & 0 \end{pmatrix} \begin{pmatrix} P(A) \\ P(B) \\ P(C) \end{pmatrix}$$

Now multiply both sides by the inverse of the square matrix in the above equation to get

$$\begin{pmatrix} 0 & 1/2 & 1/2 \\ 1/2 & 0 & 1/2 \\ 1/2 & 1/2 & 0 \end{pmatrix}^{-1} \begin{pmatrix} 0.5 \\ 0.3 \\ 0.2 \end{pmatrix} = \begin{pmatrix} P(A) \\ P(B) \\ P(C) \end{pmatrix}$$

Using R, we have

```
solve(matrix(c(0, 1, 1, 1, 0, 1, 1, 1, 0),
  byrow = TRUE,
  ncol = 3
) * 1 / 2) %*%
  matrix(c(.5, .3, .2), ncol = 1)
```

```
##       [,1]
## [1,]   0.0
## [2,]   0.4
## [3,]   0.6
```

We see that 0% of the people owe zero dollars, 40% owe between 1 and 1000 dollars, and 60% owe more than 1000 dollars.

## Exercises

---

Exercises 2.1 – 2.3 require material through Section 2.1.

**2.1.** When rolling two dice, what is the probability that one die is twice the other?

**2.2.** Consider an experiment where you roll two dice, and subtract the smaller value from the larger value (getting 0 in case of a tie).

a. What is the probability of getting 0?
b. What is the probability of getting 4?

**2.3.** A hat contains slips of paper numbered 1 through 6. You draw two slips of paper at random from the hat, without replacing the first slip into the hat.

a. Write out the sample space $S$ for this experiment.
b. Write out the event $E$, "the sum of the numbers on the slips of paper is 4."
c. Find $P(E)$.
d. Let $F$ be the event "the larger number minus the smaller number is 0." What is $P(F)$?

---

Exercises 2.4 – 2.19 require material through Section 2.2.

**2.4.** Suppose there are two boxes and each contain slips of papers numbered 1-8. You draw one number at random from each box.

a. Estimate the probability that the sum of the numbers is 8.
b. Estimate the probability that at least one of the numbers is a 2.

**2.5.** Suppose the proportion of M&M's by color is:

| *Yellow* | *Red* | *Orange* | *Brown* | *Green* | *Blue* |
|---|---|---|---|---|---|
| 0.14 | 0.13 | 0.20 | 0.12 | 0.20 | 0.21 |

a. What is the probability that a randomly selected M&M is not green?
b. What is the probability that a randomly selected M&M is red, orange, or yellow?
c. Estimate the probability that a random selection of four M&M's will contain a blue one.
d. Estimate the probability that a random selection of six M&M's will contain all six colors.

**2.6.** With the distribution from Problem 2.5, suppose you buy a bag of M&M's with 30 pieces in it. Estimate the probability of obtaining at least 9 Blue M&M's and at least 6 Orange M&M's in the bag.

**2.7.** Blood types O, A, B, and AB have the following distribution in the United States:

| Type | *A* | *AB* | *B* | *O* |
|---|---|---|---|---|
| Probability | 0.40 | 0.04 | 0.11 | 0.45 |

What is the probability that two randomly selected people have the same blood type?

**2.8.** Use simulation to estimate the probability that a 10 is obtained when two dice are rolled.

**2.9.** Estimate the probability that exactly 3 heads are obtained when 7 coins are tossed.

**2.10.** Estimate the probability that the sum of five dice is between 15 and 20, inclusive.

**2.11.** Suppose a die is tossed repeatedly, and the cumulative sum of all tosses seen is maintained. Estimate the probability that the cumulative sum ever is exactly 20. (Hint: the function `cumsum` computes the cumulative sums of a vector.)

**2.12 (Rolling two dice).** a. Simulate rolling two dice and adding their values. Perform 10,000 simulations and make a bar chart showing how many of each outcome occurred.

b. You can buy *trick* dice, which look (sort of) like normal dice. One die has numbers 5, 5, 5, 5, 5, 5. The other has numbers 2, 2, 2, 6, 6, 6. Simulate rolling the two trick dice and adding their values. Perform 10,000 simulations and make a bar chart showing how many of each outcome occurred.

c. *Sicherman dice* also look like normal dice, but have unusual numbers. One die has numbers 1, 2, 2, 3, 3, 4. The other has numbers 1, 3, 4, 5, 6, 8. Simulate rolling the two Sicherman dice and adding their values. Perform 10,000 simulations and make a bar chart showing how many of each outcome occurred. How does your answer compare to part (a)?

**2.13.** In a room of 200 people (including you), estimate the probability that at least one other person will be born on the same day as you.

**2.14.** In a room of 100 people, estimate the probability that at least two people were not only born on the same day, but also during the same hour of the same day. (For example, both were born between 2 and 3.)

**2.15.** Assuming that there are no leap-day babies and that all birthdays are equally likely, estimate the probability that at least **three** people have the same birthday in a group of 50 people. (Hint: try using `table`.)

**2.16.** If 100 balls are randomly placed into 20 urns, estimate the probability that at least one of the urns is empty.

**2.17.** A standard deck of cards has 52 cards, four each of 2,3,4,5,6,7,8,9,10,J,Q,K,A. In blackjack, a player gets two cards and adds their values. Cards count as their usual numbers, except Aces are 11 (or 1), while K, Q, J are all 10.

a. "Blackjack" means getting an Ace and a value 10 card. What is the probability of getting a blackjack?

b. What is the probability of getting 19? (The probability that the sum of your cards is 19, using Ace as 11)

Use R to simulate dealing two cards, and compute these probabilities experimentally.

**2.18.** Deathrolling in World of Warcraft works as follows. Player 1 tosses a 1000-sided die. Say they get $x_1$. Then player 2 tosses a die with $x_1$ sides on it. Say they get $x_2$. Player 1 tosses a die with $x_2$ sides on it. This pattern continues until a player rolls a 1. The player who loses is the player who rolls a 1. Estimate via simulation the probability that a 1 will be rolled on the 4th roll in deathroll.

**2.19.** In the game of Scrabble, players make words using letter tiles. The data set `fosdata::scrabble` contains all 100 tiles.

Players begin the game by drawing seven tiles from a bag of 100 tiles. Estimate the probability that a player's first seven tiles contain no vowels. (Vowels are A, E, I, O, and U.)

---

Exercises 2.20 – 2.32 require material through Section 2.3.

**2.20.** Two dice are rolled.

a. What is the probability that the sum of the numbers is exactly 10?
b. What is the probability that the sum of the numbers is at least 10?
c. What is the probability that the sum of the numbers is exactly 10, given that it is at least 10?

**2.21.** A hat contains six slips of paper with the numbers 1 through 6 written on them. Two slips of paper are drawn from the hat (without replacing), and the sum of the numbers is computed.

a. What is the probability that the sum of the numbers is exactly 10?
b. What is the probability that the sum of the numbers is at least 10?
c. What is the probability that the sum of the numbers is exactly 10, given that it is at least 10?

**2.22.** Roll two dice, one white and one red. Consider these events:

- $A$: The sum is 7.
- $B$: The white die is odd.
- $C$: The red die has a larger number showing than the white.
- $D$: The dice match (doubles).

a. Which pair(s) of events are disjoint (events $A$ and $B$ are *disjoint* if $A \cap B = \emptyset$)?
b. Which pair(s) are independent?
c. Which pair(s) are neither disjoint nor independent?

**2.23.** Suppose you do an experiment where you select ten people at random and ask their birthdays.

Here are three events:

- $A$ : all ten people were born in February.
- $B$ : the first person was born in February.
- $C$ : the second person was born in January.

a. Which pair(s) of these events are disjoint, if any?
b. Which pair(s) of these events are independent, if any?
c. What is $P(B|A)$?

**2.24.** Let $A$ and $B$ be events. Show that $P(A \cup B|B) = 1$.

**2.25.** In an experiment where you toss a fair coin twice, define events:

- $A$ : the first toss is heads.
- $B$ : the second toss is heads.
- $C$ : both tosses are the same.

Show that $A$ and $B$ are independent. Show that $A$ and $C$ are independent. Show that $B$ and $C$ are independent. Finally, show that $A$, $B$, and $C$ are **not** mutually independent.

**2.26.** Suppose a die is tossed three times. Let $A$ be the event "the first toss is a 5." Let $B$ be the event "the first toss is the largest number rolled" (the "largest" can be a tie). Determine, via simulation or otherwise, whether $A$ and $B$ are independent.

**2.27.** Suppose you have two coins that land with heads facing up with common probability $p$, where $0 < p < 1$. One coin is red and the other is white. You toss both coins. Find the probability that the red coin is heads, given that the red coin and the white coin are different. Your answer will be in terms of $p$.

**2.28.** Bob Ross was a painter with a PBS television show, "The Joy of Painting," that ran for 11 years.

a. 91% of Bob's paintings contain a tree[6] and 85% contain two or more trees. What is the probability that he painted a second tree, given that he painted a tree?
b. 18% of Bob's paintings contain a cabin. Given that he painted a cabin, there is a 35% chance the cabin is on a lake. What is the probability that a Bob Ross painting contains both a cabin and a lake?

**2.29.** Ultimate frisbee players are so poor they don't own coins. So, team captains decide which team will play offense first by flipping frisbees before the start of the game. Rather than flip one frisbee and call a side, each team captain flips a frisbee and one captain calls whether the two frisbees will land on the same side, or on different sides. Presumably, they do this instead of just flipping one frisbee because a frisbee is not obviously a fair coin - the probability of one side seems likely to be different from the probability of the other side.

a. Suppose you flip two fair coins. What is the probability they show different sides?
b. Suppose two captains flip frisbees. Assume the probability that a frisbee lands convex side up is $p$. Compute the probability (in terms of $p$) that the two frisbees match.
c. Make a graph of the probability of a match in terms of $p$.
d. One Reddit user flipped a frisbee 800 times and found that in practice, the convex side lands up 45% of the time. When captains flip, what is the probability of "same?" What is the probability of "different?"
e. What advice would you give to an ultimate frisbee team captain?
f. Is the two-frisbee flip better than a single-frisbee flip for deciding the offense?

**2.30.** Suppose there is a new test that detects whether people have a disease. If a person has the disease, then the test correctly identifies that person as being sick 99.9% of the time ( *sensitivity* of the test). If a person does not have the disease, then the test correctly identifies the person as being well 97% of the time ( *specificity* of the test). Suppose that 2% of the population has the disease. Find the probability that a randomly selected person has the disease given that they test positive for the disease.

**2.31.** Suppose that there are two boxes containing marbles.

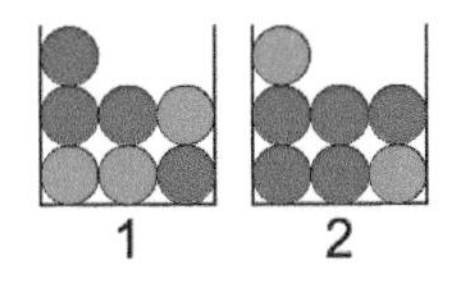


Box 1 contains 3 red and 4 blue marbles. Box 2 contains 2 red and 5 blue marbles. A single die is tossed, and if the result is 1 or 2, then a marble is drawn from box 1. Otherwise, a marble is drawn from box 2.

a. What is the probability that the marble drawn is red?
b. What is the probability that the marble came from box 1 given that the marble is red?

**2.32.** Suppose that you have 10 boxes, numbered 0-9. Box $i$ contains $i$ red marbles and $9 - i$ blue marbles.

[6]W Hickey, "A Statistical Analysis of the Work of Bob Ross," *FiveThirtyEight*, April 14, 2014, https://fivethirtyeight.com/features/a-statistical-analysis-of-the-work-of-bob-ross/.

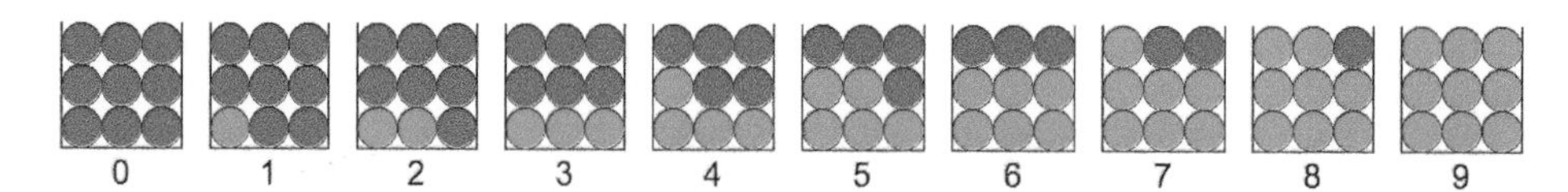


You perform the following experiment. Pick a box at random, draw a marble and record its color. Replace the marble back in the box, and draw another marble *from the same box* and record its color. Replace the marble back in the box, and draw another marble from the same box and record its color. So, all three marbles are drawn from the same box.

a. If you draw three consecutive red marbles, what is the probability that a fourth marble drawn from the same box will also be red?
b. If you draw three consecutive red marbles, what is the probability that you chose box 9?

---

Exercises 2.33 – 2.36 require material through Section 2.4.

**2.33.** How many ways are there of getting 4 heads when tossing 10 coins?

**2.34.** How many ways are there of getting 4 heads when tossing 10 coins, assuming that the 4th head came on the 10th toss?

**2.35.** Six standard six-sided dice are rolled.

a. How many outcomes are there?
b. How many outcomes are there such that all of the dice are different numbers?
c. What is the probability that you obtain six different numbers when you roll six dice?

**2.36.** A box contains 5 red marbles and 5 blue marbles. Six marbles are drawn without replacement.

a. How many ways are there of drawing the 6 marbles? Assume that getting all 5 red marbles and the first blue marble is different than getting all 5 red marbles and the second blue marble, for example.
b. How many ways are there of drawing 4 red marbles and 2 blue marbles?
c. What is the probability of drawing 4 red marbles and 2 blue marbles?

# 3

# *Discrete Random Variables*

A statistical experiment produces an outcome in a sample space, but frequently we are more interested in a number that summarizes that outcome. For example, if we randomly select a person with a fever and provide them with a dosage of medicine, the sample space might be the set of all people who currently have a fever, or perhaps the set of all possible people who could currently have a fever. However, we are more interested in the summary value of "how much did the temperature of the patient decrease." This is a random variable.

**Definition 3.1.** Let $S$ be the sample space of an experiment. A *random variable* is a function from $S$ to the real line. Random variables are usually denoted by a capital letter. Many times we will abbreviate the words *random variable* with *rv*.

Suppose $X$ is a random variable. The events of interest for $X$ are those that can be defined by a set of real numbers. For example, $X = 2$ is the event consisting of all outcomes $s \in S$ with $X(s) = 2$. Similarly, $X > 8$ is the event consisting of all outcomes $s \in S$ with $X(s) > 8$. In general, if $U \subset \mathbb{R}$:

$$X \in U \text{ is the event } \{s \in S \mid X(s) \in U\}$$

**Example 3.1.** Suppose that three coins are tossed. The sample space is

$$S = \{HHH, HHT, HTH, HTT, THH, THT, TTH, TTT\},$$

and all eight outcomes are equally likely, each occurring with probability 1/8. A natural random variable here is the number of heads observed, which we will call $X$. As a function from $S$ to the real numbers, $X$ is given by:

$$\begin{aligned} X(HHH) &= 3 \\ X(HHT) = X(HTH) = X(THH) &= 2 \\ X(TTH) = X(THT) = X(HTT) &= 1 \\ X(TTT) &= 0 \end{aligned}$$

The event $X = 2$ is the set of outcomes $\{HHT, HTH, THH\}$ and so:

$$P(X = 2) = P(\{HHT, HTH, THH\}) = \frac{3}{8}.$$

It is often easier, both notationally and for doing computations, to hide the sample space and focus only on the random variable. We will not always explicitly define the sample space of an experiment. It is easier, more intuitive and (for the purposes of this book) equivalent to just understand $P(a < X < b)$ for all choices of $a < b$. By understanding these probabilities, we can derive many useful properties of the random variable, and hence, the underlying experiment.

DOI: 10.1201/9781003004899-3

We will consider two types of random variables in this book. Discrete random variables are integers, and often come from counting something. Continuous random variables take values in an interval of real numbers, and often come from measuring something. Working with discrete random variables requires summation, while continuous random variables require integration. We will discuss these two types of random variables separately in this chapter and in Chapter 4.

## 3.1 Probability mass functions

**Definition 3.2.** A *discrete* random variable is a random variable that takes integer values.[1] A discrete random variable is characterized by its *probability mass function* (pmf). The pmf $p$ of a random variable $X$ is given by

$$p(x) = P(X = x).$$

The pmf may be given in table form or as an equation. Knowing the probability mass function determines the discrete random variable, and we will understand the random variable by understanding its pmf.

Probability mass functions satisfy the following properties:

**Theorem 3.1.** *Let $p$ be the probability mass function of $X$.*

1. $p(x) \geq 0$ *for all* $x$.
2. $\sum_x p(x) = 1$.

To check that a function is a pmf, we check that all of its values are probabilities, and that those values sum to one.

**Example 3.2.** The Eurasian lynx is a wild cat that lives in the north of Europe and Asia. When a female lynx gives birth, she may have from 1 to 4 kittens. Our statistical experiment is a lynx giving birth, and the outcome is a litter of baby lynx. Baby lynx are complicated objects, but there is a simple random variable here: the number of kittens. Call this $X$. Ecologists have estimated[2] the pmf for $X$ to be:

| $x$ | 1 | 2 | 3 | 4 |
|---|---|---|---|---|
| $p(x)$ | 0.18 | 0.51 | 0.27 | 0.04 |

In other words, the probability that a lynx mother has one kitten is 0.18, the probability that she has two kittens is 0.51, and so on. Observe that $p(1) + p(2) + p(3) + p(4) = 1$, so this is a pmf.

We can use the pmf to calculate the probability of any event defined in terms of $X$. The probability that a lynx litter has more than one kitten is:

$$P(X > 1) = P(X = 2) + P(X = 3) + P(X = 4) = 0.51 + 0.27 + 0.04 = 0.82.$$

[1] In this text, we almost exclusively consider discrete random variables with integer values, but more generally a discrete rv could take values in any subset of $\mathbb{R}$ consisting of countably many points.

[2] Jean-Michel Gaillard et al., "One Size Fits All: Eurasian Lynx Females Share a Common Optimal Litter Size," *The Journal of Animal Ecology* 83 (July 2013), https://doi.org/10.1111/1365-2656.12110.

**FIGURE 3.1** Eurasian lynx kitten. (Photo credit: Bernard Landgraf)

We can simulate the lynx litter size random variable $X$ without having to capture pregnant lynx. Recall that the R function `sample` has four arguments: the possible outcomes `x`, the sample size `size` to sample, whether we are sampling with replacement `replace`, and the probability `prob` associated with the possible outcomes. Here we generate values of $X$ for 30 lynx litters:

```
litterpmf <- c(0.18, 0.51, 0.27, 0.04)
sample(1:4, 30, replace = TRUE, prob = litterpmf)
```

```
##  [1] 1 2 2 1 2 1 1 2 3 1 3 1 1 3 3 2 1 1 3 2 2 2 3 3 3 1 2 2 3 3
```

With enough samples of $X$, we can approximate the probability $P(X > 1)$ as follows:

```
X <- sample(1:4, 10000, replace = TRUE, prob = litterpmf)
mean(X > 1)
```

```
## [1] 0.8222
```

In this code, the first line simulates the random variable. The code `X > 1` produces a vector of `TRUE` and `FALSE` which is `TRUE` when the event $X > 1$ occurs. Recall that taking the `mean` of a `TRUE/FALSE` vector gives the proportion of times that vector is `TRUE`, which will be approximately $P(X > 1)$ here.

We can also recreate the pmf by using `table` to count values and then dividing by the sample size[3]:

```
table(X) / 10000
```

```
## X
##      1      2      3      4
## 0.1778 0.5134 0.2674 0.0414
```

[3]Using the sample size as a constant in `sample` and in the computation below can lead to errors if you change the value in one place but not the other. We will see in Chapter 5 that `proportions(table(X))` is a better way to estimate probability mass functions from a sample.

**Example 3.3.** Let $X$ denote the number of heads observed when a coin is tossed three times.

In this example, we can simulate the random variable by first simulating experiment outcomes and then calculating $X$ from those. The following generates three coin flips:

```
coin_toss <- sample(c("H", "T"), 3, replace = TRUE)
```

Now we calculate how many heads were flipped, and produce one value of $X$.

```
sum(coin_toss == "H")
```

```
## [1] 2
```

Finally, we can use `replicate` to produce many samples of $X$:

```
X <- replicate(10000, {
  coin_toss <- sample(c("H", "T"), 3, replace = TRUE)
  sum(coin_toss == "H")
})
head(X, 30) # see the first 30 values of X
```

```
##  [1] 1 0 2 2 2 2 0 1 0 2 2 1 0 1 1 2 2 1 0 2 0 3 3 1 2 0 1 0 1 3
```

From the simulation, we can estimate the pmf using `table` and dividing by the number of samples:

```
table(X) / 10000
```

```
## X
##      0      1      2      3 
## 0.1279 0.3756 0.3743 0.1222
```

Instead of simulation, we could also calculate the pmf by considering the sample space, which consists of the eight equally likely outcomes HHH, HHT, HTH, HTT, THH, THT, TTH, TTT. Counting heads, we find that $X$ has the pmf:

| $x$ | 0 | 1 | 2 | 3 |
|---|---|---|---|---|
| $p(x)$ | $\frac{1}{8}$ | $\frac{3}{8}$ | $\frac{3}{8}$ | $\frac{1}{8}$ |

which matches the results of our simulation. Here is an alternative description of $p$ as a formula:

$$p(x) = \binom{3}{x}\left(\frac{1}{2}\right)^x \qquad x = 0, \ldots, 3$$

We always assume that $p$ is zero for values not mentioned; both in the table version and in the formula version.

As in the lynx example, we may simulate this random variable directly by sampling with probabilities given by the pmf. Here we sample 30 values of $X$ without "flipping" any "coins":

```
sample(0:3, 30, replace = TRUE, prob = c(0.125, 0.375, 0.375, 0.125))
```

```
##  [1] 3 1 3 0 3 1 1 2 3 0 1 2 0 0 0 2 2 1 1 3 3 1 0 2 1 1 1 1 3 1
```

**Example 3.4.** Compute the probability that we observe at least one head when three coins are tossed.

Let $X$ be the number of heads. We want to compute the probability of the event $X \geq 1$. Using the pmf for $X$,

$$\begin{aligned}
P(1 \leq X) &= P(1 \leq X \leq 3) \\
&= P(X = 1) + P(X = 2) + P(X = 3) \\
&= \frac{3}{8} + \frac{3}{8} + \frac{1}{8} \\
&= \frac{7}{8} = 0.875.
\end{aligned}$$

We could also estimate $P(X \geq 1)$ by simulation:

```
X <- replicate(10000, {
  coin_toss <- sample(c("H", "T"), 3, replace = TRUE)
  sum(coin_toss == "H")
})
mean(X >= 1)
```

```
## [1] 0.8741
```

**Example 3.5.** Suppose you toss a coin until the first time you see heads. Let $X$ denote the number of tails that you see. We will see later that the pmf of $X$ is given by

$$p(x) = \left(\frac{1}{2}\right)^{x+1} \qquad x = 0, 1, 2, 3, \ldots$$

Compute $P(X = 2)$, $P(X \leq 1)$, $P(X > 1)$, and the conditional probability $P(X = 2|X > 1)$.

1. To compute $P(X = 2)$, we just plug in $P(X = 2) = p(2) = \left(\frac{1}{2}\right)^3 = \frac{1}{8}$.
2. To compute $P(X \leq 1)$, we add $P(X = 0) + P(X = 1) = p(0) + p(1) = \frac{1}{2} + \frac{1}{4} = \frac{3}{4}$.
3. The complement of $X > 1$ is $X \leq 1$, so

$$P(X > 1) = 1 - P(X \leq 1) = 1 - \frac{3}{4} = \frac{1}{4}.$$

Alternatively, we could compute an infinite sum using the formula for the geometric series:

$$P(X > 1) = p(2) + p(3) + p(4) + \cdots = \frac{1}{8} + \frac{1}{16} + \frac{1}{32} + \cdots = \frac{1}{4}.$$

4. The formula for conditional probability gives:

$$P(X = 2|X > 1) = \frac{P(X = 2 \cap X > 1)}{P(X > 1)} = \frac{P(X = 2)}{P(X > 1)} = \frac{1/8}{1/4} = \frac{1}{2}.$$

This last answer makes sense because $X > 1$ requires the first two flips to be tails, and then there is a $\frac{1}{2}$ chance your third flip will be heads and achieve $X = 2$.

## 3.2 Expected value

Suppose you perform a statistical experiment repeatedly, and observe the value of a random variable $X$ each time. The average of these observations will (under most circumstances) converge to a fixed value as the number of observations becomes large. This value is the *expected value* of $X$, written $E[X]$. The definition looks different than this, and our intuitive explanation of the expected value is actually Theorem 3.2.

**Definition 3.3.** For a discrete random variable $X$ with pmf $p$, the *expected value* of $X$ is

$$E[X] = \sum_x xp(x),$$

provided this sum exists, where the sum is taken over all possible values of the random variable $X$.

Another word for the expected value of $X$ is the *mean* of $X$.

**Theorem 3.2 (Law of Large Numbers).** *The mean of $n$ observations of a random variable $X$ converges to the expected value $E[X]$ as $n \to \infty$, assuming $E[X]$ is defined.*

**Example 3.6.** Using simulation, we determine the expected value of a die roll. Here are 30 observations and their average:

```
rolls <- sample(1:6, 30, replace = TRUE)
rolls
```

```
##  [1] 3 3 3 3 1 5 1 2 2 3 3 6 5 1 1 6 6 5 2 6 5 2 5 2 6 1 3 6 6 5
```

```
mean(rolls)
```

```
## [1] 3.6
```

The mean appears to be somewhere between 3 and 4. Using more trials gives more accuracy:

```
rolls <- sample(1:6, 100000, replace = TRUE)
mean(rolls)
```

```
## [1] 3.4934
```

Not surprisingly, the mean value is balanced halfway between 1 and 6, at 3.5.

Using the probability distribution of a random variable $X$, one can compute the expected value $E[X]$ exactly, as in the following example.

**Example 3.7.** Let $X$ be the value of a six-sided die roll. Since the probability of each outcome is $\frac{1}{6}$, we have:

$$E[X] = 1 \cdot \frac{1}{6} + 2 \cdot \frac{1}{6} + 3 \cdot \frac{1}{6} + 4 \cdot \frac{1}{6} + 5 \cdot \frac{1}{6} + 6 \cdot \frac{1}{6} = \frac{21}{6} = 3.5$$

**Example 3.8.** Let $X$ be the number of kittens in a Eurasian lynx litter. Then

$$E[X] = 1 \cdot p(1) + 2 \cdot p(2) + 3 \cdot p(3) + 4 \cdot p(4) = 0.18 + 2 \cdot 0.51 + 3 \cdot 0.27 + 4 \cdot 0.04 = 2.17$$

This means that, on average, the Eurasian lynx has 2.17 kittens.

We can perform the computation of $E[X]$ using R:

```
litterpmf <- c(0.18, 0.51, 0.27, 0.04)
sum((1:4) * litterpmf)
```

```
## [1] 2.17
```

Alternatively, we may estimate $E[X]$ by simulation, using the Law of Large Numbers.

```
X <- sample(1:4, 10000, replace = TRUE, prob = litterpmf)
mean(X)
```

```
## [1] 2.1645
```

The expected value need not be a possible outcome associated with the random variable. You will never roll a 3.5, and a lynx mother will never have 2.17 kittens. The expected value describes the average of many observations.

**Example 3.9.** Let $X$ denote the number of heads observed when three coins are tossed. The pmf of $X$ is given by $p(x) = \binom{3}{x}(1/2)^x$, where $x = 0, \ldots, 3$. The expected value of $X$ is

$$E[X] = 0 \cdot \frac{1}{8} + 1 \cdot \frac{3}{8} + 2 \cdot \frac{3}{8} + 3 \cdot \frac{1}{8} = \frac{3}{2}.$$

We can check this with simulation:

```
X <- replicate(10000, {
  coin_toss <- sample(c("H", "T"), 3, replace = TRUE)
  sum(coin_toss == "H")
})
mean(X)
```

```
## [1] 1.4932
```

The answer is approximately 1.5, which is what our exact computation of $E[X]$ predicted.

**Example 3.10.** Consider the random variable $X$ which counts the number of tails observed before the first head when a fair coin is repeatedly tossed. The pmf of $X$ is $p(x) = 0.5^{x+1}$ for $x = 0, 1, 2, \ldots$. Finding the expected value requires summing an infinite series, which we leave as an exercise. Instead, we estimate the infinite sum via a finite sum, and we use simulation.

We assume (see below for a justification) that the infrequent results of $x \geq 100$ do not impact the expected value much. That is, we assume that $\sum_{x=0}^{99} xp(x) \approx E[X]$. We can use R to compute this:

```
xs <- 0:99
sum(xs * (0.5)^(xs + 1))
```

```
## [1] 1
```

In order to estimate the sum via simulation, we take a sample of size 10,000 and follow the same steps as in the previous example, again assuming that values of 100 or larger do not affect the expected value.

```
xs <- 0:99
probs <- .5^xs
X <- sample(0:99, 10000, replace = TRUE, prob = probs)
mean(X)
```

```
## [1] 1.0172
```

We estimate that the expected value $E[X] \approx 1$.

To justify that we do not need to include values of $x$ bigger than 99, note that

$$\begin{aligned}
E[X] - \sum_{x=0}^{99} x(1/2)^{x+1} &= \sum_{x=100}^{\infty} x(1/2)^{x+1} \\
&< \sum_{x=100}^{\infty} 2^{x/2-1}(1/2)^{x+1} \\
&= \sum_{x=100}^{\infty} 2^{-x/2} \\
&= 2^{-50}\frac{1}{2^{49}(2-\sqrt{2})} < 10^{-14}
\end{aligned}$$

So, truncating the sum at $x = 99$ introduces a negligible error. The *Loops in R* vignette at the end of this chapter shows an approach using loops that avoids truncation.

**Example 3.11.** We end this short section with an example of a discrete random variable that has expected value of $\infty$. Let $X$ be a random variable such that $P(X = 2^x) = 2^{-x}$ for $x = 1, 2, \ldots$. We see that $\sum_{x=1}^{\infty} xp(x) = \sum_{x=1}^{\infty} 2^x 2^{-x} = \infty$. If we truncated the sum associated with $E[X]$ for this random variable at any finite point, then we would introduce a very large error!

## 3.3 Binomial and geometric random variables

The binomial and geometric random variables are common and useful models for many real situations. Both involve Bernoulli trials, named after the 17th century Swiss mathematician Jacob Bernoulli.

**Definition 3.4.** A *Bernoulli trial* is an experiment that can result in two outcomes, which we will denote as "success" and "failure." The probability of a success will be denoted $p$, and the probability of failure is therefore $1 - p$.

**Example 3.12.** The following are examples of Bernoulli trials, at least approximately.

1. Toss a coin. Arbitrarily define heads to be success. Then $p = 0.5$.
2. Shoot a free throw, in basketball. Success would naturally be making the shot, failure missing the shot. Here $p$ varies depending on who is shooting. An excellent basketball player might have $p = 0.8$.
3. Ask a randomly selected voter whether they support a ballot proposition. Here success would be a yes vote, failure a no vote, and $p$ is likely unknown but of interest to the person doing the polling.
4. Roll a die, and consider success to be rolling a six. Then $p = 1/6$.

A *Bernoulli process* is a sequence (finite or infinite) of repeated, identical, independent Bernoulli trials. Repeatedly tossing a coin is a Bernoulli process. Repeatedly trying to roll a six is a Bernoulli process. Is repeated free throw shooting a Bernoulli process? There are two reasons that it might not be. First, if the person shooting is a beginner, then they would presumably get better with practice. That would mean that the probability of a success is **not** the same for each trial (especially if the number of trials is large), and the process would not be Bernoulli. Second, there is the more subtle question of whether free throws are independent. Does a free throw shooter get "hot" and become more likely to make the shot following a success? There is no way to know for sure, although research[4] suggests that repeated free throws are independent and that modeling free throws of experienced basketball players with a Bernoulli process is reasonable.

This section discusses two discrete random variables coming from a Bernoulli process: the binomial random variable which counts the number of successes in a fixed number of trials, and the geometric random variable, which counts the number of trials before the first success.

### 3.3.1 Binomial

**Definition 3.5.** A random variable $X$ is said to be a *binomial random variable* with parameters $n$ and $p$ if

$$P(X = x) = \binom{n}{x} p^x (1-p)^{n-x} \qquad x = 0, 1, \ldots, n.$$

We will sometimes write $X \sim \text{Binom}(n, p)$

The most important example of a binomial random variable comes from counting the number of successes in a Bernoulli process of length $n$. In other words, if $X$ counts the number of successes in $n$ independent and identically distributed Bernoulli trials, each with probability of success $p$, then $X \sim \text{Binom}(n, p)$. Indeed, many texts define binomial random variables in this manner, and we will use this alternative definition of a binomial random variable whenever it is convenient for us to do so.

One way to obtain a Bernoulli process is to sample *with replacement* from a population. If an urn contains 5 white balls and 5 red balls, and you count drawing a red ball as a "success," then repeatedly drawing balls will be a Bernoulli process as long as you replace the ball back in the urn after you draw it. When sampling from a *large population*, we can sample without replacement and the resulting count of successes will be approximately binomial.

**Example 3.13.** Let $X$ denote the number of heads observed when three coins are tossed. Then $X \sim \text{Binom}(3, 0.5)$ is a binomial random variable. Here $n = 3$ because there are three independent Bernoulli trials, and $p = 0.5$ because each coin has probability 0.5 of heads.

**Theorem 3.3.** *If $X$ counts the number of successes in $n$ independent and identically distributed Bernoulli trials, each with probability of success $p$, then $X \sim$ Binom$(n, p)$.*

*Proof.* There are $n$ trials. There are $\binom{n}{x}$ ways to choose $x$ of these $n$ trials as the successful trials. For each of these ways, $p^x(1-p)^{n-x}$ is the probability of having $x$ successes in the chosen spots and $n - x$ failures in the not chosen spots. ∎

**Remark.** The binomial theorem says that:

$$(a+b)^n = a^n + \binom{n}{1} a^{n-1} b^1 + \binom{n}{2} a^{n-2} b^2 + \cdots + \binom{n}{n-1} a^1 b^{n-1} + b^n$$

[4] Amos Tversky and Thomas Gilovich, "The Cold Facts about the 'Hot Hand' in Basketball," *CHANCE* 2, no. 1 (1989): 16–21, https://doi.org/10.1080/09332480.1989.11882320.

Substituting $a = p$ and $b = 1 - p$, the left-hand side becomes $(p + 1 - p)^n = 1^n = 1$, and so:

$$\begin{aligned}1 &= p^n + \binom{n}{1}p^{n-1}(1-p)^1 + \binom{n}{2}p^{n-2}(1-p)^2 + \cdots + \binom{n}{n-1}p^1(1-p)^{n-1} + (1-p)^n \\ &= P(X = n) + P(X = n-1) + P(X = n-2) + \cdots + P(X = 1) + P(X = 0),\end{aligned}$$

which shows that the pmf for the binomial distribution does sum to 1, as it must.

In R, the function `dbinom` provides the pmf of the binomial distribution:

`dbinom(x, size = n, prob = p)` gives $P(X = x)$ for $X \sim \text{Binom}(n, p)$.

If we provide a vector of values for `x`, and a single value of `size` and `prob`, then `dbinom` will compute $P(X = x)$ for all of the values in `x`. The `d` suggests "distribution," and the root `binom` identifies the binomial distribution. R uses this combination of prefix letter and root word to specify many functions for working with random variables.

A useful idiom for working with discrete probabilities is `sum(dbinom(vec, size, prob))`. If `vec` contains *distinct* values, the sum computes the probability that $X$ is in the set described by `vec`.

**Example 3.14.** For $X$ the number of heads when three coins are tossed, the pmf is

$$P(X = x) = \begin{cases} 1/8 & x = 0, 3 \\ 3/8 & x = 1, 2 \end{cases}$$

Computing with R,

```
x <- 0:3
dbinom(x, size = 3, prob = 0.5)
```

```
## [1] 0.125 0.375 0.375 0.125
```

Figure 3.2 shows sample plots of the pmf of a binomial rv for various values of $n$ when $p = .5$, while Figure 3.3 shows the binomial pmf for $n = 100$ and various $p$. In these plots of binomial pmfs, the distributions are roughly balanced around a peak. The balancing point is the expected value of the random variable, which for binomial rvs is quite intuitive: it is simply $np$.

**Theorem 3.4.** *Let $X$ be a binomial random variable with $n$ trials and probability of success $p$. Then*

$$E[X] = np$$

We will see a simple proof of this once we talk about the expected value of the sum of random variables. Here is a proof using the pmf and the definition of expected value directly.

*Proof.* The binomial theorem says that

$$(a + b)^n = a^n + \binom{n}{1}a^{n-1}b^1 + \binom{n}{2}a^{n-2}b^2 + \cdots + \binom{n}{n-1}a^1b^{n-1} + b^n.$$

Take the derivative of both sides with respect to $a$:

$$n(a+b)^{n-1} = na^{n-1} + (n-1)\binom{n}{1}a^{n-2}b^1 + (n-2)\binom{n}{2}a^{n-3}b^2 + \cdots + 1\binom{n}{n-1}a^0b^{n-1} + 0$$

and multiply both sides by $a$:

$$an(a+b)^{n-1} = na^n + (n-1)\binom{n}{1}a^{n-1}b^1 + (n-2)\binom{n}{2}a^{n-2}b^2 + \cdots + 1\binom{n}{n-1}a^1b^{n-1} + 0.$$

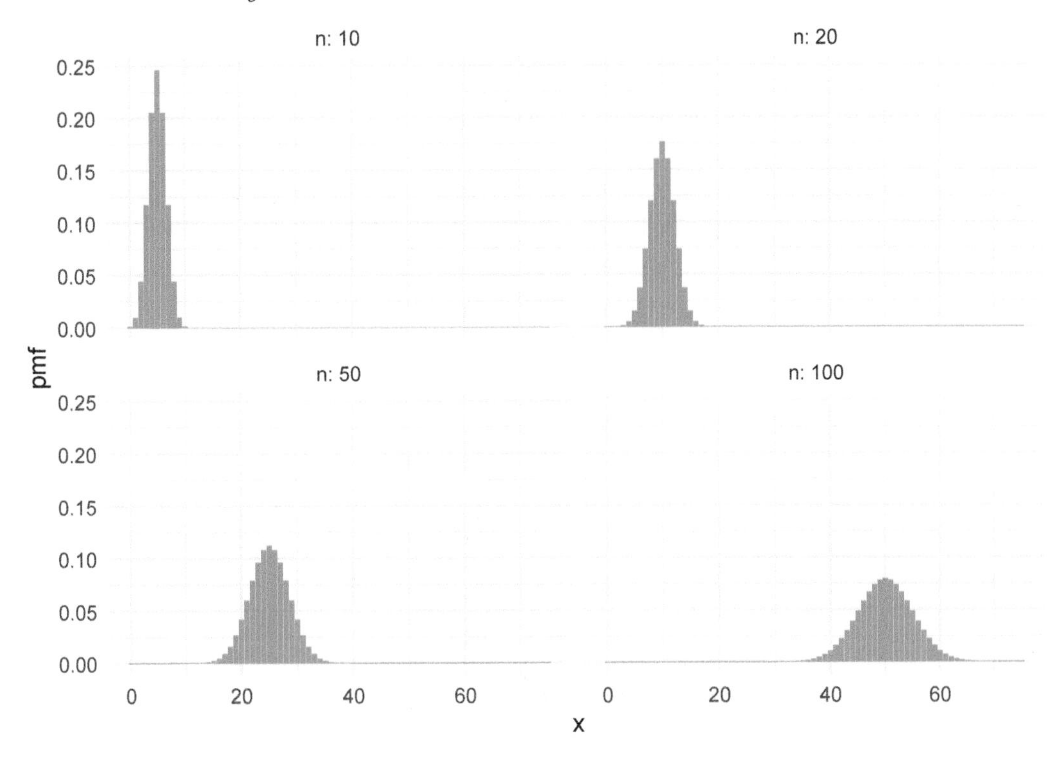


**FIGURE 3.2** Binomial distributions with $p = 0.5$ and various values of $n$.

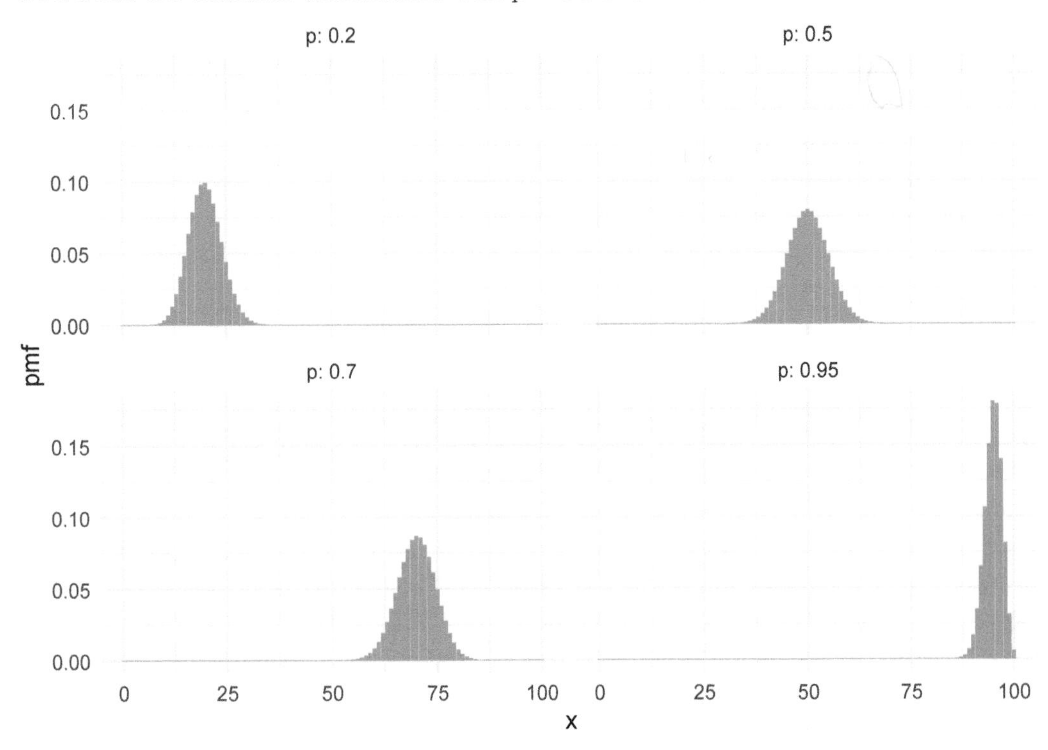


**FIGURE 3.3** Binomial distributions with $n = 100$ and various values of $p$.

Finally, substitute $a = p$, $b = 1 - p$, and note that $a + b = 1$:

$$pn = np^n + (n-1)\binom{n}{1}p^{n-1}(1-p)^1 + (n-2)\binom{n}{2}p^{n-2}(1-p)^2 + \cdots + 1\binom{n}{n-1}p^1(1-p)^{n-1} + 0(1-p)^n$$

or

$$pn = \sum_{x=0}^{n} x\binom{n}{x}p^x(1-p)^{n-x} = E[X]$$

since for $X \sim \text{Binom}(n, p)$, the pmf is given by $P(X = x) = \binom{n}{x}p^x(1-p)^{n-x}$ ■

**Example 3.15.** Suppose 100 dice are thrown. What is the expected number of sixes? What is the probability of observing 10 or fewer sixes?

We assume that the results of the dice are independent and that the probability of rolling a six is $p = 1/6$. The random variable $X$ is the number of sixes observed, and $X \sim \text{Binom}(100, 1/6)$.

Then $E[X] = 100 \cdot \frac{1}{6} \approx 16.67$. That is, we expect 1/6 of the 100 rolls to be a six.

The probability of observing 10 or fewer sixes is

$$P(X \leq 10) = \sum_{x=0}^{10} P(X = x) = \sum_{x=0}^{10} \binom{100}{x}(1/6)^x(5/6)^{100-x} \approx 0.0427.$$

In R,

```
sum(dbinom(0:10, 100, 1 / 6))
```

```
## [1] 0.04269568
```

R also provides the function `pbinom`, which is the cumulative sum of the pmf. The cumulative sum of a pmf is important enough that it gets its own name: the cumulative distribution function. We will say more about the cumulative distribution function in general in Section 4.1.

`pbinom(x, size = n, prob = p)` gives $P(X \leq x)$ for $X \sim \text{Binom}(n, p)$.

In the previous example, we could compute $P(X \leq 10)$ as

```
pbinom(10, 100, 1 / 6)
```

```
## [1] 0.04269568
```

**Example 3.16.** Suppose Peyton and Riley are running for office, and 46% of all voters prefer Peyton. A poll randomly selects 300 voters from a large population and asks their preference. What is the expected number of voters who will report a preference for Peyton? What is the probability that the poll results suggest Peyton will win?

Let "success" be a preference for Peyton, and $X$ be the random variable equal to the number of polled voters who prefer Peyton. It is reasonable to assume that $X \sim \text{Binom}(300, 0.46)$ as long as our sample of 300 voters is a small portion of the population of all voters.

We expect that $0.46 \cdot 300 = 138$ of the 300 voters will report a preference for Peyton.

For the poll results to show Peyton in the lead, we need $X > 150$. To compute $P(X > 150)$, we use $P(X > 150) = 1 - P(X \leq 150)$ and then

```
1 - pbinom(150, 300, 0.46)
```

```
## [1] 0.07398045
```

There is about a 7.4% chance the poll will show Peyton in the lead, despite their imminent defeat.

R provides the function `rbinom` to simulate binomial random variables. The first argument to `rbinom` is the number of random values to simulate, and the next arguments are `size = n` and `prob = p`. Here are 15 simulations of the Peyton vs. Riley poll:

```
rbinom(15, size = 300, prob = 0.46)
```

```
##  [1] 132 116 129 139 165 137 138 142 134 140 140 134 134 126 149
```

In this series of simulated polls, Peyton appears to be losing in all except the fifth poll where she was preferred by $165/300 = 55\%$ of the selected voters.

We can compute $P(X > 150)$ by simulation

```
X <- rbinom(10000, 300, 0.46)
mean(X > 150)
```

```
## [1] 0.0714
```

which is close to our theoretical result that Peyton should appear to be winning 7.4% of the time.

Finally, we can estimate the expected number of people in our poll who say they will vote for Peyton and compare that to the value of 138 that we calculated above.

```
mean(X) # estimate of expected value
```

```
## [1] 137.9337
```

As a final note, all of the simulations of random variables in R provide *random samples* from the specified distributions. That means that the outcomes of trials do not influence the outcomes of the other trials. See Definition 3.8 for a precise definition.

### 3.3.2 Geometric

**Definition 3.6.** A random variable $X$ is said to be a *geometric random variable* with parameter $p$ if

$$P(X = x) = (1 - p)^x p, \qquad x = 0, 1, 2, \ldots$$

**Theorem 3.5.** *Let $X$ be the random variable that counts the number of failures before the first success in a Bernoulli process with probability of success $p$. Then $X$ is a geometric random variable.*

*Proof.* The only way to achieve $X = x$ is to have the first $x$ trials result in failure and the next trial result in success. Each failure happens with probability $1 - p$, and the final success happens with probability $p$. Since the trials are independent, we multiply $1 - p$ a total of $x$ times, and then multiply by $p$.

As a check, we show that the geometric pmf does sum to one. This requires summing an infinite geometric series:

$$\sum_{x=0}^{\infty} p(1-p)^x = p\sum_{x=0}^{\infty}(1-p)^x = p\frac{1}{1-(1-p)} = 1$$

■

We defined the geometric random variable $X$ so that it is compatible with functions built into R. Some sources let $Y$ be the number of trials required for the first success, and call $Y$ geometric. In that case $Y = X + 1$, as the final success counts as one additional trial.

The functions `dgeom`, `pgeom`, and `rgeom` are available for working with a geometric random variable $X \sim \text{Geom}(p)$:

- `dgeom(x,p)` is the pmf and gives $P(X = x)$.
- `pgeom(x,p)` gives $P(X \leq x)$.
- `rgeom(N,p)` simulates $N$ random values of $X$.

**Example 3.17.** A die is tossed until the first six occurs. What is the probability that it takes 4 or more tosses?

We define success as a roll of six, and let $X$ be the number of failures before the first success. Then $X \sim \text{Geom}(1/6)$, a geometric random variable with probability of success $1/6$.

Taking 4 or more tosses corresponds to the event $X \geq 3$. Theoretically,

$$P(X \geq 3) = \sum_{x=3}^{\infty} P(X = x) = \sum_{x=3}^{\infty} \frac{1}{6} \cdot \left(\frac{5}{6}\right)^x = \frac{125}{216} \approx 0.58.$$

We cannot perform the infinite sum with `dgeom`, but we can come close by summing to a large value of $x$:

```
sum(dgeom(3:1000, 1 / 6))
```

```
## [1] 0.5787037
```

Another approach is to apply rules of probability to see that $P(X \geq 3) = 1 - P(X < 3)$. Since $X$ is discrete, $X < 3$ and $X \leq 2$ are the same event. Then $P(X \geq 3) = 1 - P(X \leq 2)$:

```
1 - sum(dgeom(0:2, 1 / 6))
```

```
## [1] 0.5787037
```

Rather than summing the pmf, we may use `pgeom`:

```
1 - pgeom(2, 1 / 6)
```

```
## [1] 0.5787037
```

The function `pgeom` has an option `lower.tail=FALSE` which makes it compute $P(X > x)$ rather than $P(X \leq x)$, leading to maybe the most concise method:

```
pgeom(2, 1 / 6, lower.tail = FALSE)
```

```
## [1] 0.5787037
```

Finally, we can use simulation to approximate the result:

```
X <- rgeom(10000, 1 / 6)
mean(X >= 3)
```

```
## [1] 0.581
```

All of these show there is about a 0.58 probability that it will take four or more tosses to roll a six.

Figure 3.4 shows the probability mass functions for geometric random variables with various $p$. Observe that for smaller $p$, we see that $X$ is likely to be larger. The lower the probability of success, the more failures we expect before our first success.

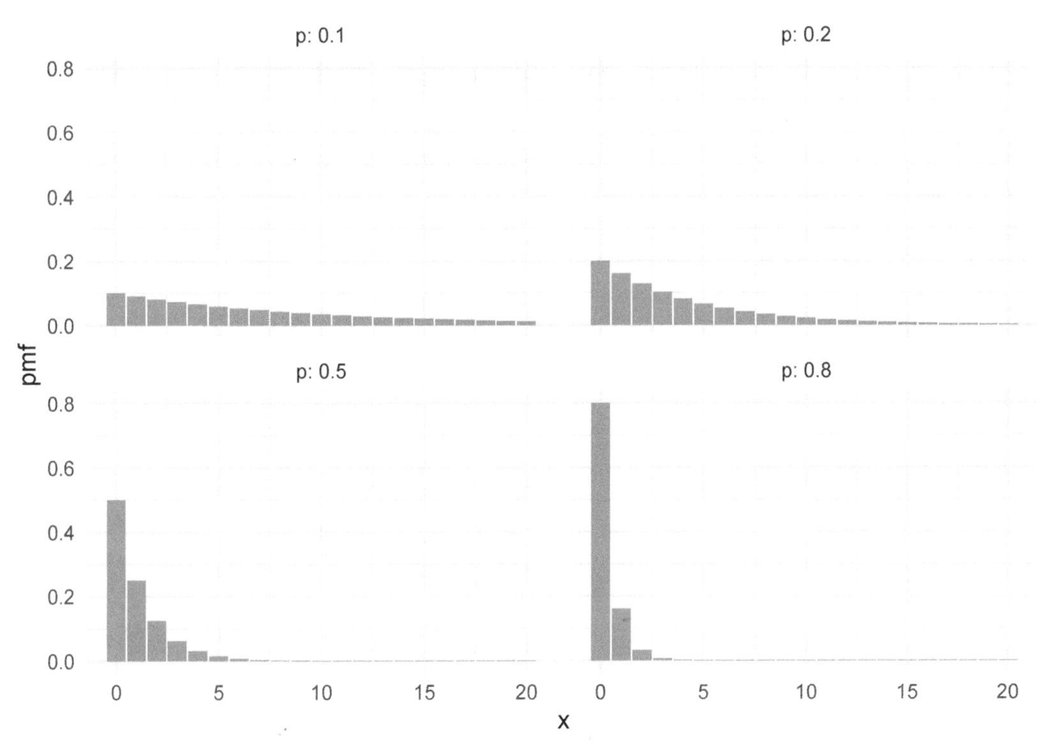


**FIGURE 3.4** PMFs for geometric random variables with various values of $p$.

**Theorem 3.6.** *Let $X$ be a geometric random variable with probability of success $p$. Then*

$$E[X] = \frac{(1-p)}{p}$$

*Proof.* Let $X$ be a geometric rv with success probability $p$. Let $q = 1 - p$ be the failure probability. We must compute $E[X] = \sum_{x=0}^{\infty} xpq^x$. Begin with a geometric series in $q$:

$$\sum_{x=0}^{\infty} q^x = \frac{1}{1-q}$$

Take the derivative of both sides with respect to $q$:

$$\sum_{x=0}^{\infty} xq^{x-1} = \frac{1}{(1-q)^2}$$

Multiply both sides by $pq$:

$$\sum_{x=0}^{\infty} xpq^x = \frac{pq}{(1-q)^2}$$

Replace $1 - q$ with $p$ and we have shown:

$$E[X] = \frac{q}{p} = \frac{1-p}{p}$$

■

**Example 3.18.** Roll a die until a six is tossed. What is the expected number of rolls?

The expected number of failures is given by $X \sim \text{Geom}(1/6)$, and so we expect $\frac{5/6}{1/6} = 5$ failures before the first success. Since the number of total rolls is one more than the number of failures, we expect 6 rolls, on average, to get a six.

**Example 3.19.** Professional basketball player Steve Nash was a 90% free-throw shooter over his career. If Steve Nash starts shooting free throws, how many would he expect to make before missing one? What is the probability that he could make 20 in a row?

Let $X$ be the random variable which counts the number of free throws Steve Nash makes before missing one. We model a Steve Nash free throw as a Bernoulli trial, but we choose "success" to be a missed free throw, so that $p = 0.1$ and $X \sim \text{Geom}(0.1)$. The expected number of "failures" is $E[X] = \frac{0.9}{0.1} = 9$, which means we expect Steve to make 9 free throws before missing one.

To make 20 in a row requires $X \geq 20$. Using $P(X \geq 20) = 1 - P(X \leq 19)$,

```
1 - pgeom(19, 0.1)
```

```
## [1] 0.1215767
```

we see that Steve Nash could run off 20 (or more) free throws in a row about 12% of the times he wants to try.

## 3.4 Functions of a random variable

Recall that a random variable $X$ is a function from the sample space $S$ to $\mathbb{R}$. Given a function $g : \mathbb{R} \to \mathbb{R}$, we can form the random variable $g \circ X$, usually written $g(X)$.

For example, the mean number of births per day in the United States is about 10300. Suppose we pick a day in February and observe $X$ the number of births on that day. We might be more interested in the deviation from the mean $X - \mu$ or the absolute deviation from the mean $|X - \mu|$ of the measurement than we are in $X$, the value of the measurement. The values $X - \mu$ and $|X - \mu|$ are examples of functions of a random variable.

As another example, if $X$ is the value from a six-sided die roll and $g(x) = x^2$, then $g(X) = X^2$ is the value of a single six-sided die roll, squared. Here $X^2$ can take the values $1, 4, 9, 16, 25, 36$, all equally likely.

The probability mass function of $g(X)$ can be computed as follows. For each $y$, the probability that $g(X) = y$ is given by $\sum p(x)$, where the sum is over all values of $x$ such that $g(x) = y$.

**Example 3.20.** Let $X$ be the value when a six-sided die is rolled and let $g(X) = X^2$. The probability mass function of $g(X)$ is given by

| $x$ | 1 | 4 | 9 | 16 | 25 | 36 |
|---|---|---|---|---|---|---|
| $p(x)$ | 1/6 | 1/6 | 1/6 | 1/6 | 1/6 | 1/6 |

This is because for each value of $y$, there is exactly one $x$ with positive probability for which $g(x) = y$.

**Example 3.21.** Let $X$ be the value when a six-sided die is rolled and let $g(X) = |X - 3.5|$ be the absolute deviation from the expected value.[5] The probability that $g(X) = .5$ is $2/6$ because $g(4) = g(3) = 0.5$, and $p(4) + p(3) = 2/6$. The full probability mass function of $g(X)$ is given by

| $x$ | 0.5 | 1.5 | 2.5 |
|---|---|---|---|
| $p(x)$ | 1/3 | 1/3 | 1/3 |

Note that we could then find the expected value of $g(X)$ by applying the definition of expected value to the probability mass function of $g(X)$. It is usually easier, however, to apply the following theorem.

**Theorem 3.7.** *Let $X$ be a discrete random variable with probability mass function $p$, and let $g$ be a function. Then,*

$$E\left[g(X)\right] = \sum g(x)p(x).$$

**Example 3.22.** Let $X$ be the value of a six-sided die roll. Then whether we use the pmf calculated in Example 3.20 or Theorem 3.7, we get that

$$E[X^2] = 1^2 \cdot 1/6 + 2^2 \cdot 1/6 + 3^2 \cdot 1/6 + 4^2 \cdot 1/6 + 5^2 \cdot 1/6 + 6^2 \cdot 1/6 = 91/6 \approx 15.2$$

Computing with simulation is straightforward:

```
X <- sample(1:6, 10000, replace = TRUE)
mean(X^2)
```

```
## [1] 15.2356
```

**Example 3.23.** Let $X$ be the number of heads observed when a fair coin is tossed three times. Compute the expected value of $(X - 1.5)^2$. Since $X$ is a binomial random variable with $n = 3$ and $p = 1/2$, it has pmf $p(x) = \binom{3}{x}(.5)^3$ for $x = 0, 1, 2, 3$. We compute

$$\begin{aligned} E[(X - 1.5)^2] &= \sum_{x=0}^{3}(x - 1.5)^2 p(x) \\ &= (0 - 1.5)^2 \cdot 0.125 + (1 - 1.5)^2 \cdot 0.375 + (2 - 1.5)^2 \cdot 0.375 + (3 - 1.5)^2 \cdot 0.125 \\ &= 0.75 \end{aligned}$$

Since `dbinom` gives the pdf for binomial rvs, we can perform this exact computation in R:

```
x <- 0:3
sum((x - 1.5)^2 * dbinom(x, 3, 0.5))
```

```
## [1] 0.75
```

[5]The careful reader will notice that this is an example of a discrete random variable that does **not** take integer values.

We conclude this section with two simple but important observations about expected values. First, expected value is linear. Second, the expected value of a constant is that constant. Stated precisely:

**Theorem 3.8.** *For random variables $X$ and $Y$, and constants $a$, $b$, and $c$:*

1. $E[aX + bY] = aE[X] + bE[Y]$
2. $E[c] = c$

The proofs follow from the definition of expected value and the linearity of summation. With these theorems in hand, we can provide a much simpler proof for the formula of the expected value of a binomial random variable.

**Example 3.24.** Let $X \sim \text{Binom}(n, p)$. Show that $E[X] = np$.

The variable $X$ is the number of successes in $n$ Bernoulli trials, each with probability $p$. Let $X_1, \ldots, X_n$ be independent Bernoulli random variables with probability of success $p$. That is, $P(X_i = 1) = p$ and $P(X_i = 0) = 1 - p$. It follows that $X = \sum_{i=1}^n X_i$. Therefore,

$$\begin{aligned} E[X] &= E\left[\sum_{i=1}^n X_i\right] \\ &= \sum_{i=1}^n E[X_i] = \sum_{i=1}^n p = np. \end{aligned}$$

## 3.5 Variance, standard deviation, and independence

The variance of a random variable measures the spread of the variable around its expected value. Random variables with large variance can be quite far from their expected values, while rvs with small variance stay near their expected value. The standard deviation is simply the square root of the variance. The standard deviation also measures spread, but in more natural units which match the units of the random variable itself.

**Definition 3.7.** Let $X$ be a random variable with expected value $\mu = E[X]$. The *variance* of $X$ is defined as

$$\text{Var}(X) = E[(X - \mu)^2]$$

The *standard deviation* of $X$ is written $\sigma(X)$ and is the square root of the variance:

$$\sigma(X) = \sqrt{\text{Var}(X)}$$

Note that the variance of an rv is always positive (in the French sense[6]), as it is the expected value of a positive function.

The next theorem gives a formula for the variance that is often easier than the definition when performing computations.

**Theorem 3.9.** $\text{Var}(X) = E[X^2] - E[X]^2$.

[6] That is, the variance is greater than or equal to zero.

*Proof.* Applying linearity of expected values (Theorem 3.8) to the definition of variance yields:

$$\begin{aligned} E[(X-\mu)^2] &= E[X^2 - 2\mu X + \mu^2] \\ &= E[X^2] - 2\mu E[X] + \mu^2 = E[X^2] - 2\mu^2 + \mu^2 \\ &= E[X^2] - \mu^2, \end{aligned}$$

as desired. ∎

**Example 3.25.** Let $X$ be the value of a single die roll. We know that $\mu = E[X] = 7/2$ and Example 3.22 showed that $E[X^2] = 91/6$. The variance of a die roll is

$$\text{Var}(X) = 91/6 - (7/2)^2 = 35/12 \approx 2.92$$

and the standard deviation is $\sigma(X) = \sqrt{35/12} \approx 1.71$.

Checking with a simulation,

```
X <- sample(1:6, 100000, replace = TRUE)
sd(X)
```

```
## [1] 1.706927
```

**Example 3.26.** Let $X \sim \text{Binom}(3, 0.5)$. Compute the standard deviation and variance of $X$.

Here $\mu = E[X] = 1.5$. In Example 3.23, we saw that $E[(X - 1.5)^2] = 0.75$. Then $\text{Var}(X) = 0.75$ and the standard deviation is $\sigma(X) = \sqrt{0.75} \approx 0.866$. We can check both of these using simulation and the built-in R functions `var` and `sd`:

```
X <- rbinom(10000, 3, 0.5)
var(X)
```

```
## [1] 0.7481625
```

```
sd(X)
```

```
## [1] 0.8649638
```

### 3.5.1 Independent random variables

We say that two random variables $X$ and $Y$ are *independent* if knowledge of the value of $X$ does not give probabilistic information about the value of $Y$ and vice versa. As an example, let $X$ be the number of days it snows at St. Louis Lambert Airport in a given month, and let $Y$ be the number of kittens in a Eurasian lynx litter. It is difficult to imagine that knowing the value of one of these random variables could give information about the other one, and it is reasonable to assume that $X$ and $Y$ are independent. On the other hand, if $X$ and $W$ are the snowy days at Lambert and the number of flights delayed that month, then knowledge of one variable could well give probabilistic information about the other. For example, if you know it snowed a lot in March, it is likely that more flights were delayed. So $X$ and $W$ are not independent.

The key result in this section is Theorem 3.10. We will usually be *assuming* that random variables are independent so that we may apply part 2 of the theorem, that the variance of a sum is the sum of variances. For readers less interested in theory, a sense of when real-world variables are independent is all that is required to follow the rest of this textbook. In Exercises 3.24, 3.27, and 4.13, you may verify Theorem 3.10, part 2 by simulation in some special cases.

We would like to formalize the notion of independence with conditional probability. The natural statement is that for any $E, F$ subsets of $\mathbb{R}$, the conditional probability $P(X \in E | Y \in F)$ is equal to $P(X \in E)$. There are several issues with formalizing the notion of independence that way, so we give a definition that is somewhat further removed from the intuition.

**Definition 3.8.** A collection of rvs $X_1, \ldots, X_n$ are called *independent* $\iff$ for all $x_1, \ldots, x_n$ the events $X_1 \le x_1$, $X_2 \le x_2$, ..., and $X_n \le x_n$ are mutually independent.

**Remark.** For two random variables $X$ and $Y$, independence means that for all $x$ and $y$,

$$P(X \le x \text{ and } Y \le y) = P(X \le x) \cdot P(Y \le y).$$

If $X$ and $Y$ are discrete, an equivalent statement is that for all $x$ and $y$,

$$P(X = x \text{ and } Y = y) = P(X = x) \cdot P(Y = y)$$

Dividing both sides by $P(Y = y)$, we see that two discrete independent random variables satisfy $P(X = x | Y = y) = P(X = x)$, which returns to the intuition that "knowing something about $Y$ tells you nothing about $X$."

**Theorem 3.10.** *1. Let $X$ be a random variable and $c$ a constant. Then*

$$\begin{aligned} Var(cX) &= c^2\, Var(X) \\ \sigma(cX) &= |c|\sigma(X) \end{aligned}$$

*2. Let $X_1, X_2, \ldots, X_n$ be **independent** random variables. Then*

$$\mathrm{Var}(X_1 + X_2 + \cdots + X_n) = \mathrm{Var}(X_1) + \cdots + \mathrm{Var}(X_n)$$

*Proof.* For part 1,

$$\begin{aligned} \mathrm{Var}(cX) =& E[(cX)^2] - E[cX]^2 = c^2 E[X^2] - (cE[X])^2 \\ =& c^2(E[X^2] - E[X]^2) = c^2 \mathrm{Var}(X) \end{aligned}$$

For part 2, we will only treat the special case of two independent discrete random variables $X$ and $Y$. We need to show that $\mathrm{Var}(X + Y) = \mathrm{Var}(X) + \mathrm{Var}(Y)$.

$$\begin{aligned} \mathrm{Var}(X+Y) &= E[(X+Y)^2] - E[X+Y]^2 \\ &= E[X^2 + 2XY + Y^2] - (E[X] + E[Y])^2 \\ &= E[X^2] - E[X]^2 + E[Y^2] - E[Y]^2 + 2E[XY] - 2E[X]E[Y] \\ &= \mathrm{Var}(X) + \mathrm{Var}(Y) + 2\,(E[XY] - E[X]E[Y]) \end{aligned}$$

The value $E[XY] - E[X]E[Y]$ is called the *covariance* of $X$ and $Y$, written $\mathrm{Cov}(X, Y)$. To finish, we need to show that the covariance is zero when $X$ and $Y$ are independent.

For any $z$, the event $XY = z$ is the disjoint union of events $(X = x) \cap (Y = y)$ where $xy = z$. Summing all $x$ and $y$ with $xy = z$ and then summing over $z$ is the same as simply summing over all possible $x$ and $y$. So,

$$\begin{aligned} E[XY] &= \sum_z zP(XY = z) = \sum_z \sum_{x,y \text{ with } xy=z} zP((X = x) \cap (Y = y)) \\ &= \sum_{x,y} xyP(X = x) \cdot P(Y = y) = \left(\sum_x xP(X = x)\right)\left(\sum_y yP(Y = y)\right) \\ &= E[X]E[Y]. \end{aligned}$$

■

**Remark.** If $X$ and $Y$ are independent rvs, then $\text{Var}(X - Y) = \text{Var}(X) + \text{Var}(Y)$.

**Example 3.27 (Variance of a binomial rv).** Let $X \sim \text{Binom}(n, p)$. We have seen that $X = \sum_{i=1}^n X_i$, where $X_i$ are independent Bernoulli random variables. Therefore,

$$\begin{aligned} \text{Var}(X) &= \text{Var}(\sum_{i=1}^n X_i) \\ &= \sum_{i=1}^n \text{Var}(X_i) \\ &= \sum_{i=1}^n p(1 - p) = np(1 - p) \end{aligned}$$

where we have used that the variance of a Bernoulli random variable is $p(1 - p)$. Indeed, $E[X_i^2] - E[X_i]^2 = p - p^2 = p(1 - p)$.

## 3.6 Poisson, negative binomial, and hypergeometric

In this section, we discuss three other commonly occurring special types of discrete random variables.

### 3.6.1 Poisson

A *Poisson process* models events that happen at random times. For example, radioactive decay is a Poisson process, where each emission of a radioactive particle is an event. Other examples modeled by the Poisson process are meteor strikes on the surface of the moon, customers arriving at a store, hits on a web page, and car accidents on a given stretch of highway.

In this section, we will discuss one natural random variable attached to a Poisson process: the *Poisson random variable*. We will see another, the *exponential random variable*, in Section 4.5.2. The Poisson random variable is discrete, and can be used to model the number of

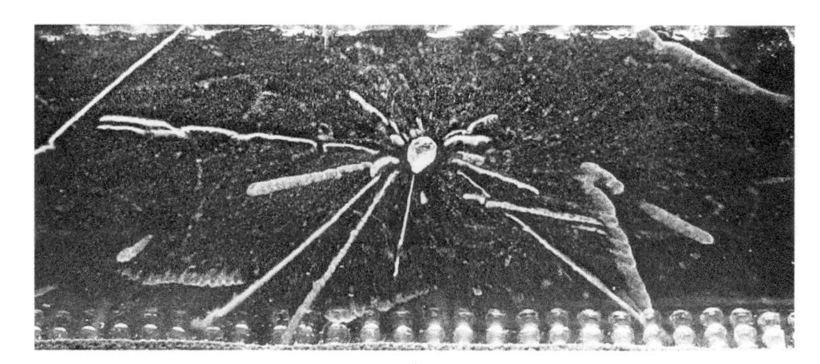

**FIGURE 3.5** Radioactivity of a Thorite mineral seen in a cloud chamber. (Photo credit: Julien Simon)

events that happen in a fixed time period. The exponential random variable models the time between events.

We begin by defining a Poisson process. Suppose events occur spread over time. The events form a Poisson process if they satisfy the following assumptions:

**Assumptions 3.1 (for a Poisson process).**

1. *The probability of an event occurring in a time interval* $[a, b]$ *depends only on the length of the interval* $[a, b]$.
2. *If* $[a, b]$ *and* $[c, d]$ *are disjoint time intervals, then the probability that an event occurs in* $[a, b]$ *is independent of whether an event occurs in* $[c, d]$. *(That is, knowing that an event occurred in* $[a, b]$ *does not change the probability that an event occurs in* $[c, d]$.*)*
3. *Two events cannot happen at the same time. (Formally, we need to say something about the probability that two or more events happens in the interval* $[a, a+h]$ *as* $h \to 0$.*)*
4. *The probability of an event occurring in a time interval* $[a, a+h]$ *is roughly* $\lambda h$*, for some constant* $\lambda$.

Property 4 says that events occur at a certain rate, which is denoted by $\lambda$.

**Definition 3.9.** The random variable $X$ is said to be a *Poisson random variable* with rate $\lambda$ if the pmf of $X$ is given by

$$p(x) = e^{-\lambda}\frac{\lambda^x}{x!} \qquad x = 0, 1, 2, \ldots$$

The parameter $\lambda$ is required to be larger than 0. We write $X \sim \text{Pois}(\lambda)$.

**Theorem 3.11.** *Let* $X$ *be the random variable that counts the number of occurrences in a Poisson process with rate* $\lambda$ *over one unit of time. Then* $X$ *is a Poisson random variable with rate* $\lambda$.

*Proof (heuristic).* We have not stated Assumption 3.1.4 precisely enough for a formal proof, so this is only a heuristic argument.

Suppose we break up the time interval from 0 to 1 into $n$ pieces of equal length. Let $X_i$ be the number of events that happen in the $i$th interval. When $n$ is big, $(X_i)_{i=1}^n$ is approximately a sequence of $n$ Bernoulli trials, each with probability of success $\lambda/n$. Therefore, if we let $Y_n$ be a binomial random variable with $n$ trials and probability of success $\lambda/n$, we have:

$$\begin{aligned}
P(X = x) &= P(\sum_{i=1}^{n} X_i = x) \\
&\approx P(Y_n = x) \\
&= \binom{n}{x}(\lambda/n)^x(1 - \lambda/n)^{(n-x)} \\
&\approx \frac{n^x}{x!}\frac{1}{n^x}\lambda^x(1 - \lambda/n)^n \\
&\to \frac{\lambda^x}{x!}e^{-\lambda} \qquad \text{as } n \to \infty
\end{aligned}$$

The key to defining the Poisson process formally correctly is to ensure that the above computation is mathematically justified.

■

We leave the proof that $\sum_x p(x) = 1$ as Exercise 3.36, along with the proof of the following Theorem.

**Theorem 3.12.** *The mean and variance of a Poisson random variable are both* $\lambda$.

Though the proof of the mean and variance of a Poisson is an exercise, we can also give a heuristic argument. From above, a Poisson with rate $\lambda$ is approximately a Binomial rv $Y_n$ with $n$ trials and probability of success $\lambda/n$ when $n$ is large. The mean of $Y_n$ is $n \times \lambda/n = \lambda$, and the variance is $n(\lambda/n)(1 - \lambda/n) \to \lambda$ as $n \to \infty$.

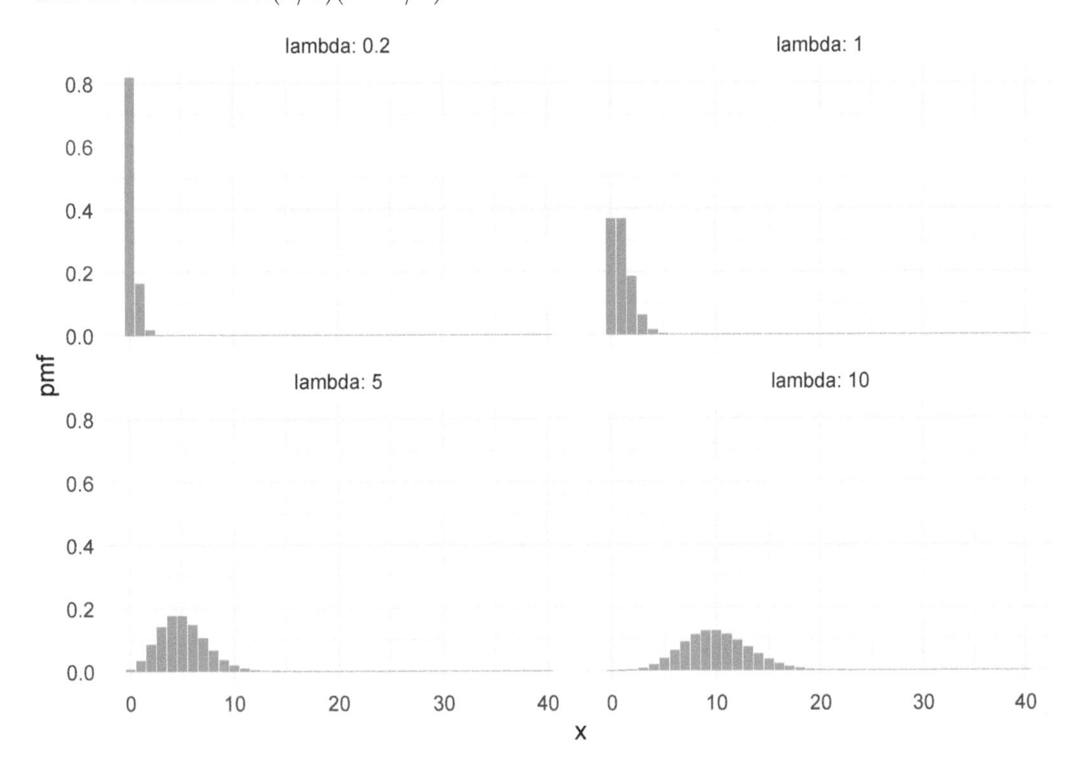


**FIGURE 3.6** Poisson pmfs with various means.

Figure 3.6 shows plots of Poisson pmfs with various means $\lambda$. Observe in Figure 3.6 that larger values of $\lambda$ correspond to more spread-out distributions, since the standard deviation of a Poisson rv is $\sqrt{\lambda}$. Also, for larger values of $\lambda$, the Poisson distribution becomes approximately normal.

The functions `dpois`, `ppois` and `rpois` are available for working with a Poisson random variable $X \sim \text{Pois}(\lambda)$:

- `dpois(x,lambda)` is the pmf, and gives $P(X = x)$.
- `ppois(x,lambda)` gives $P(X \leq x)$.
- `rpois(N,lambda)` simulates $N$ random values of $X$.

**Example 3.28.** The Taurids meteor shower is visible on clear nights in the fall and can have visible meteor rates around five per hour. What is the probability that a viewer will observe exactly eight meteors in two hours?

We let $X$ be the number of observed meteors in two hours, and model $X \sim \text{Pois}(10)$, since we expect $\lambda = 10$ meteors in our two-hour time period. Computing exactly,

$$P(X = 8) = \frac{10^8}{8!}e^{-10} \approx 0.1126.$$

Using R, and the pmf `dpois`:

```
dpois(8, 10)
```

```
## [1] 0.112599
```

Or, by simulation with `rpois`:

```
meteors <- rpois(10000, 10)
mean(meteors == 8)
```

```
## [1] 0.1118
```

We find that the probability of seeing exactly eight meteors is about 0.11.

**Example 3.29.** Suppose a typist makes typos at a rate of 3 typos per 10 pages. What is the probability that they will make at most one typo on a five-page document?

We let $X$ be the number of typos in a five-page document. Assume that typos follow the properties of a Poisson rv. It is not clear that they follow it exactly. For example, if the typist has just made a mistake, it is possible that their fingers are no longer on home position, which means another mistake is likely soon after. This would violate the independence property (2) of a Poisson process. Nevertheless, modeling $X$ as a Poisson rv is reasonable.

The rate at which typos occur *per five pages* is 1.5, so we use $\lambda = 1.5$. Then we can compute $P(X \leq 1) =$ `ppois(1,1.5)` $= 0.5578$. The typist has a 55.78% chance of making at most one typo in a five-page document.

### 3.6.2 Negative binomial

Suppose you repeatedly roll a fair die. What is the probability of getting exactly 14 non-sixes before getting your second six?

As you can see, this is an example of repeated Bernoulli trials with $p = 1/6$, but it isn't exactly geometric because we are waiting for the *second* success. This is an example of a *negative binomial* random variable.

**Definition 3.10.** Suppose that we observe a sequence of Bernoulli trials with probability of success $p$. If $X$ denotes the number of failures before the $n$th success, then $X$ is a *negative binomial random variable* with parameters $n$ and $p$. The probability mass function of $X$ is given by

$$p(x) = \binom{x+n-1}{x} p^n (1-p)^x, \qquad x = 0, 1, 2 \ldots$$

The mean of a negative binomial is $np/(1-p)$, and the variance is $np/(1-p)^2$. The root R function to use with negative binomials is `nbinom`, so `dnbinom` is how we can compute probabilities in R. The function `dnbinom` uses `prob` for $p$ and `size` for $n$ in our formula.

**Example 3.30.** Suppose you repeatedly roll a fair die. What is the probability of getting exactly 14 non-sixes before getting your second six?

```
dnbinom(x = 14, size = 2, prob = 1 / 6)
```

```
## [1] 0.03245274
```

Note that when `size = 1`, negative binomial is exactly a geometric random variable, e.g.,

```
dnbinom(x = 14, size = 1, prob = 1 / 6)
```

```
## [1] 0.01298109
```

```
dgeom(14, prob = 1 / 6)
```

```
## [1] 0.01298109
```

### 3.6.3 Hypergeometric

Consider the experiment which consists of sampling *without replacement* from a population that is partitioned into two subgroups – one subgroup is labeled a "success," and the other subgroup is labeled a "failure." The random variable that counts the number of successes in the sample is an example of a *hypergeometric* random variable.

To make things concrete, we suppose that we have $m$ successes and $n$ failures. We take a sample of size $k$ (without replacement) and we let $X$ denote the number of successes. Then

$$P(X = x) = \frac{\binom{m}{x}\binom{n}{k-x}}{\binom{m+n}{k}}$$

We also have

$$E[X] = k\left(\frac{m}{m+n}\right)$$

which is easy to remember because $k$ is the number of samples taken, and the probability of a success on any one sample (with no knowledge of the other samples) is $\frac{m}{m+n}$. The variance is similar to the variance of a binomial as well,

$$V(X) = k \cdot \frac{m}{m+n} \cdot \frac{n}{m+n} \cdot \frac{m+n-k}{m+n-1}$$

but we have the "fudge factor" of $\frac{m+n-k}{m+n-1}$, which means the variance of a hypergeometric is less than that of a binomial. In particular, when $m + n = k$, the variance of $X$ is 0. Why?

When $m + n$ is much larger than $k$, we will approximate a hypergeometric random variable with a binomial random variable with parameters $n = m + n$ and $p = \frac{m}{m+n}$. Finally, the R root for hypergeometric computations is hyper. In particular, we have the following example:

**Example 3.31.** 15 US citizens and 20 non-US citizens pass through a security line at an airport. Ten are randomly selected for further screening. What is the probability that 2 or fewer of the selected passengers are US citizens?

In this case, $m = 15$, $n = 20$, and $k = 10$. We are looking for $P(X \leq 2)$, so

```
sum(dhyper(x = 0:2, m = 15, n = 20, k = 10))
```

```
## [1] 0.08677992
```

You can find a summary of the discrete random variables together with their R commands in Section 4.6.

## Vignette: Loops in R

*Iteration* is the process of repeating a task to produce a result. In many traditional computer programming languages, iteration is done explicitly with a structure called a loop. With R, it is generally preferable and faster to use vector operations or implicit loops such as replicate. The purrr package also provides powerful tools for iterating.

However, sometimes it is easiest just to write a conventional loop, and R does provide explicit looping constructs. This vignette explains how to use while loops in the context of simulations.

A while loop repeats one or more statements until a certain condition is no longer met. It is crucial when writing a while loop that you ensure that the condition to exit the loop will eventually be satisfied! The format looks like this.

```
i <- 0 # initialize
while (i < 2) { # check condition
  print(i)
  i <- i + 1 # increment
}
```

```
## [1] 0
## [1] 1
```

R sets i to 0 and then checks whether i < 2 (it is). It then prints i and adds 1 to i. It checks again – is i < 2? Yes, so it prints i and adds 1 to it. At this point i is **not** less than 2, so R exits the loop.

As a more realistic example of while loops in action, suppose we want to estimate the expected value of a geometric random variable $X$.

Example 3.10 simulated tossing a fair coin until heads occurs, and counting the number of tails. In this case, $X$ is geometric with $p = 0.5$ and so has $E[X] = 1$. The approach was to simulate 100 coin flips and assume that heads would appear before we ran out of flips. The probability of flipping 100 tails in a row is negligible, so 100 flips was enough.

To simulate a geometric rv where $p$ is small and unknown, it may take a large number of trials before observing the first success. If each Bernoulli trial is actually a complicated calculation, we want to stop as soon as the first success occurs. Under these circumstances, it is better to use a `while` loop.

Here is a simulation of flipping a coin until heads is observed, using a `while` loop. In each iteration of the loop, we flip the coin with `sample`, print the result, and then add one to the count `num_tails`.

```
num_tails <- 0
coin_toss <- ""
while (coin_toss != "Head") {
  coin_toss <- sample(c("Head", "Tail"), 1)
  print(coin_toss)
  if (coin_toss == "Tail") {
    num_tails <- num_tails + 1
  }
}

## [1] "Tail"
## [1] "Tail"
## [1] "Head"
```

```
num_tails

## [1] 2
```

This gives one simulated value of $X$. In this case we found $X = 2$, because two tails appeared before the first head. To estimate $E[X]$, `replicate` the experiment.

```
sim_data <- replicate(10000, {
  num_tails <- 0
  coin_toss <- ""
  while (coin_toss != "Head") {
    coin_toss <- sample(c("Head", "Tail"), 1)
    if (coin_toss == "Tail") {
      num_tails <- num_tails + 1
    }
  }
  num_tails
})
mean(sim_data)

## [1] 0.9847
```

The mean of our simulated data is 0.9847, which agrees with the theoretical value $E[X] = 1$.

## Exercises

Exercises 3.1 – 3.3 require material through Section 3.1.

**3.1.** Let $X$ be a discrete random variable with probability mass function given by

$$p(x) = \begin{cases} 1/4 & x = 0 \\ 1/2 & x = 1 \\ 1/8 & x = 2 \\ 1/8 & x = 3 \end{cases}$$

a. Verify that $p$ is a valid probability mass function.
b. Find $P(X \geq 2)$.
c. Find $P(X \geq 2 \mid X \geq 1)$.
d. Find $P(X \geq 2 \cup X \geq 1)$.

**3.2.** Let $X$ be a discrete random variable with probability mass function given by

$$p(x) = \begin{cases} 1/4 & x = 0 \\ 1/2 & x = 1 \\ 1/8 & x = 2 \\ 1/8 & x = 3 \end{cases}$$

a. Use `sample` to create a sample of size 10,000 from $X$ and **estimate** $P(X = 1)$ from your sample. Your result should be close to 1/2.
b. Use `table` on your sample from part (a) to **estimate** the pmf of $X$ from your sample. Your result should be similar to the pmf given in the problem.

**3.3.** Let $X$ be a discrete random variable with probability mass function given by

$$p(x) = \begin{cases} C/4 & x = 0 \\ C/2 & x = 1 \\ C & x = 2 \\ 0 & \text{otherwise} \end{cases}$$

Find the value of $C$ that makes $p$ a valid probability mass function.

---

Exercises 3.4 – 3.15 require material through Section 3.2.

**3.4.** Give an example of a probability mass function $p$ whose associated random variable has mean 0.

**3.5.** Find the mean of the random variable $X$ given in Exercise 3.1.

**3.6.** Let $X$ be a random variable with pmf given by $p(x) = 1/10$ for $x = 1, \ldots, 10$ and $p(x) = 0$ for all other values of $x$. Find $E[X]$ and confirm your answer using a simulation.

**3.7.** Suppose you roll two ordinary dice. Calculate the expected value of their product.

**3.8.** Suppose that a hat contains slips of papers containing the numbers 1, 2, and 3. Two slips of paper are drawn without replacement. Calculate the expected value of the product of the numbers on the slips of paper.

**3.9.** Pick an integer from 0 to 999 with all possible numbers equally likely. What is the expected number of digits in your number?

**3.10.** In the summer of 2020, the U.S. was considering *pooled testing* of COVID-19.[7] This problem explores the math behind pooled testing. Since the availability of tests is limited, the testing center proposes the following pooled testing technique:

- Two samples are randomly selected and combined. The combined sample is tested.
- If the combined sample tests negative, then both people are assumed negative.
- If the combined sample tests positive, then both people need to be retested for the disease.

Suppose in a certain population, 5% of the people being tested for COVID-19 actually have COVID-19. Let $X$ be the total number of tests that are run in order to test two randomly selected people.

a. What is the pmf of $X$?
b. What is the expected value of $X$?
c. Repeat the above, but imagine that **three** samples are combined, and let $Y$ be the total number of tests that are run in order to test three randomly selected people. If the pooled test is positive, then all three people need to be retested individually.
d. If your only concern is to minimize the expected number of tests given to the population, which technique would you recommend?

**3.11.** A roulette wheel has 38 slots and a ball that rolls until it falls into one of the slots, all of which are equally likely. Eighteen slots are black numbers, eighteen are red numbers, and two are green zeros. If you bet on "red," and the ball lands in a red slot, the casino pays you your bet; otherwise, the casino wins your bet.

a. What is the expected value of a \$1 bet on red?
b. Suppose you bet \$1 on red, and if you win you "let it ride" and bet \$2 on red. What is the expected value of this plan?

**3.12.** One (questionable) roulette strategy is called bet doubling: You bet \$1 on red, and if you win, you pocket the \$1. If you lose, you double your bet so you are now betting \$2 on red, but have lost \$1. If you win, you win \$2 for a \$1 profit, which you pocket. If you lose again, you double your bet to \$4 (having already lost \$3). Again, if you win, you have \$1 profit, and if you lose, you double your bet again. This guarantees you will win \$1, unless you run out of money to keep doubling your bet.

a. Say you start with a bankroll of \$127. How many bets can you lose in a row without losing your bankroll?
b. If you have a \$127 bankroll, what is the probability that bet doubling wins you \$1?
c. What is the expected value of the bet doubling strategy with a \$127 bankroll?
d. If you play the bet doubling strategy with a \$127 bankroll, how many times can you expect to play before you lose your bankroll?

**3.13.** Flip a fair coin 10 times and let $X$ be the proportion of times that a head is followed by another head. Discard the sequence of ten tosses if you don't obtain a head in the first nine tosses. What is the expected value of $X$? (Note: this is **not** asking you to estimate the conditional probability of getting heads given that you just obtained heads.)

See Miller and Sanjurjo[8] for a connection to the "hot hand" fallacy.

---

[7] P W Cunningham, "The Health 202: The Trump Administration Is Eyeing a New Testing Strategy for Coronavirus, Anthony Fauci Says," *Washington Post*, June 26, 2020.

[8] Joshua B Miller and Adam Sanjurjo, "A Cold Shower for the Hot Hand Fallacy: Robust Evidence That Belief in the Hot Hand Is Justified," *IGIER Working Paper*, no. 518 (August 2019), https://doi.org/10.2139/ssrn.2450479.

**3.14.** To play the Missouri lottery Pick 3, you choose three digits 0-9 in order. Later, the lottery selects three digits at random, and you win if your choices match the lottery values in some way. Here are some possible bets you can play:

a. \$1 Straight wins if you correctly guess all three digits, in order. It pays \$600.
b. \$1 Front Pair wins if you correctly guess the first two digits. It pays \$60.
c. \$1 Back Pair wins if you correctly guess the last two digits. It pays \$60.
d. \$6 6-Way Combo wins if the three digits are different and you guess all three in any order. It pays \$600.
e. \$3 3-Way Combo wins if two of the three digits are the same, and you guess all three in any order. It pays \$600.
f. \$1 1-Off lets you win if some of your digits are off by 1 (9 and 0 are considered to be one off from each other). If you get the number exactly correct, you win \$300. If you have one digit off by 1, you win \$29. If you have two digits off by 1, you win \$4, and if all three of your digits are off by 1 you win \$9.

Consider the value of your bet to be your expected winnings per dollar bet. What value do each of these bets have?

**3.15.** Let $k$ be a positive integer and let $X$ be a random variable with pmf given by $p(x) = 1/k$ for $x = 1, \ldots, k$ and $p(x) = 0$ for all other values of $x$. Find $E[X]$.

---

Exercises 3.16 – 3.20 require material through Section 3.3.

**3.16.** Suppose you take a 20-question multiple choice test, where each question has four choices. You guess randomly on each question.

a. What is your expected score?
b. What is the probability you get 10 or more questions correct?

**3.17.** Steph Curry is a 91% free-throw shooter. Suppose that he shoots 10 free throws in a game.

a. What is his expected number of shots made?
b. What is the probability that he makes at least eight free throws?

**3.18.** Steph Curry is a 91% free-throw shooter. He decides to shoot free throws until his first miss. What is the probability that he shoots exactly 20 free throws (including the one he misses)?

**3.19.** In October 2020, the YouTuber called "Dream" posted a speedrun of Minecraft and was accused of cheating.

In Minecraft, when you trade with a piglin, the piglin gives you an ender pearl 4.7% of the time. Dream got 42 ender pearls after 262 trades with piglin.

a. If you trade 262 times, what is the expected number of ender pearls you receive?
b. What is the probability of getting 42 or more ender pearls after 262 trades?

When you kill a blaze, you have a 50% chance of getting a blaze rod. Dream got 211 blaze rods after killing 305 blazes.

c. If you kill 305 blazes, what is the expected number of blaze rods you receive?
d. What is the probability of getting 211 or more blaze rods after killing 305 blazes?
e. Do you think Dream was cheating?

**3.20.** Let $X \sim \text{Geom}(p)$ be a geometric rv with success probability $p$. Show that the standard deviation of $X$ is $\frac{\sqrt{1-p}}{p}$. Hint: Follow the proof of Theorem 3.6, but take the derivative *twice* with respect to $q$. This will compute $E[X(X-1)]$. Use $E[X(X-1)] = E[X^2] - E[X]$ and Theorem 3.9 to finish.

---

Exercises 3.21 – 3.22 require material through Section 3.4.

**3.21.** Let $X$ be a discrete random variable with probability mass function given by

$$p(x) = \begin{cases} 1/4 & x = 0 \\ 1/2 & x = 1 \\ 1/8 & x = 2 \\ 1/8 & x = 3 \end{cases}$$

a. Find the pmf of $Y = X - 1$.
b. Find the pmf of $U = X^2$.
c. Find the pmf of $V = (X-1)^2$.

**3.22.** Let $X$ and $Y$ be random variables such that $E[X] = 2$ and $E[Y] = 3$.

a. Find $E[4X + 5Y]$.
b. Find $E[4X - 5Y + 2]$.

---

Exercises 3.23 – 3.32 require material through Section 3.5.

**3.23.** Find the variance and standard deviation of the rv $X$ from Exercise 3.1.

**3.24.** Roll two ordinary dice and let $Y$ be their sum.

a. Compute the pmf for $Y$ exactly.
b. Compute the mean and standard deviation of $Y$.
c. Check that the variance of $Y$ is twice the variance for the roll of one die (see Example 3.25).

**3.25.** Let $X$ be a random variable such that $E[X] = 2$ and $\text{Var}(X) = 9$.

a. Find $E[X^2]$.
b. Find $E[(2X-1)^2]$.
c. Find $\text{Var}(2X - 1)$.

**3.26.** Let $X$ and $Y$ be random variables. Show that the covariance of $X$ and $Y$ can be rewritten in the following way:

$$E[XY] - E[X]E[Y] = E\left[(X - \mu_X)(Y - \mu_Y)\right].$$

**3.27.** Let $X \sim \text{Binom}(100, 0.2)$ and $Y \sim \text{Binom}(40, 0.5)$ be independent rvs.

a. Compute $\text{Var}(X)$ and $\text{Var}(Y)$ exactly.
b. Simulate the random variable $X + Y$ and compute its variance. Check that it is equal to $\text{Var}(X) + \text{Var}(Y)$.

**3.28.** In an experiment where you toss a fair coin twice, let:

- $X$ be 1 if the first toss is heads and 0 otherwise.
- $Y$ be 1 if the second toss is heads and 0 otherwise.
- $Z$ be 1 if both tosses are the same and 0 otherwise.

Show that $X$ and $Y$ are independent. Show that $X$ and $Z$ are independent. Show that $Y$ and $Z$ are independent. Finally, show that $X$, $Y$, and $Z$ are **not** mutually independent.

**3.29.** Let $X$ be a random variable with mean $\mu$ and standard deviation $\sigma$. Find the mean and standard deviation of $\frac{X-\mu}{\sigma}$.

**3.30.** Suppose that 55% of voters support Proposition A.

a. You poll 200 voters. What is the expected number that support the measure?
b. What is the margin of error for your poll (two standard deviations)?
c. What is the probability that your poll claims that Proposition A will fail?
d. How large a poll would you need to reduce your margin of error to 2%?

**3.31.** Suppose 27 people write their names down on slips of paper and put them in a hat. Each person then draws one name from the hat. Estimate the expected value and standard deviation of the number of people who draw their own name. (Assume no two people have the same name!)

**3.32.** In an experiment[9] to test whether participants have absolute pitch, scientists play notes and the participants say which of the 12 notes is being played. The participant gets 1 point for each note that is correctly identified, and 3/4 of a point for each note that is off by a half-step. (Note that if the possible guesses are 1:12, then the difference between 1 and 12 is a half-step, as is the difference between any two values that are 1 apart.)

a. If the participant hears 36 notes and randomly guesses each time, what is the expected score of the participant?
b. If the participant hears 36 notes and randomly guesses each time, what is the standard deviation of the score of the participant? Assume each guess is independent.

---

Exercises 3.33 – 3.39 require material through Section 3.6.

**3.33.** Let $X$ be a Poisson rv with mean 3.9.

a. Create a plot of the pmf of $X$.
b. What is the most likely outcome of $X$?
c. Find $a$ such that $P(a \le X \le a+1)$ is maximized.
d. Find $b$ such that $P(b \le X \le b+2)$ is maximized.

**3.34.** We stated in the text that a Poisson random variable $X$ with rate $\lambda$ is approximately a Binomial random variable $Y$ with $n$ trials and probability of success $\lambda/n$ when $n$ is large. Suppose $\lambda = 2$ and $n = 300$. What is the largest absolute value of the difference between $P(X = x)$ and $P(Y = x)$?

**3.35.** The charge $e$ on one electron is too small to measure. However, one can make measurements of the current $I$ passing through a detector. If $N$ is the number of electrons passing through the detector in one second, then $I = eN$. Assume $N$ is Poisson. Show that the charge on one electron is given by $\frac{\text{Var}(I)}{E[I]}$.

**3.36.** As stated in the text, the pdf of a Poisson random variable $X \sim \text{Pois}(\lambda)$ is

$$p(x) = \frac{1}{x!}\lambda^x e^{-\lambda}, \quad x = 0, 1, \ldots$$

[9] E Alexandra Athos et al., "Dichotomy and Perceptual Distortions in Absolute Pitch Ability," *Proceedings of the National Academy of Sciences of the United States of America* 104, no. 37 (2007): 14795–800, https://doi.org/10.1073/pnas.0703868104.

Prove the following:

a. $p$ is a pdf. (You need to show that $\sum_{x=0}^{\infty} p(x) = 1$.)
b. $E(X) = \lambda$. (Show that $\sum_{x=0}^{\infty} xp(x) = \lambda$.)
c. $\text{Var}(X) = \lambda$. (Compute $E[X(X-1)]$ by summing the infinite series $\sum_{x=0}^{\infty} x(x-1)p(x)$. Use $E[X(X-1)] = E[X^2] - E[X]$ and Theorem 3.9 to finish.)

**3.37.** Let $X$ be a negative binomial random variable with probability of success $p$ and size $n$, so that we are counting the number of failures before the $n$th success occurs.

a. Argue that $X = \sum_{i=1}^{n} X_i$ where $X_i$ are independent geometric random variables with probability of success $p$.
b. Use part (a) to show that $E[X] = \frac{np}{1-p}$ and the variance of $X$ is $\frac{np}{(1-p)^2}$.

**3.38.** Show that if $X$ is a Poisson random variable with mean $\lambda$, and $Y$ is a negative binomial random variable with $E[Y] = \frac{np}{1-p} = \lambda$, then $\text{Var}(X) \leq \text{Var}(Y)$.

**3.39.** In the game of Scrabble, players make words using letter tiles, see Exercise 2.19. The tiles consist of 42 vowels and 58 non-vowels (including blanks).

a. If a player draws 7 tiles (without replacement), what is the probability of getting 7 vowels?
b. If a player draws 7 tiles (without replacement), what is the probability of 2 or fewer vowels?
c. What is the expected number of vowels drawn when drawing 7 tiles?
d. What is the standard deviation of the number of vowels drawn when drawing 7 tiles?

**3.40 (Requires material covered in vignette or similar).** Deathrolling in World of Warcraft works as follows. Player 1 tosses a 1000-sided die. Say they get $x_1$. Then player 2 tosses a die with $x_1$ sides on it. Say they get $x_2$. Player 1 tosses a die with $x_2$ sides on it. The player who loses is the player who first rolls a 1.

a. Estimate the expected total number of rolls before a player loses.
b. Estimate the probability mass function of the total number of rolls.
c. Estimate the probability that player 1 wins.

# 4

# Continuous Random Variables

Continuous random variables are used to model random variables that can take on any value in an interval, either finite or infinite. Examples include the height of a randomly selected human or the error in measurement when measuring the height of a human. We will see that continuous random variables behave similarly to discrete random variables, except that we need to replace **sums** of the probability mass function with **integrals** of the analogous probability density function.

## 4.1 Probability density functions

**Definition 4.1.** A *probability density function* (pdf) is a function $f$ such that:

1. $f(x) \geq 0$ for all $x$.
2. $\int f(x)\,dx = 1$.

**Example 4.1.** Let

$$f(x) = \begin{cases} 2x & 0 \leq x \leq 1 \\ 0 & \text{otherwise} \end{cases}$$

We see that $f(x) \geq 0$ by inspection and $\int_{-\infty}^{\infty} f(x)\,dx = \int_0^1 2x\,dx = 1$, so $f$ is a probability density function.

**Definition 4.2.** A *continuous* random variable $X$ is a random variable described by a probability density function, in the sense that:

$$P(a \leq X \leq b) = \int_a^b f(x)\,dx.$$

whenever $a \leq b$, including the cases $a = -\infty$ or $b = \infty$.

**Definition 4.3.** The *cumulative distribution function* (cdf) associated with $X$ (either discrete or continuous) is the function $F(x) = P(X \leq x)$. Written out in terms of pdfs and pmfs:

$$F(x) = P(X \leq x) = \begin{cases} \int_{-\infty}^{x} f(t)\,dt & X \text{ is continuous} \\ \sum_{n=-\infty}^{x} p(n) & X \text{ is discrete} \end{cases}$$

DOI: 10.1201/9781003004899-4

By the Fundamental Theorem of Calculus, when $X$ is continuous, $F$ is a continuous function, hence the name continuous rv. The function $F$ is also referred to as the distribution function of $X$.

One major difference between discrete rvs and continuous rvs is that discrete rvs can take on only countably many different values, while continuous rvs typically take on values in an interval such as $[0, 1]$ or $(-\infty, \infty)$.

**Theorem 4.1.** *Let $X$ be a continuous random variable with pdf $f$ and cdf $F$.*

1. $\frac{d}{dx}F = f$.
2. $P(a \leq X \leq b) = F(b) - F(a)$.
3. $P(X \geq a) = 1 - F(a) = \int_a^\infty f(x)\,dx$.

*Proof.* All of these follow directly from the Fundamental Theorem of Calculus. ■

For a continuous rv $X$, the probability $P(X = x)$ is always zero, for any $x$. Therefore, $P(X \geq a) = P(X > a)$ and $P(X \leq b) = P(X < b)$.

**Example 4.2.** Consider the random variable $X$ with probability density function

$$f(x) = \begin{cases} 2x & 0 \leq x \leq 1 \\ 0 & \text{otherwise} \end{cases}$$

The cumulative distribution function of $X$ is given by

$$F(x) = \begin{cases} 0 & x < 0 \\ x^2 & 0 \leq x \leq 1 \\ 1 & x > 1 \end{cases}$$

We can see this is true because $F$ is a valid cumulative distribution function and $F'(x) = f(x)$ for all but finitely many $x$.

**Example 4.3.** Suppose that $X$ has pdf $f(x) = e^{-x}$ for $x > 0$, as graphed in Figure 4.1.

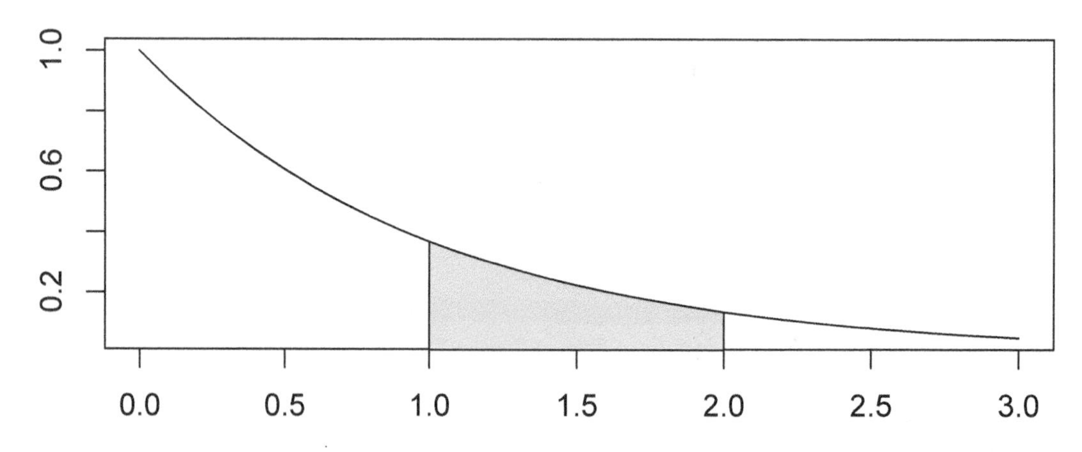


**FIGURE 4.1** $P(1 \leq X \leq 2)$ for a random variable $X$ with pdf $f(x) = e^{-x}$.

1. Find $P(1 \leq X \leq 2)$, which is the shaded area shown in the graph. By definition,

$$P(1 \leq X \leq 2) = \int_1^2 e^{-x}\,dx = -e^{-x}\Big|_1^2 = e^{-1} - e^{-2} \approx .233$$

2. Find $P(X \geq 1 | X \leq 2)$. This is the conditional probability that $X$ is greater than 1, given that $X$ is less than or equal to 2. We have

$$\begin{aligned} P(X \geq 1 | X \leq 2) &= P(X \geq 1 \cap X \leq 2)/P(X \leq 2) \\ &= P(1 \leq X \leq 2)/P(X \leq 2) \\ &= \frac{e^{-1} - e^{-2}}{1 - e^{-2}} \approx .269 \end{aligned}$$

**Example 4.4.** Let $X$ have the pdf:

$$f(x) = \begin{cases} 0 & x < 0 \\ 1 & 0 \leq x \leq 1 \\ 0 & x \geq 1 \end{cases}$$

$X$ is called the *uniform random variable* on the interval $[0, 1]$ and realizes the idea of choosing a random number between 0 and 1.

Here are some simple probability computations involving $X$:

$$P(X > 0.3) = \int_{0.3}^{1} 1\,dx = 0.7$$

$$P(0.2 < X < 0.5) = \int_{0.2}^{0.5} 1\,dx = 0.3$$

The cdf for $X$ is given by:

$$F(x) = \begin{cases} 0 & x < 0 \\ x & 0 \leq x \leq 1 \\ 1 & x \geq 1 \end{cases}$$

The pdf and cdf of the uniform random variable $X$ are implemented in R with the functions `dunif` and `punif`. Here are plots of these functions, produced by `plot` and by choosing a sequence of `x` values that cover the interesting range of the pdf and cdf:

```
x <- seq(-0.5, 1.5, 0.01)
plot(x, dunif(x))
plot(x, punif(x))
```

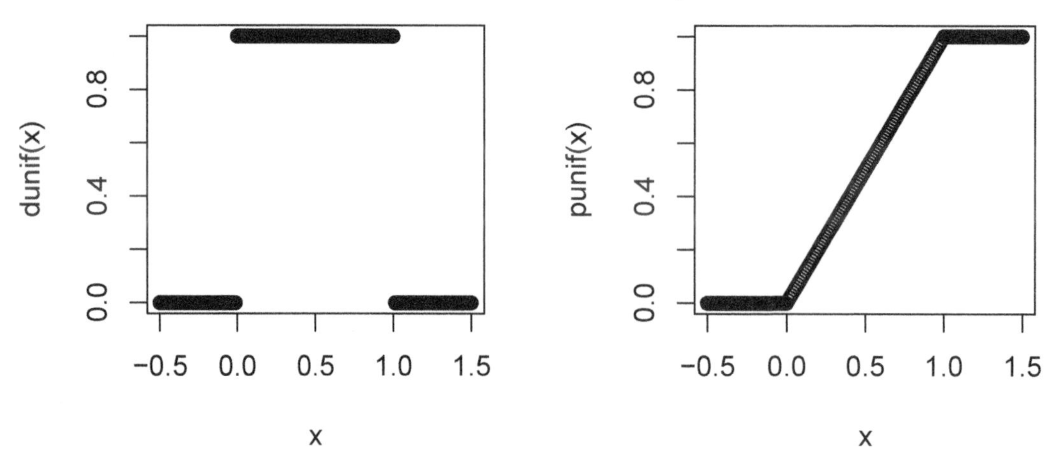


We can produce simulated values of $X$ with the function `runif` and use those to estimate probabilities:

```
X <- runif(10000)
mean(X > 0.3)
```

```
## [1] 0.7041
```

```
mean(0.2 < X & X < 0.5)
```

```
## [1] 0.3015
```

Here, we have used the **vectorized** version of the and operator, `&`.

It is important to distinguish between the operator `&` and the operator `&&` when doing simulations, or any time that you are doing logical operations on vectors. If you use `&&` on two vectors, then it ignores all of the values in each vector except the first one!

For example,

```
c(TRUE, TRUE, TRUE) && c(TRUE, FALSE, FALSE)
```

```
## [1] TRUE
```

But, compare to

```
c(TRUE, TRUE, TRUE) & c(TRUE, FALSE, FALSE)
```

```
## [1]  TRUE FALSE FALSE
```

**Example 4.5.** Suppose $X$ is a random variable with cdf given by

$$F(x) = \begin{cases} 0 & x < 0 \\ x/2 & 0 \le x \le 2 \\ 1 & x \ge 2 \end{cases}$$

1. Find $P(1 \le X \le 2)$. To do this, we note that $P(1 \le X \le 2) = F(2) - F(1) = 1 - 1/2 = 1/2$.
2. Find $P(X \ge 1/2)$. To do this, we note that $P(X \ge 1/2) = 1 - F(1/2) = 1 - 1/4 = 3/4$.
3. To find the pdf associated with $X$, we take the derivative of $F$. Since pdfs are only used via integration, it doesn't matter how we define the pdf at the places where $F$ is not

differentiable. In this case, we set $f$ equal to $1/2$ at those places to get

$$f(x) = \begin{cases} 1/2 & 0 \le x \le 2 \\ 0 & \text{otherwise} \end{cases}$$

This example is also a uniform random variable, this time on the interval $[0, 2]$.

**Example 4.6.** Let $X$ and $Y$ be independent uniform random variables on $[0, 1]$. Find the cumulative distribution function $F_W$ for the random variable $W$ which is the larger of $X$ and $Y$.

In Example 4.4 we saw that both $X$ and $Y$ have cdf given by

$$F(x) = \begin{cases} 0 & x < 0 \\ x & 0 \le x \le 1 \\ 1 & x > 1 \end{cases}.$$

Let's start with the observation that $W \le w$ exactly when *both* $X \le w$ and $Y \le w$. Therefore,

$$F_W(w) = P(W \le w) = P(X \le w \text{ and } Y \le w) = P(X \le w)P(Y \le w) = F(w)F(w) = F(w)^2.$$

The middle equality used the multiplication rule for independent events, since $X$ and $Y$ are independent. Then,

$$F_W(w) = F(w)^2 = \begin{cases} 0 & w < 0 \\ w^2 & 0 \le w \le 1 \\ 1 & w > 1 \end{cases}$$

The pdf of $W$ we compute by differentiating $F_W(w)$ to get:

$$f_W(w) = \begin{cases} 0 & w < 0 \\ 2w & 0 \le w \le 1 \\ 0 & w > 1 \end{cases}$$

We can confirm this answer by estimating the probability that the maximum of two uniform random variables is less than or equal to $2/3$. We start by taking large samples from two independent uniform distributions, and then we need to take the pairwise maximums of the vectors. Note that `max` will not work in this case, because it will simply take the maximum of the entire vector, so we need to use the *parallel* version `pmax`.

```
X <- runif(10000, 0, 1)
Y <- runif(10000, 0, 1)
W <- pmax(X, Y)
mean(W < 2 / 3)
```

```
## [1] 0.4444
```

We get that $P(W < 2/3) \approx 0.4444$, which is close to the true value $F_W(2/3) = (2/3)^2 = 4/9$.

Exercise 5.15 in Chapter 5 asks you to estimate the pdf of $W$ using simulation.

## 4.2 Expected value

**Definition 4.4.** Let $X$ be a continuous random variable with pdf $f$.

The *expected value* of $X$ is

$$E[X] = \int_{-\infty}^{\infty} x f(x)\, dx$$

**Example 4.7.** Find the expected value of $X$ when its pdf is given by $f(x) = e^{-x}$ for $x > 0$.

We compute

$$E[X] = \int_{-\infty}^{\infty} f(x)dx = \int_0^{\infty} xe^{-x}\, dx = \left(-xe^{-x} - e^{-x}\right)\Big|_0^{\infty} = 1$$

(Recall: to integrate $xe^{-x}$ you use integration by parts.)

**Example 4.8.** Find the expected value of the uniform random variable on $[0,1]$. Using integration, we get the exact result:

$$E[X] = \int_0^1 x \cdot 1\, dx = \frac{x^2}{2}\Big|_0^1 = \frac{1}{2}$$

Approximating with simulation,

```
X <- runif(10000)
mean(X)
```

```
## [1] 0.495409
```

**Remark.** Observe that the formula for the expected value of a continuous rv is the same as the formula for center of mass. If the area under the pdf $f(x)$ is cut from a thin sheet of material, the expected value is the point on the $x$-axis where this material would balance.

The balance point is at $X = 1/2$ for the uniform random variable on [0,1], since the pdf describes a square with base $[0, 1]$.

For $X$ with pdf $f(x) = e^{-x}, x \geq 0$, Figure 4.2 shows $E[X] = 1$ as the balancing point for the shaded region.

As in the discrete case, we can also define functions of random variables. In order to find the probability density function of a function of a continuous random variable, we can find the cumulative distribution function and take its derivative.

**Example 4.9.** Let $X$ be a uniform random variable on $[0, 1]$. Find the cdf of $X^2 + 1$.

Let $Y = X^2 + 1$. Since $X \in [0,1]$, we must have $Y \in [1,2]$. Then for $1 \leq y \leq 2$,

$$P(Y \leq y) = P(X^2 + 1 \leq y) = P(X \leq \sqrt{y-1}).$$

Since the cdf of $X$ is given by $F(x) = x,\ 0 \leq x \leq 1$, we have

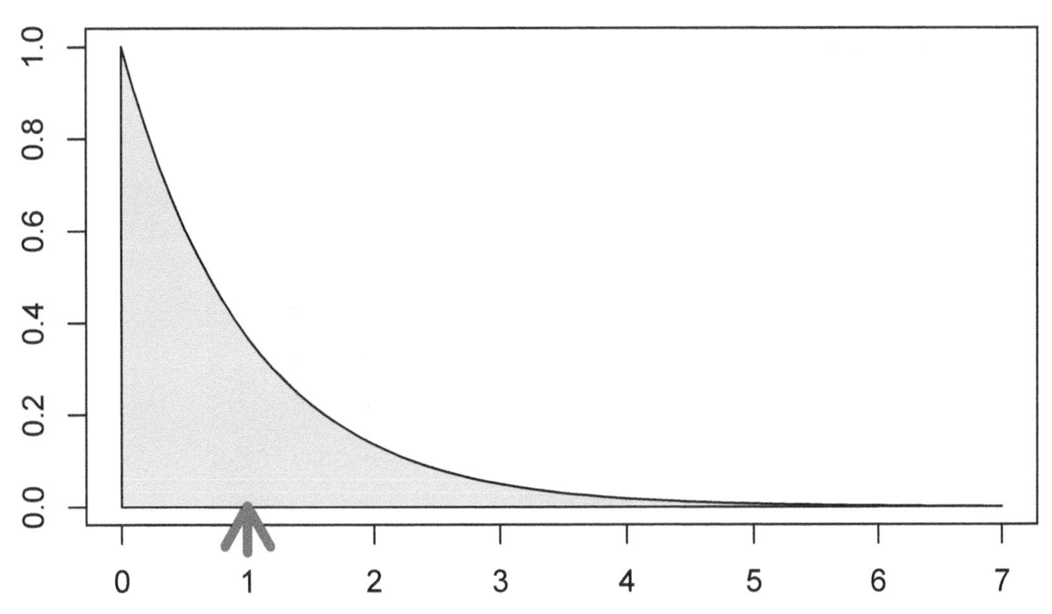


**FIGURE 4.2** The expected value of a distribution is its "balancing point."

$$P(Y \leq y) = \begin{cases} 0, & y < 1 \\ \sqrt{y-1}, & 1 \leq y \leq 2 \\ 1, & y > 2 \end{cases}$$

As in the discrete case, we will often be interested in computing expected values of functions of random variables.

**Theorem 4.2.** *Let $X$ be a continuous random variable and let $g$ be a function.*

$$E[g(X)] = \int g(x)f(x)\,dx$$

**Example 4.10.** Compute $E[X^2]$ for $X$ that has pdf $f(x) = e^{-x}$, $x > 0$.

Using integration by parts:

$$E[X^2] = \int_0^\infty x^2 e^{-x}\,dx = \left(-x^2e^{-x} - 2xe^{-x} - 2e^{-x}\right)\Big|_0^\infty = 2.$$

**Example 4.11.** Let $X$ be the uniform random variable on $[0, 1]$, and let $Y = 1/X$.

What is $P(Y < 3)$? The event $Y < 3$ is the same as $1/X < 3$ or $X > 1/3$, so $P(Y < 3) = P(X > 1/3) = 2/3$. We can check with simulation:

```
X <- runif(10000)
Y <- 1 / X
mean(Y < 3)

## [1] 0.6733
```

On the other hand, the expected value of $Y = 1/X$ is not well behaved. We compute:

$$E[1/X] = \int_0^1 \frac{1}{x} dx = \ln(x)\Big|_0^1 = \infty$$

Small values of $X$ are common enough that the huge values of $1/X$ produced cause the expected value to be infinite. Let's see what this does to simulations:

```
X <- runif(100)
mean(1 / X)
```

```
## [1] 4.116572
```

```
X <- runif(10000)
mean(1 / X)
```

```
## [1] 10.65163
```

```
X <- runif(1000000)
mean(1 / X)
```

```
## [1] 22.27425
```

Because the expected value is infinite, the simulations are not approaching a finite number as the size of the simulation increases. The reader is encouraged to try running these simulations multiple times to observe the inconsistency of the results.

Expected value is linear, which we stated as Theorem 3.8 in Chapter 3 for discrete random variables. We repeat it here for any type of random variable.

**Theorem 4.3.** *For random variables $X$ and $Y$, and constants $a$, $b$, and $c$:*

1. $E[aX + bY] = aE[X] + bE[Y]$.
2. $E[c] = c$.

## 4.3 Variance and standard deviation

The variance and standard deviation of a continuous random variable play the same role as they do for discrete random variables, that is, they measure the spread of the random variable about its mean. The definitions are unchanged from the discrete case (Definition 3.7), and Theorem 3.9 applies just as well to compute variance. For a random variable $X$ with expected value $\mu$:

$$\text{Var}(X) = E[(X - \mu)^2] = E[X^2] - E[X]^2$$
$$\sigma = \sqrt{\text{Var}(X)}$$

**Example 4.12.** Compute the variance of $X$ if the pdf of $X$ is given by $f(x) = e^{-x}$, $x > 0$.

We have already seen that $E[X] = 1$ and $E[X^2] = 2$ (Example 4.10). Therefore, the variance of $X$ is

$$\text{Var}(X) = E[X^2] - E[X]^2 = 2 - 1 = 1.$$

The standard deviation $\sigma(X) = \sqrt{1} = 1$. We interpret the standard deviation $\sigma$ as a spread around the mean, as shown in Figure 4.3.

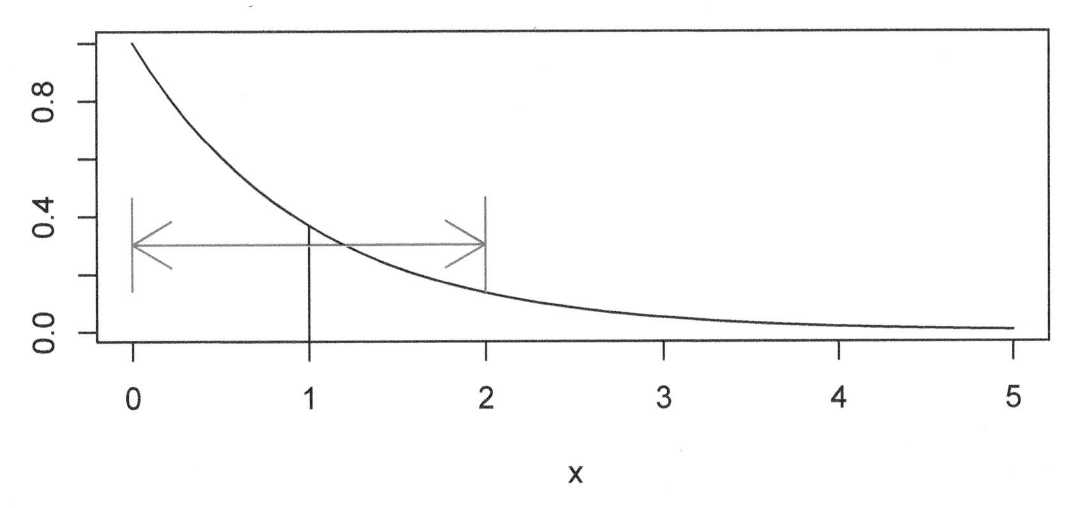


**FIGURE 4.3** One standard deviation above and below the mean $\mu = 1$.

In Figure 4.3, the mean 1 is marked with a vertical line, and the two arrows extend one standard deviation in each direction from the mean.

**Example 4.13.** Compute the standard deviation of the uniform random variable $X$ on $[0, 1]$.

$$\begin{aligned}\operatorname{Var}(X) &= E[X^2] - E[X]^2 = \int_0^1 x^2 \cdot 1\, dx - \left(\frac{1}{2}\right)^2 \\ &= \frac{1}{3} - \frac{1}{4} = \frac{1}{12} \approx 0.083.\end{aligned}$$

So the standard deviation is $\sigma(X) = \sqrt{1/12} \approx 0.289$. Figure 4.4 shows a one standard deviation spread around the mean of 1/2.

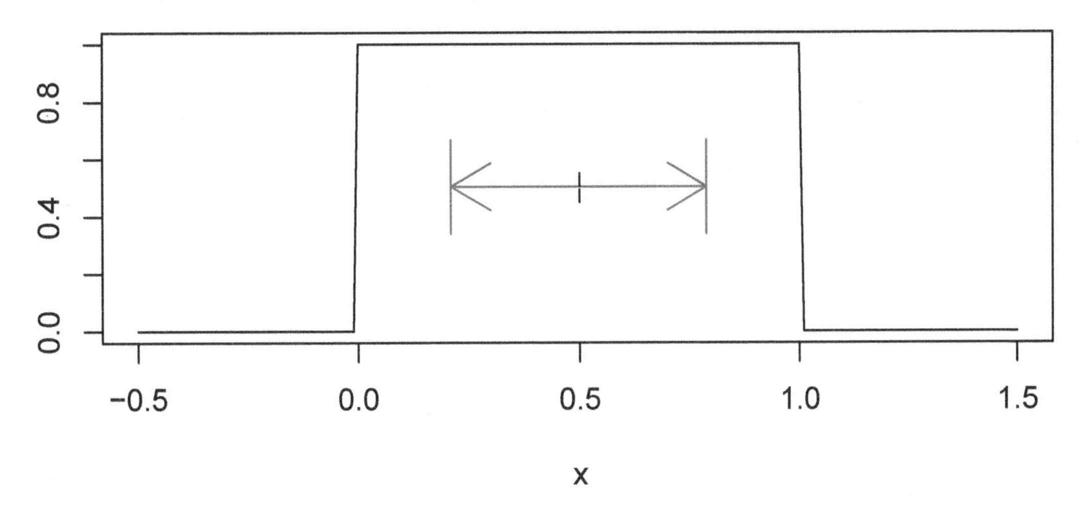


**FIGURE 4.4** One standard deviation above and below the mean $\mu = 0.5$.

For many distributions, most of the values will lie within one standard deviation of the mean, *i.e.*, within the spread shown in the example pictures. Almost all of the values will lie within two standard deviations of the mean. What do we mean by "almost all?" Well, 85% would be almost all; 15% would not be almost all. This is a very vague rule of thumb. Chebychev's Theorem is a more precise statement. It says in particular that the probability of being more than two standard deviations away from the mean is *at most* 25%.

Sometimes, you know that the data you collect will likely fall in a certain range of values. For example, if you are measuring the height in inches of 100 randomly selected adult males, you would be able to guess that your data will very likely lie in the interval 60-84. You can get a rough estimate of the standard deviation by taking the expected range of values and dividing by 6; in this case it would be $24/6 = 4$. Here, we are using the heuristic that it is very rare for data to fall more than three standard deviations from the mean. This can be useful as a quick check on your computations.

## 4.4 Normal random variables

The normal distribution is the most important in statistics. It is often referred to as the *bell curve*, because its shape resembles a bell:

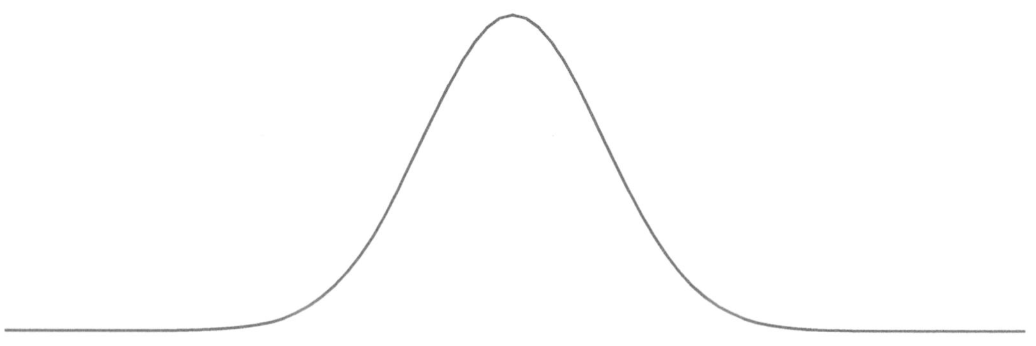

The importance of the normal distribution stems from the *Central Limit Theorem*, which implies that many random variables have normal distributions. A little more accurately, the Central Limit Theorem says that random variables which are the sum of many small independent factors are approximately normal.

For example, we might model the heights of adult females with a normal distribution. We imagine that adult height is affected by genetic contributions from generations of parents together with the sum of contributions from food eaten and other environmental factors. This is a reason to *try* a normal model for heights of adult females, and certainly should not be seen as a theoretical justification of any sort that adult female heights *must* be normal.

Many measured quantities are commonly modeled with normal distributions. Biometric measurements (height, weight, blood pressure, wingspan) are often nearly normal. Standardized test scores, economic indicators, scientific measurement errors, and variation in manufacturing processes are other examples.

The mathematical definition of the normal distribution begins with the function $h(x) = e^{-x^2}$, which produces the bell shaped curve shown above, centered at zero and with tails that decay very quickly to zero. By itself, $h(x) = e^{-x^2}$ is not a distribution since it does not have area 1 underneath the curve. In fact:

$$\int_{-\infty}^{\infty} e^{-x^2}\,dx = \sqrt{\pi}$$

This famous result is known as the *Gaussian integral.* Its proof is left to the reader in Exercise 4.14. By rescaling, we arrive at an actual pdf given by $g(x) = \frac{1}{\sqrt{\pi}} e^{-x^2}$. The distribution $g(x)$ has mean zero and standard deviation $\frac{1}{\sqrt{2}} \approx 0.707$. The inflection points of $g(x)$ are also at $\pm\frac{1}{\sqrt{2}}$ and so rescaling by 2 in the $x$ direction produces a pdf with standard deviation 1 and inflection points at $\pm 1$.

**Definition 4.5.** The *standard normal random variable* $Z$ has probability density function given by

$$f(x) = \frac{1}{\sqrt{2\pi}} e^{-x^2/2}.$$

The graph of $f(x)$ is shown in Figure 4.5.

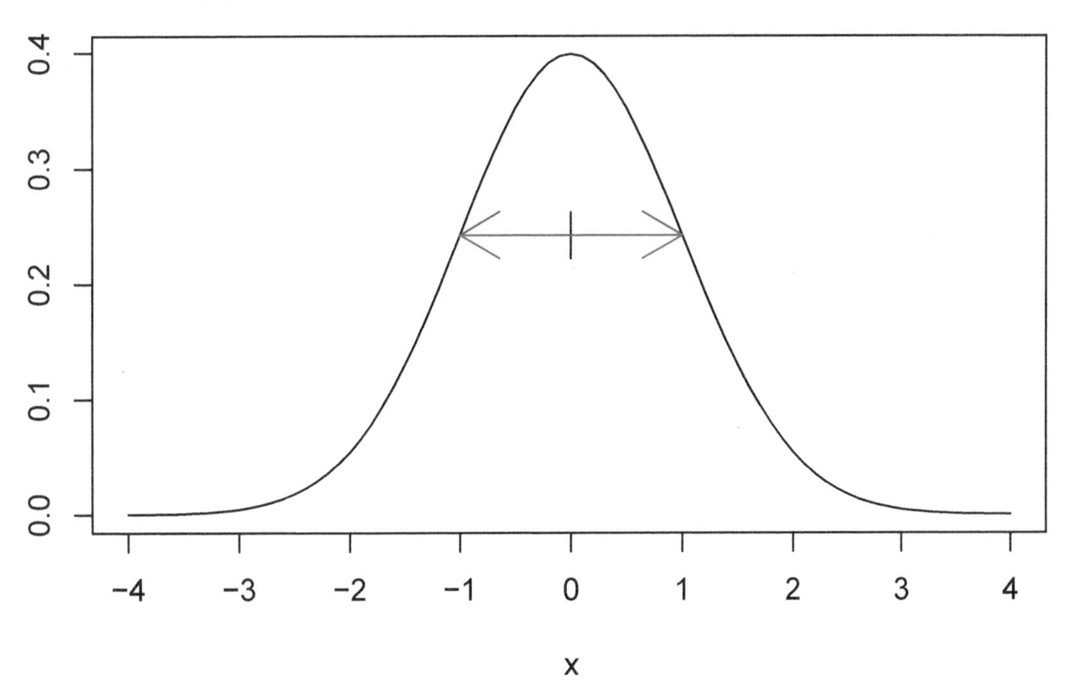


**FIGURE 4.5** The standard normal distribution (with one s.d. shown).

The R function `pnorm` computes the cdf of the normal distribution, as `pnorm(x)` $= P(Z \leq x)$. Using `pnorm`, we can compute the probability that $Z$ lies within 1, 2, and 3 standard deviations of its mean:

- $P(-1 \leq Z \leq 1) = P(Z \leq 1) - P(Z \leq -1) =$ `pnorm(1) - pnorm(-1)` $= 0.6826895$.
- $P(-2 \leq Z \leq 2) = P(Z \leq 2) - P(Z \leq -2) =$ `pnorm(2) - pnorm(-2)` $= 0.9544997$.
- $P(-3 \leq Z \leq 3) = P(Z \leq 3) - P(Z \leq -3) =$ `pnorm(3) - pnorm(-3)` $= 0.9973002$.

By shifting and rescaling $Z$, we define the normal random variable with mean $\mu$ and standard deviation $\sigma$:

**Definition 4.6.** The *normal random variable* $X$ with mean $\mu$ and standard deviation $\sigma$ is given by

$$X = \sigma Z + \mu.$$

We write $X \sim \text{Norm}(\mu, \sigma)$.

Many books write $X \sim N(\mu, \sigma^2)$, so that the second parameter in the parenthesis is the variance. We use the standard deviation and the abbreviation Norm to be consistent with R.

The pdf of a normal random variable is given in the following theorem.

**Theorem 4.4.** *Let $X$ be a normal random variable with parameters $\sigma$ and $\mu$. The probability mass function of $X$ is given by*

$$f(x) = \frac{1}{\sigma\sqrt{2\pi}} e^{-\frac{1}{2}\left(\frac{x-\mu}{\sigma}\right)^2} \qquad -\infty < x < \infty$$

The parameter names are the mean $\mu$ and the standard deviation $\sigma$.

For any normal random variable, approximately:

- 68% of the normal distribution lies within one standard deviation of the mean.
- 95% lies within two standard deviations of the mean.
- 99.7% lies within three standard deviations of the mean.

This fact is sometimes called the *empirical rule.*

Figure 4.6 shows examples of normal distributions with fixed mean $\mu = 0$ and various values of the standard deviation $\sigma$. Figure 4.7 shows normal distributions with fixed standard deviation $\sigma = 1$ and various means $\mu$.

### 4.4.1 Computations with normal random variables

R has built-in functions for working with normal distributions and normal random variables. The root name for these functions is `norm`, and as with other distributions the prefixes `d`, `p`, and `r` specify the pdf, cdf, or random sampling. We also introduce the `q` prefix here, which indicates the inverse of the cdf function. The `q` prefix is available for all random variables supported by R, but we will rarely use it for anything except normal rvs.

If $X \sim \text{Norm}(\mu, \sigma)$:

- `dnorm(x, mu, sigma)` gives the height of the pdf at $x$.
- `pnorm(x, mu, sigma)` gives $P(X \leq x)$, the cdf.
- `qnorm(p, mu, sigma)` gives the value of $x$ so that $P(X \leq x) = p$, the inverse cdf.
- `rnorm(N, mu, sigma)` simulates $N$ random values of $X$.

Here are some simple examples:

**Example 4.14.** Let $X \sim \text{Norm}(\mu = 3, \sigma = 2)$. Find $P(X \leq 4)$ and $P(0 \leq X \leq 5)$.

```
pnorm(4, mean = 3, sd = 2)
```

```
## [1] 0.6914625
```

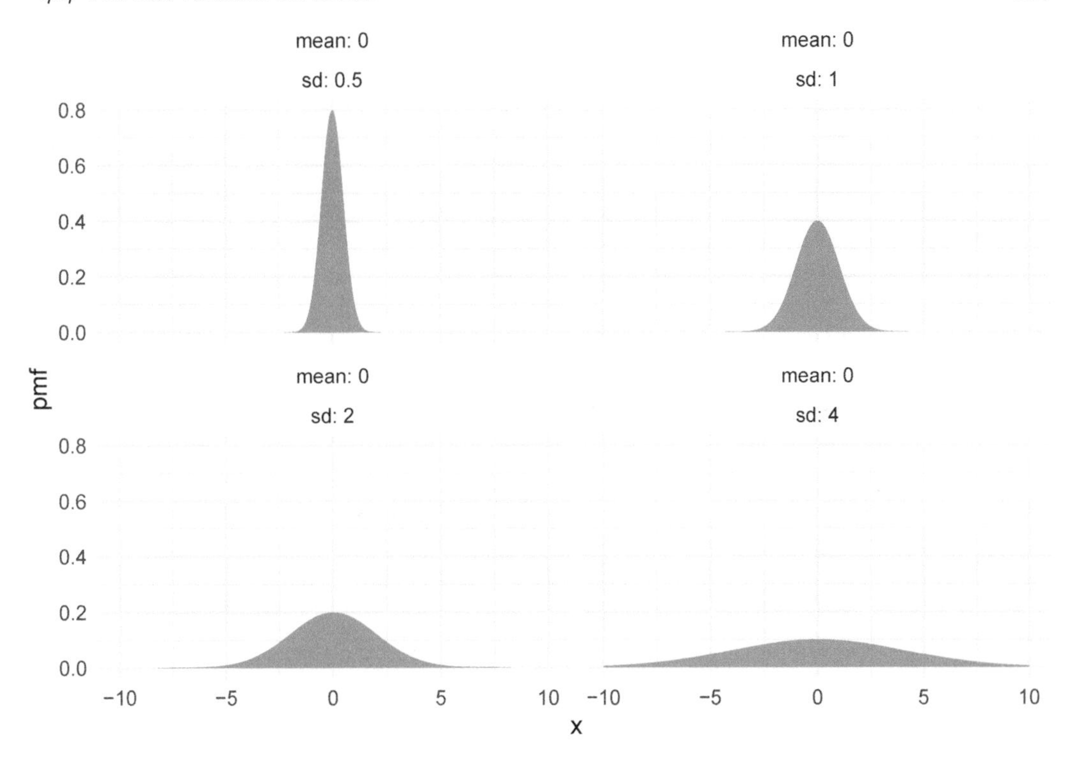


**FIGURE 4.6** Normal distributions with $\mu = 0$ and various values of $\sigma$.

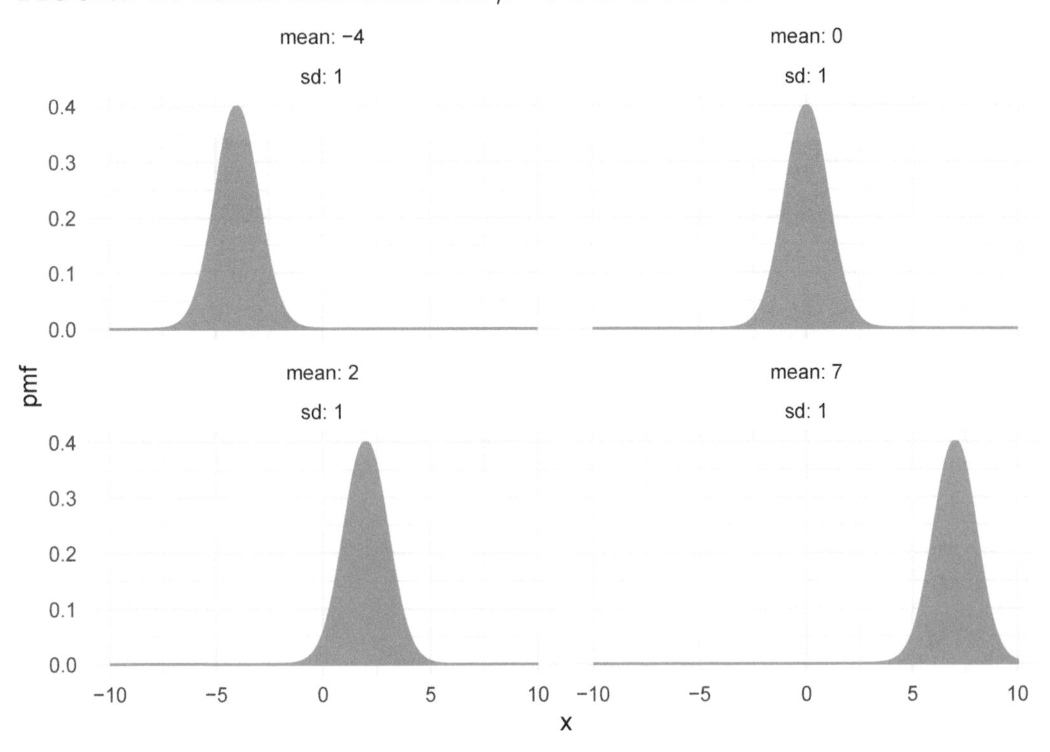


**FIGURE 4.7** Normal distributions with $\sigma = 1$ and various values of $\mu$.

```
pnorm(5, 3, 2) - pnorm(0, 3, 2)
```

```
## [1] 0.7745375
```

**Example 4.15.** Let $X \sim \text{Norm}(100, 30)$. Find the value of $q$ such that $P(X \le q) = 0.75$. One approach is to try various choices of $q$ until discovering that `pnorm(120,100,30)` is close to 0.75. However, the purpose of the `qnorm` function is to answer this exact question:

```
qnorm(0.75, 100, 30)
```

```
## [1] 120.2347
```

**Example 4.16.** The length of dog pregnancies from conception to birth varies according to a distribution that is approximately normal with mean 63 days and standard deviation 2 days.

a. What percentage of dog pregnancies last 60 days or fewer?
b. What percentage of dog pregnancies last 67 days or more?
c. What range covers the shortest 90% of dog pregnancies?
d. What is the narrowest range of times that covers 90% of dog pregnancies?

We let $X$ be the random variable which is the length of a dog pregnancy. We model $X \sim \text{Norm}(63, 2)$. Then parts (a) and (b) ask for $P(X \le 60)$ and $P(X \ge 67)$ and we can compute these with `pnorm` as follows:

```
pnorm(60, 63, 2)
```

```
## [1] 0.0668072
```

```
1 - pnorm(67, 63, 2)
```

```
## [1] 0.02275013
```

For part (c), we want $x$ so that $P(X \le x) = 0.90$. This is `qnorm(0.90,63,2)`=65.6, so 90% of dog pregnancies are shorter than 65.6 days.

For part (d), we need to use the fact that the pdfs of normal random variables are symmetric about their mean, and decreasing away from the mean. So, if we want the **shortest** interval that contains 90% of dog pregnancies, it should be centered at the mean with 5% of pregnancies to the left of the interval, and 5% of the pregnancies to the right of the interval. We get an interval of the form `[qnorm(0.05, 63, 2), qnorm(0.95, 63, 2)]`, or approximately [59.7, 66.3].

**Example 4.17.** Let $Z$ be a standard normal random variable. Find the mean and standard deviation of the variable $e^Z$.

We solve this with simulation:

```
Z <- rnorm(100000, 0, 1)
mean(exp(Z))
```

```
## [1] 1.644607
```

```
sd(exp(Z))
```

```
## [1] 2.1429
```

The mean of $e^Z$ is approximately 1.6, and the standard deviation is approximately 2.1. Note that even with 100,000 simulated values, these answers are not particularly accurate because on rare occasions $e^Z$ takes on very large values.

**Example 4.18.** Suppose you are picking seven women at random from a university to form a starting line-up in an ultimate frisbee game. Assume that women's heights at this university are normally distributed with mean 64.5 inches (5 foot, 4.5 inches) and standard deviation 2.25 inches. What is the probability that 3 or more of the women are 68 inches (5 foot, 8 inches) or taller?

To do this, we first determine the probability that a single randomly selected woman is 68 inches or taller. Let $X$ be a normal random variable with mean 64 and standard deviation 2.25. We compute $P(X \geq 68)$ using `pnorm`:

```
pnorm(68, 65, 2.25, lower.tail = FALSE)
```

```
## [1] 0.09121122
```

Now, we need to compute the probability that 3 or more of the 7 women are 68 inches or taller. Since the population of all women at a university is much larger than 7, the number of women in the starting line-up who are 68 inches or taller is binomial with $n = 7$ and $p = 0.09121122$, which we computed in the previous step. We compute the probability that at least 3 are 68 inches as

```
sum(dbinom(3:7, 7, 0.09121122))
```

```
## [1] 0.02004754
```

So, there is about a 2% chance that at least three will be 68 inches or taller.

How likely is it that the team has no players who are 68 inches or taller?

```
dbinom(0, 7, 0.09121122)
```

```
## [1] 0.5119655
```

According to our model, this happens about 50% of the time. None of the 2019 national champion UC San Diego Psychos were 68 inches or taller, although it is unlikely that ultimate frisbee players are drawn randomly from the population of women.

**Example 4.19.** Throwing a dart at a dartboard with the bullseye at the origin, model the location of the dart with independent coordinates $X \sim \text{Norm}(0, 3)$ and $Y \sim \text{Norm}(0, 3)$ (both in inches). What is the expected distance from the bullseye?

The distance from the bullseye is given by the Euclidean distance formula $d = \sqrt{X^2 + Y^2}$. We simulate the $X$ and $Y$ random variables and then compute the mean of $d$:

```
X <- rnorm(10000, 0, 3)
Y <- rnorm(10000, 0, 3)
d <- sqrt(X^2 + Y^2)
mean(d)
```

```
## [1] 3.76416
```

We expect the dart to land about 3.8 inches from the bullseye, on average.

### 4.4.2 Normal approximation to the binomial

The value of a binomial random variable is the sum of independent factors: the Bernoulli trials. A special case of the Central Limit Theorem is that a binomial random variable can be well approximated by a normal random variable when the number of trials is large.

First, we need to understand the standard deviation of a binomial random variable.

**Theorem 4.5.** *Let $X \sim Binom(n, p)$. The variance and standard deviation of $X$ are given by:*

$$Var(X) = np(1-p) \tag{4.1}$$

$$\sigma(X) = \sqrt{np(1-p)} \tag{4.2}$$

The proof of this theorem was given in Example 3.27. There is also an instructive proof that is similar to the proof of Theorem 3.3, except that we take the derivative of the binomial theorem two times and compute $E[X(X-1)]$. The result follows from $E[X^2] = E[X(X-1)] + E[X]$ and Theorem 3.9.

Now the binomial rv $X$ can be approximated by a random normal variable with the same mean and standard deviation as $X$:

**Theorem 4.6.** *Fix $p$. For large $n$, the binomial random variable $X \sim Binom(n, p)$ is approximately normal with mean $\mu = np$ and standard deviation $\sigma = \sqrt{np(1-p)}$.*

The size of $n$ required to make the normal approximation accurate depends on the accuracy required and also depends on $p$. Binomial distributions with $p$ close to 0 or 1 are not as well approximated by the normal distribution as those with $p$ near 1/2.

This normal approximation was traditionally used to work with binomial random variables, since calculating the binomial distribution exactly requires quite a bit of computation. Probabilities for the normal distribution were readily available in tables, and so were easier to use. With R, the `pbinom` function makes it easy to work with binomial pmfs directly.

**Example 4.20.** Let $X \sim \text{Binom}(300, 0.46)$. Compute $P(X > 150)$.

Computing exactly, $P(X > 150) =$ `1 - pbinom(150,300,0.46)` $= 0.0740$.

To use the normal approximation, we calculate that $X$ has mean $300 \cdot 0.46 = 138$ and standard deviation $\sqrt{300 \cdot 0.46 \cdot 0.54} \approx 8.63$. Then $P(X > 150) \approx$ `1 - pnorm(150,138,8.63)` $=$ 0.0822.

As an improvement, notice that the continuous normal variable can take values in between 150 and 151, but the discrete binomial variable cannot. To account for this, we use a *continuity correction* and assign each integer value of the binomial variable to the one-unit wide interval centered at that integer. Then 150 corresponds to the interval (145.5,150.5) and 151 corresponds to the interval (150.5,151.5). To approximate $X > 150$, we want our normal random variable to be larger than 150.5. The normal approximation with continuity correction gives $P(X > 150) \approx$ `1 - pnorm(150.5,138,8.63)` $= 0.0737$, much closer to the actual value of 0.0740.

## 4.5 Uniform and exponential random variables

### 4.5.1 Uniform random variables

Uniform random variables may be discrete or continuous. A *discrete uniform variable* may take any one of finitely many values, all equally likely. The classic example is the die roll, which is uniform on the numbers 1,2,3,4,5,6. Another example is a coin flip, where we assign 1 to heads and 0 to tails. Unlike most other named random variables, R has no special functions for working with discrete uniform variables. Instead, we use `sample` to simulate these.

**Definition 4.7.** A continuous uniform random variable $X$ on the interval $[a, b]$ has pdf given by

$$f(x) = \begin{cases} \frac{1}{b-a} & a \le x \le b \\ 0 & \text{otherwise} \end{cases}$$

The graph of $f(x)$ is shown in Figure 4.8.

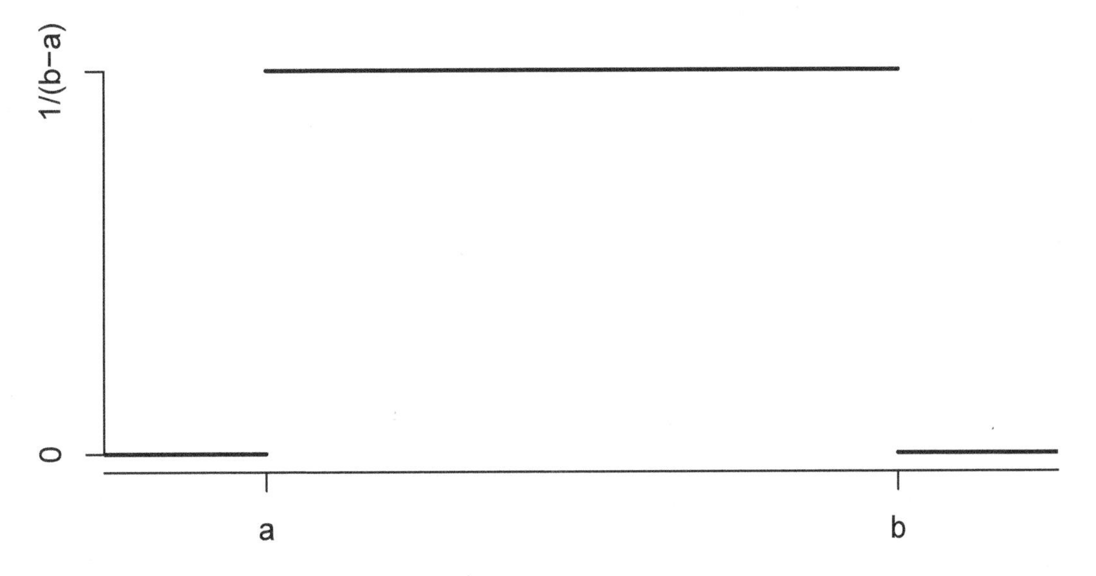


**FIGURE 4.8** PDF of a continuous uniform random variable on $[a, b]$.

A continuous uniform rv $X$ is characterized by the property that for any interval $I \subset [a, b]$, the probability $P(X \in I)$ depends only on the length of $I$. We write $X \sim \text{Unif}(a, b)$ if $X$ is continuous uniform on the interval $[a, b]$.

One example of a random variable that could be modeled by a continuous uniform random variable is a round-off error in measurements. Say we measure height and record only feet and inches. It is a reasonable first approximation that the error associated with rounding to the nearest inch is uniform on the interval $[-1/2, 1/2]$. There may be other sources of measurement error which might not be well modeled by a uniform random variable, but the round-off is uniform.

Another example is related to the Poisson process. If you observe a Poisson process after some length of time $T$ and see that exactly one event has occurred, then the time that the event occurred in the interval $[0, T]$ is uniformly distributed.

**Example 4.21.** A random real number is chosen uniformly in the interval 0 to 10. What is the probability that it is bigger than 7, given that it is bigger than 6?

Let $X \sim \text{Unif}(0, 10)$ be the random real number. Then

$$P(X > 7 \mid X > 6) = \frac{P(X > 7 \cap X > 6)}{P(X > 6)} = \frac{P(X > 7)}{P(X > 6)} = \frac{3/10}{4/10} = \frac{3}{4}.$$

Alternately, we can compute with the `punif` function, which gives the cdf of a uniform random variable.

```
(1 - punif(7, 0, 10)) / (1 - punif(6, 0, 10))
```

```
## [1] 0.75
```

The conditional density of a uniform over the interval $[a, b]$ given that it is in the subset $[c, d]$ is uniformly distributed on the interval $[c, d]$. Applying that fact to Example 4.21, we know that $X$ given $X > 6$ is uniform on the interval $[6, 10]$. Therefore, the probability that $X$ is larger than 7 is simply 3/4. Note that this only works for *uniform* random variables! For other random variables, you need to compute conditional probabilities as in Example 4.21.

We finish this section with a computation of the mean and variance of a uniform random variable $X$. Not surprisingly, the mean is exactly halfway along the interval of possible values for $X$.

**Theorem 4.7.** *For $X \sim$ Unif$(a, b)$, the expected value of $X$ is*

$$E[X] = \frac{b + a}{2}$$

*and the variance is*

$$Var(X) = \frac{(b - a)^2}{12}$$

*Proof.* We compute the mean of $X$ as follows:

$$\begin{aligned} E[X] &= \int_a^b \frac{x}{b - a}\, dx = \left. \frac{x^2}{2(b - a)} \right|_{x=a}^{x=b} \\ &= \frac{b^2}{2(b - a)} - \frac{a^2}{2(b - a)} = \frac{(b - a)(b + a)}{2(b - a)} \\ &= \frac{a + b}{2}. \end{aligned}$$

For the variance, first calculate $E[X^2] = \int_a^b \frac{x^2}{b-a} dx$. Then

$$\text{Var}(X) = E[X^2] - E[X]^2 = E[X^2] - \left(\frac{a + b}{2}\right)^2.$$

Working the integral and simplifying Var$(X)$ is left as Exercise 4.20. ■

### 4.5.2 Exponential random variables

**Definition 4.8.** An *exponential random variable* $X$ with rate $\lambda$ has pdf

$$f(x) = \lambda e^{-\lambda x}, \qquad x > 0$$

We write $X \sim \text{Exp}(\lambda)$. Figure 4.9 shows the graph of $f(x)$ for various values of the rate $\lambda$.

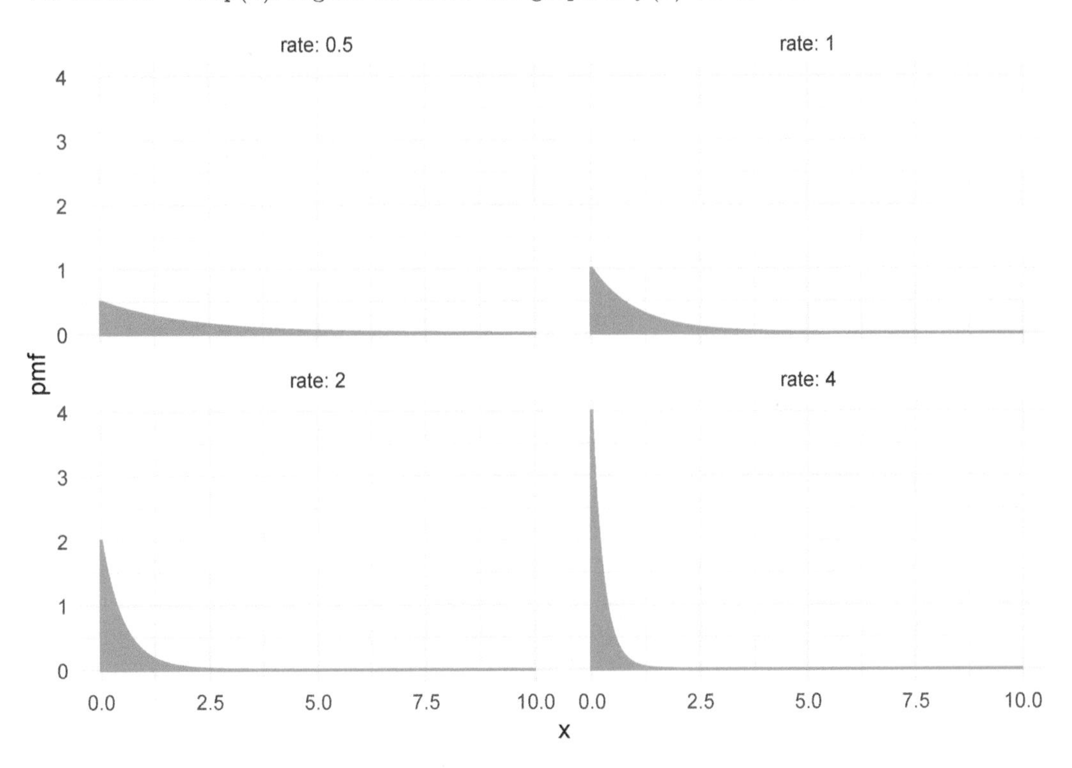


**FIGURE 4.9** Exponential distributions with various values of $\lambda$.

Exponential random variables measure the waiting time until the first event occurs in a Poisson process, see Section 3.6.1. The waiting time until an electronic component fails could be exponential. In a store, the time between customers could be modeled by an exponential random variable by starting the Poisson process at the moment the first customer enters. Observe in Figure 4.9 that the higher the rate, the smaller $X$ will be. This is because we generally wait less time for events that occur more frequently.

The exponential distribution is a *skew distribution*, which means it is not symmetric. Because it has a long tail on the right, we say it has *right skew.*

**Theorem 4.8.** *Let $X \sim Exp(\lambda)$ be an exponential random variable with rate $\lambda$. Then the mean and variance of $X$ are:*

$$E[X] = \frac{1}{\lambda} \tag{4.3}$$

$$Var(X) = \frac{1}{\lambda^2} \tag{4.4}$$

*Proof.* We compute the mean of an exponential random variable with rate $\lambda$ using integration by parts as follows:

$$\begin{aligned} E[X] &= \int_{-\infty}^{\infty} x f(x)\, dx \\ &= \int_{0}^{\infty} x\lambda e^{-\lambda x} + \int_{-\infty}^{0} x \cdot 0\, dx \\ &= -xe^{-\lambda x} - \frac{1}{\lambda} e^{-\lambda x} \Big|_{x=0}^{x=\infty} \\ &= \frac{1}{\lambda} \end{aligned}$$

For the variance, we compute the rather challenging integral:

$$E[X^2] = \int_{0}^{\infty} x^2 \lambda e^{-\lambda x} dx = \frac{2}{\lambda^2}$$

Then

$$\text{Var}(X) = E[X^2] - E[X]^2 = \frac{2}{\lambda^2} - \left(\frac{1}{\lambda}\right)^2 = \frac{1}{\lambda^2}.$$

■

**Example 4.22.** When watching the Taurids meteor shower, meteors arrive at a rate of five per hour. How long do you expect to wait for the first meteor? How long should you wait to have a 95% change of seeing a meteor?

Here, $X$ is the length of time before the first meteor. We model the meteors as a Poisson process with rate $\lambda = 5$ (and time in hours). Then $E[X] = \frac{1}{5} = 0.2$ hours, or 12 minutes.

For a 95% chance, we are interested in finding $x$ so that $P(X < x) = 0.95$. One way to approach this is by playing with values of $x$ in the R function `pexp(x,5)`. Some effort will yield `pexp(0.6,5) = 0.95`, so that we should plan on waiting 0.6 hours, or 36 minutes to be 95% sure of seeing a meteor.

A more straightforward approach is to use the inverse cdf function `qexp`, which gives

```
qexp(0.95, 5)
```

```
## [1] 0.5991465
```

and the exact waiting time of 0.599 hours.

The *memoryless property* of exponential random variables is the equation

$$P(X > s + t \,|\, X > s) = P(X > t)$$

for any $s, t > 0$. It helps to interpret this equation in the context of a Poisson process, where $X$ measures waiting time for some event. The left-hand side of the equation is the probability that we wait $t$ units longer, given that we have already waited $s$ units. The right-hand side is the probability that we wait $t$ units, from the beginning. Because these two probabilities are the same, it means that waiting $s$ units has gotten us no closer to the occurrence of the event. The Poisson process has no memory that you have "already waited" $s$ units.

We prove the memoryless property here by computing the probabilities involved. The cdf of an exponential random variable with rate $\lambda$ is given by

$$F(x) = \int_0^\infty e^{-\lambda x} dx = 1 - e^{-\lambda x}$$

for $x > 0$. Then $P(X > x) = 1 - F(x) = e^{-\lambda x}$, and

$$\begin{aligned} P(X > s + t \mid X > s) &= \frac{P(X > s + t \cap X > s)}{P(X > s)} \\ &= \frac{P(X > s + t)}{P(X > s)} \\ &= e^{-\lambda(s+t)}/e^{-\lambda s} \\ &= e^{-\lambda t} \\ &= P(X > t) \end{aligned}$$

## 4.6 Summary

This chapter and Chapter 3 introduced the notion of a random variable, and the associated notion of a probability distribution. For any random variable, we might be interested in answering probability questions either exactly or through simulation. Usually, these questions involve knowledge of the probability distribution. For some commonly occurring types of random variable, the probability distribution functions are well understood.

The following table provides a quick reference for random variables introduced so far, together with pmf/pdf, expected value, variance and root R function.

TABLE 4.1: Properties of common probability distributions.

| RV | PMF/PDF | Range | Mean | Variance | R Root |
|---|---|---|---|---|---|
| Binomial | $\binom{n}{x} p^x (1-p)^{n-x}$ | $0 \leq x \leq n$ | $np$ | $np(1-p)$ | `binom` |
| Geometric | $p(1-p)^x$ | $x \geq 0$ | $\frac{1-p}{p}$ | $\frac{1-p}{p^2}$ | `geom` |
| Poisson | $\frac{1}{x!} \lambda^x e^{-\lambda}$ | $x \geq 0$ | $\lambda$ | $\lambda$ | `pois` |
| Hypergeometric | $\frac{\binom{m}{x}\binom{n}{k-x}}{\binom{m+n}{k}}$ | $x = 0, \ldots, k$ | $kp$ | $kp(1-p)\frac{m+n-k}{m+n-1}$ | `hyper` |
| Negative Binomial | $\binom{x+n-1}{x} p^n (1-p)^x$ | $x \geq 0$ | $n\frac{1-p}{p}$ | $n\frac{1-p}{p^2}$ | `nbinom` |
| Uniform | $\frac{1}{b-a}$ | $a \leq x \leq b$ | $\frac{a+b}{2}$ | $\frac{(b-a)^2}{12}$ | `unif` |
| Exponential | $\lambda e^{-\lambda x}$ | $x \geq 0$ | $1/\lambda$ | $1/\lambda^2$ | `exp` |
| Normal | $\frac{1}{\sigma\sqrt{2\pi}} e^{-(x-\mu)^2/(2\sigma^2)}$ | $-\infty < x < \infty$ | $\mu$ | $\sigma^2$ | `norm` |

When modeling a count of something, you often need to choose between binomial, geometric, and Poisson. The binomial and geometric random variables both come from Bernoulli trials, where there is a sequence of individual trials each resulting in success or failure. In the Poisson process, events are spread over a time interval, and appear at random.

The normal random variable is a good starting point for continuous measurements that have a central value and become less common away from that mean. Exponential variables show up when waiting for events to occur. Continuous uniform variables sometimes occur as the location of an event in time or space, when the event is known to have happened on some fixed interval.

R provides these random variables (and many more!) through a set of four functions for each known distribution. The four functions are determined by a *prefix*, which can be `p`, `d`, `r`, or `q`. The *root* determines which distribution we are talking about. Each distribution function takes a single argument first, determined by the prefix, and then some number of parameters, determined by the root. The general form of a distribution function in R is:

```
[prefix][root] ( argument, parameter1, parameter2, ..)
```

The available prefixes are:

- `p` compute the cumulative distribution function $P(X < x)$, and the argument is $x$.
- `d` compute the pdf or pmf $f$. The value is $f(x)$, and the argument is $x$. In the discrete case, this is the probability $P(X = x)$.
- `r` sample from the rv. The argument is $N$, the number of samples to take.
- `q` quantile function, the inverse cdf. This computes $x$ so that $P(X < x) = q$, and the argument is $q$.

The distributions we have introduced so far, with their parameters, are:

`binom`
: binomial, parameters are $n$, number of trials and $p$, probability of success.

`geom`
: geometric, parameter is $p$, probability of success.

`pois`
: Poisson, parameter is $\lambda$, the rate at which events occur, or the mean number of events over the time interval.

`nbinom`
: negative binomial, parameters are `size` which is the number of successes, and `prob`.

`hyper`
: hypergeometric with parameters $m$, number of white balls, $n$, number of black balls, $k$, number of balls drawn without replacement.

`unif`
: uniform, parameters are $a, b$.

`norm`
: normal, parameters are $\mu$, the mean, and $\sigma$, the standard deviation.

`exp`
: exponential, parameter is $\lambda$, the rate.

There will be many more distributions to come, and the four prefixes work the same way for all of them.

## Exercises

Exercises 4.1 – 4.5 require material through Section 4.1.

**4.1.** Let $X$ be a random variable with pdf given by $f(x) = 2x$ for $0 \le x \le 1$ and $f(x) = 0$ otherwise.

a. Find $P(X \ge 1/2)$.
b. Find $P(X \ge 1/2 | X \ge 1/4)$.

**4.2.** Let $X$ be a random variable with pdf

$$f(x) = \begin{cases} Cx^2 & 0 \le x \le 1 \\ C(2-x)^2 & 1 \le x \le 2 \end{cases}$$

Find $C$.

**4.3.** For each of the following functions, decide whether the function is a valid pdf, a valid cdf or neither.

a. $h(x) = \begin{cases} 1 & 0 \le x \le 2 \\ -1 & 2 \le x \le 3 \\ 0 & \text{otherwise} \end{cases}$
b. $h(x) = \sin(x) + 1$
c. $h(x) = \begin{cases} 1 - e^{-x^2} & x \ge 0 \\ 0 & x < 0 \end{cases}$
d. $h(x) = \begin{cases} 2xe^{-x^2} & x \ge 0 \\ 0 & x < 0 \end{cases}$

**4.4.** Provide an example of a pdf $f$ for a random variable $X$ such that there exists an $x$ for which $f(x) > 1$. Is it possible to have $f(x) > 1$ for all values of $x$?

**4.5.** Is there a function which is both a valid pdf **and** a valid cdf? If so, give an example. If not, explain why not.

---

Exercises 4.6 – 4.8 require material through Section 4.3.

**4.6.** Let $X$ be a random variable with pdf $f(x) = 3(1-x)^2$ when $0 \le x \le 1$, and $f(x) = 0$ otherwise.

a. Verify that $f$ is a valid pdf.
b. Find the mean and variance of $X$.
c. Find $P(X \le 1/2)$.
d. Find $P(X \le 1/2 \mid X \ge 1/4)$.

**4.7.** If $\text{Var}(X) = 3$, what is $\text{Var}(2X + 1)$?

**4.8.** Let $X$ be a random variable whose pdf is given by the plot in Figure 4.10. Assume that the pdf is zero outside of the interval given in the plot.

a. Estimate the mean of $X$.
b. Estimate the standard deviation of $X$.
c. For which $a$ is $P(a \le X \le a + 2)$ maximized?
d. Estimate $P(0 \le X \le 2)$.

---

Exercises 4.9 – 4.14 require material through Section 4.4.

**4.9.** Compare the pdfs of three normal random variables, one with mean 1 and standard

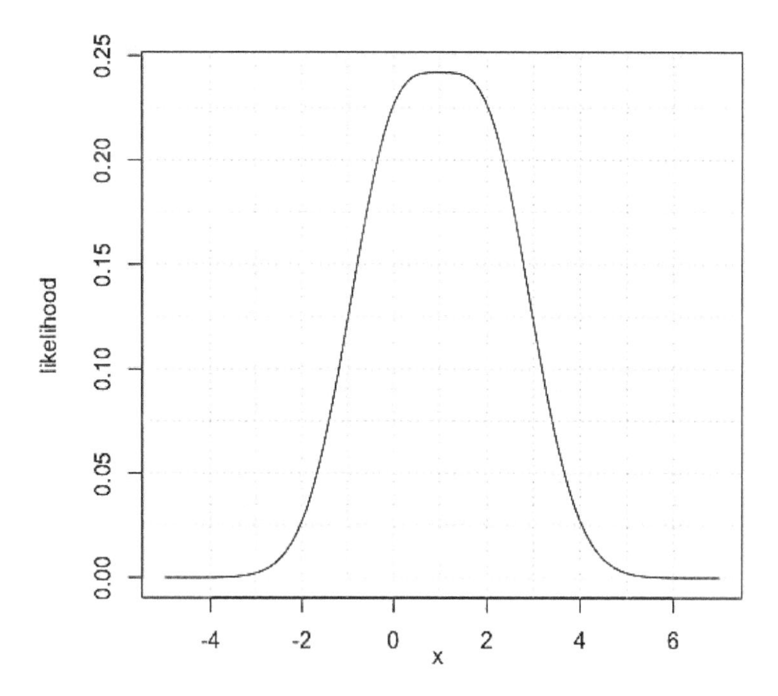


**FIGURE 4.10** PDF of X

deviation 1, one with mean 1 and standard deviation 10, and one with mean -4 and standard deviation 1.

**4.10.** Let $X$ be a normal rv with mean 1 and standard deviation 2.

a. Find $P(a \leq X \leq a+2)$ when $a = 3$.
b. Sketch the graph of the pdf of $X$, and indicate the region that corresponds to your answer in the previous part.
c. Find the value of $a$ such that $P(a \leq X \leq a+2)$ is the largest.

**4.11.** Suppose that scores on an exam are normally distributed with mean 80 and standard deviation 5, and that scores are not rounded.

a. What is the probability that a student scores higher than 85 on the exam?
b. Assume that exam scores are independent and that 10 students take the exam. What is the probability that 4 or more students score 85 or higher on the exam?

**4.12.** Climbing rope will break if pulled hard enough. Experiments show that 10.5 mm dynamic nylon rope has a mean breaking point of 5036 lbs with a standard deviation of 122 lbs. Assume breaking points of rope are normally distributed.

a. Sketch the distribution of breaking points for this rope.
b. What proportion of ropes will break with 5000 lbs of load?
c. At what load will 95% of all ropes break?

**4.13.** In this exercise, we verify Theorem 3.10 in a special case. Let $X$ be normal with mean 0 and standard deviation 2, and let $Y$ be normal with mean 0 and standard deviation 1. Assume $X$ and $Y$ are independent. Use simulation to estimate the variance of $X + 3Y$, and compare to $1^2 \times 4 + 3^2 \times 1$.

**4.14 (Gaussian integral).** There is no elementary antiderivative for the function $e^{x^2}$.

However, the Gaussian integral $\int_{-\infty}^{\infty} e^{-x^2}\, dx$ can be computed exactly. Begin with the following:

$$\left(\int_{-\infty}^{\infty} e^{-x^2}\, dx\right)^2 = \int_{-\infty}^{\infty} e^{-x^2}\, dx \int_{-\infty}^{\infty} e^{-y^2}\, dy = \int_{-\infty}^{\infty}\int_{-\infty}^{\infty} e^{-(x^2+y^2)}\, dx\, dy$$

Now switch to polar coordinates and show that the Gaussian integral is equal to $\sqrt{\pi}$

---

Exercises 4.15 – 4.28 require material through Section 4.5.

**4.15.** Plot the pdf and cdf of a uniform random variable on the interval $[0, 1]$.

**4.16.** Compare the cdf and pdf of an exponential random variable with rate $\lambda = 2$ with the cdf and pdf of an exponential rv with rate 1/2. (If you wish to read ahead in Chapter 7, you can learn how to put plots on the same axes, with different colors.)

**4.17.** Let $X$ be an exponential rv with rate $\lambda = 1/4$.

a. What is the mean of $X$?
b. Find the value of $a$ such that $P(a \le X \le a + 1)$ is maximized. Is the mean contained in the interval $[a, a + 1]$?

**4.18.** Suppose the time (in seconds) it takes your professor to set up their computer to start class is uniformly distributed on the interval $[0, 30]$. Suppose also that it takes you 5 seconds to send your mom a nice, quick text that you are thinking of her. You only text her if you can complete it during the time your professor is setting up their computer. If you try to text your mom every day in class, what is the probability that she will get a text on 3 consecutive days?

**4.19.** Suppose the time to failure (in years) for a particular component is distributed as an exponential random variable with rate $\lambda = 1/5$. For better performance, the system has two components installed, and the system will work as long as either component is functional. Assume the time to failure for the two components is independent. What is the probability that the system will fail before 10 years have passed?

**4.20.** Verify that a uniform random variable on the interval $[a, b]$ has variance given by $\sigma^2 = \frac{(b-a)^2}{12}$.

**4.21 (Memoryless property).** Let $X$ be an exponential random variable with rate $\lambda$. If $a$ and $b$ are positive numbers, then

$$P(X > a + b \mid X > b) = P(X > a)$$

a. Explain why this is called the *memoryless property*.
b. Show that for an exponential rv $X$ with rate $\lambda$, $P(X > a) = e^{-a\lambda}$.
c. Use the result in (b) to prove the memoryless property for exponential random variables.

**4.22.** For each of the following descriptions of a random variable, indicate whether it can best be modeled by binomial, geometric, Poisson, uniform, exponential, or normal. Answer the associated questions. Note that not all of the experiments yield random variables that are *exactly* of the type listed above, but we are asking about reasonable modeling.

a. Let $Y$ be the random variable that counts the number of sixes which occur when a die is tossed 10 times. What type of random variable is $Y$? What is $P(Y = 3)$? What is the expected number of sixes? What is $\text{Var}(Y)$?

b. Let $U$ be the random variable which counts the number of accidents which occur at an intersection in one week. What type of random variable is $U$? Suppose that, on average, 2 accidents occur per week. Find $P(U = 2)$, $E(U)$ and $\text{Var}(U)$.
c. Suppose a stop light has a red light that lasts for 60 seconds, a green light that lasts for 30 seconds, and a yellow light that lasts for 5 seconds. When you first observe the stop light, it is red. Let $X$ denote the time until the light turns green. What type of rv would be used to model $X$? What is its mean?
d. Customers arrive at a teller's window at a uniform rate of 5 per hour. Let $X$ be the length in minutes of time that the teller has to wait until they see their first customer after starting their shift. What type of rv is $X$? What is its mean? Find the probability that the teller waits less than 10 minutes for their first customer.
e. A coin is tossed until a head is observed. Let $X$ denote the total number of tails observed during the experiment. What type of rv is $X$? What is its mean? Find $P(X \leq 3)$.
f. Let $X$ be the recorded body temperature of a healthy adult in degrees Fahrenheit. What type of rv is $X$? Estimate its mean and standard deviation, based on your knowledge of body temperatures.

**4.23.** Suppose you turn on a soccer game and see that the score is 1-0 after 30 minutes of play. Let $X$ denote the time (in minutes from the start of the game) that the goal was scored. What type of rv is $X$? What is its mean?

**4.24.** Let $X_1, X_2, X_3$ be independent uniform random variables on the interval $[0, 1]$. Find the cdf of the random variable $Z$ which is the maximum of $X_1, X_2$ and $X_3$. (Hint: the event $Z \leq z$ is the same as the event $(X_1 \leq z) \cap (X_2 \leq z) \cap (X_3 \leq z)$, and $F(z) = P(Z \leq z)$.)

**4.25.** There exist naturally occurring random variables that are neither discrete nor continuous. Suppose a group of people is waiting for one more person to arrive before starting a meeting. Suppose that the arrival time of the person is exponential with mean 4 minutes, and that the meeting will start either when the person arrives, or after 5 minutes, whichever comes first. Let $X$ denote the length of time the group waits before starting the meeting.

a. Find $P(0 \leq X \leq 4)$.
b. Find $P(X = 5)$.

**4.26.** Suppose you pick 4 numbers $x_1, \ldots, x_4$ uniformly in the interval $[0, 1]$ and you create four intervals of length 1/2 centered at the $x_i$; namely, $[x_i - 1/4, x_i + 1/4]$. Note that these intervals need not be contained in $[0, 1]$.

Estimate via simulation the probability that the union of the 4 intervals is an **interval** of length at least 1.[1]

**4.27.** Suppose that you have two infinite, horizontal parallel lines that are one unit apart. You drop a needle of length 1/2 so that its center between the two lines is uniform on $[0, 1]$, and the angle that the needle forms relative to the parallel lines is uniform on $[0, \pi]$.

Estimate the probability that the needle touches one of the parallel lines, and confirm that your answer is approximately $1/\pi$.

**4.28.** This problem was reported to be a Google interview question. Suppose you have a stick of length one meter. You randomly select two points on the stick, and break the stick at those two places. Estimate the probability that the resulting three pieces of stick can be used to form a triangle.

[1] This problem is a special case of the arc covering problem, which asks the probability that $n$ randomly placed arcs of length $a$ cover a circle of circumference 1, which was solved by Stevens in 1939. For general $a$, we would need to reformulate the problem to be equivalent to the arc question.

# 5

# *Simulation of Random Variables*

In this chapter we discuss simulation related to random variables. After a review of probability simulation, we turn to the estimation of pdfs and pmfs of random variables. These simulations provide a foundation for understanding the fundamental concepts of statistical inference: sampling distributions, point estimators, and the Central Limit Theorem.

## 5.1 Estimating probabilities

In order to run simulations with random variables, we use R's built-in random generation functions. These functions all take the form `r`*distname*, where *distname* is the root name of the distribution. Normal random variables have root `norm`, so the random generation function for normal rvs is `rnorm`. Other root names we have encountered so far are `unif`, `geom`, `pois`, `exp`, and `binom`.

The first argument to all of R's random generation functions is **always** the number of samples to create. Random generation functions take additional parameters that fully describe the distribution. For example, if we want to create a random sample of size 100 from a normal random variable with mean 5 and sd 2, we would use `rnorm(100, mean = 5, sd = 2)`, or simply `rnorm(100, 5, 2)`. Without the additional arguments, `rnorm` gives the standard normal random variable, with mean 0 and sd 1. For example, `rnorm(10000)` gives 10,000 independent random values of the standard normal random variable $Z$.

Our strategy for estimating probabilities of events involving random variables is as follows:

- Sample the random variable using the appropriate random generation function.
- Evaluate the event to get a vector of true/false values.
- Use the `mean` function to compute the proportion of times that the event occurs.

If the sample size is reasonably large, then the true probability that the event occurs should be close to the percentage of times that the event occurs in our sample. The larger the sample size is, the closer we expect our estimate to be to the true value. In practice, a sample of size 10,000 gives a good balance between speed and accuracy. Estimates vary, and rare events are harder to sample, so you should always repeat your estimate a few times to see what kind of variation you obtain in your answers.

**Example 5.1.** Let $Z$ be a standard normal random variable. Estimate $P(Z > 1)$.

We begin by creating a large random sample from a normal random variable.

```
Z <- rnorm(10000)
head(Z)
```

```
## [1]  0.8826500 -0.5235463 -0.9814831  1.1042537  0.5727897  0.3843779
```

DOI: 10.1201/9781003004899-5

Next, we evaluate the event $Z > 1$. We create a vector `bigZ` that is `TRUE` when the sample is greater than one, and `FALSE` when the sample is not greater than one.

```
bigZ <- Z > 1
head(bigZ)
```

```
## [1] FALSE FALSE FALSE  TRUE FALSE FALSE
```

Now, we count the number of times we see a `TRUE`, and divide by the length of the vector.

```
sum(bigZ == TRUE) / 10000
```

```
## [1] 0.1588
```

There are a few improvements we can make. First, `bigZ == TRUE` is just the same as `bigZ`, so we can compute:

```
sum(bigZ) / 10000
```

```
## [1] 0.1588
```

We note that `sum` converts `TRUE` to one and `FALSE` to zero. Adding up those values and dividing by the length is the **same thing** as taking the mean. So, we can simply do:

```
mean(bigZ)
```

```
## [1] 0.1588
```

For simple problems like this one, there is no real need to store the event in a variable, so we may go directly to the computation:

```
mean(Z > 1)
```

```
## [1] 0.1588
```

We estimate that $P(Z > 1) = 0.1588$.

**Example 5.2.** Let $Z$ be a standard normal rv. Estimate $P(Z^2 > 1)$.

```
Z <- rnorm(10000)
mean(Z^2 > 1)
```

```
## [1] 0.322
```

We can also easily estimate means and standard deviations of random variables. To do so, we create a large random sample from the distribution in question, and we take the mean or standard deviation of the large sample. If the sample is large, then we expect the sample mean to be close to the true mean, and the sample standard deviation to be close to the true standard deviation. Let's begin with an example where we know what the correct answer is, in order to check that the technique is working.

**Example 5.3.** Let $Z$ be a normal random variable with mean 0 and standard deviation 1. Estimate the mean and standard deviation of $Z$.

```
Z <- rnorm(10000)
mean(Z)
```

```
## [1] 0.001501601
```

```
sd(Z)
```

```
## [1] 1.003491
```

We see that we are reasonably close to the correct answers of 0 and 1.

**Example 5.4.** Estimate the mean and standard deviation of $Z^2$.

```
Z <- rnorm(10000)
mean(Z^2)
```

```
## [1] 0.9873908
```

```
sd(Z^2)
```

```
## [1] 1.416691
```

We will see later in this chapter that $Z^2$ is a $\chi^2$ random variable with one degree of freedom. It is known that a $\chi^2$ random variable with $\nu$ degrees of freedom has mean $\nu$ and standard deviation $\sqrt{2\nu}$. Thus, the answers above are pretty close to the correct answers.

**Example 5.5.** Let $X$ and $Y$ be independent standard normal random variables. Estimate $P(XY > 1)$.

There are two ways to do this. The first is:

```
X <- rnorm(10000)
Y <- rnorm(10000)
mean(X * Y > 1)
```

```
## [1] 0.105
```

A technique that is closer to what we will be doing below is the following. We want to create a random sample from the random variable $W = XY$. To do so, we would use

```
W <- rnorm(10000) * rnorm(10000)
```

Note that R multiplies the vectors component-wise. Then, we compute the percentage of times that the sample is greater than 1 as before.

```
mean(W > 1)
```

```
## [1] 0.0995
```

The two methods give slightly different answers because simulations are random and only estimate the true values.

Notice that $P(XY > 1)$ is *not* the same as the answer we got for $P(Z^2 > 1)$ in Example 5.2.

## 5.2 Estimating discrete distributions

In this section, we show how to estimate the pmf of a discrete random variable via simulation. Let's begin with an example.

**Example 5.6.** Suppose that two dice are rolled, and their sum is denoted as $X$. Estimate the pmf of $X$ via simulation.

To estimate $P(X = 5)$, for example, we would use

```
X <- replicate(10000, {
  dieRoll <- sample(1:6, 2, TRUE)
  sum(dieRoll)
})
mean(X == 5)
```

```
## [1] 0.1107
```

It is possible to repeat this approach for each value $2, 3, \ldots, 12$, but that would take a long time. A more efficient method is to keep track of all observations of the random variable, and divide each by the total number of times the rv was observed. We will use `table` for this. Recall, `table` gives a vector of counts of each unique element in a vector. That is,

```
table(c(1, 1, 1, 1, 1, 2, 2, 3, 5, 1))
```

```
##
## 1 2 3 5
## 6 2 1 1
```

indicates that there are 6 occurrences of "1," 2 occurrences of "2," and 1 occurrence each of "3" and "5." To apply this to the die rolling, we create a vector of length 10,000 that has all observations of the random variable $X$:

```
X <- replicate(10000, {
  dieRoll <- sample(1:6, 2, TRUE)
  sum(dieRoll)
})
table(X)
```

```
## X
##    2    3    4    5    6    7    8    9   10   11   12
##  300  547  822 1080 1412 1655 1418 1082  846  562  276
```

We then divide each entry of the table by 10,000 to estimate of the pmf of $X$:

```
table(X) / 10000
```

```
## X
##      2      3      4      5      6      7      8      9     10     11
## 0.0300 0.0547 0.0822 0.1080 0.1412 0.1655 0.1418 0.1082 0.0846 0.0562
##     12
## 0.0276
```

We don't want to hard-code the 10,000 value in the above command, because it can be a source of error if we change the number of replications and forget to change the denominator here. There is a base R function `proportions`[1] which is useful for dealing with tabled data. In particular, if the tabled data is of the form above, then it computes the proportion of each value.

```
proportions(table(X))
```

```
## X
##      2      3      4      5      6      7      8      9     10     11
## 0.0300 0.0547 0.0822 0.1080 0.1412 0.1655 0.1418 0.1082 0.0846 0.0562
```

[1]The R function `proportions` is new to R 4.0.1 and is recommended as a drop-in replacement for the unfortunately named `prop.table`.

```
##      12
## 0.0276
```

And, there is our estimate of the pmf of $X$. For example, we estimate the probability of rolling an 11 to be 0.0562.

A simple way to visualize a pmf is to plot the table:

```
plot(proportions(table(X)),
  main = "Rolling two dice", ylab = "Probability"
)
```

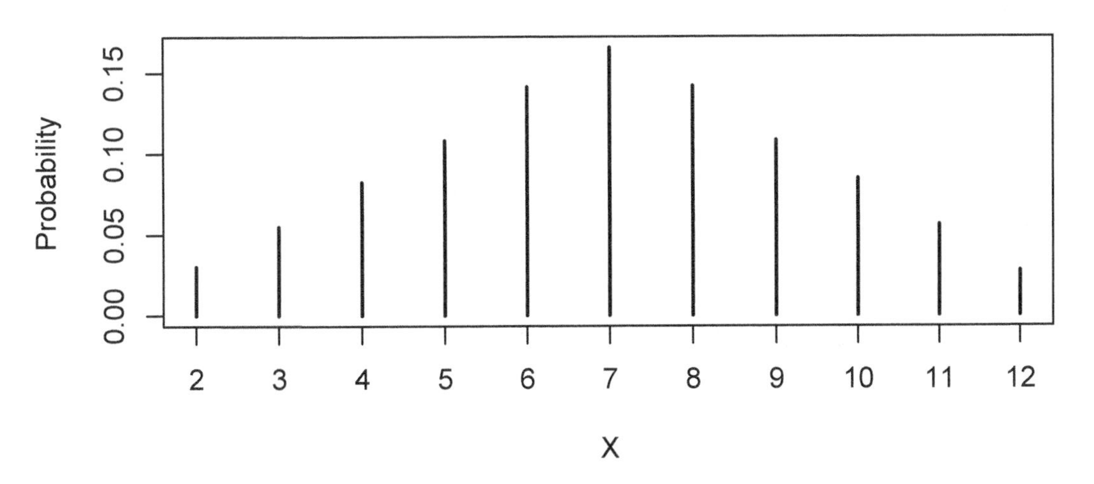


**Example 5.7.** Suppose 50 randomly chosen people are in a room. Let $X$ denote the number of people in the room who have the same birthday as someone else in the room. Estimate the pmf of $X$.

> **Try It Yourself.**
> Before reading further, what do you think the most likely outcome of $X$ is?

We will simulate birthdays by taking a random sample from `1:365` and storing it in a vector. The tricky part is counting the number of elements in the vector that are repeated somewhere else in the vector. We will create a table and add up all of the values that are bigger than 1. Like this:

```
birthdays <- sample(1:365, 50, replace = TRUE)
table(birthdays)
```

```
## birthdays
##   8   9  13  20  24  41  52  64  66  70  92  98  99 102 104 119 123 126
##   2   1   1   2   2   1   1   1   1   1   1   1   1   1   1   1   1   1
## 151 175 179 182 185 222 231 237 240 241 249 258 259 276 279 285 287 313
##   1   1   1   1   1   1   1   1   1   1   1   1   2   2   1   1   1   1
## 317 323 324 327 333 344 346 364 365
##   1   1   1   1   1   1   1   1   1
```

Look through the table. Anywhere there is a number bigger than 1, there are that many people who share that birthday. We can use `table(birthdays) > 1` to detect the multiple birthday entries, and then use that to index into the table to select those days:

```
table(birthdays)[table(birthdays) > 1]
```

```
## birthdays
##   8  20  24 259 276
##   2   2   2   2   2
```

Finally we sum to count the number of people who share a birthday with someone else, producing one observation of the random variable $X$.

```
sum(table(birthdays)[table(birthdays) > 1])
```

```
## [1] 10
```

Now, we replicate to produce many observations of $X$.

```
X <- replicate(10000, {
  birthdays <- sample(1:365, 50, replace = TRUE)
  sum(table(birthdays)[table(birthdays) > 1])
})
pmf <- proportions(table(X))
pmf
```

```
## X
##      0      2      3      4      5      6      7      8      9     10
## 0.0309 0.1196 0.0059 0.1982 0.0219 0.2146 0.0309 0.1688 0.0293 0.0917
##     11     12     13     14     15     16     17     18     19     21
## 0.0205 0.0361 0.0097 0.0137 0.0034 0.0034 0.0006 0.0006 0.0001 0.0001
```

Let's plot it.

```
plot(pmf,
  main = "Fifty people in a room",
  xlab = "Number of people sharing a birthday",
  ylab = "Probability"
)
```

Looking at the pmf (in Figure 5.1), the most likely outcome is that 6 people in the room share a birthday with someone else, followed closely by 4, and then 8. Note that it is impossible for exactly one person to share a birthday with someone else in the room!

**Example 5.8.** You toss a coin 100 times. After each toss, either there have been more heads, more tails, or the same number of heads and tails. Let $X$ be the number of times in the 100 tosses that there were more heads than tails. Estimate the pmf of $X$.

> **Try It Yourself.**
> Before looking at the solution, guess whether the pmf of $X$ will be centered around 50, or not.

We start by doing a single run of the experiment. The function `cumsum` accepts a numeric vector and returns the *cumulative sum* of the vector. So, `cumsum(c(1, 3, -1))` would return `c(1, 4, 3)`.

```
# flip 100 coins
coin_flips <- sample(c("H", "T"), 100, replace = TRUE)
coin_flips[1:10]
```

```
##  [1] "H" "H" "T" "T" "T" "T" "H" "H" "H" "T"
```

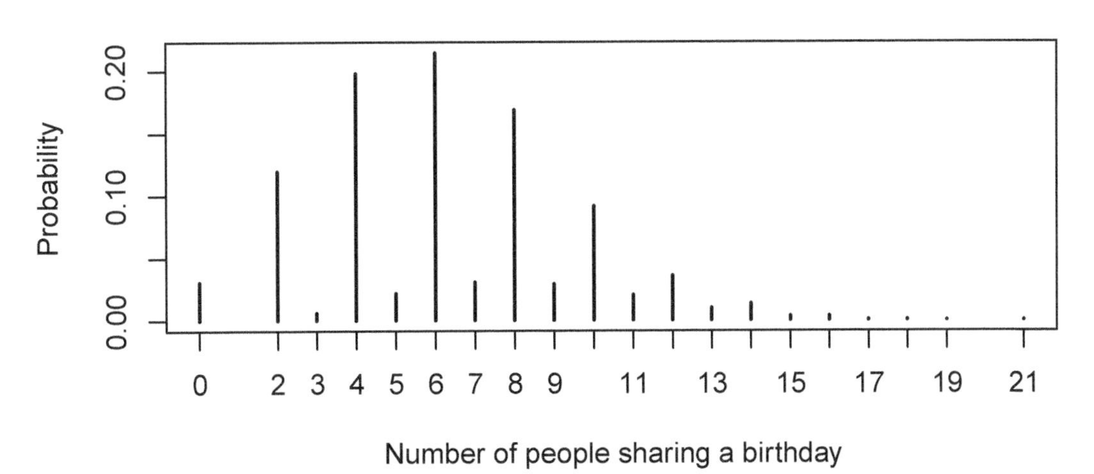


**FIGURE 5.1** Probability mass function for the number of people who share a birthday with another person in a room with 50 people.

```
# calculate the cumulative number of heads or tails so far
num_heads <- cumsum(coin_flips == "H")
num_tails <- cumsum(coin_flips == "T")
num_heads[1:10]
```

```
##  [1] 1 2 2 2 2 2 3 4 5 5
```

```
num_tails[1:10]
```

```
##  [1] 0 0 1 2 3 4 4 4 4 5
```

```
# calculate the number of times there were more heads than tails
sum(num_heads > num_tails)
```

```
## [1] 53
```

Now, we put that inside of replicate.

```
X <- replicate(100000, {
  coin_flips <- sample(c("H", "T"), 100, replace = TRUE)
  num_heads <- cumsum(coin_flips == "H")
  num_tails <- cumsum(coin_flips == "T")
  sum(num_heads > num_tails)
})
pmf <- proportions(table(X))
```

When we have this many possible outcomes, it is easier to view a plot of the pmf than to look directly at the table of probabilities.

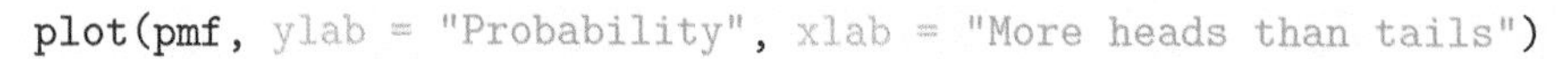

```
plot(pmf, ylab = "Probability", xlab = "More heads than tails")
```

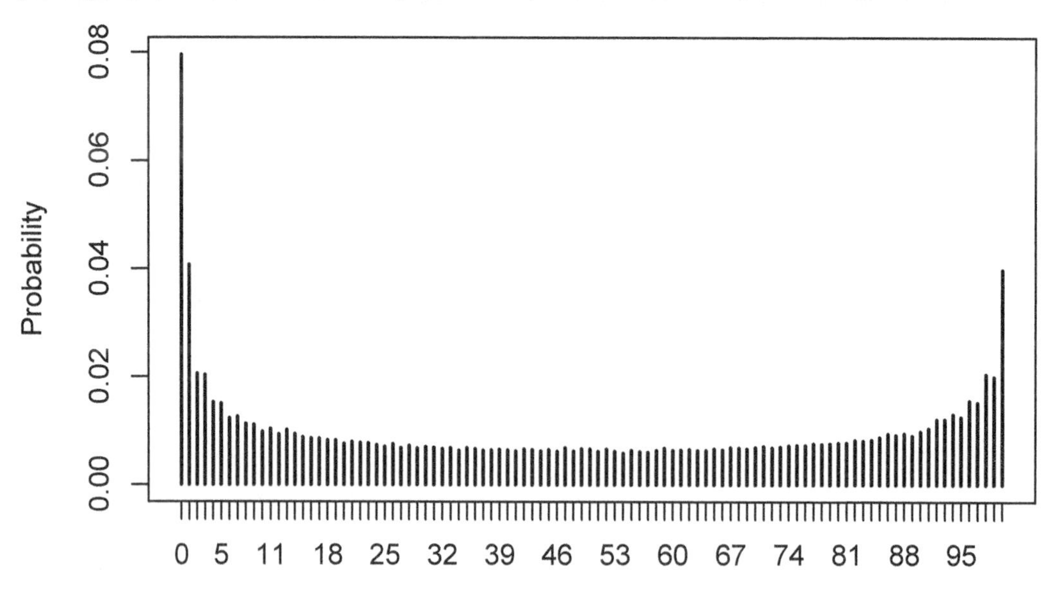


The most likely outcome (by far) is that there are **never** more heads than tails in the 100 tosses of the coin. This result can be surprising, even to experienced mathematicians.

**Example 5.9.** Suppose you have a bag full of marbles; 50 are red and 50 are blue. You are standing on a number line, and you draw a marble out of the bag. If you get red, you go left one unit. If you get blue, you go right one unit. This is called a *random walk*. You draw marbles up to 100 times, each time moving left or right one unit. Let $X$ be the number of marbles drawn from the bag until you return to 0 for the first time. The rv $X$ is called the *first return time* since it is the number of steps it takes to return to your starting position.

Estimate the pmf of $X$.

First, simulate the steps of the walk, with 1 and -1 representing steps right and left.

```
movements <- sample(rep(c(1, -1), times = 50), 100, replace = FALSE)
movements[1:10]
```

```
##  [1]  1  1 -1 -1 -1 -1  1  1  1 -1
```

Next, use `cumsum` to calculate the cumulative sum of the steps. This vector gives the position of the walk at each step.

```
cumsum(movements)[1:10]
```

```
##  [1]  1  2  1  0 -1 -2 -1  0  1  0
```

The values where the cumulative sum is zero represent a return to the origin. Using `which`, we learn which steps of the walk were zero, and then find the first of these with `min`.

```
which(cumsum(movements) == 0)
```

```
## [1]   4   8  10  12  56  78  82  94 100
```

```
min(which(cumsum(movements) == 0))
```

```
## [1] 4
```

This results in a single value of the rv $X$, in this case 4. To finish, we replicate the code and `table` it to compute the pmf.

```
X <- replicate(10000, {
  movements <- sample(rep(c(1, -1), times = 50), 100, replace = FALSE)
  min(which(cumsum(movements) == 0))
})
pmf <- proportions(table(X))
plot(pmf,
  main = "First return time for a 100 step random walk",
  xlab = "Steps to return",
  ylab = "Probability"
)
```

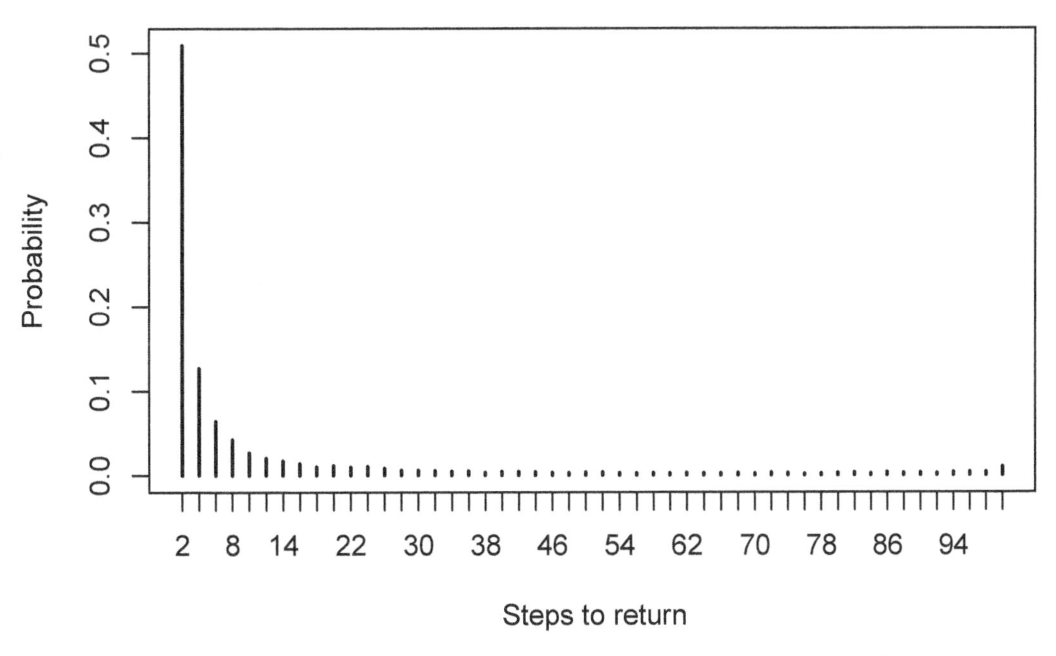


Only even return times are possible (why?). Half the time the first return is after two draws, one of each color. There is a slight bump near 100, when the bag of marbles empties out and the number of red and blue marbles drawn are forced to equalize.

## 5.3 Estimating continuous distributions

In this section, we show how to estimate the pdf of a continuous rv $X$ via simulation. For discrete rvs, we used `table` to produce a count of each value that occurred in our simulation.

These counts approximated the pmf. However, continuous rvs essentially never take the same value twice, so `table` is not helpful.

Instead, we divide the range of $X$ into segments and count the number of values in each segment that appear in the simulation. These counts can be visualized by drawing a vertical bar over each segment of the range, with height corresponding to the count or proportion of values that appeared in that segment. The resulting graph is called a *histogram* and is easily produced with the R command `hist`.

For distributions where the pdf is continuous, we may also use *density estimation*. The height of the density estimation is a weighted sum of the distances to all of the data points in the sample. Places which are close to many data points will have higher density estimates, while places that are far from most data points will have lower density estimates. The R command for density estimation is `density`, and in this book we will only use the result of density estimation to produce a plot. The plot results in a smooth curve whose height approximates the pdf of the random variable.

**Example 5.10.** Estimate the pdf of $2Z$ when $Z$ is a standard normal random variable.

We simulate 10,000 values of a standard normal rv $Z$ and then multiply by 2 to get 10,000 values sampled from $2Z$.

```
Z <- rnorm(10000)
twoZ <- 2 * Z
```

We then use `density` to estimate and plot the pdf of the data:

```
plot(density(twoZ),
  main = "Density of 2Z",
  xlab = "2Z"
)
```

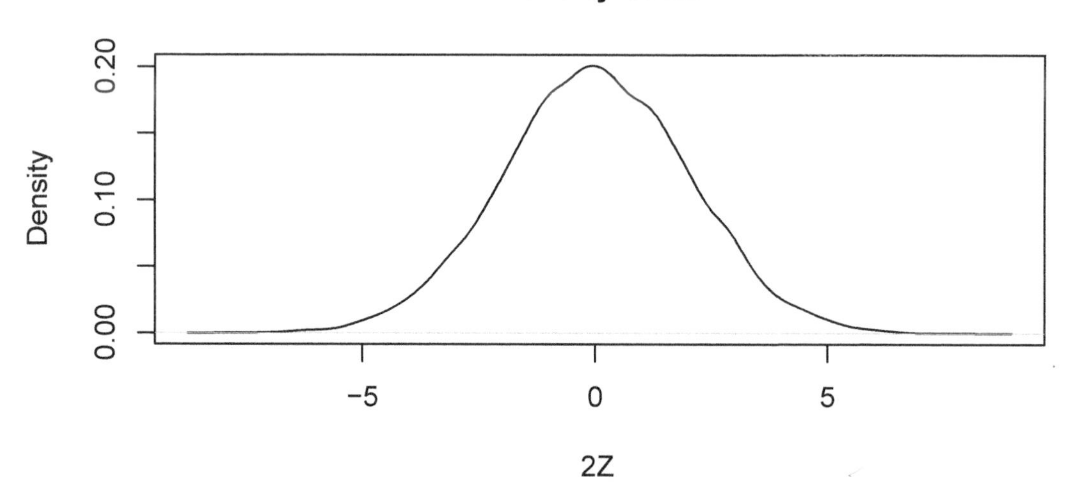


Notice that the most likely outcome of $2Z$ seems to be 0, just as it is for $Z$, but that the spread of the distribution of $2Z$ is twice as wide as the spread of $Z$.

Alternatively, we could create a histogram of the data, using `hist`. We set `probability = TRUE` to adjust the scale on the $y$-axis so that the area of each rectangle in the histogram is the probability that the random variable falls in the interval given by the base of the rectangle.

```
hist(twoZ,
  probability = TRUE,
  main = "Histogram of 2Z",
  xlab = "2Z"
)
```

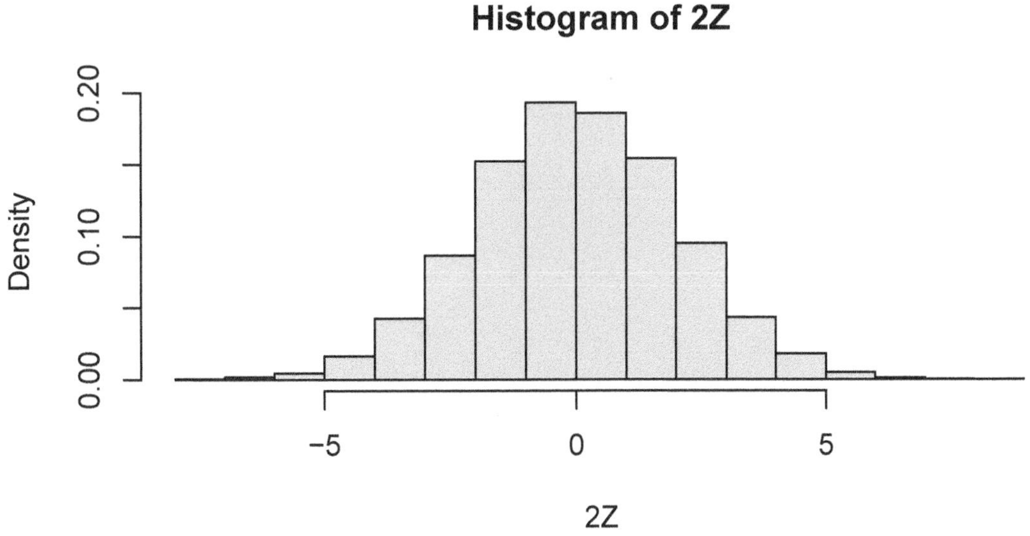

Given experimental information about a probability density, we wish to compare it to a known theoretical pdf. A direct way to make this comparison is to plot the estimated pdf on the same graph as the theoretical pdf. To add a curve to a histogram, density plot, or any plot, use the R function `curve`.

**Example 5.11.** Compare the pdf of $2Z$, where $Z \sim \text{Norm}(0, 1)$ to the pdf of a normal random variable with mean 0 and standard deviation 2.

We already saw how to estimate the pdf of $2Z$, we just need to plot the pdf of $\text{Norm}(0, 2)$ on the same graph. We show how to do this using both the histogram and the density plot approach. The pdf $f(x)$ of $\text{Norm}(0, 2)$ is given in R by the function $f(x) = \text{dnorm}(x, 0, 2)$.

```
hist(twoZ,
  probability = TRUE,
  main = "Density and histogram of 2Z",
  xlab = "2Z"
)
curve(dnorm(x, 0, 2), add = TRUE, col = "red")
```

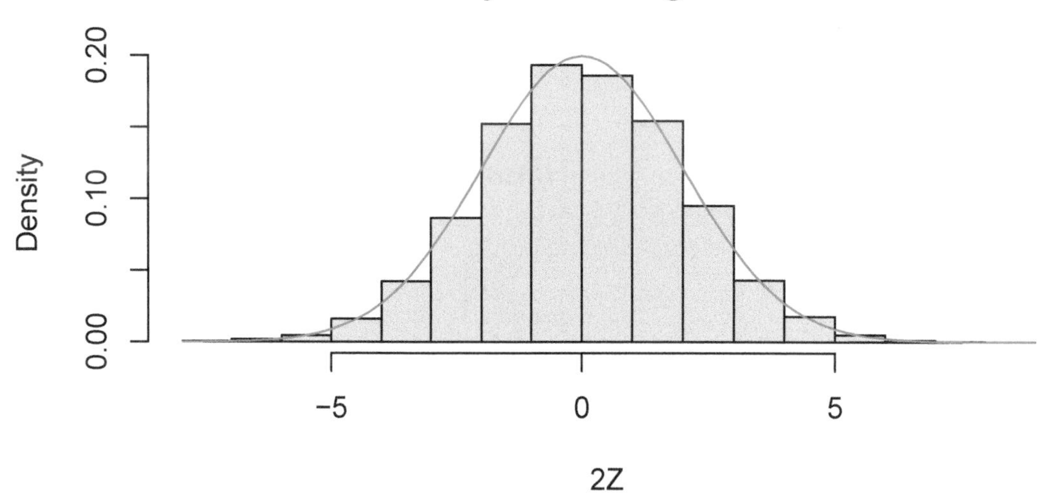


Since the area of each rectangle in the histogram is approximately the same as the area under the curve over the same interval, this is evidence that $2Z$ is normal with mean 0 and standard deviation 2. Next, let's look at the density estimation together with the true pdf of a normal rv with mean 0 and $\sigma = 2$.

```
plot(density(twoZ),
  xlab = "2Z",
  main = "Density 2Z and Norm(0, 2)"
)
curve(dnorm(x, 0, 2), add = TRUE, col = "red")
```

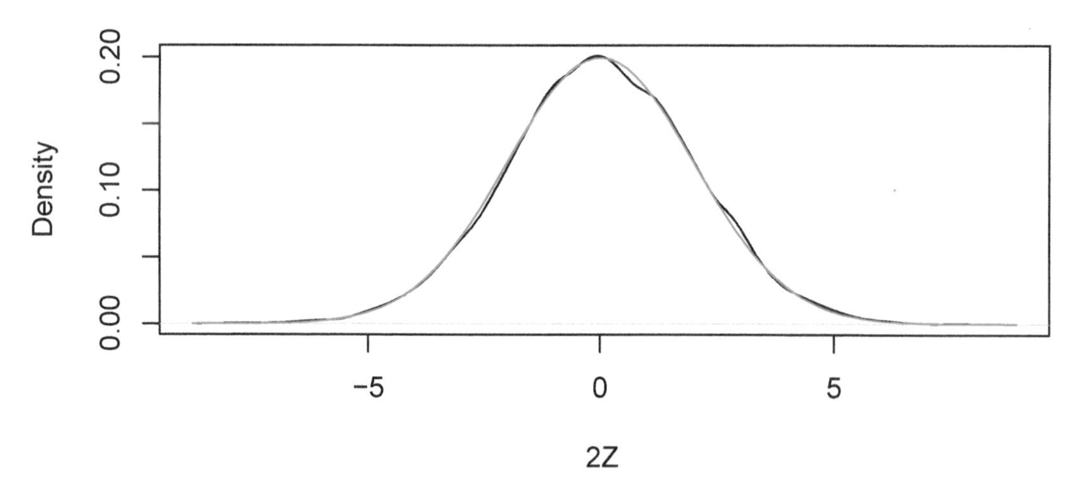


Wow! Those look really close to the same thing! This is again evidence that $2Z \sim \text{Norm}(0, 2)$.

It would be a good idea to label the two curves in our plots, but that sort of finesse is easier with ggplot, which will be discussed in Chapter 7. Also in Chapter 7, we introduce *qq plots*, which are a more accurate (but less intuitive) way to compare continuous distributions.

**Example 5.12.** Estimate via simulation the pdf of $W = \log(|Z|)$ when $Z$ is standard normal.

```
W <- log(abs(rnorm(10000)))
plot(density(W),
  main = "Density of log|Z|",
  xlab = "log|Z|"
)
```

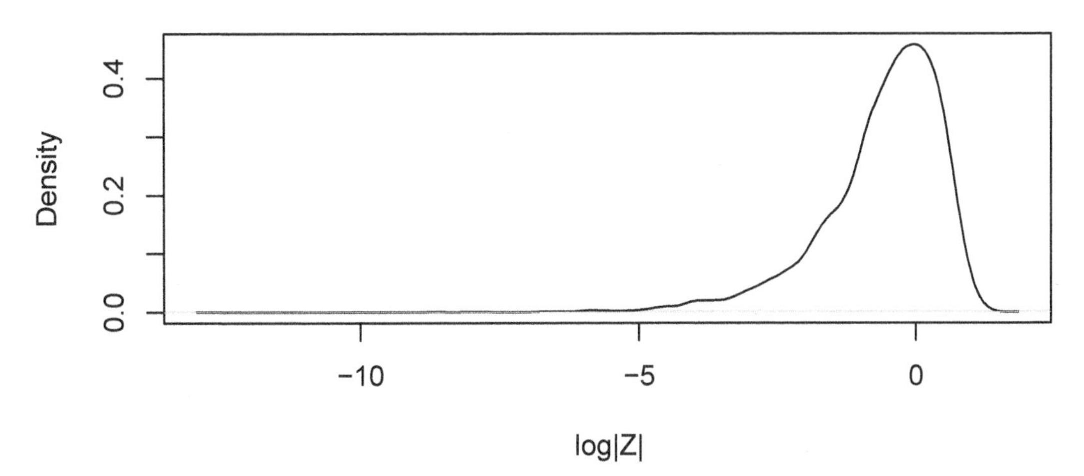


The simulations in Examples 5.10 and 5.12 worked well because the pdfs of $2Z$ and of $\log(|Z|)$ are continuous functions. For pdfs with discontinuities, density estimation of this type will misleadingly smooth out the jumps in the function.

**Example 5.13.** Estimate the pdf of $X$ when $X$ is uniform on the interval $[-1, 1]$.

```
X <- runif(10000, -1, 1)
plot(density(X),
  main = "Density of Unif[-1,1]",
  xlab = "X"
)
curve(dunif(x, -1, 1), add = TRUE, col = "red")
```

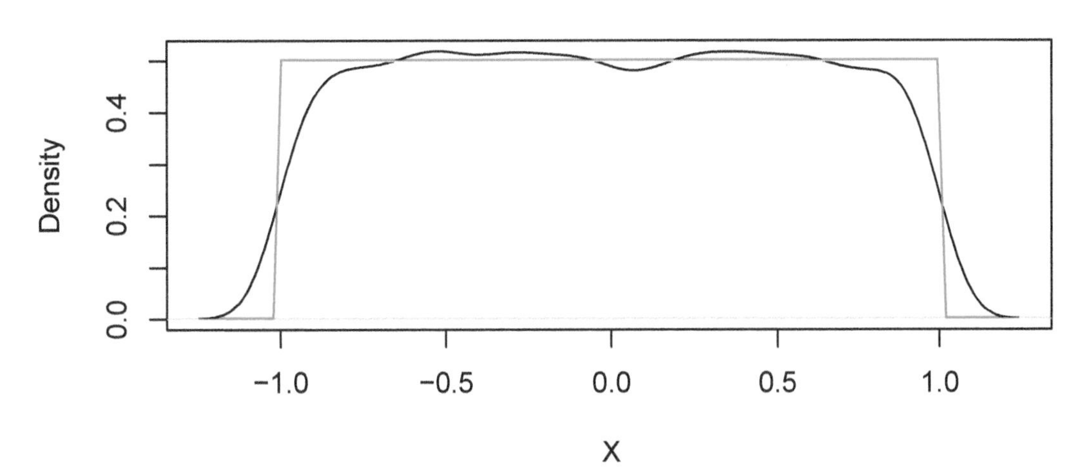

This distribution should be exactly 0.5 between -1 and 1, and zero everywhere else, as shown by the theoretical distribution in red. The simulation is reasonable in the middle, but the discontinuities at the endpoints are not shown well. Increasing the number of data points helps, but it does not fix the problem.

For this distribution, which has a jump discontinuity, using a histogram works better.

```
hist(X,
  probability = TRUE,
  main = "Histogram of Unif[-1,1]",
  xlab = "X"
)
```

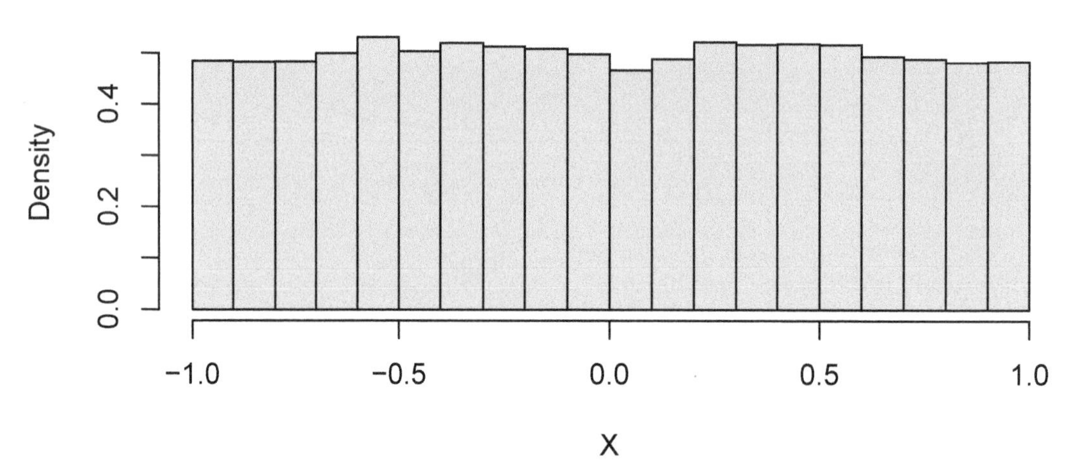


We now use density estimation to give evidence of an important fact: the sum of normal random variables is still normal.

**Example 5.14.** Estimate the pdf of $Z_1 + Z_2$ where $Z_1$ and $Z_2$ are independent standard normal random variables.

```
Z1 <- rnorm(10000)
Z2 <- rnorm(10000)
plot(density(Z1 + Z2),
  main = "Sum of two standard normal rvs",
  xlab = expression(Z[1] + Z[2])
)
curve(dnorm(x, 0, sqrt(2)), add = TRUE, col = "red")
```

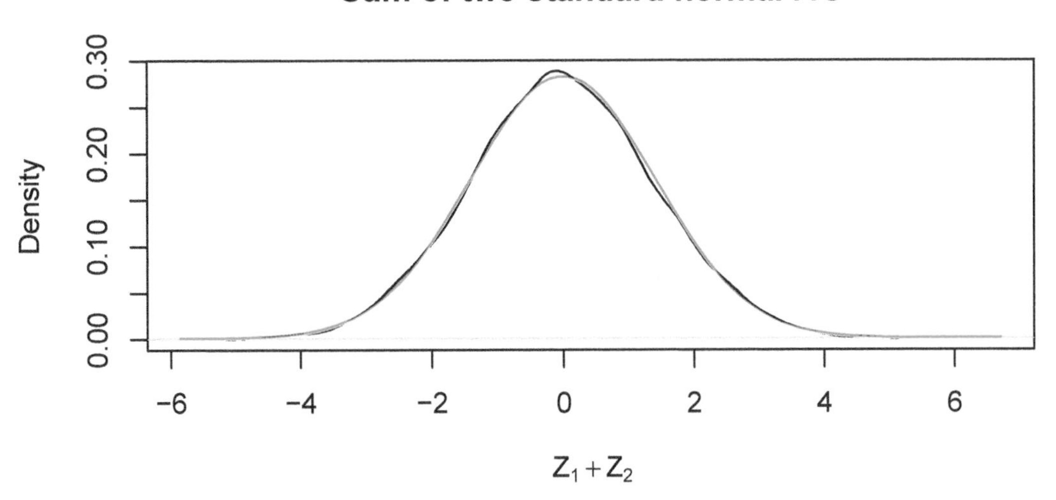


In the plot, the pdf for a Norm(0, $\sqrt{2}$) rv is included. It appears that the sum of two standard normal random variables is again a normal random variable. The standard deviation is $\sqrt{2}$ since $\text{Var}(Z_1 + Z_2) = \text{Var}(Z_1) + \text{Var}(Z_2) = 2$ by Theorem 3.10.

> **Try It Yourself.**
> For $Z_1$,$Z_2$ independent standard normal, estimate the pdf of $Z_1 - Z_2$ and see that it is also Norm(0, $\sqrt{2}$).

Exercise 5.16 asks you to investigate the sum of two normal rvs which are not standard mean 0, sd 1, and see that their sum is normal as well. We will not prove that this is always true, but state it as the following:

**Theorem 5.1.** *If $X \sim \text{Norm}(\mu_X, \sigma_X)$ and $Y \sim \text{Norm}(\mu_Y, \sigma_Y)$ are independent, then $X+Y$ is normal with mean $\mu_X + \mu_Y$ and variance $\sigma_X^2 + \sigma_Y^2$.*

More generally,

**Theorem 5.2.** *Let $X_1, \ldots, X_n$ be mutually independent normal random variables with means $\mu_1, \ldots, \mu_n$ and standard deviations $\sigma_1, \ldots, \sigma_n$. The random variable*

$$\sum_{i=1}^{n} a_i X_i$$

*is a normal random variable with mean $\sum_{i=1}^{n} a_i \mu_i$ and standard deviation $\sqrt{\sum_{i=1}^{n} a_i^2 \sigma_i^2}$.*

Let's meet an important distribution, the $\chi^2$ (chi-squared) distribution. In fact, $\chi^2$ is a family of distributions controlled by a parameter called the *degrees of freedom*, usually abbreviated df. The root name for a $\chi^2$ rv is `chisq`, and `dchisq` requires the degrees of freedom to be specified using the `df` parameter.

**Example 5.15.** Let $Z$ be a standard normal rv. Find the pdf of $Z^2$ and compare it to the pdf of a $\chi^2$ rv with one df on the same plot.

```
Z <- rnorm(10000)
hist(Z^2,
  probability = T,
  xlab = expression(Z^2)
```

```
)
curve(dchisq(x, df = 1), add = TRUE, col = "red")
```

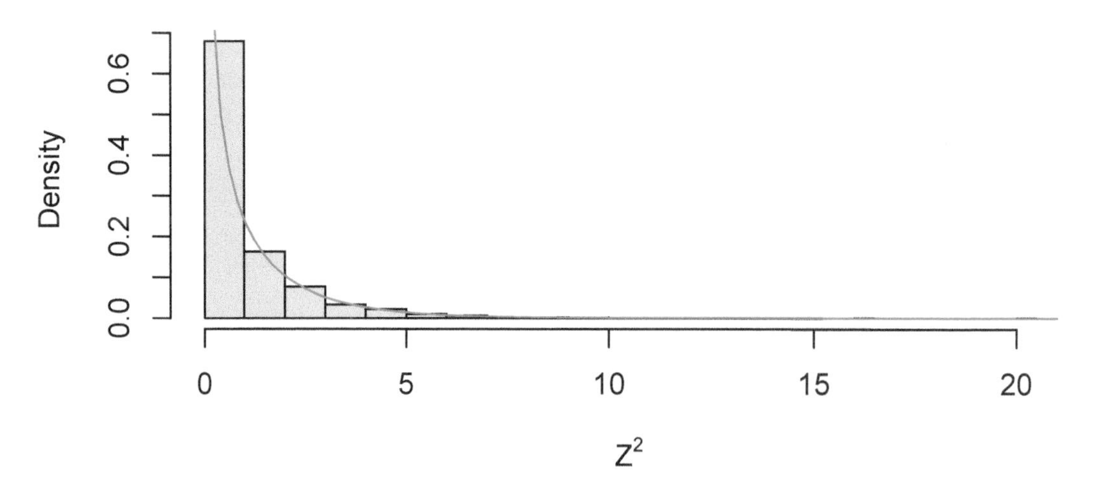


**FIGURE 5.2** Estimated pdf of $Z^2$ with the exact pdf of a $\chi^2$ with 1 df overlaid in red.

As you can see in Figure 5.2, the estimated density follows the true histogram quite well. This is evidence that $Z^2$ is, in fact, chi-squared. Notice that $\chi^2$ with one df has a vertical asymptote at $x = 0$.

**Example 5.16.** The sum of exponential random variables follows what is called a *gamma distribution*. The gamma distribution is represented in R via the root name `gamma` together with the typical prefixes `dpqr`. A gamma random variable has two parameters, the `shape` and the `rate`. It turns out (see Exercise 5.27) that:

1. The sum of independent gamma random variables with shapes $\alpha_1$ and $\alpha_2$ and the same rate $\beta$ is again a gamma random variable with shape $\alpha_1 + \alpha_2$ and rate $\beta$.
2. A gamma random variable with shape 1 is an exponential random variable.

We check this in a simple case when $X$ and $Y$ are independent exponential random variables with rate 2. The sum $W = X + Y$ should have a gamma distribution with shape 2 and rate 2.

We estimate the pdf of $W = X + Y$, plot it, and overlay the pdf of a gamma random variable in Figure 5.3.

```
W <- replicate(10000, {
  X <- rexp(1, 2)
  Y <- rexp(1, 2)
  X + Y
})
hist(W,
  probability = TRUE,
  main = "Sum of exponentials is gamma",
  xlab = "X + Y",
  ylim = c(0, .72) # so the top of the curve is not clipped
```

```
)
curve(dgamma(x, shape = 2, rate = 2), add = TRUE, col = "red")
```

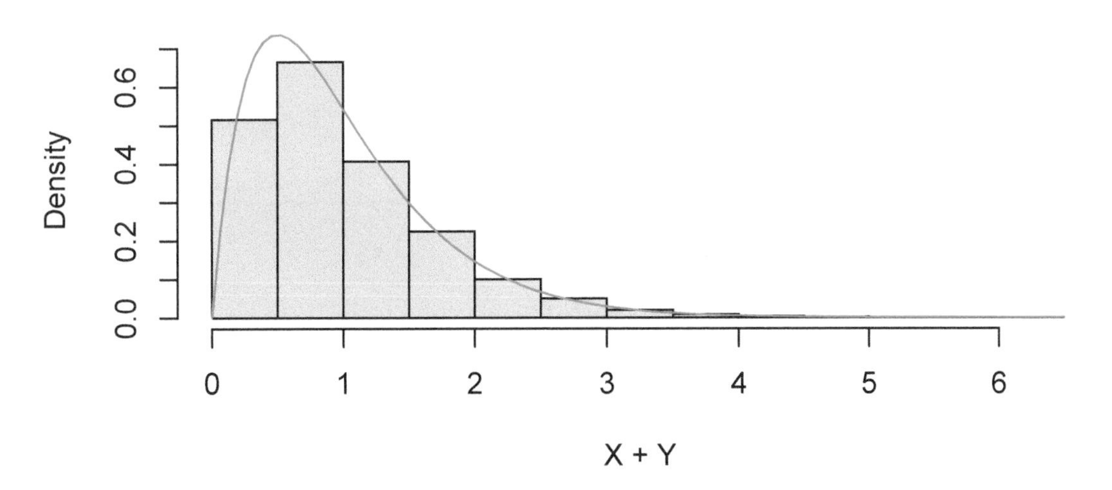


**FIGURE 5.3** Histogram of the sum of two exponential variables $X$ and $Y$ with rate 2. The pdf for a gamma random variable with shape 2 and rate 2 is overlaid in red.

**Example 5.17.** Estimate the density of $X_1 + X_2 + \cdots + X_{20}$ when all of the $X_i$ are independent exponential random variables with rate 2.

This one is trickier and is our first example where it is easier to use `replicate` to create the data. Let's build up the experiment that we are replicating from the ground up.

Here's the experiment of summing 20 exponential rvs with rate 2:

```
sum(rexp(20, 2))
```

```
## [1] 8.325553
```

Now replicate it (10 times to test):

```
replicate(10, sum(rexp(20, 2)))
```

```
##  [1] 11.898357  6.407845  8.360866  9.297432 12.212909 11.200338
##  [7]  8.746138 10.663090 11.940880  8.289535
```

Of course, we don't want to just replicate it 10 times; we need about 10,000 data points to get a good density estimate.

```
sumExpData <- replicate(10000, sum(rexp(20, 2)))
plot(density(sumExpData),
  main = "Density of sum of 20 exponentials",
  xlab = "X1 + ... + X20"
)
curve(dgamma(x, shape = 20, rate = 2), add = TRUE, col = "red")
```

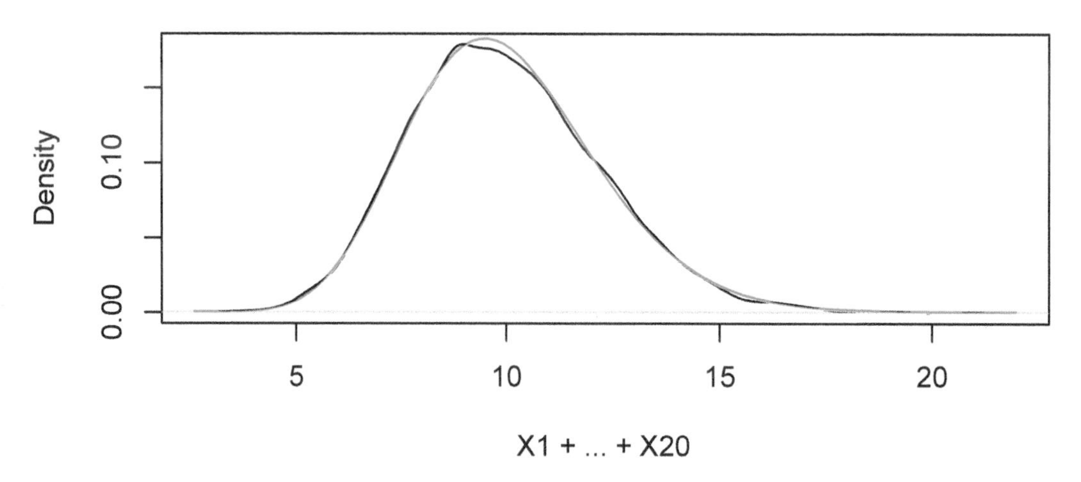


As explained in Example 5.16, this is exactly a gamma random variable with rate 2 and shape 20, and it is also starting to look like a normal random variable! Exercise 5.17 asks you to fit a normal curve to this distribution.

## 5.4 Central Limit Theorem

**Definition 5.1.** Random variables $X_1, \ldots, X_n$ are called *independent and identically distributed* or *iid* if the variables are mutually independent, and each $X_i$ has the same probability distribution.

All of R's random generation functions (`rnorm`, `rexp`, etc.) produce samples from iid random variables. In the real world, iid rvs are produced by taking a *random sample* of size $n$ from a population. Ideally, this sample would be obtained by numbering all members of the population and then using a random number generator to select members for the sample. As long as the population size is much larger than $n$, the sampled values will be independent, and all will have the same probability distribution as the population distribution. However, in practice it is quite difficult to produce a true random sample from a real population, and failure to sample randomly can introduce error into mathematical results that depend on the assumption that random variables are iid.

**Definition 5.2.** Let $X_1, \ldots, X_n$ be a random sample. A random variable $Y = h(X_1, \ldots, X_n)$ that is derived from the random sample is called a *sample statistic* or sometimes just a *statistic*. The probability distribution of $Y$ is called a *sampling distribution.*

Three of the most important statistics are the sample mean, sample variance, and sample standard deviation.

**Definition 5.3.** Assume $X_1, \ldots, X_n$ are independent and identically distributed. Define:

- The *sample mean*
$$\overline{X} = \frac{X_1 + \cdots + X_n}{n}.$$

- The *sample variance*
$$S^2 = \frac{1}{n-1}\sum_{i=1}^{n}(X_i - \overline{X})^2$$
- The *sample standard deviation* is $S$, the square root of the sample variance.

These three statistics are computed in R with the commands `mean`, `var`, and `sd` that we have been using all along.

A primary goal of statistics is to learn something about a population (the distribution of the $X_i$'s) by studying sample statistics. We can make precise statements about a population from a random sample because it is possible to describe the sampling distributions of statistics like $\overline{X}$ and $S$ even in the absence of information about the distribution of the $X_i$. The most important example of this phenomenon is the Central Limit Theorem, which is fundamental to statistical reasoning.

We first establish basic properties of the distribution of the sample mean.

**Proposition 5.1.** *If $X_1, \ldots, X_n$ are iid with mean $\mu$ and standard deviation $\sigma$, then the sample mean $\overline{X}$ has mean and variance given by*

$$E[\overline{X}] = \mu$$
$$Var(\overline{X}) = \sigma^2/n$$

*Proof.* From linearity of expected value (Theorem 3.8),

$$E[\overline{X}] = E\left[\frac{X_1 + \cdots X_n}{n}\right] = \frac{1}{n}\Big(E[X_1] + \cdots + E[X_n]\Big) = \frac{1}{n}(n\mu) = \mu.$$

Since the $X_i$ are mutually independent, Theorem 3.10 applies and:

$$\text{Var}[\overline{X}] = \text{Var}\left(\frac{1}{n}\sum_i X_i\right) = \frac{1}{n^2}\text{Var}\left(\sum_i X_i\right) = \frac{1}{n^2}\sum_i \text{Var}(X_i) = \frac{1}{n^2}\cdot n\sigma^2 = \frac{\sigma^2}{n}$$

■

From Proposition 5.1, the random variable $\frac{\overline{X}-\mu}{\sigma/\sqrt{n}}$ has mean 0 and standard deviation 1. When the population is normally distributed, Theorem 5.1 implies the following:

**Proposition 5.2.** *If $X_1, \ldots, X_n$ are iid normal, then $\frac{\overline{X}-\mu}{\sigma/\sqrt{n}} = Z$, where $Z$ is a standard normal rv.*

Remarkably, for large sample sizes we can still describe the distribution of $\frac{\overline{X}-\mu}{\sigma/\sqrt{n}}$ even when we don't know that the population is normal. This is the Central Limit Theorem.

**Theorem 5.3 (Central Limit Theorem).** *Let $X_1, \ldots, X_n$ be iid rvs with finite mean $\mu$ and standard deviation $\sigma$. Then*

$$\frac{\overline{X}-\mu}{\sigma/\sqrt{n}} \to Z \qquad \text{as } n \to \infty$$

*where $Z$ is a standard normal rv.*

We will not prove Theorem 5.3, but we will do simulations for several examples. They will all follow a similar format.

**Example 5.18.** Let $X_1, \ldots, X_{30}$ be independent Poisson random variables with rate 2. From our knowledge of the Poisson distribution, each $X_i$ has mean $\mu = 2$ and standard deviation $\sigma = \sqrt{2}$. Assuming $n = 30$ is a large enough sample size, the Central Limit Theorem says that

$$Z = \frac{\overline{X} - 2}{\sqrt{2}/\sqrt{30}}$$

will be approximately normal with mean 0 and standard deviation 1. Let us check this with a simulation.

This is a little bit more complicated than our previous examples, but the idea is still the same. We create an experiment which computes $\overline{X}$ and then transforms it by subtracting 2 and dividing by $\sqrt{2}/\sqrt{30}$.

Here is a single experiment:

```
Xbar <- mean(rpois(30, 2))
(Xbar - 2) / (sqrt(2) / sqrt(30))
```

```
## [1] 0.1290994
```

Now, we replicate and plot:

```
Z <- replicate(10000, {
  Xbar <- mean(rpois(30, 2))
  (Xbar - 2) / (sqrt(2) / sqrt(30))
})

plot(density(Z),
  main = "Standardized sum of 30 Poisson rvs", xlab = "Z"
)
curve(dnorm(x), add = TRUE, col = "red")
```

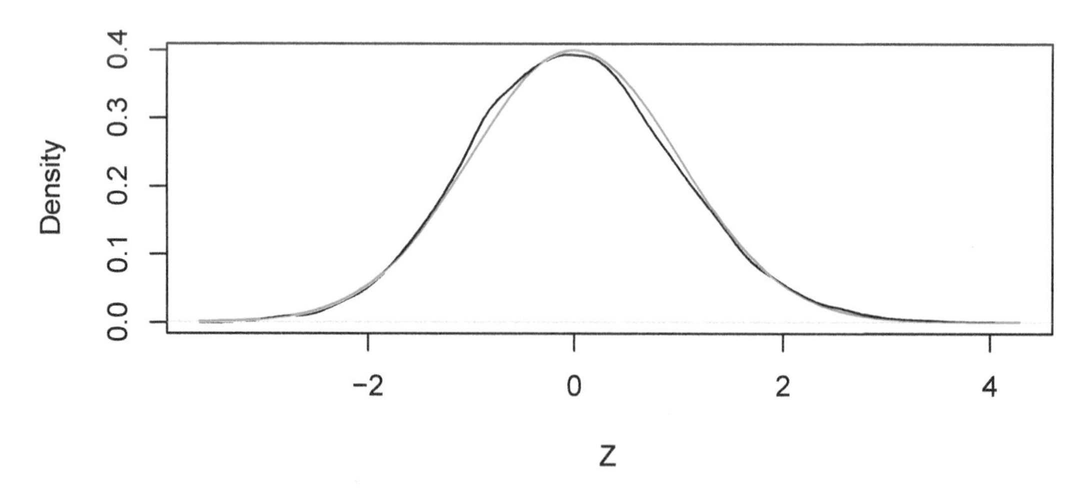


**FIGURE 5.4** Standardized sum of 30 Poisson random variables compared to a standard normal rv.

In Figure 5.4 we see very close agreement between the simulated density of $Z$ and the standard normal density curve.

**Example 5.19.** Let $X_1, \ldots, X_{50}$ be independent exponential random variables with rate 1/3. From our knowledge of the exponential distribution, each $X_i$ has mean $\mu = 3$ and standard deviation $\sigma = 3$. The Central Limit Theorem says that

$$Z = \frac{\overline{X} - 3}{3/\sqrt{n}}$$

is approximately normal with mean 0 and standard deviation 1 when $n$ is large. We check this with a simulation in the case $n = 50$. The resulting plot, in Figure 5.5, shows that even with $n = 50$ a sum of exponential random variables is still slightly skew to the right.

```
Z <- replicate(10000, {
  Xbar <- mean(rexp(50, 1 / 3))
  (Xbar - 3) / (3 / sqrt(50))
})

plot(density(Z),
  main = "Standardized sum of 50 exponential rvs", xlab = "Z"
)
curve(dnorm(x), add = TRUE, col = "red")
```

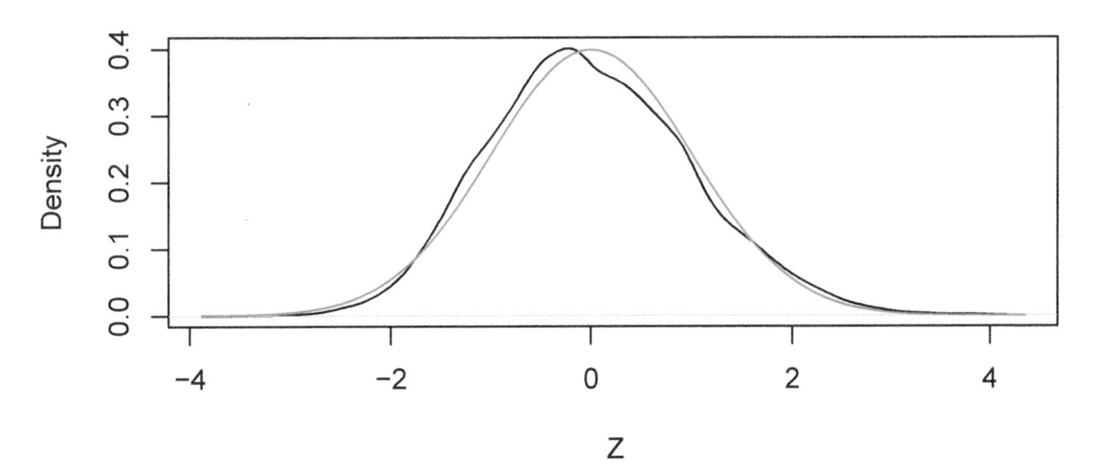


**FIGURE 5.5** Standardized sum of 50 exponential random variables compared to a standard normal rv.

The Central Limit Theorem says that, if you take the mean of a large number of independent samples, then the distribution of that mean will be approximately normal. How large does $n$ need to be in practice?

- If the population you are sampling from is symmetric with no outliers, a good approximation to normality appears after as few as 15-20 samples.
- If the population is moderately skewed, such as exponential or $\chi^2$, then it can take between 30-50 samples before getting a good approximation.
- Data with extreme skewness, such as some financial data where most entries are 0, a few are small, and even fewer are extremely large, may not be appropriate for the Central Limit Theorem even with 1000 samples (see Example 5.20).

There are versions of the Central Limit Theorem available when the $X_i$ are not iid, but outliers of sufficient size will cause the distribution of $\overline{X}$ to not be normal, see Exercise 5.34.

**Example 5.20.** A distribution for which sample size of $n = 1000$ is not sufficient for good approximation via normal distributions.

We create a distribution that consists primarily of zeros, but has a few modest sized values and a few large values:

```
# Start with lots of zeros
skewdata <- replicate(2000, 0)
# Add a few moderately sized values
skewdata <- c(skewdata, rexp(200, 1 / 10))
# Add a few large values
skewdata <- c(skewdata, seq(100000, 500000, 50000))

mu <- mean(skewdata)
sig <- sd(skewdata)
```

We use `sample` to take a random sample of size $n$ from this distribution. We take the mean, subtract the true mean of the distribution, and divide by $\sigma/\sqrt{n}$. We replicate that 10,000 times to estimate the sampling distribution. Here is the code we use to generate the sample and produce the plots shown in Figure 5.6.

```
Z <- replicate(10000, {
  Xbar <- mean(sample(skewdata, 100, TRUE))
  (Xbar - mu) / (sig / sqrt(100))
})
hist(Z,
  probability = TRUE,
  main = "100 Samples",
  xlab = "Z"
)
curve(dnorm(x), add = TRUE, col = "red")
```

Even with a sample size of 1000, the density still fails to be normal, especially the lack of left tail. Of course, the Central Limit Theorem is still true, so $\overline{X}$ *must* become approximately normal if we choose $n$ large enough. When $n = 5000$ there is still a slight skewness in the distribution, but it is finally close to normal.

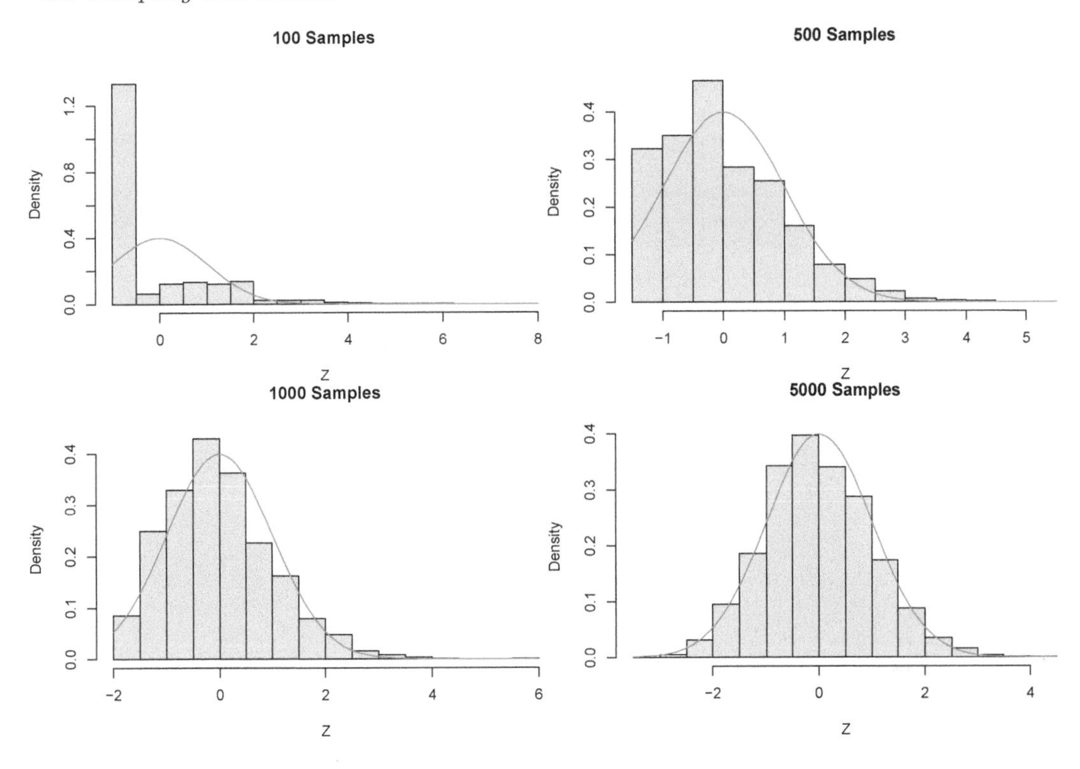


**FIGURE 5.6** The Central Limit Theorem in action for an extremely skew population.

## 5.5 Sampling distributions

We describe the $\chi^2$, $t$, and $F$ distributions, which are examples of distributions that are derived from random samples from a normal distribution. The emphasis is on understanding the relationships between the random variables and how they can be used to describe distributions related to the sample statistics $\overline{X}$ and $S$. Your goal should be to get comfortable with the idea that sample statistics have known distributions. You may not be able to *prove* relationships that you see later in this book, but with careful study of this chapter you won't be surprised, either. We will use density estimation extensively to illustrate the relationships.

### 5.5.1 The $\chi^2$ distribution

**Definition 5.4.** Let $Z$ be a standard normal random variable. An rv with the same distribution as $Z^2$ is called a *Chi-squared random variable with one degree of freedom.*

Let $Z_1, \ldots, Z_n$ be $n$ iid standard normal random variables. An rv with the same distribution as $Z_1^2 + \cdots + Z_n^2$ is called a *Chi-squared random variable with $n$ degrees of freedom.*

**Remark.** The sum of $n$ independent $\chi^2$ rvs with 1 degree of freedom is a $\chi^2$ rv with $n$ degrees of freedom. More generally, the sum of a $\chi^2$ with $\nu_1$ degrees of freedom and a $\chi^2$ rv with $\nu_2$ degrees of freedom is $\chi^2$ rv with $\nu_1 + \nu_2$ degrees of freedom.

Let's check by estimating pdfs that the sum of a $\chi^2$ rv with 2 df and a $\chi^2$ rv with 3 df is a $\chi^2$ rv with 5 df.

```
X <- rchisq(10000, 2)
Y <- rchisq(10000, 3)
hist(X + Y,
  probability = TRUE,
  main = "Sum of chi^2 random variables"
)
curve(dchisq(x, df = 5), add = TRUE, col = "red")
```

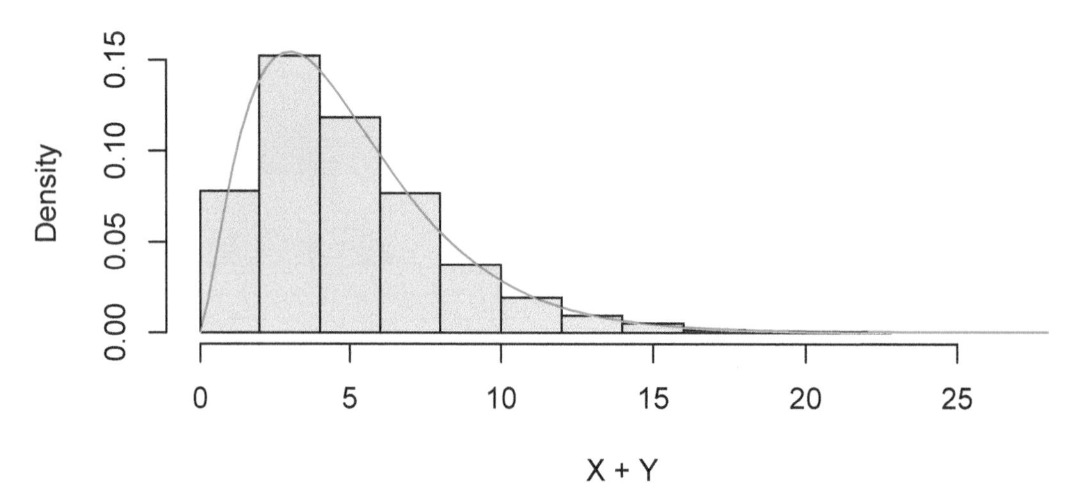


The $\chi^2$ distribution is important for understanding the sample variance $S^2$.

**Theorem 5.4.** *If $X_1, \ldots, X_n$ are iid normal rvs with mean $\mu$ and standard deviation $\sigma$, then*

$$\frac{n-1}{\sigma^2} S^2$$

*has a $\chi^2$ distribution with $n-1$ degrees of freedom.*

We provide a heuristic and a simulation as evidence that Theorem 5.4 is true. Additionally, in Exercise 5.40 you are asked to prove Theorem 5.4 in the case that $n = 2$.

Heuristic argument: Note that

$$\frac{n-1}{\sigma^2} S^2 = \frac{1}{\sigma^2} \sum_{i=1}^{n} \left(X_i - \overline{X}\right)^2 = \sum_{i=1}^{n} \left(\frac{X_i - \overline{X}}{\sigma}\right)^2$$

Now, if we had $\mu$ in place of $\overline{X}$ in the above equation, we would have exactly a $\chi^2$ with $n$ degrees of freedom. Replacing $\mu$ by its estimate $\overline{X}$ reduces the degrees of freedom by one, but it remains $\chi^2$.

For the simulation, we estimate the density of $\frac{n-1}{\sigma^2} S^2$ and compare it to that of a $\chi^2$ with $n-1$ degrees of freedom, in the case that $n = 4$, $\mu = 5$ and $\sigma = 9$.

```
S2 <- replicate(10000, 3 / 81 * sd(rnorm(4, 5, 9))^2)
hist(S2,
  probability = TRUE,
```

```
  main = "Sample variance compared to chi^2",
  xlab = expression(S^2),
  ylim = c(0, .25)
)
curve(dchisq(x, df = 3), add = TRUE, col = "red")
```

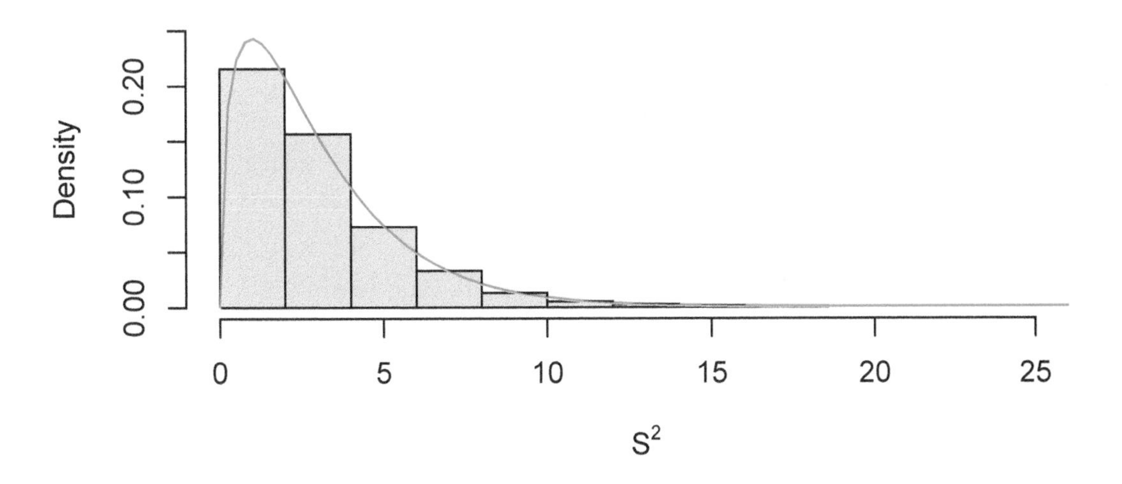


### 5.5.2 The $t$ distribution

This section introduces Student's $t$ distribution. The distribution is named after William Gosset, who published under the pseudonym "Student" while working at the Guinness Brewery in the early 1900's.

**Definition 5.5.** If $Z$ is a standard normal random variable, $\chi^2_\nu$ is a Chi-squared rv with $\nu$ degrees of freedom, and $Z$ and $\chi^2_\nu$ are independent, then

$$t_\nu = \frac{Z}{\sqrt{\chi^2_\nu/\nu}}$$

is distributed as a *$t$ random variable with $\nu$ degrees of freedom.*

The $t$ distributions are symmetric around zero and bump-shaped, like the normal distribution, but the $t$ distributions are heavy tailed. The heavy tails mean that $t$ random variables have higher probability of being far from 0 than normal random variables do, in the sense that

$$\lim_{M\to\infty} \frac{P(|T| > M)}{P(|Z| > M)} = \infty$$

for any normal rv $Z$ and $T$ having any $t$ distribution.

The $t$ distribution will play an important role in Chapter 8, where it appears as the result of random sampling. Theorem 5.5 establishes the connection between a random sample $X_1, \ldots, X_n$ and the $t$ distribution.

**Theorem 5.5.** *If $X_1, \ldots, X_n$ are iid normal rvs with mean $\mu$ and sd $\sigma$, then*

$$\frac{\overline{X} - \mu}{S/\sqrt{n}}$$

*is t with $n-1$ degrees of freedom.*

*Proof.*

$$\frac{Z}{\sqrt{\chi^2_{n-1}/n}} = \frac{\overline{X} - \mu}{\sigma/\sqrt{n}} \cdot \sqrt{\frac{\sigma^2(n-1)}{S^2(n-1)}} = \frac{\overline{X} - \mu}{S/\sqrt{n}}$$

Where we have used that $(n-1)S^2/\sigma^2$ is $\chi^2$ with $n-1$ degrees of freedom. ∎

Since the mean and standard deviation of $\overline{X}$ are $\mu$ and $\sigma/\sqrt{n}$, the random variable $\frac{\overline{X}-\mu}{\sigma/\sqrt{n}}$ is Norm$(0, 1)$. Replacing the denominator $\sigma/\sqrt{n}$ with the random variable $S/\sqrt{n}$ changes the distribution from normal to $t$.

**Definition 5.6.** The random variable $S/\sqrt{n}$ is the *standard error* of $\overline{X}$.

**Example 5.21.** Estimate the pdf of $\frac{\overline{X}-\mu}{S/\sqrt{n}}$ in the case that $X_1, \ldots, X_6$ are iid normal with mean 3 and standard deviation 4, and compare it to the pdf of the appropriate $t$ random variable.

```
tData <- replicate(10000, {
  X <- rnorm(6, 3, 4)
  (mean(X) - 3) / (sd(X) / sqrt(6))
})
hist(tData,
  probability = TRUE,
  ylim = c(0, .37),
  main = "Histogram of t",
  xlab = "t"
)
curve(dt(x, df = 5), add = TRUE, col = "red")
```

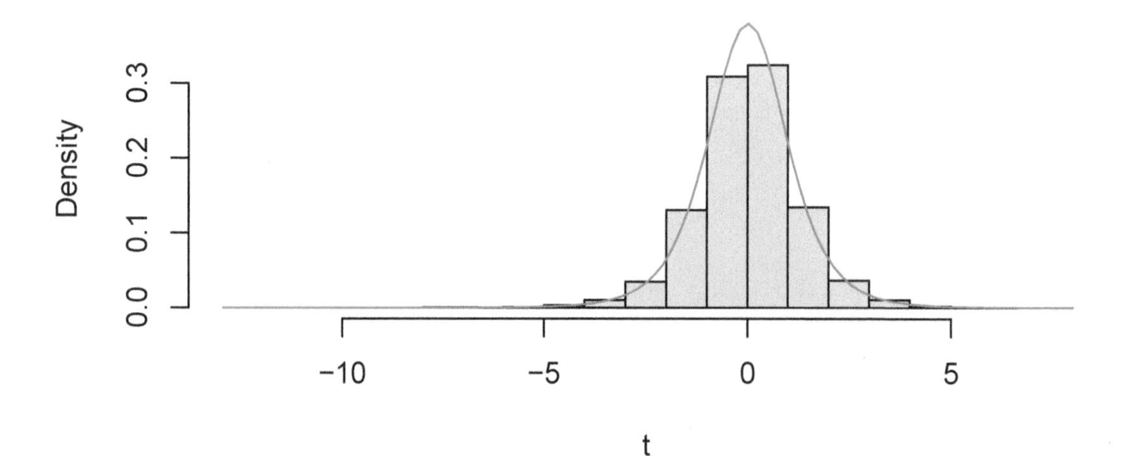


We see a very good agreement.

### 5.5.3 The $F$ distribution

An $F$ *distribution* has the same density function as

$$F_{\nu_1,\nu_2} = \frac{\chi^2_{\nu_1}/\nu_1}{\chi^2_{\nu_2}/\nu_2}$$

when $\chi^2_{\nu_1}$ and $\chi^2_{\nu_2}$ are independent. We say $F$ has $\nu_1$ numerator degrees of freedom and $\nu_2$ denominator degrees of freedom.

One example of this type is:

$$\frac{S_1^2/\sigma_1^2}{S_2^2/\sigma_2^2}$$

where $X_1, \ldots, X_{n_1}$ are iid normal with standard deviation $\sigma_1$ and $Y_1, \ldots, Y_{n_2}$ are iid normal with standard deviation $\sigma_2$.

### 5.5.4 Summary

This chapter has introduced the key variables associated with sampling distributions. Here, we summarize the relationships among these variables for ease of reference.

Suppose $X$ is normal with mean $\mu$ and sd $\sigma$. Then:

- $aX + b$ is normal with mean $a\mu + b$ and standard deviation $a\sigma$.
- $\frac{X - \mu}{\sigma} \sim Z$, where $Z$ is standard normal.

Given a random sample $X_1, \ldots, X_n$ with mean $\mu$ and standard deviation $\sigma$,

- $\frac{\overline{X} - \mu}{\sigma/\sqrt{n}} \sim Z$ if the $X_i$ are normal.
- $\lim_{n\to\infty} \frac{\overline{X} - \mu}{\sigma/\sqrt{n}} \sim Z$ if the $X_i$ have finite variance (the Central Limit Theorem).

To understand the sample mean $\overline{X}$ and standard deviation $S$, we introduced:

- $\chi^2_\nu$ the chi-squared random variable with $\nu$ degrees of freedom.
- $t_\nu$ the $t$ random variable with $\nu$ degrees of freedom.
- $F_{\nu,\mu}$ the $F$ random variable with $\nu$ and $\mu$ degrees of freedom.

These satisfy:

- $Z^2 \sim \chi^2_1$.
- $\chi^2_\nu + \chi^2_\eta \sim \chi^2_{\nu+\eta}$ if independent.
- $\frac{(n-1)S^2}{\sigma^2} \sim \chi^2_{n-1}$ when $X_1, \ldots, X_n$ iid normal.
- $\frac{Z}{\sqrt{\chi^2_\nu/\nu}} \sim t_\nu$ if independent.
- $\frac{\overline{X} - \mu}{S/\sqrt{n}} \sim t_{n-1}$ when $X_1, \ldots, X_n$ iid normal.
- $\frac{\chi^2_\nu/\nu}{\chi^2_\mu/\mu} \sim F_{\nu,\mu}$ if independent.
- $\frac{S_1^2/\sigma_1^2}{S_2^2/\sigma_2^2} \sim F_{n_1-1,n_2-1}$ if $X_1, \ldots, X_{n_1}$ iid normal with sd $\sigma_1$ and $Y_1, \ldots, Y_{n_2}$ iid normal with sd $\sigma_2$.

## 5.6 Point estimators

This section explores the definition of sample mean and (especially) sample variance, in the more general context of point estimators and their properties.

Suppose that a random variable $X$ depends on some parameter $\theta$. In many instances we do not know $\theta$, but are interested in estimating it from observations of $X$.

**Definition 5.7.** If $X_1, \ldots, X_n$ are iid with the same distribution as $X$, then a *point estimator* for $\theta$ is a function $\hat{\theta} = \hat{\theta}(X_1, \ldots, X_n)$.

Typically the value of $\hat{\theta}$ represents our best estimate of the parameter $\theta$, based on the data. Note that $\hat{\theta}$ is a random variable, and since $\hat{\theta}$ is derived from $X_1, \ldots, X_n$, $\hat{\theta}$ is a sample statistic and has a sampling distribution.

The two point estimators that we examine in this section are the sample mean and sample variance. The sample mean $\overline{X}$ is a point estimator for the *mean* of a random variable, and the sample variance $S^2$ is a point estimator for the *variance* of an rv.

Suppose we want to understand a random variable $X$. We collect a random sample of data $x_1, \ldots, x_n$ from the $X$ distribution. We then imagine a new discrete random variable $X_0$ which samples uniformly from these $n$ data points.

The variable $X_0$ satisfies $P(X_0 = x_i) = 1/n$, and so its mean is

$$E[X_0] = \frac{1}{n}\sum_{i=1}^{n} x_i = \overline{x}.$$

Since $X_0$ incorporates everything we know about $X$, it is reasonable to use the mean of $X_0$ as a point estimator for $\mu$, the mean of $X$:

$$\hat{\mu} = \overline{X} = \frac{1}{n}\sum_{i=1}^{n} X_i$$

Continuing in the same way,

$$\sigma^2 = E[(X_0 - \mu)^2] = \frac{1}{n}\sum_{i=1}^{n}(x_i - \mu)^2$$

This works just fine *as long as $\mu$ is known*. However, most of the time, we do not know $\mu$ and we must replace the true value of $\mu$ with our estimate from the data. There is a heuristic "degrees of freedom" that each time you replace a parameter with an estimate of that parameter, you divide by one less. Following that heuristic, we obtain

$$\hat{\sigma}^2 = S^2 = \frac{1}{n-1}\sum_{i=1}^{n}(x_i - \overline{x})^2$$

### 5.6.1 Properties of point estimators

The point estimators $\overline{X}$ for $\mu$ and $S^2$ for $\sigma^2$ are random variables themselves, since they are computed using a random sample from a distribution. As such, they also have distributions, means and variances. One desirable property of point estimators is that they be *unbiased.*

**Definition 5.8.** A point estimator $\hat{\theta}$ for $\theta$ is an *unbiased estimator* if $E[\hat{\theta}] = \theta$.

Intuitively, "unbiased" means that the estimator does not consistently underestimate or overestimate the parameter it is estimating. If we were to estimate the parameter over and over again, the average value of the estimates would converge to the correct value of the parameter.

**Example 5.22.** Let $X_1, \ldots, X_5$ be a random sample from a normal distribution with mean 1 and variance 4. Compute $\overline{X}$ based on the random sample.

```
mean(rnorm(5, 1, 2))
```

```
## [1] 0.866061
```

You can see the estimate is pretty close to the correct value of 1, but not exactly 1. However, if we repeat this experiment 10,000 times and take the average value of $\overline{x}$, that should be very close to 1.

```
mean(replicate(10000, mean(rnorm(5, 1, 2))))
```

```
## [1] 1.012652
```

**Proposition 5.3.** *For a random sample from a population with mean $\mu$, the sample mean $\overline{X}$ is an unbiased estimator of $\mu$.*

*Proof.* Let $X_1, \ldots, X_n$ be a random sample from the population, so $E[X_i] = \mu$. Then

$$E[\overline{X}] = E\left[\frac{X_1 + \cdots + X_n}{n}\right] = \frac{1}{n}\left(E[X_1] + \cdots + E[X_n]\right) = \mu.$$

■

Now consider the sample variance $S^2$. The formula

$$S^2 = \frac{1}{n-1}\sum_{i=1}^{n}(X_i - \overline{X})^2$$

contains a somewhat surprising division by $n - 1$, since the sum has $n$ terms. We explore this with simulation.

**Example 5.23.** Let $X_1, \ldots, X_5$ be a random sample from a normal distribution with mean 1 and variance 4. Estimate the variance using $S^2$.

```
sd(rnorm(5, 1, 2))^2
```

```
## [1] 3.393033
```

We see that our estimate is not ridiculous, but not the best either. Let's repeat this 10,000 times and take the average. (We are estimating $E[S^2]$.)

```
mean(replicate(10000, sd(rnorm(5, 1, 2))^2))
```

```
## [1] 3.992801
```

> **Try It Yourself.**
> Examples 5.22 and 5.23 use simulations to estimate the expected value of the random variables $\overline{X}$ and $S^2$. Repeat these simulations a few times and convince yourself that they are not consistently overestimating or underestimating the correct value.

**Proposition 5.4.** *For a random sample from a population with mean $\mu$ and variance $\sigma^2$, the sample variance $S^2$ is an unbiased estimator of $\sigma^2$.*

The proof of Proposition 5.4 is left as Exercise 5.46.

**Example 5.24.** In this example, we see that dividing by $n$ would lead to a *biased* estimator. That is, $\frac{1}{n}\sum_{i=1}^n (X_i - \overline{X})^2$ is *not* an unbiased estimator of $\sigma^2$.

We use replicate twice. On the inside, it is running 10,000 simulations. On the outside, it is repeating the overall simulation 7 times.

```
replicate(
  7,
  mean(replicate(10000, {
    X <- rnorm(5, 1, 2)
    1 / 5 * sum((X - mean(X))^2)
  }))
)
```

```
## [1] 3.242466 3.212477 3.231688 3.187677 3.157526 3.218117 3.227146
```

In each of the 7 trials, $\frac{1}{n}\sum_{i=1}^n (X_i - \overline{X})^2$ underestimates the correct value $\sigma^2 = 4$ by quite a bit.

### 5.6.2 Variance of unbiased estimators

We showed that $\overline{X}$ and $S^2$ are unbiased estimators for $\mu$ and $\sigma^2$, respectively. However, there are other reasonable estimators for the mean and variance.

The *median* of a collection of numbers is the middle number, after sorting (or the average of the two middle numbers). When a population is symmetric about its mean $\mu$, the median is an unbiased estimator of $\mu$. If the population is not symmetric, the median is not typically an unbiased estimator for $\mu$, see Exercise 5.41 and Exercise 5.42.

If $X_1, \ldots, X_n$ is a random sample from a normal rv, then the median of the $X_i$ is an unbiased estimator for the mean. Moreover, the median seems like a perfectly reasonable thing to use to estimate $\mu$, and in many cases is actually preferred to the mean.

There is one way, however, in which the sample mean $\overline{X}$ is definitely better than the median, and that is that the mean has a lower variance than the median. This means that $\overline{X}$ does not deviate from the true mean as much as the median will deviate from the true mean, as measured by variance.

Let's do a simulation to illustrate. Suppose you wish to compute the mean $\overline{X}$ based on a random sample of size 11 from a normal rv with mean 0 and standard deviation 1. We estimate the variance of $\overline{X}$.

```
var(replicate(10000, mean(rnorm(11, 0, 1))))
```

```
## [1] 0.08961579
```

Now we repeat for the median.

```
var(replicate(10000, median(rnorm(11, 0, 1))))
```

```
## [1] 0.1400155
```

We see that the variance of the mean is lower than the variance of the median. This is a good reason to use the sample mean to estimate the mean of a normal rv rather than using the median, absent other considerations such as outliers.

### 5.6.3 MSE and bias-variance decomposition

Let $\hat{\theta}$ be a point estimator for $\theta$. One way to quantify how well $\hat{\theta}$ estimates $\theta$ is to compute the *mean squared error* (MSE) given by the formula

$$\text{MSE}(\hat{\theta}) = E[(\hat{\theta} - \theta)^2].$$

The *bias* of the point estimator $\hat{\theta}$ is defined to be the difference between the expected value of $\hat{\theta}$ and $\theta$,

$$\text{Bias}(\hat{\theta}) = E[\hat{\theta}] - \theta.$$

In particular, $\hat{\theta}$ is unbiased exactly when $\text{Bias}(\hat{\theta}) = 0$.

**Theorem 5.6.** *Let $\hat{\theta}$ be a point estimator for $\theta$.*

$$\text{MSE}(\hat{\theta}) = \left(\text{Bias}(\hat{\theta})\right)^2 + \text{Var}(\hat{\theta})$$

*Proof.* Note that $\text{Bias}(\hat{\theta}) = E[\hat{\theta}] - \theta$ is a constant. Then

$$\begin{aligned}
E\left[\left(\hat{\theta} - \theta\right)^2\right] &= E\left[\left(\hat{\theta} - E[\hat{\theta}] + E[\hat{\theta}] - \theta\right)^2\right] \\
&= E\left[\left(\hat{\theta} - E[\hat{\theta}] + \text{Bias}(\hat{\theta})\right)^2\right] \\
&= E\left[\left(\hat{\theta} - E[\hat{\theta}]\right)^2\right] + 2E\left[\left(\hat{\theta} - E[\hat{\theta}]\right) \cdot \text{Bias}(\hat{\theta})\right] + E\left[\text{Bias}(\hat{\theta})^2\right] \\
&= \text{Var}(\hat{\theta}) + 2\text{Bias}(\hat{\theta})E[\left(\hat{\theta} - E[\hat{\theta}]\right)] + \left(\text{Bias}(\hat{\theta})\right)^2 \\
&= \left(\text{Bias}(\hat{\theta})\right)^2 + \text{Var}(\hat{\theta})
\end{aligned}$$

since $E\left[\hat{\theta} - E[\hat{\theta}]\right] = 0$. ■

**Example 5.25.** Suppose a population is normal with mean 1 and standard deviation 2, and we take a random sample of size $n = 10$. Use simulation to calculate the MSE of the sample standard deviation $S$ as a point estimate of $\sigma$.

We solve this problem in two ways. First, by estimating the MSE directly, then by computing the variance and bias of $S$ and applying Theorem 5.6.

We start by estimating the MSE directly.

```
S <- replicate(10000, {
  x <- rnorm(10, 1, 2)
  sd(x)
})
mean((S - 2)^2)
```

```
## [1] 0.2173475
```

Now, we estimate the variance and bias separately and add the bias squared to the variance.

```
bias_estimate <- mean(replicate(10000, {
  x <- rnorm(10, 1, 2)
  s <- sd(x)
  (s - 2)
}))
variance_estimate <- var(replicate(10000, {
  x <- rnorm(10, 1, 2)
  s <- sd(x)
  s
}))
variance_estimate + bias_estimate^2
```

```
## [1] 0.2169428
```

We see that we get approximately the same answer for the estimate of the MSE, which confirms Theorem 5.6.

It is often possible to replace an unbiased estimator with a biased estimator that has lower variance and lower mean squared error. This is the *bias-variance tradeoff*.

For example, the sample variance $S^2$ is an unbiased estimator for $\sigma^2$. What if we divide by $n$ instead of $n-1$ in the formula for sample variance?

$$\hat{\sigma}^2 = \frac{1}{n}\sum_{i=1}^{n}\left(X_i - \overline{X}\right)^2.$$

Division by $n$ makes $\hat{\sigma}^2$ smaller than $S^2$. This means that the variance of $\hat{\sigma}^2$ will be less than the variance of $S^2$. On the other hand, since $S^2$ is an unbiased estimator of $\sigma^2$, our new estimator $\hat{\sigma}^2$ will be consistently smaller than $\sigma^2$. It it no longer unbiased.

The smaller variance offsets the bias, and in fact $\hat{\sigma}^2$ has a better MSE when estimating $\sigma^2$ than $S^2$ does.

**Example 5.26.** Let $X_1, \ldots, X_{20}$ be iid normal with mean 0 and standard deviation $\sigma = 2$. Compare the MSE of $S^2$ and $\hat{\sigma}^2$ when estimating $\sigma^2 = 4$.

First, estimate the MSE of $S^2$:

```
S2 <- replicate(10000, {
  x <- rnorm(20, 0, 2)
  var(x)
})
mean((S2 - 4)^2)
```

```
## [1] 1.713287
```

Since $S^2$ is unbiased, its MSE is just its variance (Theorem 5.6)

```
var(S2)
```

```
## [1] 1.713379
```

We get slightly different estimates when we estimate the variance of $S^2$ or when we estimate the MSE directly, but either way we get an estimate for the MSE of about 1.71.

For $\hat{\sigma}^2$, we can estimate the variance, bias and MSE:

```
hat_sigma2 <- replicate(10000, {
  x <- rnorm(20, 0, 2)
  xbar <- mean(x)
  1 / 20 * sum((x - xbar)^2)
})
var(hat_sigma2)

## [1] 1.576474
mean(hat_sigma2 - 4) # estimated bias

## [1] -0.1638332
var(hat_sigma2) + mean(hat_sigma2 - 4)^2 # MSE = var + bias^2

## [1] 1.603315
mean((hat_sigma2 - 4)^2) # direct estimate of MSE

## [1] 1.603157
```

The estimated variance of $\hat{\sigma}^2$, 1.58 is lower than the estimated variance of $S^2$, which was 1.71. The bias of $\hat{\sigma}^2$ is negative, since it consistently underestimates $\sigma^2$.

Overall, the MSE of $\hat{\sigma}^2$ is lower than the MSE of $S^2$.

---

## Vignette: Stein's paradox

In this book, we mainly consider random variables whose probability density functions are one-dimensional. When working with multiple random variables at once, their joint probability distribution can be thought of as a multivariate function. In this vignette, we imagine three random variables as together determining a distribution on three-dimensional space, and provide a shocking fact about estimating its mean.

Let's start with a normal distribution with standard deviation 1 and unknown mean $\mu$. Suppose we are only allowed a single sample $x$ from it. What would the best guess of the mean be? Well, it seems clear that it would be tough to beat $x$.

We will measure the error of an estimate by taking the difference between the estimate and the true value $\mu$ and then squaring it. Here is an example when the true mean $\mu$ is zero:

```
mu <- 0
x <- rnorm(1, mu, 1)
err <- x - mu
err^2

## [1] 1.457008
```

In this case, $(x - \mu)^2 = 1.457$. If we replicate this a bunch of times and take the average, we get the mean squared error of the estimate,

```
mu <- 0
err <- replicate(10000, {
  x <- rnorm(1, mu, 1)
  estimate <- x
  estimate - mu
})
mean(err^2)
```

```
## [1] 0.9753385
```

We see that the mean squared error is about 0.98 when estimating the true mean as $x$ given one data point $x$. What if instead we try estimating the true mean by $x/2$?

```
mu <- 0
err <- replicate(10000, {
  x <- rnorm(1, mu, 1)
  estimate <- x / 2
  estimate - mu
})
mean(err^2)
```

```
## [1] 0.2529804
```

It's a better estimate, since the mean squared error is lower. However, that trick only works when $\mu = 0$, because dividing by two moves every number closer to zero. Using $x/2$ to estimate the mean is worse when the true mean is $\mu = 5$. Here is an estimate of the mean of a normal distribution with $\mu = 5$ using $x$:

```
mu <- 5
err <- replicate(10000, {
  x <- rnorm(1, mu, 1)
  estimate <- x
  estimate - mu
})
mean(err^2)
```

```
## [1] 0.9985144
```

And now an estimate using $x/2$:

```
mu <- 5
err <- replicate(10000, {
  x <- rnorm(1, mu, 1)
  estimate <- x / 2
  estimate - mu
})
mean(err^2)
```

```
## [1] 6.47556
```

Using our single data point $x$ to estimate the mean works much better than using $x/2$ here.

In general, if you want to estimate the unknown mean of a normal random variable, and you only have one observation $x$, the best you can do is to guess that the mean is $x$.

*Stein's paradox* involves the three-dimensional version of this estimation problem. Imagine you have three independent normal random variables $X$, $Y$, $Z$, with standard deviation 1 and unknown means $\mu, \nu, \lambda$. Take single observations $x$ of $X$, $y$ of $Y$, and $z$ of $Z$, respectively. This determines a point $\mathbf{p} = (x, y, z)$ in three-dimensional Euclidean space. The point $\mathbf{p}$ is more likely to be near the point $(\mu, \nu, \lambda)$, and less likely to be further away.

We ask: what is the best estimate for the three means? Is the point $\mathbf{p}$?

Let's explore with a simulation where the means are $\mu = 1$, $\nu = 2$, $\lambda = -1$.

```
means <- c(1, 2, -1)
err <- replicate(10000, {
  p <- rnorm(3, means, 1)
  estimate <- p
  estimate - means
})
mean(err^2)
```

```
## [1] 1.003518
```

When estimating the means with the point $\mathbf{p}$ itself, the mean squared error is about 1. Actually it's always exactly 1 whatever the means are.

Here's another way to estimate, called the James-Stein estimator. We take the sample point $\mathbf{p}$ and move it towards the origin by a distance of the reciprocal of $\|\mathbf{p}\|$. Mathematically, our estimate for the means will be

$$\mathbf{p} - \frac{1}{\|\mathbf{p}\|}\frac{\mathbf{p}}{\|\mathbf{p}\|} = \left(1 - \frac{1}{\|\mathbf{p}\|^2}\right)\mathbf{p}$$

Here's the James-Stein estimate in action:

```
means <- c(1, 2, -1)
err <- replicate(10000, {
  p <- rnorm(3, means, 1)
  estimate <- (1 - 1 / sum(p^2)) * p
  estimate - means
})
mean(err^2)
```

```
## [1] 0.9214803
```

We see that, in this case, the James-Stein estimator has a lower mean squared error than the ordinary estimator $\mathbf{p}$!

> **Try It Yourself.**
> Compare the James-Stein estimator with the ordinary estimator for a variety of means $(\mu, \nu, \lambda)$. You will see (though you might need to take a **lot** of samples) that the James-Stein estimator is always at least as good as the ordinary estimator no matter what the means are.

The fact that the James-Stein estimator is better than the ordinary estimator is Stein's paradox. In general, if you are trying to estimate a bunch of variables (more than 2), then it is not necessarily best to estimate each variable separately as well as possible.

## Exercises

---

Exercises 5.1 – 5.2 require material through Section 5.1.

**5.1.** Let $Z$ be a standard normal random variable. Estimate via simulation $P(Z^2 < 2)$.

**5.2.** Let $X$ and $Y$ be independent exponential random variables with rate 3. Let $Z = \max(X, Y)$ be the maximum of $X$ and $Y$.

a. Estimate via simulation $P(Z < 1/2)$.
b. Estimate the mean and standard deviation of $Z$.

---

Exercises 5.3 – 5.13 require material through Section 5.2.

**5.3.** Five coins are tossed and the number of heads $X$ is counted. Estimate via simulation the pmf of $X$.

**5.4.** Three dice are tossed and their sum $X$ is observed. Use simulation to estimate and plot the pmf of $X$.

**5.5.** Seven balls numbered 1-7 are in an urn. Two balls are drawn from the urn without replacement and the sum $X$ of the numbers is computed.

a. Estimate via simulation the pmf of $X$.
b. What are the least likely outcomes of $X$?

**5.6.** Five six-sided dice are tossed and their **product** is observed. Use the estimated pmf to find the most likely outcome. (The R function `prod` computes the product of a vector.)

**5.7.** Fifty people put their names in a hat. They then all randomly choose one name from the hat. Let $X$ be the number of people who get their own name. Estimate and plot the pmf of $X$.

**5.8.** Consider an experiment where 20 balls are placed randomly into 10 urns. Let $X$ denote the number of urns that are empty.

a. Estimate via simulation the pmf of $X$.
b. What is the most likely outcome?
c. What is the least likely outcome that has positive probability?

**5.9 (Hard).** Suppose 6 people, numbered 1-6, are sitting at a round dinner table with a big plate of spaghetti, and a bag containing 5005 red marbles and 5000 blue marbles. Person 1 takes the spaghetti, serves themselves, and chooses a marble. If the marble is red, they pass the spaghetti to the left (person number 2). If it is blue, they pass the spaghetti to the right (person number 6). The guests continue doing this until the last person receives the plate of spaghetti. Let $X$ denote the number of the person holding the spaghetti at the end of the experiment. Estimate the pmf of $X$.

**5.10.** Suppose you roll a die until the first time you obtain an even number. Let $X_1$ be the total number of times you roll a 1, and let $X_2$ be the total number of times that you roll a 2.

a. Is $E[X_1] = E[X_2]$? (Hint: Use simulation. It is **extremely** unlikely that you will roll 30 times before getting the first even number, so you can safely assume that the first even occurs inside of 30 rolls.)

b. Is the pmf of $X_1$ the same as the pmf of $X_2$?
c. Estimate the variances of $X_1$ and $X_2$.

**5.11.** Simulate creating independent uniform random numbers in [0,1] and summing them until your cumulative sum is larger than 1. Let $N$ be the random variable which is how many numbers you needed to sample. For example, if your numbers were 0.35, 0.58, 0.22 you would have $N = 3$ since the sum exceeds 1 after the third number. What is the expected value of $N$?

**5.12.** Recall Example 5.8 in the text, where we show that the most likely number of times you have more heads than tails when a coin is tossed 100 times is zero. Suppose you toss a coin 100 times.

a. Let $X$ be the number of times in the 100 tosses that you have exactly the same number of heads as tails. Estimate the expected value of $X$.
b. Let $Y$ be the number of tosses for which you have more heads than tails. Estimate the expected value of $Y$.

**5.13.** Suppose there are two candidates in an election. Candidate A receives 52 votes and Candidate B receives 48 votes. You count the votes one at a time, keeping a running tally of who is ahead. At each point, either A is ahead, B is ahead, or they are tied. Let $X$ be the number of times that Candidate B is ahead in the 100 tallies.

a. Estimate the pmf of $X$ and plot it.
b. Estimate $P(X > 50)$.

---

Exercises 5.14 – 5.29 require material through Section 5.3.

**5.14.** Let $X$ and $Y$ be independent uniform random variables on the interval $[0, 1]$. Estimate the pdf of $X + Y$ and plot it.

**5.15.** Let $X$ and $Y$ be independent uniform random variables on the interval $[0, 1]$. Let $Z$ be the maximum of $X$ and $Y$.

a. Plot the pdf of $Z$.
b. From your answer to (a), decide whether $P(0 \leq Z \leq 1/3)$ or $P(1/3 \leq Z \leq 2/3)$ is larger.

**5.16.** Let $X$ and $Y$ be independent normal random variables with means $\mu_X = 0, \mu_Y = 8$ and standard deviations $\sigma_X = 3$ and $\sigma_Y = 4$.

a. What are the mean and variance of $X + Y$?
b. Simulate the distribution of $X + Y$ and plot it. Add a normal pdf to your plot with mean and standard deviation to match the density of $X + Y$.
c. What are the mean and standard deviation of $5X - Y/2$?
d. Simulate the distribution of $5X - Y/2$ and plot it. Add a normal pdf to your plot with mean and standard deviation to match the density of $5X - 2Y$.

**5.17.** Example 5.17 plotted the probability density function for the sum of 20 independent exponential random variables with rate 2. The density appeared to be approximately normal.

a. What are $\mu$ and $\sigma$ for the exponential distribution with rate 2?

b. What should $\mu$ and $\sigma$ be for the sum of 20 independent rvs which are exponential with rate 2?
c. Plot the density for the sum of exponentials and add the pdf of Normal$(\mu, \sigma)$ to your plot. How well does it seem to fit?

**5.18.** Estimate the value of $a$ such that $P(a \leq Y \leq a+1)$ is maximized when $Y$ is the *maximum* value of two independent exponential random variables with mean 2.

**5.19.** Is the product of two normal rvs normal? Estimate via simulation the pdf of the product of $X$ and $Y$, when $X$ and $Y$ are independent normal random variables with mean 0 and standard deviation 1. How can you determine whether $XY$ is normal? (Hint: you will need to estimate the mean and standard deviation.)

**5.20.** The minimum of two independent exponential rvs with mean 2 is an exponential rv. Use simulation to determine what the rate is.

**5.21.** The sum of two independent chi-squared rvs with 2 degrees of freedom is either exponential or chi-squared. Which one is it? What is the parameter associated with your answer?

**5.22.** Richard Feynman said, "I couldn't claim that I was smarter than sixty-five other guys–but the average of sixty-five other guys, certainly!" Assume that "intelligence" is normally distributed with mean 0 and standard deviation 1 and that the 65 people Feynman is referring to are drawn at random.

a. Estimate via simulation the pdf of the maximum intelligence from among 65 people.
b. Estimate via simulation the pdf of the mean intelligence of 65 people.
c. (Open-ended) About how many standard deviations above the mean in intelligence did Feynman think he was?

**5.23.** Consider the Log Normal distribution, whose root in R is `lnorm`. The log normal distribution takes two parameters, `meanlog` and `sdlog`.

a. Graph the pdf of a Log Normal random variable with `meanlog = 0` and `sdlog = 1`. The pdf of a Log Normal rv is given by `dlnorm(x, meanlog, sdlog)`.
b. Let $X$ be a Log Normal rv with `meanlog = 0` and `sdlog = 1`. Estimate via simulation the density of $\log(X)$, and compare it to a standard normal random variable.

**5.24.** The *beta distribution* plays an important role in Bayesian statistics. It has two parameters, called *shape parameters.*

Let $X$ and $Y$ be independent uniform rvs on the interval $[0, 1]$. Estimate via simulation the pdf of the maximum of $X$ and $Y$, and compare it to the pdf of a beta distribution with parameters `shape1 = 2` and `shape2 = 1`. (Use `dbeta()`.)

**5.25.** Estimate the density of the maximum of 7 independent uniform random variables on the interval $[0, 1]$. This density is also beta. What shape parameters are required for the beta distribution to match your estimated density?

**5.26.** Let $X_1, \ldots, X_7$ be iid uniform random variables on the interval $[0, 1]$, and let $Y$ be the **second largest** of the $X_1, \ldots, X_7$.

a. Estimate the pdf of $Y$.
b. This pdf is also beta. What are the shape parameters?

**5.27.** Let $X_1$ and $X_2$ be independent gamma random variables with rate 2 and shapes 3 and 4. Confirm via simulation that $X_1 + X_2$ is gamma with rate 2 and shape 7.

**5.28.** Plot the pdf of a gamma random variable with shape 1 and rate 3, and confirm that is the same pdf as an exponential random variable with rate 3.

**5.29.** If $X$ is a gamma random variable with rate $\beta$ and shape $\alpha$, use simulation to show that $\frac{1}{c}X$ is a gamma random variable with rate $c\beta$ and shape $\alpha$ for two different choices of positive $\alpha$, $\beta$, and $c$.

---

Exercises 5.30 – 5.36 require material through Section 5.4.

**5.30.** Let $X_1, \ldots, X_n$ be independent uniform rvs on the interval $(0, 1)$.

a. What are the mean $\mu$ and the sd $\sigma$ of a uniform rv on the interval $(0, 1)$?
b. How large does $n$ need to be before the pdf of $\frac{\overline{X}-\mu}{\sigma/\sqrt{n}}$ is approximately that of a standard normal rv?

(Note: there is no "right" answer, as it depends on what is meant by "approximately." You should try various values of $n$ until you find the lowest one where the estimate of the $\overline{X}$ distribution is still close to standard normal.)

**5.31.** Let $X_1, \ldots, X_n$ be independent exponential rvs with rate $\lambda = 10$.

a. What are the mean $\mu$ and the sd $\sigma$ of an exponential rv with rate 10?
b. How large does $n$ need to be before the pdf of $\frac{\overline{X}-\mu}{\sigma/\sqrt{n}}$ is approximately that of a standard normal rv?

**5.32.** Let $X_1, \ldots, X_n$ be independent chi-squared rvs with 2 degrees of freedom.

a. What are the mean $\mu$ and the sd $\sigma$ of a chi-squared rv with 2 degrees of freedom?
b. How large does $n$ need to be before the pdf of $\frac{\overline{X}-\mu}{\sigma/\sqrt{n}}$ is approximately that of a standard normal rv?

**5.33.** Let $X_1, \ldots, X_n$ be independent binomial rvs with $n = 10$ and $p = 0.8$.

a. What are the mean $\mu$ and the sd $\sigma$ of the Binom$(10, 0.8)$ distribution?
b. How large does $n$ need to be before the pdf of $\frac{\overline{X}-\mu}{\sigma/\sqrt{n}}$ is approximately that of a standard normal rv?

**5.34.** Let $X_1, \ldots, X_{1000}$ be a random sample from a uniform random variable on the interval $[-1, 1]$. Suppose this sample is contaminated by a single outlier that is chosen uniformly in the interval $[200, 300]$, so that there are in total 1001 samples. Plot the estimated pdf of $\overline{X}$ and give convincing evidence that it is not normal.

**5.35.** The Central Limit Theorem describes the distribution for the **mean** of iid random variables. This exercise explores the distribution of the **median** of iid random variables. Suppose $X_1, \ldots, X_n$ are iid continuous random variables with pdf $f(x)$, median $\theta$, and with the additional condition that the cdf $F$ has derivative greater than 0 at the median, i.e., $F'(\theta) > 0$. Then for $n$ large, the median is approximately normal with mean $\theta$ and standard deviation $1/(2\sqrt{n}f(\theta))$. You will check this for some examples using simulation. The R function to compute the median of a sample is `median`.

a. Let $X_1, \ldots, X_n$ be independent uniform random variables on $[-1, 1]$. The median is 0 and $f(0) = 1/2$. Show that for large $n$, the median of $X_1, \ldots, X_n$ is approximately normal with mean 0 and standard deviation $1/\sqrt{n}$.
b. Let $X_1, \ldots, X_n$ be iid exponential random variables with rate 1. Find an $n$ so that the pdf of the *median* of $X_1, \ldots, X_n$ is approximately normally distributed. The true median of an exponential random variable with rate 1 is $\ln 2$.

c. Let $X_1, \ldots, X_n$ be iid binomial random variables with size 3 and $p = 1/2$. Note this does **not** satisfy the hypotheses that would guarantee that the median is approximately normal. Choose $n = 10, 100$ and $n = 1000$, and see whether the median appears to be approximately normally distributed as $n$ gets bigger. The true median of a binomial with parameters 3 and 1/2 is 1.5.

**5.36.** In this exercise, we investigate the importance of the assumption of finite mean and variance in the statement of the Central Limit Theorem. Let $X_1, \ldots, X_n$ be iid random variables with a $t$ distribution with one degree of freedom, also called the *Cauchy distribution*. You can sample from such a $t$ random variable using `rt(N, df = 1)`.

a. Use `dt(x,1)` to plot the pdf of a $t$ random variable with one degree of freedom.
b. Confirm for $N = 100, 1000$ or $10000$ that `mean(rt(N, 1))` does not give consistent results. This is because $\int_{-\infty}^{\infty} |x| \mathrm{dt}(x, 1)\, dx = \infty$, so the mean of a $t$ random variable with 1 degree of freedom does not exist.
c. Estimate by simulation the pdf of $\overline{X}$ for $N = 100, 1000, 10000$. To visualize this distribution, use a histogram with `breaks = c(-Inf, -20:20, Inf)` and `xlim = c(-20,20)`. Check by adding a curve that $\overline{X}$ has the $t$ distribution with 1 df no matter what $N$ you choose.
d. Does the Central Limit Theorem hold for this distribution?

---

Exercises 5.37 – 5.40 require material through Section 5.5.

**5.37.** Let $X_1, \ldots, X_9$ be independent identically distributed normal random variables with mean 2 and standard deviation 3. For what constants $a$ and $b$ is

$$\frac{\overline{X} - a}{b}$$

a standard normal random variable?

**5.38.** Let $X_1, \ldots, X_8$ be independent normal random variables with mean 2 and standard deviation 3. Show using simulation that

$$\frac{\overline{X} - 2}{S/\sqrt{8}}$$

is a $t$ random variable with 7 degrees of freedom.

**5.39.** Let $X_1, \ldots, X_{10}$ and $Y_1, \ldots, Y_{20}$ be independent normal random variables with means $\mu_{X_i} = 0, \mu_{Y_i} = 1$ and standard deviations $\sigma_X = 1$ and $\sigma_Y = 2$. Show using simulation that

$$\frac{S_X^2/1^2}{S_Y^2/2^2}$$

is an $F$ random variable with 9 and 19 degrees of freedom.

**5.40.** Prove Theorem 5.4 in the case $n = 2$. That is, suppose that $X_1$ and $X_2$ are independent, identically distributed normal random variables. Prove that

$$\frac{(n-1)S^2}{\sigma^2} = \frac{1}{\sigma^2}\sum_{i=1}^{2}(X_i - \overline{X})^2$$

is a $\chi^2$ random variable with one degree of freedom.

---

Exercises 5.41 – 5.47 require material through Section 5.6.

**5.41.** In this exercise, we show through simulation that the median is an unbiased estimator for the population mean when the population is symmetric.

a. Use 20 replicates to estimate the expected value of the median of five independent normal random variables with mean 2 and standard deviation 4. Repeat 5 times and write down the 5 values.
b. Use 1000 replicates to estimate the expected value of the median of five independent normal random variables with mean 2 and standard deviation 4. Repeat 5 times and write down the 5 values.
c. Use 20000 replicates to estimate the expected value of the median of five independent normal random variables with mean 2 and standard deviation 4. Repeat 5 times and write down the 5 values.
d. Does it appear that the estimate for the expected value of the median is getting closer to 2 as the number of replicates increases?

**5.42.** Show through simulation that the median is a biased estimator for the mean of an exponential rv with $\lambda = 1$. Assume a random sample of size 8.

**5.43.** Determine via simulation whether $\frac{1}{n}\sum_{i=1}^n (x_i - \mu)^2$ is a biased estimator for $\sigma^2$ when $n = 10$ and $x_1, \ldots, x_{10}$ is a random sample from a normal rv with mean 1 and variance 9.

**5.44.** Show through simulation that $S^2 = \frac{1}{n-1}\sum_{i=1}^n \left(x_i - \overline{x}\right)^2$ is an unbiased estimator for $\sigma^2$ when $x_1, \ldots, x_{10}$ are iid normal rvs with mean 0 and variance 4.

**5.45.** Show through simulation that $S = \sqrt{S^2}$, where $S$ is defined in the previous problem, is a *biased* estimator for $\sigma$ when $x_1, \ldots, x_{10}$ are iid normal random variables with mean 0 and variance 4, i.e., standard deviation 2.

**5.46.** This exercise walks step-by-step through the mathematical proof of Proposition 5.4, that $S^2$ is an unbiased estimator of the population variance $\sigma^2$. Suppose $X_1, \ldots, X_n$ are an independent random sample from a population with mean $\mu$ and variance $\sigma^2$.

a. Use Theorem 3.10 to show that $\text{Var}(\overline{X}) = \sigma^2/n$.
b. Use the computational formula for variance, Theorem 3.9, to show that $E[X_i^2] = \mu^2 + \sigma^2$ and that $E[\overline{X}^2] = \mu^2 + \sigma^2/n$.
c. Show that $E[\sum_i X_i \overline{X}] = n\mu^2 + \sigma^2$
d. Show that $E[\sum_i (X_i - \overline{X})^2] = E[\sum_i X_i^2] - 2E[\sum_i X_i \overline{X}] + E[\sum_i \overline{X}^2] = (n-1)\sigma^2$.
e. Conclude that $E[S^2] = \sigma^2$.

**5.47.** Determine whether $\hat{\sigma}_1^2 = \frac{1}{n}\sum_{i=1}^n \left(x_i - \overline{x}\right)^2$ or $\hat{\sigma}_2^2 = \frac{1}{n+1/2}\sum_{i=1}^n \left(x_i - \overline{x}\right)^2$ has lower mean squared error when estimating $\sigma^2$ when $x_1, \ldots, x_{10}$ are iid normal random variables with mean 0 and variance 4.

# 6

# *Data Manipulation*

In this chapter we introduce the *tidyverse*. The tidyverse consists of a collection of R packages designed to work together, organized around common principles. The central organizing principle is that data should be *tidy*: Data should be rectangular, each row should correspond to one observation, and each column should correspond to one observed variable. Data should not be stored in the names of variables.

The tidyverse tools we will use is in this chapter, in order of amount used, are the following:

`dplyr`
: Pronounced "dee - ply - er," this is the main package that we are learning about in this chapter.

`stringr`
: For basic *string* operations.

`tidyr`
: For the `pivot` functions.

`lubridate`
: For dealing with times and dates.

`janitor`
: To clean the names of variables in data frames.

To use these packages, you will need to install them. The simplest method is to install the entire `tidyverse` package, with

```
install.packages("tidyverse")
```

Then you can load all the tidyverse tools at once with

```
library(tidyverse)
```

Alternately, you can choose to be more selective and install and load each package as you need it. This chapter requires `dplyr`[1]:

```
library(dplyr)
```

The advantage to using individual packages is that the `tidyverse` package is large and has many effects you may not understand. Using `dplyr` by itself keeps change to a minimum, and it also helps you learn the correct location of the tools used in this chapter.

[1] Loading `dplyr` produces warning messages because the `dplyr` package contains commands that override some base R commands (notably the `filter` command). This book will not show these warnings. Use `suppressMessages(library(dplyr))` if you want to load the package quietly.

DOI: 10.1201/9781003004899-6

## 6.1 Data frames and tibbles

Data in R is usually stored in data frames, with one row per observation and one column per variable. The tidyverse tools work naturally with data frames but prefer a new data format called a *tibble*. Sometimes tidyverse tools will automatically convert data frames into tibbles. Or, you can make the conversion yourself using the `as_tibble` function:

```
as_tibble(mtcars)
```

```
## # A tibble: 32 x 11
##      mpg   cyl  disp    hp  drat    wt  qsec    vs    am  gear  carb
##    <dbl> <dbl> <dbl> <dbl> <dbl> <dbl> <dbl> <dbl> <dbl> <dbl> <dbl>
##  1  21       6  160    110  3.9   2.62  16.5     0     1     4     4
##  2  21       6  160    110  3.9   2.88  17.0     0     1     4     4
##  3  22.8     4  108     93  3.85  2.32  18.6     1     1     4     1
##  4  21.4     6  258    110  3.08  3.22  19.4     1     0     3     1
##  5  18.7     8  360    175  3.15  3.44  17.0     0     0     3     2
##  6  18.1     6  225    105  2.76  3.46  20.2     1     0     3     1
##  7  14.3     8  360    245  3.21  3.57  15.8     0     0     3     4
##  8  24.4     4  147.    62  3.69  3.19  20       1     0     4     2
##  9  22.8     4  141.    95  3.92  3.15  22.9     1     0     4     2
## 10  19.2     6  168.   123  3.92  3.44  18.3     1     0     4     4
## # ... with 22 more rows
```

As usual, this does not change `mtcars` into a tibble unless we store the result back into the variable via `mtcars <- as_tibble(mtcars)`. Most of the time, you don't need to worry about the difference between a data frame and a tibble, but tibbles have a couple of advantages over data frames:

1. Printing. By now you have probably noticed that if you print a data frame with thousands of rows, you get thousands of rows of output. This is rarely desirable. Printing a tibble will only show you the first 10 rows of data, and will shorten or remove columns if they do not fit in your output window. Printing a tibble also shows the size of the data and the types of the variables automatically, as you can see in the example above.

2. Type consistency. If you select two columns of a tibble, you get a tibble. If you select one column of a tibble, you get a tibble. This is type consistency. Data frames do not have this behavior, since selecting two columns gives a data frame but selecting one column gives a vector:

```
mtcars[1:5, c("am", "gear", "carb")] # data frame
```

```
##                   am gear carb
## Mazda RX4          1    4    4
## Mazda RX4 Wag      1    4    4
## Datsun 710         1    4    1
## Hornet 4 Drive     0    3    1
## Hornet Sportabout  0    3    2
```

```
mtcars[1:5, "carb"] # vector
```

```
## [1] 4 4 1 1 2
```

In this chapter, we will primarily use a data set consisting of user generated numerical ratings of movies. This data comes from the website MovieLens[2] (movielens.org), a non-commercial site for movie recommendations. The data is collected and distributed by GroupLens research of the University of Minnesota and made available for research and educational use.

The MovieLens data contains the variables **userId**, **movieId**, **rating**, **timestamp**, **title** and **genres**. The data set consists of 100,836 observations taken at a random starting point, and it includes the reviews of 610 consecutively numbered users.

In addition to being available on the MovieLens web page, we have made it available in the package `fosdata` associated with this book. If you haven't yet done so, you will need to install `fosdata` as follows.

```
install.packages("remotes") # if you don't already have this package
remotes::install_github(repo = "speegled/fosdata")
```

We recommend using `fosdata::<data_set>` to reference data sets individually, rather than loading them all with `library(fosdata)`.

```
movies <- fosdata::movies
```

Let's convert it to a `tibble` so that when we print things out to the screen, we aren't overwhelmed with information.

```
movies <- as_tibble(movies)
movies
```

```
## # A tibble: 100,836 x 6
##    userId movieId rating timestamp title                        genres
##     <int>   <int>  <dbl>     <int> <chr>                        <chr>
##  1      1       1      4 964982703 Toy Story (1995)             Adventur~
##  2      1       3      4 964981247 Grumpier Old Men (1995)      Comedy|R~
##  3      1       6      4 964982224 Heat (1995)                  Action|C~
##  4      1      47      5 964983815 Seven (a.k.a. Se7en) (1995)  Mystery|~
##  5      1      50      5 964982931 Usual Suspects, The (1995)   Crime|My~
##  6      1      70      3 964982400 From Dusk Till Dawn (1996)   Action|C~
##  7      1     101      5 964980868 Bottle Rocket (1996)         Adventur~
##  8      1     110      4 964982176 Braveheart (1995)            Action|D~
##  9      1     151      5 964984041 Rob Roy (1995)               Action|D~
## 10      1     157      5 964984100 Canadian Bacon (1995)        Comedy|W~
## # ... with 100,826 more rows
```

## 6.2 dplyr verbs

The `dplyr` package is organized around commands called *verbs*. Each verb takes a tibble (or data frame) as its first argument, and possibly additional arguments.

[2]F Maxwell Harper and Joseph A Konstan, "The MovieLens Datasets: History and Context," *ACM Trans. Interact. Intell. Syst.* 5, no. 4 (December 2015), https://doi.org/10.1145/2827872.

The first verb we will meet is `filter`,[3] which forms a new data frame consisting of rows that satisfy certain *filter*ing conditions.

Here we create a new data frame with all 218 reviews of the 1999 film *Fight Club*:

```
filter(movies, title == "Fight Club (1999)")

## # A tibble: 218 x 6
##     userId movieId rating  timestamp title             genres
##      <int>   <int>  <dbl>      <int> <chr>             <chr>
##  1       1    2959    5    964983282 Fight Club (1999) Action|Crime|Dram~
##  2       4    2959    2    945078528 Fight Club (1999) Action|Crime|Dram~
##  3      10    2959    0.5 1455356582 Fight Club (1999) Action|Crime|Dram~
##  4      15    2959    2.5 1510571747 Fight Club (1999) Action|Crime|Dram~
##  5      16    2959    3.5 1377476874 Fight Club (1999) Action|Crime|Dram~
##  6      17    2959    4.5 1305696867 Fight Club (1999) Action|Crime|Dram~
##  7      18    2959    4.5 1455049351 Fight Club (1999) Action|Crime|Dram~
##  8      19    2959    5    965703109 Fight Club (1999) Action|Crime|Dram~
##  9      21    2959    2   1441392954 Fight Club (1999) Action|Crime|Dram~
## 10      22    2959    3.5 1268726211 Fight Club (1999) Action|Crime|Dram~
## # ... with 208 more rows
```

Here are two more examples of the `filter` command. Find all user ratings of 1 or less:

```
filter(movies, rating <= 1)

## # A tibble: 4,181 x 6
##     userId movieId rating  timestamp title                  genres
##      <int>   <int>  <dbl>      <int> <chr>                  <chr>
##  1       1    3176    1    964983504 Talented Mr. Ripley, ~ Drama|Myster~
##  2       3      31    0.5 1306463578 Dangerous Minds (1995) Drama
##  3       3     527    0.5 1306464275 Schindler's List (199~ Drama|War
##  4       3     647    0.5 1306463619 Courage Under Fire (1~ Action|Crime~
##  5       3     688    0.5 1306464228 Operation Dumbo Drop ~ Action|Adven~
##  6       3     720    0.5 1306463595 Wallace & Gromit: The~ Adventure|An~
##  7       3     914    0.5 1306463567 My Fair Lady (1964)    Comedy|Drama~
##  8       3    1093    0.5 1306463627 Doors, The (1991)      Drama
##  9       3    1124    0.5 1306464216 On Golden Pond (1981)  Drama
## 10       3    1263    0.5 1306463569 Deer Hunter, The (197~ Drama|War
## # ... with 4,171 more rows
```

All reviews of 1 or less for *Fight Club*:

```
filter(movies, title == "Fight Club (1999)", rating <= 1)

## # A tibble: 3 x 6
##   userId movieId rating  timestamp title             genres
##    <int>   <int>  <dbl>      <int> <chr>             <chr>
## 1     10    2959    0.5 1455356582 Fight Club (1999) Action|Crime|Drama~
## 2    153    2959    0.5 1525548681 Fight Club (1999) Action|Crime|Drama~
## 3    308    2959    0.5 1421374757 Fight Club (1999) Action|Crime|Drama~
```

[3]If filter reports an error such as `Error in filter(movies, rating <= 1) : object 'rating' not found` that you don't understand, it is likely that you forgot to load the `dplyr` package. Load it with `library(dplyr)`.

It turns out that there are only three users in this data set who really disliked *Fight Club*!

Now that we have some basic idea of how verbs work, let's look at an overview of some of the ones that we will use in this chapter. We will use the following verbs regularly when working with data:

`filter()`
: Form a new data frame consisting of rows that satisfy certain *filter*ing conditions.

`select()`
: Form a new data frame with *select*ed columns.

`arrange()`
: Form a new data frame with row(s) *arrange*d in a specified order.

`slice_max()`, `slice_min()`
: Filter the *max*imum or *min*imum rows according to some ranking.

`summarize()`
: *Summarize* a data frame into a single row.

`distinct()`
: Collapse identical data to produce a single row for each *distinct* value

`mutate()`
: Create new variables by computation.

**Try It Yourself.**
Here are examples of how to use dplyr verbs with the MovieLens data.

`select(movies, rating, movieId)`
: Select the columns `rating` and `movieId` (in that order).

`arrange(movies, timestamp)`
: Arrange the ratings by the date they were reviewed.

`arrange(movies, desc(rating))`
: Arrange the ratings in descending order of rating.

`slice_max(movies, n=5, timestamp)`
: Find the last five reviews (by timestamp) in the data.

`filter(movies, rating != round(rating))`
: Filter to find all the half-star ratings.

`summarize(movies, mean(rating))`
: Summarize by finding the mean of all ratings in the data set.

`distinct(movies, userId)`
: Form a data frame consisting of the unique User ID's.

`mutate(movies, when = lubridate::as_datetime(timestamp))`
: Mutate the timestamp to a human readable date and time in the new variable called `when`. Note that this uses a function from the `lubridate` package, part of the tidyverse.

## 6.3 dplyr pipelines

Verb commands are simple, but are designed to be used together to perform more complicated operations. The *pipe operator* is the dplyr method for combining verbs. Pipe is the three-character symbol `%>%`, which you can also type using the three-key combination ctrl-shift-m. Pipe works by taking the value produced on its left and feeding it as the first argument of the function on its right.

For example, we can sort ratings of the 1958 movie *Vertigo* by timestamp using arrange and filter together.

```
movies %>%
  filter(title == "Vertigo (1958)") %>%
  arrange(timestamp)
```

```
## # A tibble: 60 x 6
##     userId movieId rating timestamp title          genres
##      <int>   <int>  <dbl>     <int> <chr>          <chr>
##  1     385     903      4 865023813 Vertigo (1958) Drama|Mystery|Romance~
##  2     171     903      5 866905882 Vertigo (1958) Drama|Mystery|Romance~
##  3     372     903      3 874414948 Vertigo (1958) Drama|Mystery|Romance~
##  4     412     903      4 939115095 Vertigo (1958) Drama|Mystery|Romance~
##  5     597     903      3 940362409 Vertigo (1958) Drama|Mystery|Romance~
##  6     199     903      3 940379738 Vertigo (1958) Drama|Mystery|Romance~
##  7     383     903      4 943571272 Vertigo (1958) Drama|Mystery|Romance~
##  8     554     903      5 944898646 Vertigo (1958) Drama|Mystery|Romance~
##  9      45     903      4 951756950 Vertigo (1958) Drama|Mystery|Romance~
## 10      59     903      5 953610229 Vertigo (1958) Drama|Mystery|Romance~
## # ... with 50 more rows
```

This has the same effect as the harder to read command:

```
arrange(filter(movies, title == "Vertigo (1958)"), timestamp)
```

With pipelines, we imagine the data (movies) flowing into the pipe then passing through the verbs in sequence, first being filtered and then being arranged. Pipelines also make it natural to break up long commands into multiple lines after each pipe operator, although this is not required.

**Example 6.1.** Find the mean[4] rating of *Toy Story*, which has movieId one.

```
movies %>%
  filter(movieId == 1) %>%
  summarize(mean(rating))
```

```
## # A tibble: 1 x 1
##   `mean(rating)`
##            <dbl>
## 1           3.92
```

The filter() command creates a data frame that consists solely of the observations of *Toy Story*, and summarize() computes the mean rating.

Pipelines can be used with any R function, not just dplyr verbs. One handy trick, especially if your data is not a tibble, is to pipe into head.

**Example 6.2.** Find all movies that some user rated 5.

[4]Rating should almost certainly have been reclassified as an ordered factor, and it is not clear that taking the mean is really a valid thing to do.

```
movies %>%
  filter(rating == 5) %>%
  select(title) %>%
  distinct() %>%
  head(n = 5)
```

```
## # A tibble: 5 x 1
##   title
##   <chr>
## 1 Seven (a.k.a. Se7en) (1995)
## 2 Usual Suspects, The (1995)
## 3 Bottle Rocket (1996)
## 4 Rob Roy (1995)
## 5 Canadian Bacon (1995)
```

The `filter` picks out the 5-star movies, `select` picks only the title variable, `distinct` removes duplicate titles, and then `head` shows only the first five distinct titles.

### 6.3.1 Group by and summarize

The MovieLens data has one observation for each user review. However, we are often interested in working with the movies themselves. The tool to do this is the `dplyr` verb `group_by`, which groups data to perform tasks by groups. By itself, `group_by` has little effect; in fact, the only visual indication that it has done anything is a line at the top of the tibble output noting that there are groups. The following example shows how to find the mean rating of each movie in the MovieLens data.

**Example 6.3.** In order to find the mean rating of each movie, we will use the `group_by()` function.

```
movies %>%
  group_by(title)
```

```
## # A tibble: 100,836 x 6
## # Groups:   title [9,719]
##    userId movieId rating timestamp title                        genres
##     <int>   <int>  <dbl>     <int> <chr>                        <chr>
##  1      1       1      4 964982703 Toy Story (1995)             Adventur~
##  2      1       3      4 964981247 Grumpier Old Men (1995)      Comedy|R~
##  3      1       6      4 964982224 Heat (1995)                  Action|C~
##  4      1      47      5 964983815 Seven (a.k.a. Se7en) (1995)  Mystery|~
##  5      1      50      5 964982931 Usual Suspects, The (1995)   Crime|My~
##  6      1      70      3 964982400 From Dusk Till Dawn (1996)   Action|C~
##  7      1     101      5 964980868 Bottle Rocket (1996)         Adventur~
##  8      1     110      4 964982176 Braveheart (1995)            Action|D~
##  9      1     151      5 964984041 Rob Roy (1995)               Action|D~
## 10      1     157      5 964984100 Canadian Bacon (1995)        Comedy|W~
## # ... with 100,826 more rows
```

In essentially all uses, we follow `group_by` with another operation on the groups, most commonly `summarize`:

```
movies %>%
  group_by(title) %>%
  summarize(mean(rating))
```

```
## # A tibble: 9,719 x 2
##    title                                   `mean(rating)`
##    <chr>                                            <dbl>
##  1 ¡Three Amigos! (1986)                             3.13
##  2 ...All the Marbles (1981)                         2
##  3 ...And Justice for All (1979)                     3.17
##  4 '71 (2014)                                        4
##  5 'burbs, The (1989)                                3.18
##  6 'Hellboy': The Seeds of Creation (2004)           4
##  7 'night Mother (1986)                              3
##  8 'Round Midnight (1986)                            3.5
##  9 'Salem's Lot (2004)                               5
## 10 'Til There Was You (1997)                         4
## # ... with 9,709 more rows
```

This produced a new tibble with one row for each group and created the new variable with the awkward name `mean(rating)` which records the mean rating for each movie. We could give that variable an easier name to type by assigning it as part of the summarize operation: `summarize(rating = mean(weight))`.

```
movies %>%
  group_by(title) %>%
  summarize(rating = mean(rating))
```

```
## # A tibble: 9,719 x 2
##    title                                   rating
##    <chr>                                    <dbl>
##  1 ¡Three Amigos! (1986)                     3.13
##  2 ...All the Marbles (1981)                 2
##  3 ...And Justice for All (1979)             3.17
##  4 '71 (2014)                                4
##  5 'burbs, The (1989)                        3.18
##  6 'Hellboy': The Seeds of Creation (2004)   4
##  7 'night Mother (1986)                      3
##  8 'Round Midnight (1986)                    3.5
##  9 'Salem's Lot (2004)                       5
## 10 'Til There Was You (1997)                 4
## # ... with 9,709 more rows
```

**Example 6.4.** The built-in `chickwts` records the weight of chickens fed on different diets. Calculate the number of chickens in each feed group, and the mean weight of chickens in that group.

```
chickwts %>%
  group_by(feed) %>%
  summarize(mw = mean(weight), count = n())
```

```
## # A tibble: 6 x 3
##   feed          mw count
```

```
##   <fct>        <dbl> <int>
## 1 casein        324.    12
## 2 horsebean     160.    10
## 3 linseed       219.    12
## 4 meatmeal      277.    11
## 5 soybean       246.    14
## 6 sunflower     329.    12
```

This example illustrates the special function `n()`, which gives the number of rows in each group.

The final example of `group_by` in this section illustrates the use of two grouping variables. After using `summarize`, the data will **still** be grouped by default, so be careful.

**Example 6.5.** The `storms` data set in the `dplyr` package has information about Atlantic hurricanes, with one observation every six hours during the lifetime of a storm. The variable `wind` records the maximum sustained wind speed for each six-hour observation window. Find the four storms that attained the highest overall wind speed.

To do this, we will need to group the observations by storm and find the maximum of `wind` for each storm. There is no single variable which identifies a storm, but `name` and `year` together do.[5] If we group by both `name` and `year`, and then `summarize`, we will get one entry per storm. However, the resulting tibble will remain **grouped** by `name`. Before we can find the top four, we must remove the grouping with the `ungroup` function.

```
dplyr::storms %>%
  group_by(name, year) %>%
  summarize(max_wind = max(wind)) %>%
  ungroup() %>%
  slice_max(max_wind, n = 4)
```

```
## # A tibble: 4 x 3
##   name      year max_wind
##   <chr>    <dbl>    <int>
## 1 Gilbert   1988      160
## 2 Wilma     2005      160
## 3 Mitch     1998      155
## 4 Rita      2005      155
```

We see that Hurricanes Gilbert (1988), Wilma (2005), Mitch (1998), and Rita (2005) had the highest maximum wind speeds.

**Try It Yourself.**
Try the code in the previous chunk after **removing** the `ungroup` line. Can you explain what is happening?

[5]This is a lie. There was one storm, Zeta, which formed on December 30, 2005 and lasted until January 6, 2006. It and the unusual storm Eta, also from 2005, are the subject of an XKCD comic, https://xkcd.com/1126/.

## 6.4 The power of dplyr

With a small collection of verbs and the pipe operator, you are well equipped to perform complicated data analysis. This section shows some of the techniques you can use to learn answers from data.

**Example 6.6.** Create a data frame consisting of the observations associated with movies that have an average rating of 5 stars. That is, each rating of the movie was 5 stars.

In order to do this, we will use the `group_by()` function to find the mean rating of each movie, as above.

```
movies %>%
  group_by(title) %>%
  summarize(rating = mean(rating))

## # A tibble: 9,719 x 2
##    title                                      rating
##    <chr>                                       <dbl>
##  1 ¡Three Amigos! (1986)                        3.13
##  2 ...All the Marbles (1981)                    2
##  3 ...And Justice for All (1979)                3.17
##  4 '71 (2014)                                   4
##  5 'burbs, The (1989)                           3.18
##  6 'Hellboy': The Seeds of Creation (2004)      4
##  7 'night Mother (1986)                         3
##  8 'Round Midnight (1986)                       3.5
##  9 'Salem's Lot (2004)                          5
## 10 'Til There Was You (1997)                    4
## # ... with 9,709 more rows
```

Now, we can filter out those whose mean rating is 5.

```
movies %>%
  group_by(title) %>%
  summarize(rating = mean(rating)) %>%
  filter(rating == 5)

## # A tibble: 296 x 2
##    title                                  rating
##    <chr>                                   <dbl>
##  1 'Salem's Lot (2004)                         5
##  2 12 Angry Men (1997)                         5
##  3 12 Chairs (1976)                            5
##  4 20 Million Miles to Earth (1957)            5
##  5 61* (2001)                                  5
##  6 7 Faces of Dr. Lao (1964)                   5
##  7 9/11 (2002)                                 5
##  8 A Detective Story (2003)                    5
##  9 A Flintstones Christmas Carol (1994)        5
## 10 A Perfect Day (2015)                        5
```

```
## # ... with 286 more rows
```

And we see that there are 296 movies which have mean rating of 5.

**Example 6.7.** Which movie that received only 5-star ratings has the most ratings?

```
movies %>%
  group_by(title) %>%
  summarize(mr = mean(rating), numRating = n()) %>%
  arrange(desc(mr), desc(numRating))
```

```
## # A tibble: 9,719 x 3
##    title                                                   mr numRating
##    <chr>                                                <dbl>     <int>
##  1 Belle époque (1992)                                      5         2
##  2 Come and See (Idi i smotri) (1985)                       5         2
##  3 Enter the Void (2009)                                    5         2
##  4 Heidi Fleiss: Hollywood Madam (1995)                     5         2
##  5 Jonah Who Will Be 25 in the Year 2000 (Jonas qui aur~    5         2
##  6 Lamerica (1994)                                          5         2
##  7 Lesson Faust (1994)                                      5         2
##  8 'Salem's Lot (2004)                                      5         1
##  9 12 Angry Men (1997)                                      5         1
## 10 12 Chairs (1976)                                         5         1
## # ... with 9,709 more rows
```

There are seven movies that have a perfect mean rating of 5 and have been rated twice. No movie with a perfect mean rating has been rated three or more times.

**Example 6.8.** Out of movies with a lot of ratings, which has the highest rating? Well, we need to decide what "a lot of ratings" means. Let's see how many ratings some of the movies had.

```
movies %>%
  group_by(title) %>%
  summarize(count = n()) %>%
  arrange(desc(count))
```

```
## # A tibble: 9,719 x 2
##    title                                     count
##    <chr>                                     <int>
##  1 Forrest Gump (1994)                         329
##  2 Shawshank Redemption, The (1994)            317
##  3 Pulp Fiction (1994)                         307
##  4 Silence of the Lambs, The (1991)            279
##  5 Matrix, The (1999)                          278
##  6 Star Wars: Episode IV - A New Hope (1977)   251
##  7 Jurassic Park (1993)                        238
##  8 Braveheart (1995)                           237
##  9 Terminator 2: Judgment Day (1991)           224
## 10 Schindler's List (1993)                     220
## # ... with 9,709 more rows
```

It seems like 100 ratings could classify as "a lot." Let's see which movie with at least 100 ratings has the highest mean rating.

```
movies %>%
  group_by(title) %>%
  summarize(count = n(), meanRating = mean(rating)) %>%
  filter(count >= 100) %>%
  slice_max(meanRating)

## # A tibble: 1 x 3
##   title                             count meanRating
##   <chr>                             <int>      <dbl>
## 1 Shawshank Redemption, The (1994)    317       4.43
```

Lots of people like *The Shawshank Redemption.* If we want to see all of the highest rated movies, we can arrange by `meanRating`. Tibbles only print 10 rows by default, so to get a longer list we pipe to `print` and specify the number of movies we wish to see.[6]

```
movies %>%
  group_by(title) %>%
  summarize(count = n(), meanRating = mean(rating)) %>%
  filter(count > 100) %>%
  arrange(desc(meanRating)) %>%
  select(title, meanRating) %>%
  print(n = 35)

## # A tibble: 134 x 2
##    title                                                           meanRating
##    <chr>                                                                <dbl>
##  1 Shawshank Redemption, The (1994)                                      4.43
##  2 Godfather, The (1972)                                                 4.29
##  3 Fight Club (1999)                                                     4.27
##  4 Godfather: Part II, The (1974)                                        4.26
##  5 Departed, The (2006)                                                  4.25
##  6 Goodfellas (1990)                                                     4.25
##  7 Dark Knight, The (2008)                                               4.24
##  8 Usual Suspects, The (1995)                                            4.24
##  9 Princess Bride, The (1987)                                            4.23
## 10 Star Wars: Episode IV - A New Hope (1977)                             4.23
## 11 Schindler's List (1993)                                               4.22
## 12 Apocalypse Now (1979)                                                 4.22
## 13 American History X (1998)                                             4.22
## 14 Star Wars: Episode V - The Empire Strikes Back (1980)                 4.22
## 15 Raiders of the Lost Ark (Indiana Jones and the Raiders of~            4.21
## 16 One Flew Over the Cuckoo's Nest (1975)                                4.20
## 17 Reservoir Dogs (1992)                                                 4.20
## 18 Pulp Fiction (1994)                                                   4.20
## 19 Matrix, The (1999)                                                    4.19
## 20 Amelie (Fabuleux destin d'Amélie Poulain, Le) (2001)                  4.18
## 21 Forrest Gump (1994)                                                   4.16
## 22 Monty Python and the Holy Grail (1975)                                4.16
```

[6]If you definitely want to see **all** of the movies, you can pipe to `print(n = nrow(.))`

```
## 23 Silence of the Lambs, The (1991)                               4.16
## 24 Eternal Sunshine of the Spotless Mind (2004)                   4.16
## 25 Green Mile, The (1999)                                         4.15
## 26 Saving Private Ryan (1998)                                     4.15
## 27 Star Wars: Episode VI - Return of the Jedi (1983)              4.14
## 28 Memento (2000)                                                 4.12
## 29 Lord of the Rings: The Return of the King, The (2003)          4.12
## 30 Fargo (1996)                                                   4.12
## 31 Lord of the Rings: The Fellowship of the Ring, The (2001)      4.11
## 32 Taxi Driver (1976)                                             4.11
## 33 Blade Runner (1982)                                            4.10
## 34 Full Metal Jacket (1987)                                       4.10
## 35 Shining, The (1980)                                            4.08
## # ... with 99 more rows
```

You just got your summer movie watching list. You're welcome!

**Try It Yourself.**
Use dplyr to determine:

- The 10 worst rated movies that were rated at least 100 times.
- Which user made the most ratings.
- Which genre had the highest rating.

Continue reading to check your work.

- Of movies rated at least 100 times, the list of the worst starts with *Waterworld*, *Batman Forever*, and *Home Alone*.
- User number 414 rated 2698 movies!
- There are many genres tied for highest rating, with a perfect 5 rating and only 1 user. Your data science instincts should tell you to restrict to movies rated a lot. The restriction that we chose led to concluding that the genre "Crime|Drama" movies are the highest rated.

When `group_by` is followed by `summarize`, the resulting output has one row per group. It is also reasonable to follow `group_by` with `mutate`, which can be used to calculate a value for each row based off of data from the groups. In this case, you can add `ungroup` to the pipeline to remove the grouping before using the data further.

**Example 6.9.** Find the "worst opinions" in the MovieLens data set. We are interested in finding the movie that received the most ratings while also receiving exactly one 5-star rating, and in finding the user who had that bad take.

```
movies %>%
  group_by(title) %>%
  mutate(
    num5 = sum(rating == 5),
    numRating = n()
  ) %>%
  ungroup() %>%
  filter(rating == 5, num5 == 1) %>%
  select(userId, title, numRating) %>%
  slice_max(numRating, n = 5)
```

```
## # A tibble: 5 x 3
##   userId title                                    numRating
##    <int> <chr>                                        <int>
## 1    246 Addams Family Values (1993)                     84
## 2    313 Charlie's Angels (2000)                         72
## 3    587 Lost World: Jurassic Park, The (1997)           67
## 4     43 Coneheads (1993)                                63
## 5     45 Signs (2002)                                    63
```

User number 246 has the worst opinion in the data set, as none of 83 other people thought *Addams Family Values* was all that good.

## 6.5 Working with character strings

This section introduces the `stringr` package for basic string manipulation. As far as R is concerned, strings are variables of type character, or `chr`. For example:

```
my_name <- "Darrin Speegle"
str(my_name)
```

```
##  chr "Darrin Speegle"
```

Recall that `str` returns the `str`ucture of the variable, together with the contents if it is not too long. Even finding out how long a string is can be challenging for newcomers, because `length` gives the number of strings in the variable `my_name`:

```
length(my_name)
```

```
## [1] 1
```

One might very reasonably be interested in the number of spaces (a proxy for the number of words), the capital letters (a person's initials), or in sorting a vector of characters containing names by the last name of the people. All of these tasks (and more) are made easier using the `stringr` package.

```
library(stringr)
```

Most functions from `stringr` begin with the prefix `str_`. The function `str_length` accepts one argument, a string, which can either be a single string or a vector of strings. It returns the length(s) of the string(s). For example,

```
str_length(my_name)
```

```
## [1] 14
```

If we had a vector of two names, then it works like this:

```
our_names <- c("Darrin Speegle", "Bryan Clair")
str_length(our_names)
```

```
## [1] 14 11
```

Next, we consider the function `str_count`, which has two arguments, `string` and `pattern`. The first is a string, and the second is a *regular expression* that you use to indicate the pattern that you are counting. To count the number of spaces, we will pass a single character to the pattern, namely the space character.

```
str_count(our_names, " ")
```

```
## [1] 1 1
```

There is one space in each of the two strings in the variable `our_names`. Regular expressions are very useful and powerful. If you end up doing much work with text data, you will not regret learning more about them. For example, if you want to match lower case vowels, you could use

```
str_count(our_names, "[aeiou]")
```

```
## [1] 5 3
```

Note the use of the grouping symbols `[` and `]`. If we want to match any lower case letter, we use `[a-z]`, any upper case letter is `[A-Z]`, and any digit is `[0-9]`. So, to count the number of upper case letters in the strings, we would use

```
str_count(our_names, "[A-Z]")
```

```
## [1] 2 2
```

Next, we examine `str_extract` and its sibling function `str_extract_all`. Suppose we want to extract the initials of the people in the vector `our_names`. Our first thought might be to use `str_extract`, which takes two parameters, `string` which contains the string(s) we wish to extract from, and `pattern` which is the pattern we wish to match, in order to extract.

```
str_extract(our_names, "[A-Z]")
```

```
## [1] "D" "B"
```

Note that this only extracts the *first* occurrence of the pattern in each string. To get all of them, we use `str_extract_all`

```
str_extract_all(our_names, "[A-Z]")
```

```
## [[1]]
## [1] "D" "S"
##
## [[2]]
## [1] "B" "C"
```

Note the format of the output. It is a *list* of character vectors. Let's check using `str`:

```
initials <- str_extract_all(our_names, "[A-Z]")
str(initials)
```

```
## List of 2
##  $ : chr [1:2] "D" "S"
##  $ : chr [1:2] "B" "C"
```

This is because `str_extract_all` doesn't know how many matches there are going to be in each string, and because it doesn't know that we are thinking of these as initials and want them all in the same string. That is, we would like a vector that looks like `c("DS", "BC")`. After learning about `sapply` in Section 6.7, we could apply `str_flatten` to each element of the list, but that seems like a long road to walk for this purpose. Perhaps it is easier to do the opposite and remove all of the things that don't match capital letters! To do this, we use `str_remove_all`, which again takes two arguments, `string` and `pattern` to remove. We need to know that regular expressions can be told to *not* match characters by including ^ inside the brackets:

```
str_remove_all(our_names, "[^A-Z]")
```

```
## [1] "DS" "BC"
```

The last two functions that we are looking at are `str_detect` and `str_split`. These are most useful within a data analysis flow. The function `str_detect` accepts `string` and `pattern` and returns `TRUE` if the pattern is detected, and `FALSE` if the pattern is not detected. For example, we could look for the pattern `an` in `our_names`:

```
str_detect(our_names, "an")
```

```
## [1] FALSE  TRUE
```

Note that if we had put `"an"` inside a bracket like this `"[an]"`, then `str_detect` would have looked for a pattern consisting of either an `a` or an `n`, and we would have received two `TRUE` values. Finally, suppose we want to split the strings into first name and last name. We use `str_split` or its sibling function `str_split_fixed`. The function `str_split` takes two arguments, a string and a pattern on which to split the string. Every time the function sees the pattern, it splits off a piece of the string. For example,

```
str_split(our_names, " ")
```

```
## [[1]]
## [1] "Darrin"  "Speegle"
##
## [[2]]
## [1] "Bryan" "Clair"
```

Note that, once again, the function returns a list of vectors of strings. If we *know* that we only want to split into at most 2 groups, say, then we can use `str_split_fixed` to indicate that. The first two arguments are the same, but `str_split_fixed` has a third argument, `n` which is the number of strings that we want to split each string into.

```
str_split_fixed(our_names, " ", n = 2)
```

```
##      [,1]     [,2]
## [1,] "Darrin" "Speegle"
## [2,] "Bryan"  "Clair"
```

**Example 6.10.** Let's apply string manipulation to understand the `Genres` variable of the `fosdata::movies` data set.

What are the best comedies? We look for the highest-rated movies that have been rated at least 50 times and include "Comedy" in the genre list.

```
movies %>%
  filter(str_detect(genres, "Comedy")) %>%
  group_by(title) %>%
  summarize(rating = mean(rating), count = n()) %>%
  filter(count >= 50) %>%
  slice_max(rating, n = 10)
```

```
## # A tibble: 10 x 3
```

```
##    title                                                      rating count
##    <chr>                                                       <dbl> <int>
##  1 Dr. Strangelove or: How I Learned to Stop Worrying and ~    4.27    97
##  2 Princess Bride, The (1987)                                  4.23   142
##  3 Pulp Fiction (1994)                                         4.20   307
##  4 Amelie (Fabuleux destin d'Amélie Poulain, Le) (2001)        4.18   120
##  5 Forrest Gump (1994)                                         4.16   329
##  6 Monty Python and the Holy Grail (1975)                      4.16   136
##  7 Snatch (2000)                                               4.16    93
##  8 Life Is Beautiful (La Vita è bella) (1997)                  4.15    88
##  9 Fargo (1996)                                                4.12   181
## 10 Toy Story 3 (2010)                                          4.11    55
```

**Example 6.11.** Find the highest-rated movies of 1999.

The challenge here is that the year that the movie was released is hidden at the end of the title of the movie. Let's pull out the year of release and move it into its own variable. We use regular expressions to do this. The character `$` inside a regular expression means that the pattern occurs at the end of a string.

Let's extract 4 digits, followed by a right parenthesis, at the end of a string. The pattern for this is `[0-9]{4}\\)$`. The {4} indicates that we are matching the digit pattern four times, we have to escape the right parenthesis using `\\` because `)` is a reserved character in regular expressions, and the `$` indicates that we are anchoring this to the end of the string. Let's test it out:

```
pattern <- "[0-9]{4}\\)$"
str_extract(movies$title, pattern) %>%
  head()
```

```
## [1] "1995)" "1995)" "1995)" "1995)" "1995)" "1996)"
```

Hmmm, even better would be to get rid of the right parenthesis. We can use the look-ahead operator `?` to only pull out 4 digits if they are followed by a right parenthesis and then the end of string. Since parentheses are special characters in regular expressions, we have to escape the right parenthesis using `\\)`.

```
year_pattern <- "[0-9]{4}(?=\\)$)"
str_extract(movies$title, year_pattern) %>%
  head()
```

```
## [1] "1995" "1995" "1995" "1995" "1995" "1996"
```

Sweet! Let's convert to numeric and do a histogram, just to make sure no weird stuff snuck in.

```
str_extract(movies$title, year_pattern) %>%
  as.numeric() %>%
  hist(main = "MovieLens movies", xlab = "Release year")
```

The histogram (in Figure 6.1) looks pretty good. Let's create a new variable called `year` that contains the year of release of the movie.

```
movies <- mutate(movies,
  year = str_extract(title, year_pattern) %>% as.numeric()
```

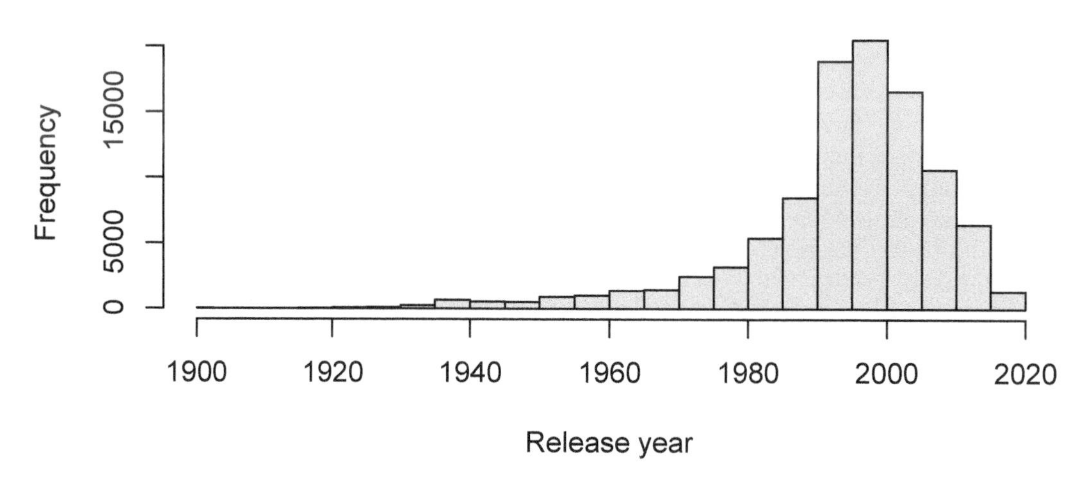


**FIGURE 6.1** Distribution of release years in the movies data set.

```
)
summary(movies$year)
```

```
##    Min. 1st Qu.  Median    Mean 3rd Qu.    Max.    NA's
##    1902    1990    1997    1994    2003    2018      30
```

Whoops! We have 30 NA years. Let's take a look and see what happened.

```
movies %>%
  filter(is.na(year)) %>%
  select(title)
```

```
## # A tibble: 30 x 1
##    title
##    <chr>
##  1 "Black Mirror"
##  2 "Runaway Brain (1995) "
##  3 "The Adventures of Sherlock Holmes and Doctor Watson"
##  4 "Maria Bamford: Old Baby"
##  5 "Generation Iron 2"
##  6 "Ready Player One"
##  7 "Babylon 5"
##  8 "Ready Player One"
##  9 "Nocturnal Animals"
## 10 "Guilty of Romance (Koi no tsumi) (2011) "
## # ... with 20 more rows
```

Aha! Some of the movies don't have years, so those will have to remain NA. A few other movies have an extra space after the end parentheses. We can handle the extra space with `\\s*`, which matches zero or more spaces.

```
year_pattern <- "[0-9]{4}(?=\\)\\s*$)"
movies <- mutate(movies,
```

```
  year = str_extract(title, year_pattern) %>% as.numeric()
)
summary(movies$year)
```

```
##    Min. 1st Qu.  Median    Mean 3rd Qu.    Max.    NA's
##    1902    1990    1997    1994    2003    2018      17
```

Finally, what were the five highest rated movies of 1999 that had at least 50 ratings?

```
movies %>%
  filter(year == 1999) %>%
  group_by(title) %>%
  summarize(mean = mean(rating), count = n()) %>%
  filter(count >= 50) %>%
  slice_max(mean, n = 5)
```

```
## # A tibble: 5 x 3
##   title                      mean count
##   <chr>                     <dbl> <int>
## 1 Fight Club (1999)          4.27   218
## 2 Matrix, The (1999)         4.19   278
## 3 Green Mile, The (1999)     4.15   111
## 4 Office Space (1999)        4.09    94
## 5 American Beauty (1999)     4.06   204
```

One final common usage of regular expressions is within `select`. If your data frame has a lot of variables, and you want to select out some that follow a certain pattern, then knowing regular expressions can be very helpful.

**Example 6.12.** Consider the `accelerometer` data in the `fosdata` package. This data set will be described in more detail in Chapter 11. Here we list the names of all the variables in this data set:

```
accelerometer <- fosdata::accelerometer
names(accelerometer)
```

```
##  [1] "participant"
##  [2] "machine"
##  [3] "set"
##  [4] "contraction_mode"
##  [5] "time_video_rater_cv_ms"
##  [6] "time_video_rater_dg_ms"
##  [7] "time_smartphone_1_ms"
##  [8] "time_smartphone_2_ms"
##  [9] "video_rater_mean_ms"
## [10] "smartphones_mean_ms"
## [11] "relative_difference"
## [12] "difference_video_smartphone_ms"
## [13] "mean_video_smartphone_ms"
## [14] "contraction_mode_levels"
## [15] "difference_video_raters_ms"
## [16] "difference_smartphones_ms"
## [17] "video_smartphone_difference_outlier"
## [18] "rater_difference_outlier"
```

```
## [19] "smartphone_difference_outlier"
## [20] "normalized_error_smartphone"
## [21] "participant_age_levels"
## [22] "participant_age_years"
## [23] "participant_height_cm"
## [24] "participant_weight_kg"
## [25] "participant_gender"
```

Look at all of those variables! Most of the data sets that we see in this book have been condensed, but it is not at all uncommon for experimenters to collect data on dozens or even hundreds of values. Let's suppose you wanted to create a new data frame that had all of the time measurements in it. Those are the variables whose names *end* with `ms`. We can do this by combining `select` with the regular expression `ms$`, which matches the letters ms and the end of the string.

```
select(accelerometer, matches("ms$")) %>%
  print(n = 5)
```

```
## # A tibble: 12,245 x 10
##   time_video_rater~ time_video_rater~ time_smartphone_~ time_smartphone~
##               <dbl>             <dbl>             <dbl>            <dbl>
## 1              1340              1360              1650             1700
## 2              1160              1180              1350             1350
## 3              1220              1240              1400             1350
## 4              1260              1280              1400             1350
## 5              1560              1180              1350             1300
## # ... with 12,240 more rows, and 6 more variables:
## #   video_rater_mean_ms <dbl>, smartphones_mean_ms <dbl>,
## #   difference_video_smartphone_ms <dbl>,
## #   mean_video_smartphone_ms <dbl>, difference_video_raters_ms <dbl>,
## #   difference_smartphones_ms <dbl>
```

To build a data frame that includes both variables that start with `smartphone` or end with `ms`, use the regular expression | character for "or."

```
select(accelerometer, matches("^smartphone|ms$")) %>%
  print(n = 5)
```

```
## # A tibble: 12,245 x 11
##   time_video_rater~ time_video_rater~ time_smartphone_~ time_smartphone~
##               <dbl>             <dbl>             <dbl>            <dbl>
## 1              1340              1360              1650             1700
## 2              1160              1180              1350             1350
## 3              1220              1240              1400             1350
## 4              1260              1280              1400             1350
## 5              1560              1180              1350             1300
## # ... with 12,240 more rows, and 7 more variables:
## #   video_rater_mean_ms <dbl>, smartphones_mean_ms <dbl>,
## #   difference_video_smartphone_ms <dbl>,
## #   mean_video_smartphone_ms <dbl>, difference_video_raters_ms <dbl>,
## #   difference_smartphones_ms <dbl>,
## #   smartphone_difference_outlier <lgl>
```

## 6.6 Structure of data

### 6.6.1 Tidy data: pivoting

The tidyverse is designed for tidy data, and a key feature of tidy data is that all data should be stored in a rectangular array, with each row an observation and each column a variable. In particular, no information should be stored in variable names. As an example, consider the built-in data set `WorldPhones` which gives the number of telephones in each region of the world, back in the day when telephones were still gaining popularity as a device for making phone calls.

The `WorldPhones` data is stored as a matrix with row names, so we convert it to a tibble called `phones` and preserve the row names in a new variable called Year. We also clean up the names into a standard format using `janitor::clean_names`.

```
phones <- as_tibble(WorldPhones, rownames = "year") %>%
  janitor::clean_names()
phones
```

```
## # A tibble: 7 x 8
##   year  n_amer europe  asia s_amer oceania africa mid_amer
##   <chr>  <dbl>  <dbl> <dbl>  <dbl>   <dbl>  <dbl>    <dbl>
## 1 1951   45939  21574  2876   1815    1646     89      555
## 2 1956   60423  29990  4708   2568    2366   1411      733
## 3 1957   64721  32510  5230   2695    2526   1546      773
## 4 1958   68484  35218  6662   2845    2691   1663      836
## 5 1959   71799  37598  6856   3000    2868   1769      911
## 6 1960   76036  40341  8220   3145    3054   1905     1008
## 7 1961   79831  43173  9053   3338    3224   2005     1076
```

Notice that the column names are regions of the world, so that there is useful information stored in those names.

Every entry in this data set gives the value for a year and a region, so the tidy format should instead have three variables: year, region, and telephones. Making this change will cause this data set to become much longer. Instead of 7 rows and 7 columns, we will have 49 rows, one for each of the $7 \times 7$ year and region combinations. The tool to make this change is the `pivot_longer` function from the `tidyr` package (part of the tidyverse):

```
library(tidyr)

phones %>%
  pivot_longer(cols = !year, names_to = "region", values_to = "telephones")
```

```
## # A tibble: 49 x 3
##    year  region   telephones
##    <chr> <chr>         <dbl>
##  1 1951  n_amer        45939
##  2 1951  europe        21574
##  3 1951  asia           2876
##  4 1951  s_amer         1815
##  5 1951  oceania        1646
```

```
## 6 1951  africa           89
## 7 1951  mid_amer        555
## 8 1956  n_amer        60423
## 9 1956  europe        29990
## 10 1956  asia          4708
## # ... with 39 more rows
```

This function used four arguments. The first was the data frame we wanted to pivot, in this case `phones`. Second, we specified which columns to use, with the expression `cols = !year`, which meant "all columns except year." Finally, we told the function that the names of the columns should become a new variable called "region," and the values in those columns should become a new variable called "telephones."

**Example 6.13.** As a more complex example, let's look at the `billboard` data set provided with the `tidyr` package. This contains the weekly *Billboard* rankings for each song that entered the Top 100 during the year 2000. Observe that each row is a song (`track`) and the track's place on the charts is stored in up to 76 columns named `wk1` through `wk76`. There are many `NA` in the data since none of these tracks were ranked for 76 consecutive weeks.

```
billboard %>% head(n = 5)
```

```
## # A tibble: 5 x 79
##   artist  track    date.entered   wk1   wk2   wk3   wk4   wk5   wk6   wk7
##   <chr>   <chr>    <date>       <dbl> <dbl> <dbl> <dbl> <dbl> <dbl> <dbl>
## 1 2 Pac   Baby D~  2000-02-26      87    82    72    77    87    94    99
## 2 2Ge+her The Ha~  2000-09-02      91    87    92    NA    NA    NA    NA
## 3 3 Door~ Krypto~  2000-04-08      81    70    68    67    66    57    54
## 4 3 Door~ Loser    2000-10-21      76    76    72    69    67    65    55
## 5 504 Bo~ Wobble~  2000-04-15      57    34    25    17    17    31    36
## # ... with 69 more variables: wk8 <dbl>, wk9 <dbl>, wk10 <dbl>,
## #   wk11 <dbl>, wk12 <dbl>, wk13 <dbl>, wk14 <dbl>, wk15 <dbl>,
## #   wk16 <dbl>, wk17 <dbl>, wk18 <dbl>, wk19 <dbl>, wk20 <dbl>,
## #   wk21 <dbl>, wk22 <dbl>, wk23 <dbl>, wk24 <dbl>, wk25 <dbl>,
## #   wk26 <dbl>, wk27 <dbl>, wk28 <dbl>, wk29 <dbl>, wk30 <dbl>,
## #   wk31 <dbl>, wk32 <dbl>, wk33 <dbl>, wk34 <dbl>, wk35 <dbl>,
## #   wk36 <dbl>, wk37 <dbl>, wk38 <dbl>, wk39 <dbl>, wk40 <dbl>, ...
```

This data is not tidy, since the week column names contain information. To do any sort of analysis on this data, the week needs to be a variable. We use `pivot_longer` to replace all 76 week columns with two columns, one called "week" and another called "rank":

```
long.bill <- billboard %>%
  pivot_longer(
    cols = wk1:wk76,
    names_to = "week", values_to = "rank",
    values_drop_na = TRUE
  )
long.bill
```

```
## # A tibble: 5,307 x 5
##    artist track                   date.entered week   rank
##    <chr>  <chr>                   <date>       <chr> <dbl>
##  1 2 Pac  Baby Don't Cry (Keep... 2000-02-26   wk1      87
##  2 2 Pac  Baby Don't Cry (Keep... 2000-02-26   wk2      82
```

```
## 3 2 Pac    Baby Don't Cry (Keep... 2000-02-26  wk3       72
## 4 2 Pac    Baby Don't Cry (Keep... 2000-02-26  wk4       77
## 5 2 Pac    Baby Don't Cry (Keep... 2000-02-26  wk5       87
## 6 2 Pac    Baby Don't Cry (Keep... 2000-02-26  wk6       94
## 7 2 Pac    Baby Don't Cry (Keep... 2000-02-26  wk7       99
## 8 2Ge+her The Hardest Part Of ... 2000-09-02  wk1       91
## 9 2Ge+her The Hardest Part Of ... 2000-09-02  wk2       87
## 10 2Ge+her The Hardest Part Of ... 2000-09-02  wk3       92
## # ... with 5,297 more rows
```

The `values_drop_na = TRUE` argument removes the `NA` values from the rank column while pivoting.

Which 2000 song spent the most weeks on the charts? This question would have been near impossible to answer without tidying the data. Now we can group by track and count the number of times it was ranked:

```
long.bill %>%
  group_by(track) %>%
  summarize(weeks_on_chart = n()) %>%
  slice_max(weeks_on_chart)
```

```
## # A tibble: 1 x 2
##   track  weeks_on_chart
##   <chr>           <int>
## 1 Higher             57
```

*Higher*, by Creed, entered the charts in 2000 and spent 57 weeks in the top 100 (though not consecutively!)

The `tidyr` package also provides a function `pivot_wider` which performs the opposite transformation to `pivot_longer`. You will need to provide a `names_from` value, which contains the variable name to get the column names from, and a `values_from` value or values, which gives the columns to get the cell values from. You often will also need to provide an `id_cols` which uniquely identifies each observation. Let's see how to use it in an example.

**Example 6.14.** Consider the `babynames` data set in the `babynames` package. This consists of all names of babies born in the USA from 1880 through 2017, together with the sex assigned to the baby, the number of babies of that sex given that name in that year, and the proportion of babies with that sex in that year that were given the name. Only names that were given to five or more babies are included.

```
library(babynames)
head(babynames)
```

```
## # A tibble: 6 x 5
##    year sex   name          n   prop
##   <dbl> <chr> <chr>     <int>  <dbl>
## 1  1880 F     Mary       7065 0.0724
## 2  1880 F     Anna       2604 0.0267
## 3  1880 F     Emma       2003 0.0205
## 4  1880 F     Elizabeth  1939 0.0199
## 5  1880 F     Minnie     1746 0.0179
## 6  1880 F     Margaret   1578 0.0162
```

Convert the `babynames` data from year 2000 into wide format, where each row is a name together with the number of male babies and female babies with that name. Note that we need to either select down to the variables of interest or we would need to provide the id columns.

```
babynames %>%
  filter(year == 2000) %>%
  pivot_wider(id_cols = name, names_from = sex, values_from = n)
```

```
## # A tibble: 27,512 x 3
##    name          F     M
##    <chr>     <int> <int>
##  1 Emily     25953    30
##  2 Hannah    23080    25
##  3 Madison   19967   138
##  4 Ashley    17997    82
##  5 Sarah     17697    26
##  6 Alexis    17629  2714
##  7 Samantha  17266    21
##  8 Jessica   15709    27
##  9 Elizabeth 15094    22
## 10 Taylor    15078  2853
## # ... with 27,502 more rows
```

Not every name had both a male and a female entry. Those are listed as `NA` in the data frame. If we would prefer them to be 0, we can add `values_fill = 0`.

Now that we have the data in a nice format, let's determine the most common name for which the number of male babies was within 200 of the number of female babies.

```
babynames %>%
  filter(year == 2000) %>%
  pivot_wider(
    id_cols = name, names_from = sex,
    values_from = n, values_fill = 0
  ) %>%
  filter(M > F - 200, M < F + 200) %>%
  mutate(total_names = F + M) %>%
  slice_max(n = 5, total_names)
```

```
## # A tibble: 5 x 4
##   name         F     M total_names
##   <chr>    <int> <int>       <int>
## 1 Peyton    1967  2001        3968
## 2 Skyler    1284  1472        2756
## 3 Jessie     719   533        1252
## 4 Justice    477   656        1133
## 5 Sage       557   392         949
```

We see by this criteria, Peyton was the most common gender neutral name in 2000, followed by Skyler, Jessie, Justice, and Sage.

### 6.6.2 Using join to merge data frames

In this section, we introduce the `join` family of functions, part of the `dplyr` package. We will focus on the function `left_join`, which **joins** two data frames together into one. The syntax is `left_join(x, y, by = NULL)`, where `x` and `y` are data frames and `by` is a list of columns that you want to join the data frames by (there are additional arguments that we won't be using). The way to remember what `left_join` does is that it adds the columns of `y` to the columns of `x` by matching columns with the same names.

**Example 6.15.** The `band_members` and `band_instruments` data sets are both included in the `dplyr` package.

```
band_members
```

```
## # A tibble: 3 x 2
##   name  band
##   <chr> <chr>
## 1 Mick  Stones
## 2 John  Beatles
## 3 Paul  Beatles
```

```
band_instruments
```

```
## # A tibble: 3 x 2
##   name  plays
##   <chr> <chr>
## 1 John  guitar
## 2 Paul  bass
## 3 Keith guitar
```

Left join adds instrument information to the band member data.

```
left_join(band_members, band_instruments)
```

```
## Joining, by = "name"
```

```
## # A tibble: 3 x 3
##   name  band    plays
##   <chr> <chr>   <chr>
## 1 Mick  Stones  <NA>
## 2 John  Beatles guitar
## 3 Paul  Beatles bass
```

**Example 6.16.** What happens if Paul learns how to play the drums as well?

```
new_band_instruments
```

```
## # A tibble: 4 x 2
##   name  plays
##   <chr> <chr>
## 1 John  guitar
## 2 Paul  bass
## 3 Keith guitar
## 4 Paul  drums
```

```
left_join(band_members, new_band_instruments)

## Joining, by = "name"

## # A tibble: 4 x 3
##   name  band    plays
##   <chr> <chr>   <chr>
## 1 Mick  Stones  <NA>
## 2 John  Beatles guitar
## 3 Paul  Beatles bass
## 4 Paul  Beatles drums
```

When there are multiple matches to `name`, then `left_join` makes multiple rows in the new data frame.

Here is a more compelling example where we combine data from multiple sources into a single data frame.

**Example 6.17.** Consider the data sets `pres_election` available in the `fosdata` package, and `unemp` available in the `mapproj` package. These data sets give the by county level presidential election results from 2000-2016, and the population and unemployment rate of all counties in the US. Some packages, such as mapproj, require you to use a command like `data("unemp"))` to access data sets.

```
pres_election <- fosdata::pres_election
library(mapproj)

## Loading required package: maps

data("unemp")
```

Suppose we want to create a new data frame that contains **both** the election results and the unemployment data. That is a job for `left_join`.

```
combined_data <- left_join(pres_election, unemp, by = c("FIPS" = "fips"))
```

Because the two data frames didn't have any column names in common, the join required `by =` to specify which columns to join by.

```
head(combined_data)

##   year   state state_po  county FIPS       candidate      party
## 1 2000 Alabama       AL Autauga 1001         Al Gore   democrat
## 2 2000 Alabama       AL Autauga 1001 George W. Bush republican
## 3 2000 Alabama       AL Autauga 1001    Ralph Nader      green
## 4 2000 Alabama       AL Autauga 1001          Other       <NA>
## 5 2000 Alabama       AL Baldwin 1003         Al Gore   democrat
## 6 2000 Alabama       AL Baldwin 1003 George W. Bush republican
##   candidatevotes totalvotes   pop unemp
## 1           4942      17208 23288   9.7
## 2          11993      17208 23288   9.7
## 3            160      17208 23288   9.7
## 4            113      17208 23288   9.7
## 5          13997      56480 81706   9.1
## 6          40872      56480 81706   9.1
```

Word to the wise: when dealing with spatial data that is organized by county, it is much easier to use the FIPS code than the county name, because there are variances in capitalization and other issues that come up when trying to use county names.

As a final example, we show how the `movies` data for `fosdata` was constructed using `left_join`.

**Example 6.18.** MovieLens distributes their movie rating information in two separate CSV files, which you may access as follows:

```
movies_orig <-
  read.csv("https://stat.slu.edu/~speegle/data/ml-small/movies.csv")
ratings_orig <-
  read.csv("https://stat.slu.edu/~speegle/data/ml-small/ratings.csv")
```

The `movies_orig` data contains the movie names and genres, while the `ratings_orig` data contains the user ID, the rating, and the timestamp.

```
head(movies_orig)
```

```
##   movieId                              title                          genres
## 1       1                   Toy Story (1995) Adventure|Animation|Chi...
## 2       2                     Jumanji (1995) Adventure|Children|Fantasy
## 3       3            Grumpier Old Men (1995)              Comedy|Romance
## 4       4           Waiting to Exhale (1995)        Comedy|Drama|Romance
## 5       5 Father of the Bride Part II (1995)                      Comedy
## 6       6                        Heat (1995)       Action|Crime|Thriller
```

```
head(ratings_orig)
```

```
##   userId movieId rating timestamp
## 1      1       1      4 964982703
## 2      1       3      4 964981247
## 3      1       6      4 964982224
## 4      1      47      5 964983815
## 5      1      50      5 964982931
## 6      1      70      3 964982400
```

We combine the two data sets using `left_join` to get the `movies` data that we used throughout the chapter.

```
movies <- left_join(movies_orig, ratings_orig, by = "movieId")
```

## 6.7 The apply family

This book focuses on `dplyr` and the tidyverse for data manipulation. However, it can be useful to know the base R tools for doing some of the same tasks. At a minimum, it can be useful to know these tools when reading other people's code!

The `apply` family consists of quite a few functions, including `apply`, `sapply`, `lapply`, `vapply`, `tapply`, `mapply`, and even `replicate`! While `tapply` and its variants might be the closest in spirit to how we used `dplyr` in this chapter, we will focus on `apply` and `sapply`.

All of the functions in the apply family are implicit loops, like replicate. The typical usage of apply is apply(X, MARGIN, FUN). The function apply applies the function FUN to either the rows or columns of X, depending on whether MARGIN is 1 (rows) or 2 (columns). So, the argument X is typically a matrix or data frame and MARGIN is either 1 or 2. The argument FUN is a function, such as sum or mean, or a custom function defined by the user.

**Example 6.19.** Consider the data set USJudgeRatings, which contains the ratings of US judges by lawyers on various facets.

```
head(USJudgeRatings)
```

```
##                  CONT INTG DMNR DILG CFMG DECI PREP FAMI ORAL WRIT PHYS RTEN
## AARONSON,L.H.     5.7  7.9  7.7  7.3  7.1  7.4  7.1  7.1  7.1  7.0  8.3  7.8
## ALEXANDER,J.M.    6.8  8.9  8.8  8.5  7.8  8.1  8.0  8.0  7.8  7.9  8.5  8.7
## ARMENTANO,A.J.    7.2  8.1  7.8  7.8  7.5  7.6  7.5  7.5  7.3  7.4  7.9  7.8
## BERDON,R.I.       6.8  8.8  8.5  8.8  8.3  8.5  8.7  8.7  8.4  8.5  8.8  8.7
## BRACKEN,J.J.      7.3  6.4  4.3  6.5  6.0  6.2  5.7  5.7  5.1  5.3  5.5  4.8
## BURNS,E.B.        6.2  8.8  8.7  8.5  7.9  8.0  8.1  8.0  8.0  8.0  8.6  8.6
```

Which judge had the highest mean rating? Which rating scale had the highest mean across all judges?

Using apply across rows computes the mean rating for each judge.

```
head(apply(USJudgeRatings, MARGIN = 1, mean))
```

```
##  AARONSON,L.H. ALEXANDER,J.M. ARMENTANO,A.J.    BERDON,R.I.
##       7.291667       8.150000       7.616667       8.458333
##   BRACKEN,J.J.     BURNS,E.B.
##       5.733333       8.116667
```

To pull the highest mean rating:

```
max(apply(USJudgeRatings, MARGIN = 1, mean))
```

```
## [1] 8.858333
```

```
which.max(apply(USJudgeRatings, MARGIN = 1, mean))
```

```
## CALLAHAN,R.J.
##             7
```

We see that judge seven, R.J. Callahan, had the highest rating of 8.858333.

Using apply across columns computes the mean rating of all of the judges in each category.

```
apply(USJudgeRatings, MARGIN = 2, mean)
```

```
##     CONT     INTG     DMNR     DILG     CFMG     DECI     PREP     FAMI
## 7.437209 8.020930 7.516279 7.693023 7.479070 7.565116 7.467442 7.488372
##     ORAL     WRIT     PHYS     RTEN
## 7.293023 7.383721 7.934884 7.602326
```

Judges scored highest on the "integrity" scale.

The sapply function has as its typical usage sapply(X, FUN), where X is a vector, and FUN is the function that we wish to apply to each element of X. You may create custom functions with the function keyword.

**Example 6.20.** Approximate the sd of a $t$ random variable with 5, 10, 15, 20, 25, and 30 degrees of freedom.

To do this for 5 df, we simulate 10,000 random values of $t$ and take the standard deviation:

```
sd(rt(10000, df = 5))
```

```
## [1] 1.328793
```

To repeat this computation for the other degrees of freedom, we turn the simulation into a function we choose to call `sim_t_sd`. The function takes an argument we call `n` and runs the simulation. In a function, the last value before the closing } is the returned value of the function.

```
sim_t_sd <- function(n) {
  sd(rt(10000, df = n))
}
sim_t_sd(5) # check it for n=5
```

```
## [1] 1.260566
```

We create a sequence of values $5, 10, \ldots, 30$ and then use `sapply` to feed each of these values to the `sim_t_sd` function.

```
sapply(X = seq(5, 30, 5), FUN = sim_t_sd)
```

```
## [1] 1.292638 1.117897 1.071840 1.050075 1.047600 1.033202
```

`sapply` took each element of the vector `X`, applied the `sim_t_sd` function to it, and returned the results as a vector.

Unfortunately, `sapply` is not type consistent in the way it simplifies results, sometimes returning a vector, sometimes a matrix. This gives `sapply` a bad reputation, but it is still useful for exploratory analysis. The function `replicate` is a "convenience wrapper" for a common usage of `sapply`, so that `replicate(100, {expr})` is equivalent to `sapply(1:100, function(x) {expr})`, where in this case, the function inside of `sapply` will have no dependence on `x`.

## Vignette: dplyr murder mystery

> There has been a murder in dplyr City! You have lost your notes, but you remember that the murder took place on January 15, 2018. All of the clues that you will need to solve the murder are contained in the `dplyrmurdermystery` package once you load it into your environment.

The `dplyr` murder mystery is a whodunit, where you use your `dplyr` skillz to analyze data sets that lead to the solution of a murder mystery. The original caper of this sort was the Command Line Murder Mystery,[7] which was recast as SQL Murder Mystery.[8] We have created an R package containing a mystery that is not entirely unlike the SQL Murder Mystery for you to enjoy. You can download the package via

[7] https://github.com/veltman/clmystery
[8] https://github.com/NUKnightLab/sql-mysteries

```
remotes::install_github(repo = "speegled/dplyrmurdermystery")
```

Once you install the package, you can load the data into your environment by typing

```
library(dplyrmurdermystery)
data("dplyr_murder_mystery")
```

This loads quite a bit of data into your environment, so you may want to make sure you are starting with a clean environment before you do it.

## Vignette: Data and gender

In most studies with human subjects, the investigators collect demographic information on their subjects to control for lurking variables that may interact with the variables under investigation. One of the most commonly collected demographic variables is gender. In this vignette, we discuss current best practices for collecting gender information in surveys and studies.

### Why not male or female?

The term *transgender* refers to people whose gender expression defies social expectations. More narrowly, the term *transgender* refers to people whose gender identity or gender expression differs from expectations associated with the sex assigned at birth.[9]

Approximately 1 in 250 adults in the US identify as transgender, with that proportion expected to rise as survey questions improve and stigma decreases.[10] For teenagers, this proportion may be as high as 0.7%.[11]

The traditional "Male or Female?" survey question fails to measure the substantial transgender population, and the under- or non-representation of transgender individuals is a barrier to understanding social determinants and health disparities faced by this population.

### How to collect gender information

Best practices[12] in collecting gender information are to use a two-step approach, with the following questions:

1. Assigned sex at birth: What sex were you assigned at birth, on your original birth certificate?

- Male
- Female

2. Current gender identity: How do you describe yourself? (check one)

[9] Marla Berg-Weger, *Social Work and Social Welfare: An Invitation* (Routledge, 2016).

[10] Esther L Meerwijk and Jae M Sevelius, "Transgender Population Size in the United States: A Meta-Regression of Population-Based Probability Samples," *American Journal of Public Health* 107, no. 2 (2017): e1–8, https://doi.org/10.2105/AJPH.2016.303578.

[11] N Chokshi, "One in Every 137 Teenagers Would Identify as Transgender, Report Says," *The New York Times*, February 23, 2017.

[12] The GenIUSS Group, "Best Practices for Asking Questions to Identify Transgender and Other Gender Minority Respondents on Population-Based Surveys," ed. J. L. Herman (Los Angeles, CA: The Williams Institute, 2014).

- Male
- Female
- Transgender
- Do not identify as female, male, or transgender

These questions serve as guides. Question 1 has broad agreement, while there is still discussion and research being done as to what the exact wording of Question 2 should be. It is also important to talk to people who have expertise in or familiarity with the specific population you wish to sample from, to see whether there are any adjustments to terminology or questions that should be made for that population.

In general, questions related to sex and gender are considered "sensitive." Respondents may be reluctant to answer sensitive questions or may answer inaccurately. For these questions, privacy matters. When possible, place sensitive questions on a self-administered survey. Placing gender and sexual orientation questions at the start of a survey may also decrease privacy, since those first questions are encountered at the same time for all subjects and are more visible on paper forms.

### How to incorporate gender in data analysis

The relatively small samples of transgender populations create challenges for analysis. One type of error is called a *specificity error* and occurs when subjects mistakenly indicate themselves as transgender or another gender minority. The transgender population is less than 1% of the overall population. So, if even one in one hundred subjects misclassifies themselves, then the subjects identifying as transgender in the sample will be a mix of half transgender individuals and half misclassified individuals. The best way to combat specificity errors is with carefully worded questions and prepared language to explain the options if asked.

Small samples lead to a large margin of error and make it difficult to detect statistically significant differences within groups.[13] For an individual survey, this may prevent analysis of fine-grained gender information. Aggregating data over time, space, and across surveys can allow analysis of gender minority groups.

### Example

The data `fosdata::gender` was collected by Sell, Goldberg, and Conron[14] to determine if it is feasible to sample from rare and dispersed populations using the online service "Google Android Panel." They use the two-step approach to gender, with a sex-at-birth question and a gender identity question that is broadly worded and that allows respondents to select multiple options. This multiple-option question is coded as seven different T/F variables beginning with `gender_`.

```
gender <- fosdata::gender
str(gender)
```

```
## 'data.frame':    20305 obs. of  10 variables:
##  $ gender_male    : logi  FALSE FALSE FALSE FALSE FALSE TRUE ...
##  $ gender_female  : logi  TRUE TRUE TRUE TRUE TRUE FALSE ...
##  $ gender_trans   : logi  FALSE FALSE FALSE TRUE FALSE FALSE ...
##  $ gender_queer   : logi  FALSE FALSE FALSE FALSE FALSE FALSE ...
```

[13]These terms are defined in Chapter 8. If these terms are new to you, just think of it for now as meaning that it is hard to get information from small amounts of data.

[14]Randall Sell, Shoshana Goldberg, and Kerith Conron, "The Utility of an Online Convenience Panel for Reaching Rare and Dispersed Populations," *PLOS One* 10 (December 2015): e0144011, https://doi.org/10.1371/journal.pone.0144011.

```
##  $ gender_not_sure: logi  FALSE FALSE FALSE FALSE FALSE FALSE ...
##  $ gender_unclear : logi  FALSE FALSE FALSE FALSE FALSE FALSE ...
##  $ gender_na      : logi  FALSE FALSE FALSE FALSE FALSE FALSE ...
##  $ sex_at_birth   : Factor w/ 2 levels "Female","Male": 1 1 1 1 1 2 1 ..
##  $ hispanic       : Factor w/ 3 levels "Don't know","No",..: 2 2 2 2 2..
##  $ race           : chr  "White" "White" "Black or African American" "..
```

As a first step, let's explore how many respondents selected more than one option:

```
gender <- gender %>%
  mutate(responses = rowSums(across(matches("^gender"))))
table(gender$responses)
```

```
##
##     1     2     3     4     5     6
## 20050   200    34     1     3    17
```

Here we used dplyr's `across` function to say we want to take row sums of responses across all variables that start with "gender." When summing, `TRUE` values count as 1.

Observe that the vast majority of the respondents selected only one value. However, the prevalence of respondents with two or more values is not small when compared

with the number of non-cisgender individuals (only 757 as we will see shortly). In particular, 17 individuals answered yes to six of the seven gender questions available, and those answers seem likely to be due to user confusion or error possibly leading to specificity error in our results.

Following Sell et al., we compute a "transgender status" summary variable which labels respondents reporting:

1. current gender identity of male and being assigned male at birth as "Male (cisgender)"
2. current gender identity of female and being assigned female at birth as "Female (cisgender)"
3. current gender identity of male, transgender, or genderqueer/gender non-conforming and being assigned female at birth as "Male,Trans,GenQ/Female@Birth"
4. current gender identity of female, transgender, or genderqueer/gender non-conforming and being assigned male at birth as "Female,Trans,GenQ/Male@Birth"

We use dplyr's `case_when` statement to select the cases.

```
gender <- gender %>%
  mutate(trans_status = case_when(
    sex_at_birth == "Male" & gender_male & responses == 1 ~
    "Male (cisgender)",
    sex_at_birth == "Female" & gender_female & responses == 1 ~
    "Female (cisgender)",
    sex_at_birth == "Female" & gender_male | gender_trans | gender_queer ~
    "Male,Trans,GenQ/Female@Birth",
    sex_at_birth == "Male" & gender_female | gender_trans | gender_queer ~
    "Female,Trans,GenQ/Male@Birth"
  ))
gender %>%
  group_by(trans_status) %>%
  summarize(n())
```

```
## # A tibble: 5 x 2
```

```
##   trans_status                  `n()`
##   <chr>                         <int>
## 1 Female (cisgender)             9733
## 2 Female,Trans,GenQ/Male@Birth     47
## 3 Male (cisgender)               9815
## 4 Male,Trans,GenQ/Female@Birth    396
## 5 <NA>                            314
```

We finish this example by making a table of trans status gender by race. For simplicity, we only look at respondents with a single answer to the race question and ignore the `hispanic` variable.

```
gender %>%
  filter(!stringr::str_detect(race, ",")) %>%
  mutate(trans = stringr::str_detect(trans_status, "Trans")) %>%
  group_by(race) %>%
  summarize(
    respondents = n(),
    trans_pct = round(100 * mean(trans, na.rm = TRUE), 1)
  )
```

```
## # A tibble: 6 x 3
##   race                               respondents trans_pct
##   <chr>                                    <int>     <dbl>
## 1 American Indian or Alaskan Native          170       9.1
## 2 Asian                                     1647       1.5
## 3 Black or African American                 1212       2.1
## 4 None of the above                          626       3.2
## 5 Some other race                            842       3.6
## 6 White                                   13443       1.5
```

# Exercises

Exercises 6.1 – 6.2 require material through Section 6.2.

**6.1.** The built-in data set `iris` is a data frame containing measurements of the sepals and petals of 150 iris flowers. Convert this data to a tibble with new variables `Sepal.Area` and `Petal.Area` which are the product of the corresponding length and width measurements.

**6.2.** Consider the `austen` data set in the `fosdata` package. This data frame contains the complete texts of *Emma* and *Pride and Prejudice*, with additional information which you can read about in the help page for the data set. Each of the following tasks corresponds to using a single `dplyr` verb.

a. Create a new data frame that consists only of the observations in *Emma*.
b. Create a new data frame that contains only the variables `word`, `word_length` and `novel`.
c. Create a new data frame that has the words in both books arranged in descending word length.

d. Create a new data frame that contains only the longest words that appeared in either of the books.
e. What was the mean word length in the two books together?
f. Create a new data frame that consists only of the distinct words found in the two books, together with the word length and sentiment score variables. (Hint: use `distinct`).

---

Exercises 6.3 – 6.26 require material through Section 6.4.

**6.3.** Consider the `mpg` data set in the `ggplot2` package. For the purposes of this question, consider each observation a different car.

a. Which car(s) had the highest highway gas mileage?
b. Compute the mean city mileage for compact cars.
c. Compute the mean city mileage for each class of cars, and arrange in decreasing order.
d. Which cars have the smallest absolute difference between highway mileage and city mileage?
e. Compute the mean highway mileage for each year, and arrange in decreasing order.

**6.4.** This question uses the `DrinksWages` from the `HistData` package. This data, gathered in 1910, was a survey of people working in various trades (bakers, plumbers, goldbeaters, etc.). The trades are assigned class values of A, B, or C based on required skill. For each trade, the number of workers who drink (`drinks`), the number of sober workers (`sober`), and wage information (`wage`) was recorded. There is also a column `n = drinks + sober` which is the total number of workers surveyed for each trade.

a. Compute the mean wages for each class, A, B, and C.
b. Find the three trades with the highest proportion of drinkers. Consider only trades with 10 or more workers in the survey.

**6.5.** Consider the data set `oly12` from the `VGAMdata` package (that you will probably need to install). It has individual competitor information from the Summer 2012 London Olympic Games.

a. According to this data, which country won the most medals? How many did that country win? (You need to sum Gold, Silver, and Bronze.)
b. Which countries were the heaviest? Compute the mean weight of male athletes for all countries with at least 10 competitors, and report the top three.

---

Exercises 6.6 – 6.10 all use the `movies` data set from the `fosdata` package.

**6.6.** What is the movie with the highest mean rating that has been rated at least 30 times?

**6.7.** Which genre has been rated the most? (For the purpose of this, consider Comedy and Comedy|Romance as completely different genres, for example.)

**6.8.** Which movie in the genre Comedy|Romance that has been rated at least 50 times has the lowest mean rating? Which has the highest mean rating?

**6.9.** Which movie that has a mean rating of 4 or higher has been rated the most times?

**6.10.** Which user gave the highest mean ratings?

---

Exercises 6.11 – 6.16 all use the `Batting` data set from the `Lahman` package. This gives the batting statistics of every player who has played baseball from 1871 through the present day.

**6.11.** Which player has been hit-by-pitch the most number of times?

**6.12.** How many doubles were hit in 1871?

**6.13.** Which team has the most total number of home runs, all time?

**6.14.** Which player who has played in at least 500 games has scored the most number of runs per game?

**6.15.** a. Which player has the most lifetime at bats without ever having hit a home run?
b. Which active player has the most lifetime at bats without ever having hit a home run? (An active player is someone with an entry in the most recent year of the data set).

**6.16.** a. Verify that Curtis Granderson hit the most triples in a single season since 1960.
b. In which season did the major league leader in triples have the fewest triples?
c. In which season was there the biggest difference between the major league leader in stolen bases (SB) and the player with the second most stolen bases?

---

Exercises 6.17 – 6.24 all use the `Pitching` data set from the `Lahman` package. This gives the pitching statistics of every pitcher who has played baseball from 1871 through the present day.

**6.17.** a. Which pitcher has won (W) the most number of games?
b. Which pitcher has lost (L) the most number of games?

**6.18.** Which pitcher has hit the most opponents with a pitch (HBP)?

**6.19.** Which year had the most number of complete games (CG)?

**6.20.** Among pitchers who have won at least 100 games, which has the highest winning percentage? (Winning percentage is wins divided by wins + losses.)

**6.21.** Among pitchers who have struck out at least 500 batters, which has the highest strikeout to walk ratio? (Strikeout to walk ratio is SO/BB.)

**6.22.** List the pitchers for the St Louis Cardinals (SLN) in 2006 with at least 30 recorded outs (IPouts), sorted by ERA from lowest to highest.

**6.23.** A balk (BK) is a subtle technical mistake that a pitcher can make when throwing. What were the top five years with the most balks in major league baseball history? Why was 1988 known as "the year of the balk?"

**6.24.** Which pitcher has the most outs pitched (IPouts) of all the pitchers who have more bases on balls (BB) than hits allowed (H)? Who has the second-most?

**6.25.** Consider the `storms` data set in the `dplyr` package, from Example 6.5. Recall that `name` and `year` together identify all storms except Zeta (2005-2006).

a. Which name(s) was/were given to the most storms?
b. Which year(s) had the most named storms?
c. The second strongest storm named Lili had maximum wind speed of 100. Which name's second strongest storm in terms of maximum wind speed was the strongest among all names' second strongest storms? The `dplyr` function `nth` may be useful for doing this problem.

**6.26.** M&M's are small pieces of candy that come in various colors and are sold in various sizes of bags. One popular size for Halloween is the "Fun Size," which typically contains between 15 and 18 M&M's. For the purposes of this problem, we assume the mean number of candies in a bag is 17 and the standard deviation is 1. According to Rick Wicklin[15], the proportion of M&M's of various colors produced in the New Jersey M&M factory are as follows.

| Color | Proportion |
|---|---|
| Blue | 25.0 |
| Orange | 25.0 |
| Green | 12.5 |
| Yellow | 12.5 |
| Red | 12.5 |
| Brown | 12.5 |

Suppose that you purchase a big bag containing 200 Fun Size M&M packages. The purpose of this exercise it to estimate the probability that each of your bags has a different distribution of colors.

a. We will model the number of M&M's in a bag with a binomial random variable. Find values of $n$ and $p$ such that a binomial random variable with parameters $n$ and $p$ has mean approximately 17 and variance approximately 1.
b. (Hard) Create a data frame that has its rows being bags and columns being colors, where each entry is the number of M&M's in that bag of that color. For example, the first entry in the first column might be 2, the number of Blue M&M's in the first bag.
c. Use `nrows` and `dplyr::distinct` to determine whether each row in the data frame that you created is distinct.
d. Put inside `replicate` and estimate the probability that all bags will be distinct. This could take some time to run, depending on how you did parts (b) and (c), so start with 500 replicates.

---

Exercises 6.27 – 6.33 require material through Section 6.5.

**6.27.** The data set `words` is built into the `stringr` package.

a. How many words in this data set contain "ff?"
b. What percentage of these words start with "s?"

**6.28.** The data set `sentences` is built into the `stringr` package.

a. What percentage of sentences contain the string "the?"
b. What percentage of sentences contain the word "the" (so, either "the" or "The")?
c. What percentage of sentences start with the word "The?"
d. Find the one sentence that has both an "x" and a "q" in it.
e. Which words are the most common words that a sentence *ends* with?

**6.29.** The data set `fruit` is built into the `stringr` package.

a. How many fruits have the word "berry" in their name?
b. Some of these fruits have the word "fruit" in their name. Find these fruit and remove the word "fruit" to create a list of words that can be made into fruit. (Hint: use `str_remove`)

---

[15] Purtill, "A Statistician Got Curious about M&M Colors and Went on an Endearingly Geeky Quest for Answers."

Exercises 6.30 – 6.32 require the `babynames` data frame from the `babynames` package.

**6.30.** Say that a name is popular if it was given to 1000 or more babies of a single sex. How many popular female names were there in 2015? What percentage of these popular female names ended in the letter 'a?'

**6.31.** Consider the `babynames` data set. Restrict to babies born in 2003. We'll consider a name to be gender neutral if the number of male babies given that name is within plus or minus 20% of the number of girl babies given that name. What were the 5 most popular gender neutral names in 2003?

**6.32.** The `phonics` package has a function `metaphone` that gives a rough phonetic transcription of English words. Restrict to babies born in 2003 and create a new variable called `phonetic`, which gives the phonetic transcription of the names.

a. Filter the data set so that the year is 2003 and so that each phonetic transcription has at least two distinct names associated with it.
b. Filter the data so that it only contains the top 2 most given names for each phonetic transcription.
c. Among the pairs of names for girls obtained in the previous part with the same metaphone representation and such that each name occurs less than 120% of the times that the other name occurs, which pair of names was given most frequently?
d. We wouldn't consider the most common pair to actually be the "same name." Look through the list sorted by total number of times the pair of names occurs, and state which pair of names that you would consider to be the same name is the most common. That pair is, in some sense, the hardest to spell common name from 2003.
e. What is the most common hard to spell *triple* of girls names from 2003? (Answers may vary, depending on how you interpret this. To get interesting answers, you may need to loosen the 120% rule from the previous part.)

---

**6.33.** In this exercise, we examine the expected number of coin tosses until various patterns occur. For each pattern given below, find the expected number of tosses until the pattern occurs. For example, if the pattern is HTH and you toss HHTTHTTHTH, then it would be 10 tosses until HTH occurs. With the same sequence of tosses, it would be 5 tosses until TTH occurs. (Hint: `paste(coins, collapse = "")` converts a vector containing H and T into a character string on which you can use `stringr` commands such as `str_locate`.)

a. What is the expected number of tosses until HTH occurs?
b. Until HHT occurs?
c. Until TTTT occurs?
d. Until THHH occurs?

---

Exercises 6.34 – 6.37 require material through Section 6.6.

**6.34.** This exercise uses the `billboard` data from the `tidyr` package.

a. Which artist had the most tracks on the charts in 2000?
b. Which track from 2000 spent the most weeks at #1? (This requires tidying the data as described in Section 6.6.1.)

**6.35.** Consider the `scrabble` and `letter_frequency` data sets in the `fosdata` package.

a. Create a new data frame that contains the variables `letter`, `points`, `number`, and `frequency` for all English letters a through z. `number` should be the number of tiles in a Scrabble set with that letter, and `frequency` should be the letter frequency from `letter_frequency`

b. Which letter has the largest absolute difference between the percentage of tiles that are that letter and the frequency of that letter occurring in `letter_frequency`? Note that the frequency of tiles in Scrabble should be related to the frequency of letters in the **dictionary**. The data set `letter_frequency` gives the frequency in texts, so we would not expect the frequencies to match.

**6.36.** Consider the `scotland_births` data set in the `fosdata` package. This data gives the number of births by the age of the mother in Scotland for each year from 1945-2019. This data is in wide format. (Completion of this exercise will be helpful for Exercise 7.28.)

a. Convert the data into long format with three variable names: `age`, `year` and `births`, where each observation is the number of `births` in `year` to mothers that are `age` years old.
b. Convert the year to integer by removing the `x` and using `as.integer`.
c. Which year had the most babies born to mothers 20-years-old or younger?

**6.37.** The data set `world_cup` from `fosdata` has the results of all games in the 2014 and 2015 FIFA World Cup soccer finals. From this data, create a data frame which has the total number of goals scored by each team in the 2015 World Cup. Your data frame should have only two variables, `team` and `goals`, and 24 rows, one for each team. Display the entire data frame in descending order of goals scored.

---

Exercise 6.38 requires material through Section 6.7.

**6.38.** Suppose you sample five numbers from a uniform distribution on the interval $[0, 1]$. Use simulation to show that the expected value of the $k$th smallest of the five values is $\frac{k}{6}$. That is, the minimum of the five values has expected value $1/6$, the second smallest of the values has expected value $2/6$, and so on.

# 7

# *Data Visualization with ggplot*

Data visualization is a critical aspect of statistics and data science. Visualization is crucial for **communication** because it presents the essence of the underlying data in a way that is immediately understandable. Visualization is also a tool for **exploration** that may provide insights into the data that lead to new discoveries.

This chapter introduces the package `ggplot2` to create visualizations that are more expressive and better looking than base R plots. `ggplot2` has a powerful yet simple syntax based on a conceptual framework called the *grammar of graphics*.

```
library(ggplot2)
```

Creating visuals often requires transforming, reorganizing or summarizing the data prior to plotting. The `ggplot2` package is designed to work well with `dplyr`, and an understanding of the `dplyr` tools introduced in Chapter 6 will be assumed throughout this chapter.

## 7.1 ggplot fundamentals

In this section, we discuss the grammar of graphics. We introduce the scatterplot as a concrete example of how to use the grammar of graphics, and we discuss how the presentation of data is affected by its structure.

### 7.1.1 The grammar of graphics

The *grammar of graphics* is a theory describing the creation of data visualizations, developed by Wilkinson.[1] Much like the grammar of a language allows the combination of words in ways that express complicated ideas, the grammar of graphics allows the combination of simpler components in ways that express complicated visuals.

The most important aspect of a visualization is the **data**. Without data, there is nothing to visualize. Wilkinson describes three types of data: empirical, abstract, and meta. Empirical data is data that comes from observations, abstract data comes from formal models, and meta data is data on data. We will be concerned with empirical data in this chapter. So, we assume that we have a collection of observations of variables from some experiment that is stored in a data frame.

Next, we need a mapping from variables in the data to **aesthetics**. An aesthetic is a graphical property that can be altered to convey information. Common aesthetics are location in terms

[1] L Wilkinson et al., *The Grammar of Graphics*, Statistics and Computing (Springer New York, 2005).

DOI: 10.1201/9781003004899-7

of $x$- or $y$-coordinates, size, shape, and color. **Scales** control the details of how variables are mapped to aesthetics. For example, we may want to transform the coordinates or modify the default color assignment with a scale.

Variables are visualized through the aesthetic mapping and through **geometries** which are put in **layers** on a **coordinate system**. Examples of geometries are points, lines, or areas. A typical coordinate system would be the regular $x$-$y$ plane from mathematics. **Statistics** of the variables are computed inside the geometries, depending on the type of visualization desired. Some examples of statistics are counting the number of occurrences in a region, tabling the values of a variable, or the identity statistic which does nothing to the variable. **Facets** divide a single visualization into multiple graphs. Finally, a **theme** is applied to the overall visualization.

Creating a graph using ggplot2 is an iterative process. We will start with the data and aesthetics, then add geoms, scales, facets, and a theme, if desired. The data and aesthetics are typically inherited at each step, with a few exceptions. One notable exception is a class of **annotations**, which are layers that do not inherit global settings. An example might be if we wish to add a reference line whose slope and intercept are not computed from the data.

Don't worry if some of these words and concepts are unfamiliar to you right now. We will explain them further with examples throughout the text. We recommend returning to the vocabulary of the grammar of graphics introduced in this section after reading through the rest of the chapter.

### 7.1.2 Basic plot creation

The purpose of this example is to connect the terminology given in the previous section to a visualization. We will go into much more detail on this throughout the chapter.

Consider the built-in data set, CO2, which will be the **data** in this visualization. This data set gives the carbon dioxide uptake of six plants from Quebec and six plants from Mississippi. The uptake is measured at different levels of carbon dioxide concentration, and half of the plants of each type were chilled overnight before the experiment was conducted.

We will choose the variables conc, uptake and eventually Type as the variables that we wish to map to aesthetics. We start by mapping conc to the $x$-coordinate and uptake to the $y$-coordinate. The geometry that we will add is a scatterplot, given by geom_point.

```
ggplot(CO2, aes(x = conc, y = uptake)) +
  geom_point()
```

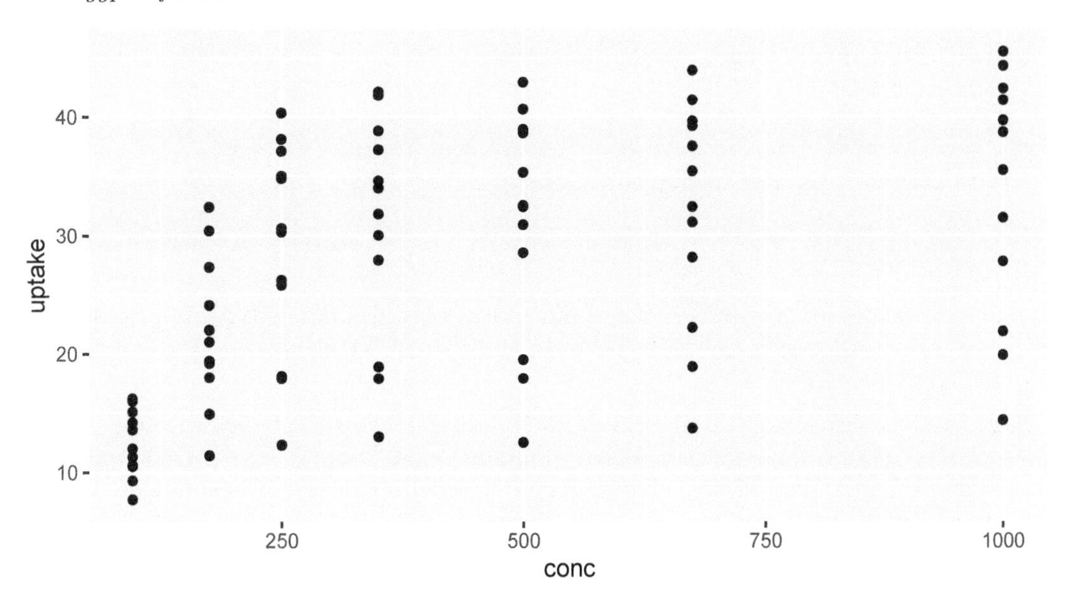


The function ggplot[2] takes as its first argument the data frame that we are working with, and as its second argument the aesthetic mappings between variables and visual properties. In this case, we are telling ggplot that the aesthetic "$x$-coordinate" is to be associated with the variable conc, and the aesthetic "$y$-coordinate" is to be associated with the variable uptake. Let's see what that command does all by itself:

```
ggplot(CO2, aes(x = conc, y = uptake))
```

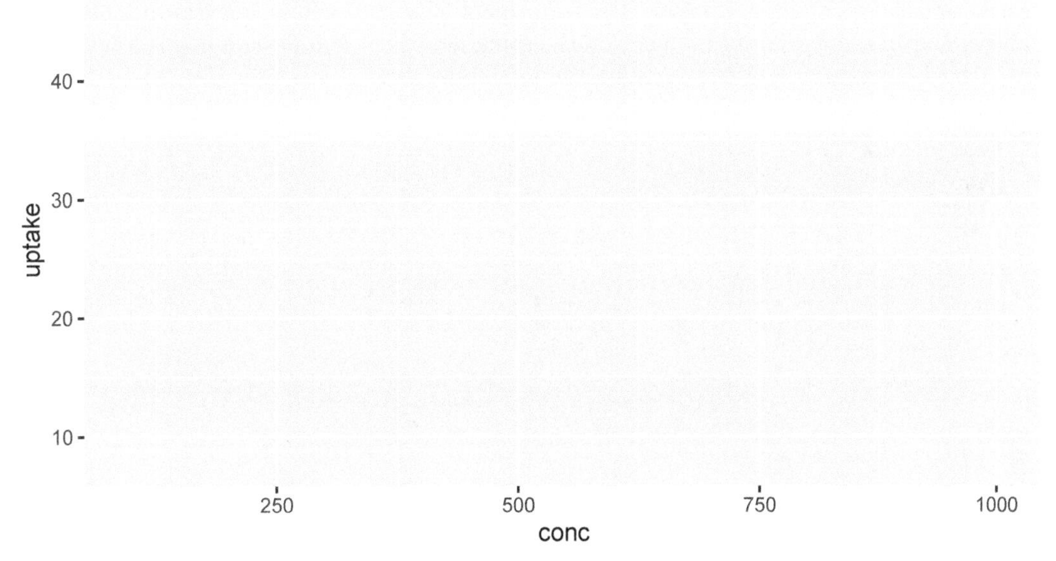


The ggplot function has set up the $x$-coordinates and $y$-coordinates for conc and uptake. Now, we just need to tell it what we want to do with those coordinates. That's where geom_point comes in. The function geom_point() inherits the x- and $y$-coordinates from ggplot, and plots them as points. To display a curve fit to the data, we can use the geometry geom_smooth().

[2]The workhorse function in the ggplot2 package is ggplot without the 2.

```
ggplot(CO2, aes(x = conc, y = uptake)) +
  geom_smooth()

## `geom_smooth()` using method = 'loess' and formula 'y ~ x'
```

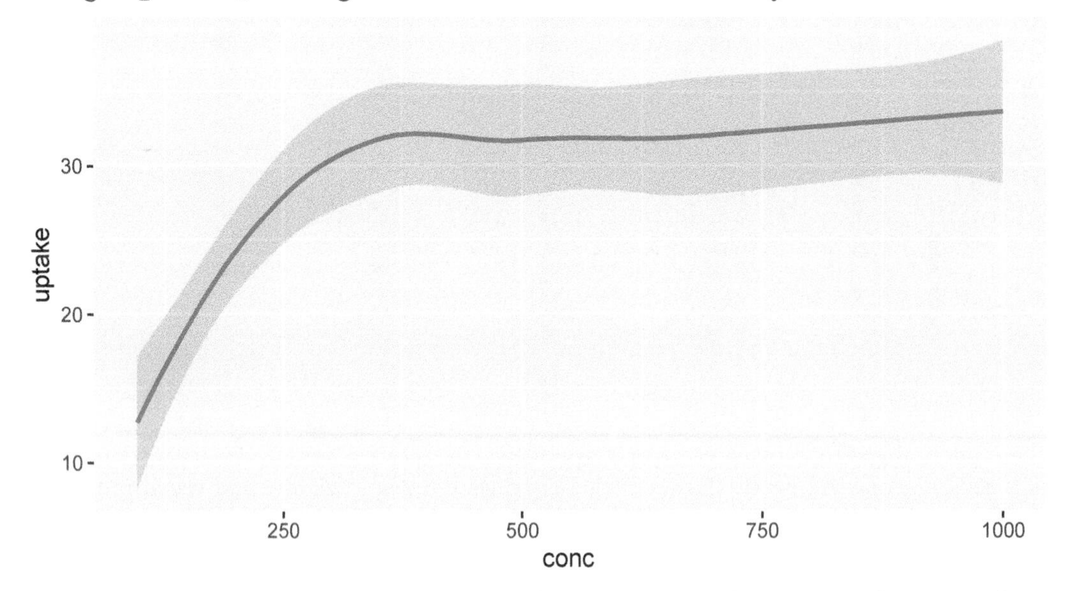


Notice that `geom_smooth` produced a message describing exactly how it went about smoothing the data. That can be helpful, but this book will suppress those messages from here on.

Each geometry in a plot is a **layer**. The '+' symbol is used to add new layers to a plot, and allows multiple geometries to appear on the same plot. For example, we might want to see the data points and the fitted curve on the same plot.

```
ggplot(CO2, aes(x = conc, y = uptake)) +
  geom_point() +
  geom_smooth()
```

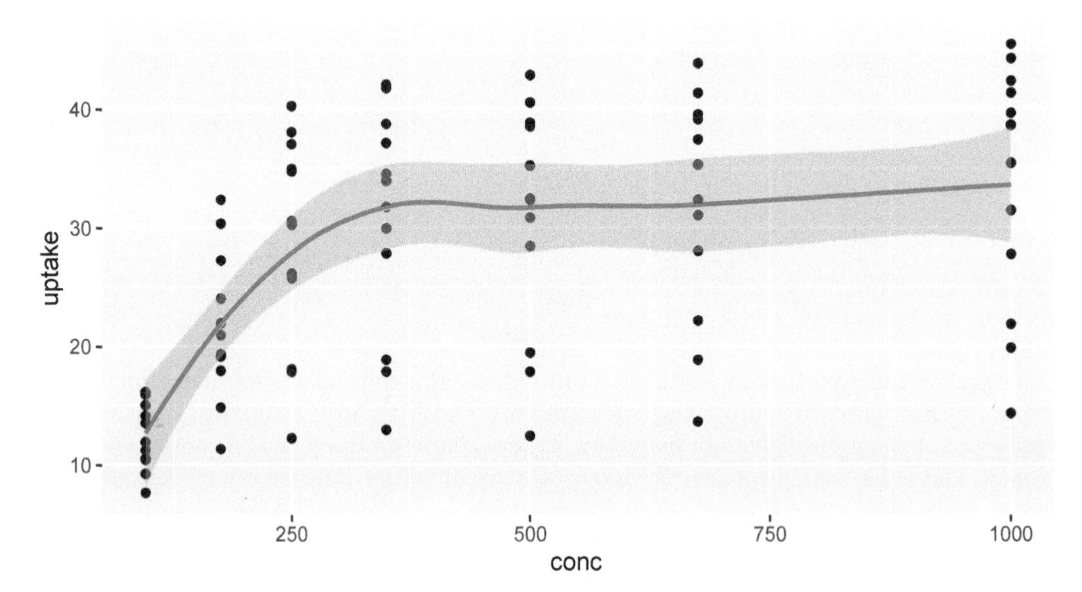


To add more information to the plot, we use **aesthetics** to map additional variables to visual properties of the graph. Different geometries support different aesthetics. `geom_point` requires both `x` and `y` and supports these:

**x** $x$ position

**y** $y$ position

**alpha**
transparency

**color**
color, or outline color

**fill**
fill color

**group**
grouping variable for fitting lines and curves

**shape**
point shape

**size**
point size

**stroke**
outline thickness

Let's use color to display the `Type` variable in the `CO2` data set, which shows whether readings are from Mississippi or Quebec. Both the `color` aesthetic and the `group` aesthetic assign groups to the data, which affect the results of geometries like `geom_smooth`, producing a smoothed curve for each group.

```
ggplot(CO2, aes(x = conc, y = uptake, color = Type)) +
  geom_point() +
  geom_smooth()
```

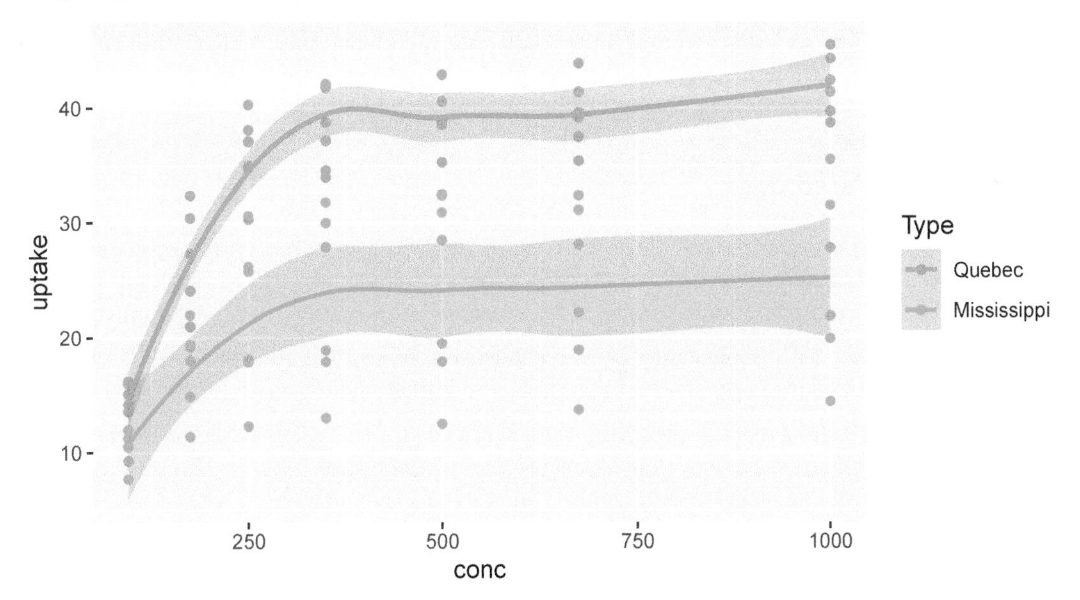


That plot is very useful! We see that the Quebecois plants have a larger uptake of $CO_2$ at each concentration level than do the Mississippians.

**Try It Yourself.**
Experiment with different aesthetics to display the other variables in the `CO2` data set.

The **ggplot** command produces a ggplot object, which is normally displayed immediately. It is also possible to store the ggplot object in a variable and then continue to modify that variable until the plot is ready for display. This technique is useful for building complicated plots in stages.

```
# store the plot without displaying it
co2plot <- ggplot(CO2, aes(x = conc, y = uptake, color = Type)) +
  geom_point() +
  geom_smooth()
```

Let's complete this plot by adding better labels, a **scale** to customize the colors assigned to Type, and a **theme** that will change the overall look and feel of the plot. See Figure 7.1.

```
co2plot <- co2plot +
  labs(
    title = "Carbon Dioxide Uptake in Grass Plants",
    caption = "Data from Potvin, Lechowicz, Tardif (1990)",
    x = "Ambient CO2 concentration (mL/L)",
    y = "CO2 uptake rates (umol/m^2 sec)",
    color = "Origin"
  )
co2plot <- co2plot +
  scale_color_manual(values = c("blue", "red")) +
  theme_minimal()
co2plot # display the completed plot
```

> **Try It Yourself.**
> The data in CO2 was obtained from measurements on only six plants total. Each plant was measured at six different concentration levels and chilled/nonchilled. Map **Plant** to the color aesthetic and create a scatterplot of uptake versus concentration.

### 7.1.3 Structured data

Plotting with ggplot requires you to provide clean, tidy data, with properly typed variables and factors in order. Many traditional plotting programs (such as spreadsheets) will happily generate charts from unstructured data. This contrast in philosophy can be summed up:

- To change how data is presented by a spreadsheet, you adjust the chart.
- To change how data is presented by ggplot, you adjust the structure of the data.

The advantage to the ggplot approach is that it forces you to reckon with the structure of your data, and understand how that structure is represented directly in the visualization. Visualizations with ggplot are reproducible, which means that new or changed data is easily dealt with, and chart designs can be consistently applied to different data sets.

With the spreadsheet approach, adjusting a chart after creating it requires the user to click through menus and dialog boxes. This method is not reproducible, and frequently leads to tedious repetition of effort. However, if you are used to editing charts in a spreadsheet, then the ggplot philosophy takes some getting used to.

The data manipulation tools from Chapter 6 are well suited to work with ggplot. The first

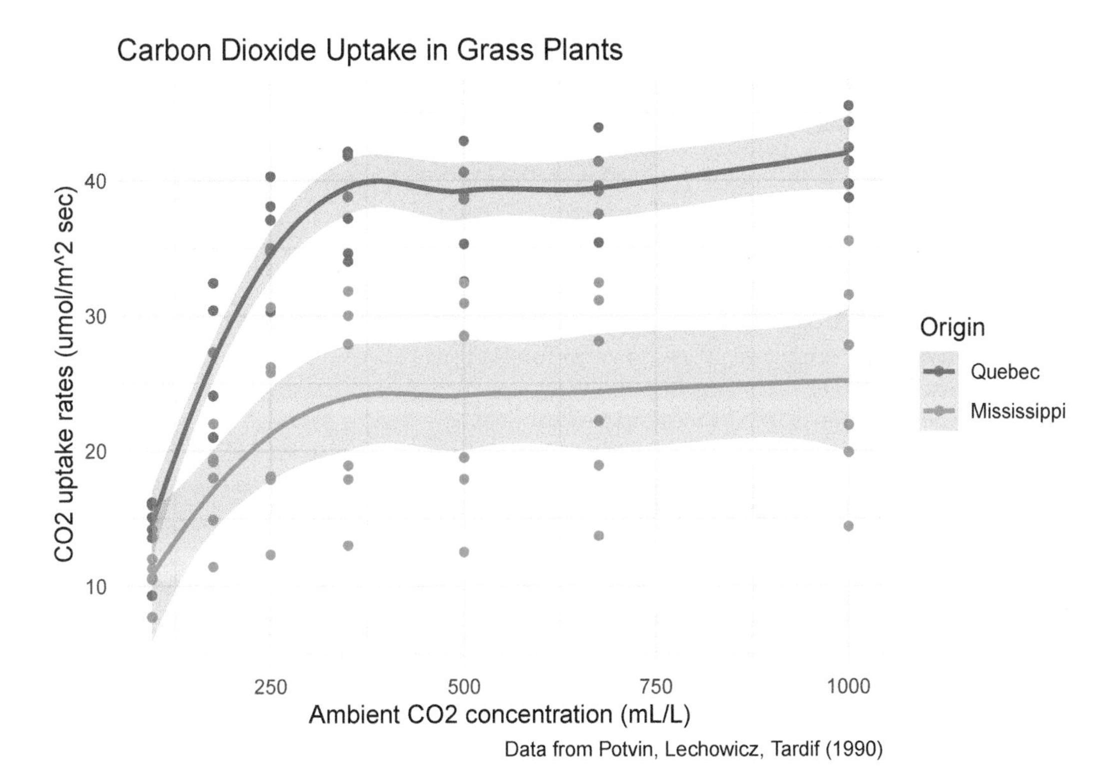


**FIGURE 7.1** A finished plot.

argument to any ggplot command is the data, which means that ggplot can be placed at the end of a dplyr pipeline.

For example, you may want to visualize something that isn't directly represented as a variable in the data frame. We can use `mutate` to create the variable that we want to visualize, and then use `ggplot` to visualize it.

**Example 7.1.** Consider the `houses` data set in the `fosdata` package. This data set gives the house prices and characteristics of houses sold in King County (home of Seattle) from May 2014 through May 2015.

Suppose we want to get a scatterplot of price per square foot versus the log of the lot size for houses in zip code 98001. The variables `price`, `sqft_living`, and `sqft_lot` are given in the data frame, but we will need to compute the price per square foot and take the log of the square footage of the lot ourselves.[3] We will also need to `filter` the data frame so that only houses in zip code 98001 appear. We do this using `dplyr` tools, and then we can pipe directly into `ggplot2` without saving the modified data frame.

```
fosdata::houses %>%
  filter(zipcode == 98001) %>%
  mutate(
    price_per_sf = price / sqft_living,
    log_lot_size = log(sqft_lot)
  ) %>%
  ggplot(aes(x = log_lot_size, y = price_per_sf)) +
  geom_point() +
  labs(title = "Home prices in zip code 98001")
```

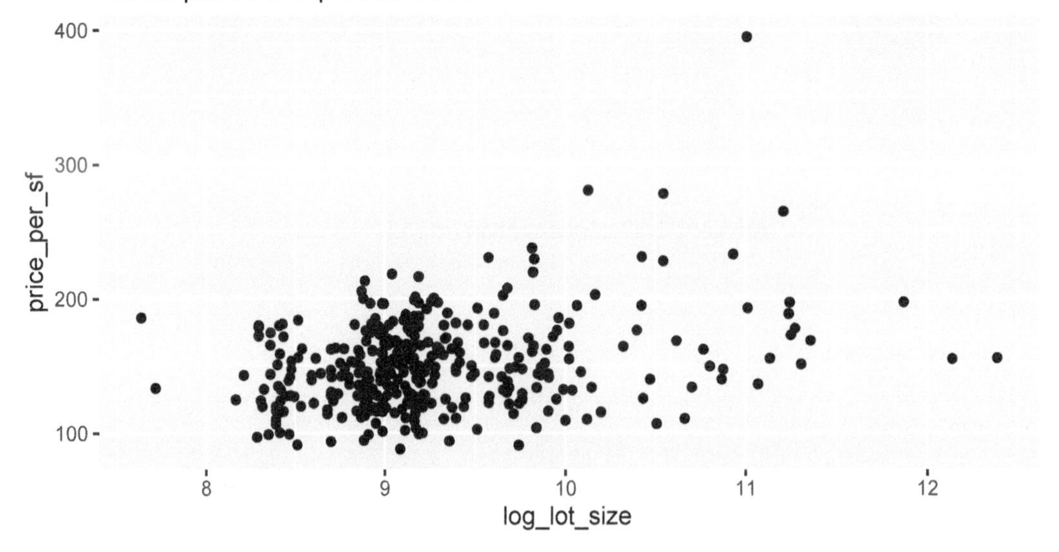


It appears as though an increase in the logarithm of the lot size is associated with an increase in the price per square foot of houses in the 98001 zip code.

Data for ggplot *must* be stored as a data frame (or equivalent structure, such as a tibble),

[3] `ggplot2` will also allow you to make simple computations inside the aesthetic mapping, but we choose not to do so in order to illustrate the `dplyr` technique.

and usually needs to be tidy. A common issue is when a data set contains multiple columns of data that you would like to deal with in the same way.

**Example 7.2.** The `weight_estimate` data in `fosdata` comes from an experiment where children watched actors lift objects of various weights. We would like to plot the actual weight of the object on the $x$-axis and the child's estimate on the $y$-axis. However, the actual weights are stored in the names of the four column variables `mean100`, `mean200`, `mean300`, and `mean400`:

```
head(fosdata::weight_estimate)
```

```
##        id height mean100 mean200 mean300 mean400 age
## 1 6YO_S1    124     150     240     270     380   6
## 2 6YO_S2    118     240     280     330     400   6
## 3 6YO_S3    121     170     370     360     380   6
## 4 6YO_S4    122     160     140     250     390   6
## 5 6YO_S5    116     170     300     360     310   6
## 6 6YO_S6    118     210     230     300     320   6
```

We use `pivot_longer` to tidy the data.

```
weight_tidy <- fosdata::weight_estimate %>%
  tidyr::pivot_longer(
    cols = matches("^mean"),
    names_to = "actual_weight",
    names_prefix = "mean",
    values_to = "estimated_weight"
  )
head(weight_tidy)
```

```
## # A tibble: 6 x 5
##   id     height age   actual_weight estimated_weight
##   <chr>   <int> <fct> <chr>                    <dbl>
## 1 6YO_S1    124 6     100                        150
## 2 6YO_S1    124 6     200                        240
## 3 6YO_S1    124 6     300                        270
## 4 6YO_S1    124 6     400                        380
## 5 6YO_S2    118 6     100                        240
## 6 6YO_S2    118 6     200                        280
```

Now we can provide the actual weight as a variable for plotting:

```
ggplot(weight_tidy, aes(x = actual_weight, y = estimated_weight)) +
  geom_point() +
  labs(title = "Child estimates of weights lifted by actors")
```

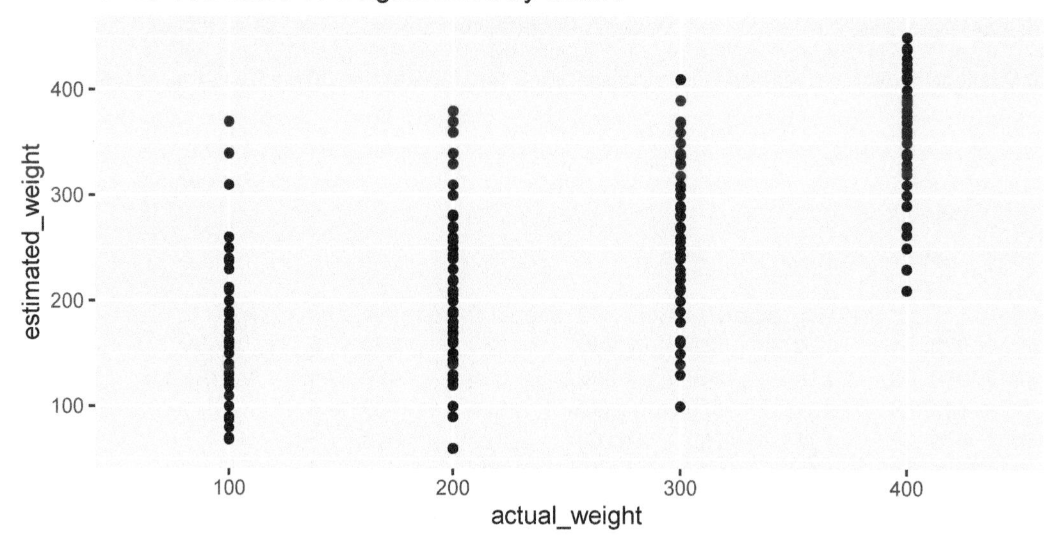


Another common issue when plotting with ggplot is the ordering of factor type variables. When ggplot deals with a character or factor type variable, it will revert to alphabetic order unless the variable is a factor with appropriately set levels.

**Example 7.3.** The `ecars` data from `fosdata` gives information about electric cars charging at workplace charging stations.

Let's make a barplot showing the number of charges for each day of the week.

```
ecars <- fosdata::ecars
ggplot(ecars, aes(x = weekday)) +
  geom_bar()
```

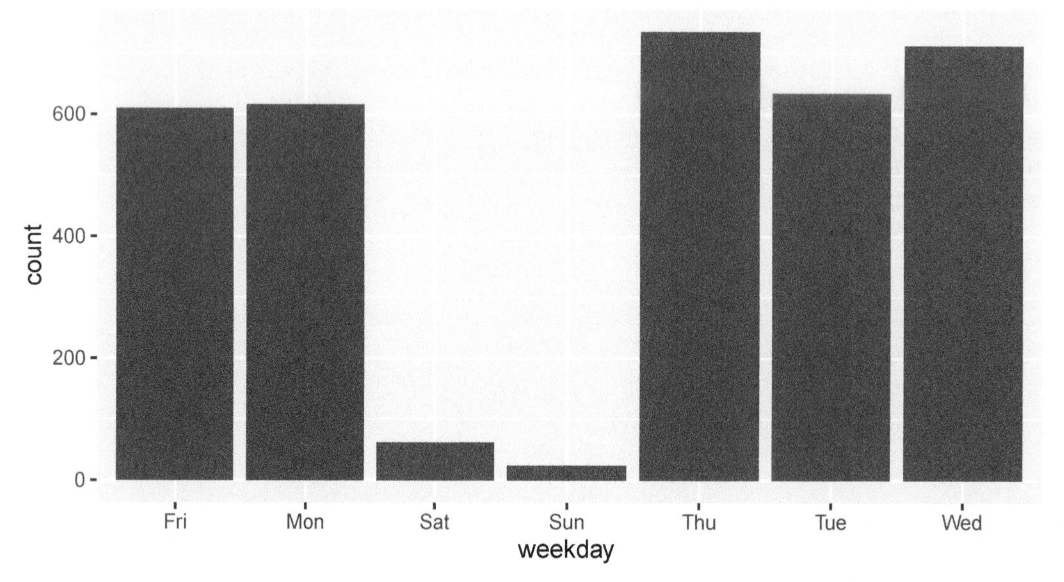


Observe that the days of the week are in alphabetical order, which is not helpful. To correct the chart, we need to correct the data. We do this by making the `weekday` variable into a factor with the correct order of levels.[4] After that, the plot shows the bars in the correct order.

```
days <- c("Sun", "Mon", "Tue", "Wed", "Thu", "Fri", "Sat")
ecars <- mutate(ecars, weekday = factor(weekday, levels = days))
ggplot(ecars, aes(x = weekday)) +
  geom_bar()
```

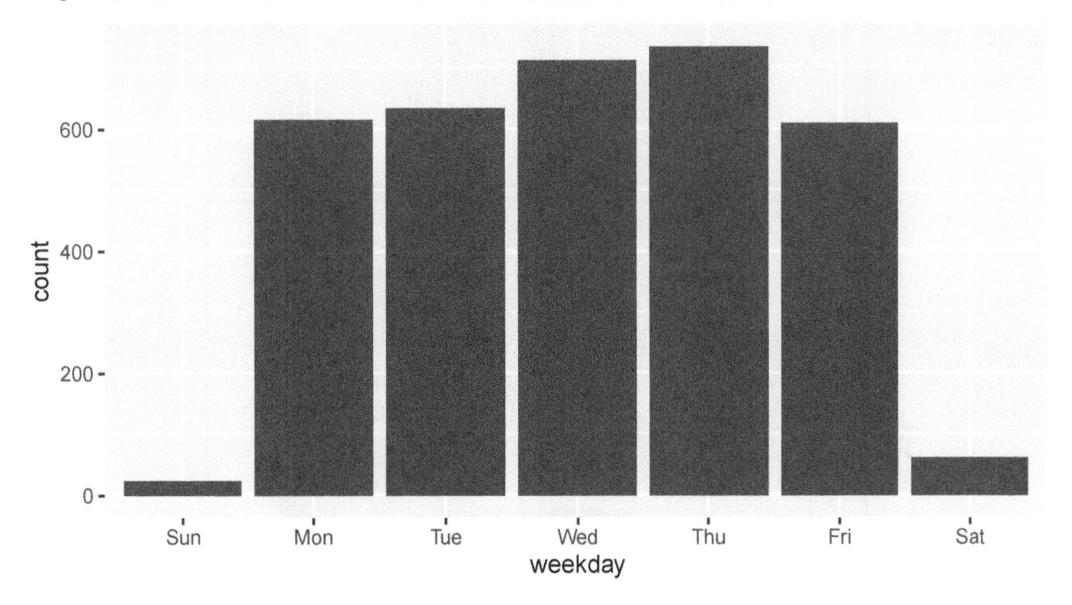


## 7.2 Visualizing a single variable

In this section, we present several ways of visualizing data that is contained in a single numeric or categorical variable. We have already seen three ways of doing this in Chapter 5, namely histograms, barplots, and density plots. We will see how to do these three types of plots using ggplot2, and we will also introduce qq plots and boxplots.

### 7.2.1 Histograms

A histogram displays the distribution of numerical data. Data is divided into equal-width ranges called *bins* that cover the full range of the variable. The histogram geometry draws rectangles whose bases are the bins and whose heights represent the number of data points falling into that bin. This representation assigns equal area to each point in the data set.

**Example 7.4.** The Bechdel test[5] is a simple measure of the representation of women in fiction. A movie passes the Bechdel test when it has two female characters who have a conversation about something other than a man.

---

[4]There is some dispute in the world about whether the calendar week should begin on Sunday or Monday, but we all agree the days come in order.

[5]See https://en.wikipedia.org/wiki/Bechdel_test

*Image credit: Branson Reese.*

The `bechdel` data set in the `fosdata` package contains information about 1794 movies, their budgets, and whether or not they passed the Bechdel test. This data set was used in the *FiveThirtyEight* article, "The Dollar-And-Cents Case Against Hollywood's Exclusion of Women."[6]

You can load the data via

```
bechdel <- fosdata::bechdel
```

The variable `binary` is FAIL/PASS as to whether the movie passes the Bechdel test. Let's see how many movies failed the Bechdel test.

```
summary(bechdel$binary)
```

```
## FAIL PASS
##  991  803
```

In 991 out of the 1794 movies, there was no conversation between two women about something other than a man. There are several interesting things that one could do here, and some of them are explored in the exercises. For now, let's look at a histogram of the **budgets** of all movies in 2013 dollars. The geometry `geom_histogram` has required aesthetics of `x` and `y`, but `y` is computed within the function itself and does not need to be supplied.

[6] W Hickey, "The Dollar-and-Cents Case Against Hollywood's Exclusion of Women," *FiveThirtyEight*, April 1, 2014, http://fivethirtyeight.com/features/the-dollar-and-cents-case-against-hollywoods-exclusion-of-women/.

```
ggplot(bechdel, aes(x = budget_2013)) +
  geom_histogram()
```

```
## `stat_bin()` using `bins = 30`. Pick better value with `binwidth`.
```

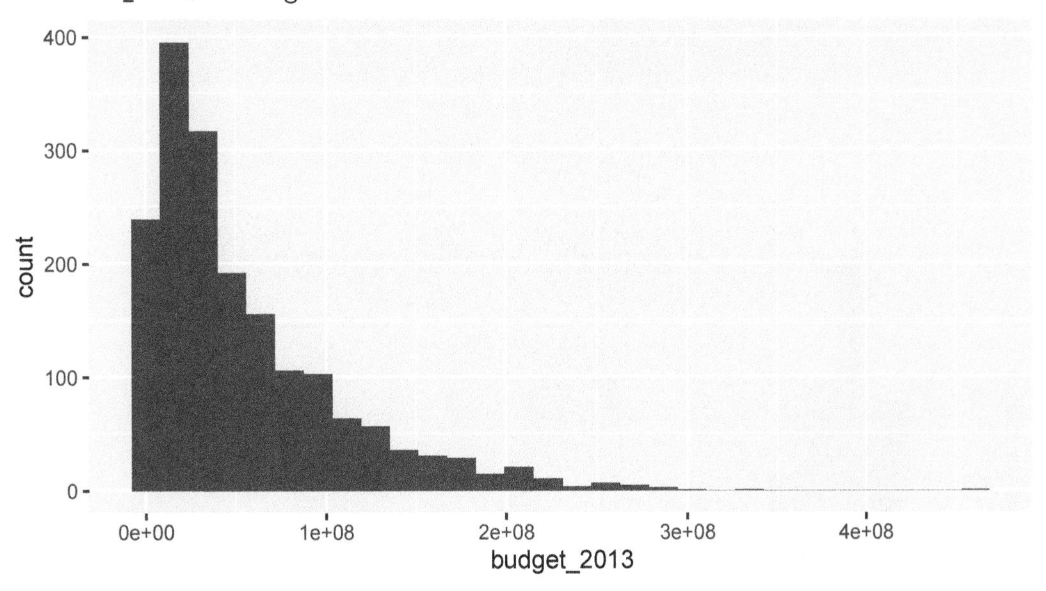


The function **geom_histogram** always uses 30 bins in its default setting and produces a message about it. In contrast, the **hist** command in base R has an algorithm that tries to provide a default number of bins that will work well for your data. With ggplot, adjust the binwidth or number of bins to see how those impact the general shape of the distribution. We will not show the message indicating that the number of bins is 30 further in this text.

```
ggplot(bechdel, aes(x = budget_2013)) +
  geom_histogram(bins = 15) # 15 bins
```

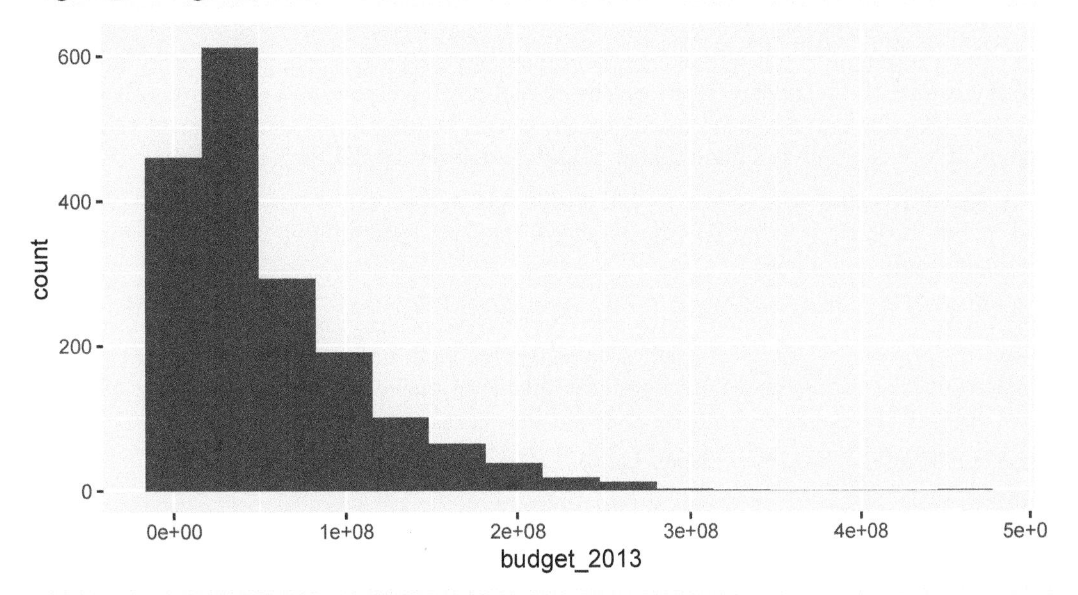


```
ggplot(bechdel, aes(x = budget_2013)) +
  geom_histogram(binwidth = 20e+6) + # binwidth of 20 million dollars
```

```
  scale_x_continuous(
    labels =
      scales::unit_format(unit = "M", scale = 1e-6)
  ) +
  labs(
    title = "Hollywood Movie Budgets",
    x = "Film budget, 2013 US dollars",
    y = "Number of films"
  )
```

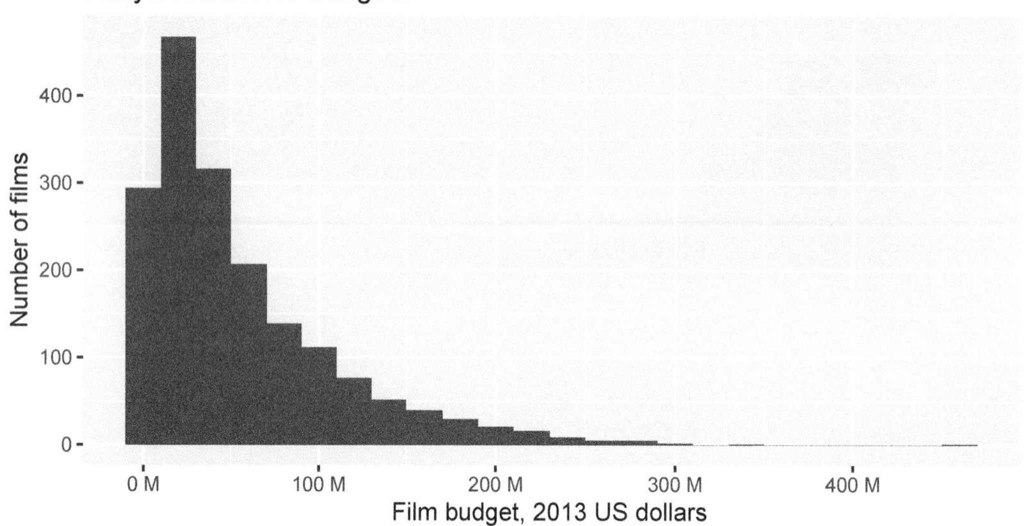


### 7.2.2 Barplots

Barplots are similar to histograms in appearance, but do not perform any binning. A barplot simply counts the number of occurrences at each level (resp. value) of a categorical (resp. integer) variable. It plots the count along the $y$-axis and the levels of the variable along the $x$-axis. It works best when the number of different levels in the variable is small.

The `ggplot2` geometry for creating barplots is `geom_bar`, which has required aesthetics `x` and `y`, but again the `y` value is computed from the data by default so is not always included in the aesthetic mapping.

**Example 7.5.** In the `bechdel` data set, the variable `clean_test` contains the levels 'ok,' 'dubious,' 'nowomen,' 'notalk,' 'men.' The values 'ok' and 'dubious' are for movies that pass the Bechdel test, while the other values describe how a movie failed the test. Create a barplot that displays the number of movies that fall into each level of `clean_test`.

```
ggplot(bechdel, aes(x = clean_test)) +
  geom_bar()
```

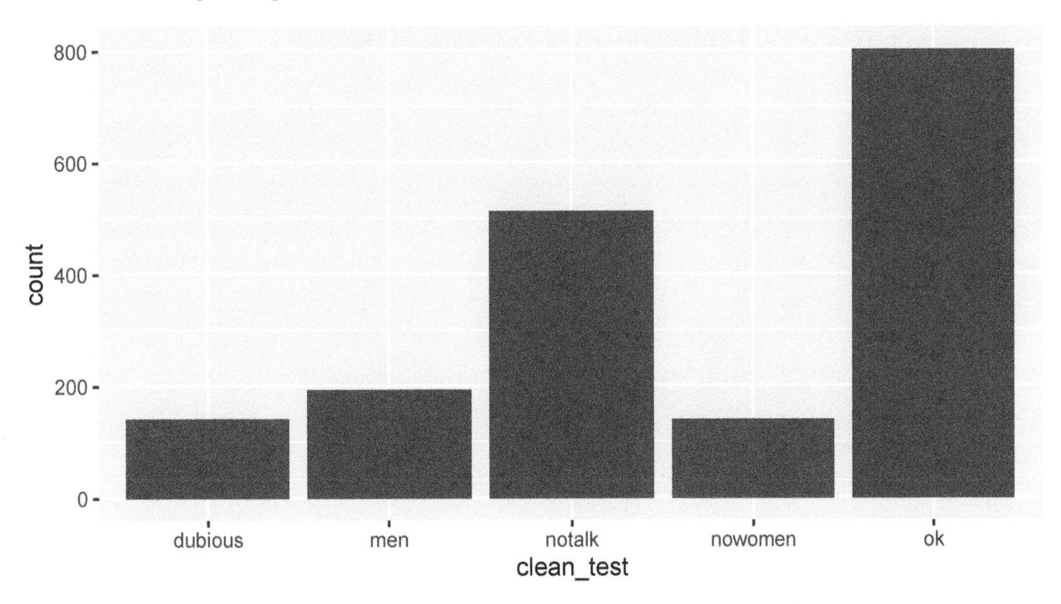


This visualization leaves something to be desired, as it fails to show the natural grouping of `ok` with `dubious`, and then `men`, `notalk` and `nowomen` in another group. See Section 7.4.1 for a better visualization.

We can also use `geom_bar` when the variable in question is integer valued rather than categorical.

**Example 7.6.** Consider the `austen` data set in the `fosdata` package. This data set contains the complete texts of *Emma* and *Pride and Prejudice*, by Jane Austen. The format is that each observation is a word, together with sentence, chapter, novel, length, and sentiment score information about the word.

Create a barplot showing the number of words of each length (in the two novels combined).

```
austen <- fosdata::austen
ggplot(austen, aes(x = word_length)) +
  geom_bar() +
  labs(title = "Length of words in Emma and Pride and Prejudice")
```

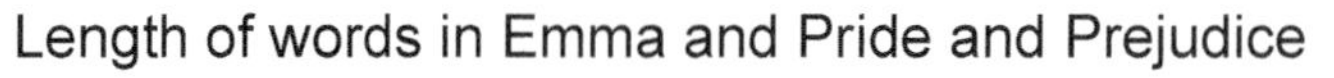


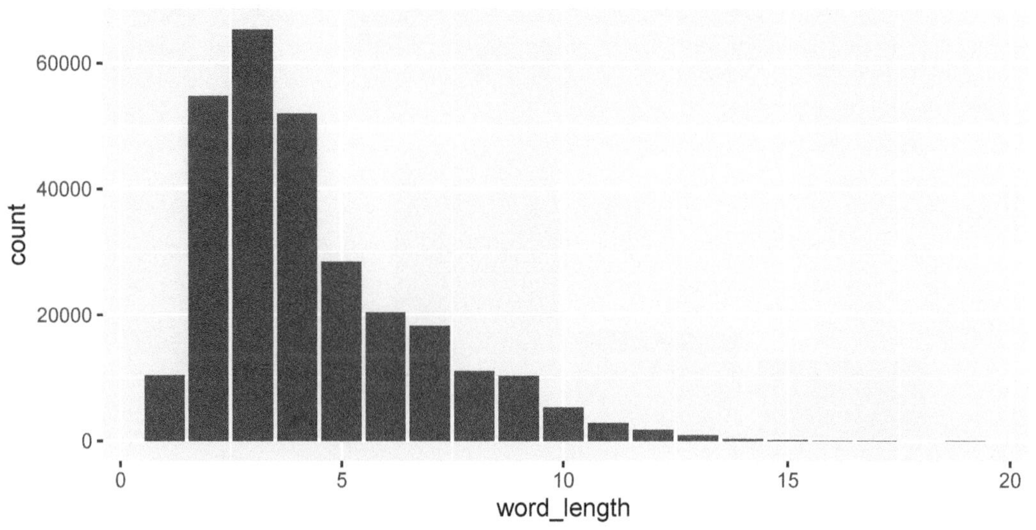


Sometimes, we wish to create a barplot where the $y$ values are something other than the number of occurrences of a variable. In this case, we provide the y aesthetic and use `geom_col` instead of `geom_bar`.

**Example 7.7.** In the `austen` data set, plot the mean sentiment score of words in *Pride and Prejudice* versus the chapter the words appear in using a barplot. We will need to do some manipulation to the data before piping it into `ggplot`.

```
austen %>%
  filter(novel == "Pride and Prejudice") %>%
  group_by(chapter) %>%
  summarize(mean_sentiment = mean(sentiment_score)) %>%
  ggplot(aes(x = chapter, y = mean_sentiment)) +
  geom_col() +
  labs(title = "Sentiment by chapter in Pride and Prejudice")
```

The resulting plot is in Figure 7.2.

### 7.2.3 Density plots

Density plots display the distribution of a continuous random variable. Rather than binning, as histograms do, a density plot shows a curve whose height is the mean of the pdfs of normal random variables centered at the data points. The standard deviations are controlled by the bandwidth in much the same way that binwidth controls width of histogram bins. Density plots work best when values of the variable we are plotting aren't spread out too far. It can be a good idea to try a few different bandwidths in order to see the effect that it has on the density estimation.

In base R, we used `plot(density(data))` to produce a density plot of `data`. In `ggplot2`, we use the `geom_density` geometry. As in the other geoms we have seen so far, `geom_density` requires x and y as aesthetics, but will compute the y values if they are not provided.

**Example 7.8.** Consider the `chimps` data set in the `fosdata` package. The amount of grey

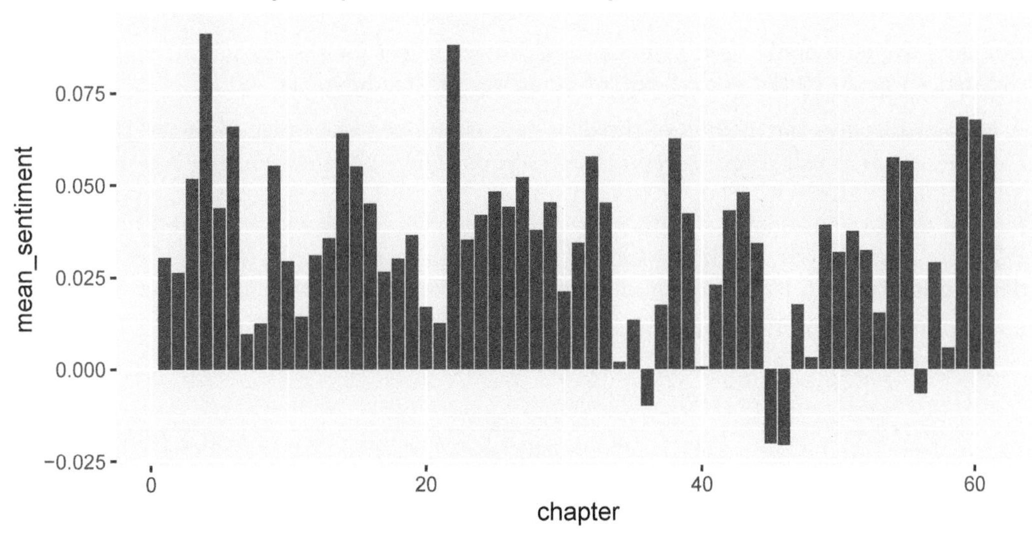


**FIGURE 7.2** Using geom_col to plot bars with heights set by a y aesthetic.

in a chimpanzee's hair was rated by 158 humans, and the mean value of their rankings is stored in the variable `grey_score_avg`. Plot the density of the average grey hair score for chimpanzees.

```
chimps <- fosdata::chimps
chimps %>%
  ggplot(aes(x = grey_score_avg)) +
  geom_density()
```

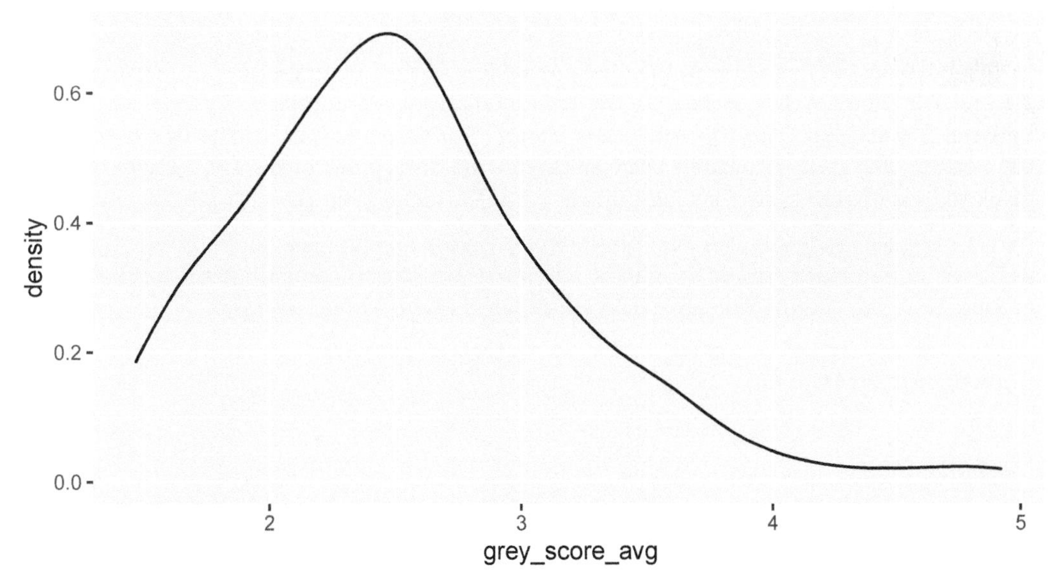


We interpret this plot as an estimate of the plot of the pdf of the random variable `grey_score_avg`. For example, the most likely outcome is that the mean grey score will be about 2.5.

### 7.2.4 Boxplots

Boxplots are commonly used visualizations to get a feel for the median and spread of a variable. In ggplot2 the geom_boxplot geometry creates boxplots.

A boxplot displays the median of the data as a dark line. The range from the 25th to the 75th percentile is called the *interquartile range* or *IQR*, and is shown as a box or "hinges." Vertical lines called "whiskers" extend 1.5 times the IQR, or are truncated if the whisker reaches the last data point. Points that fall outside the whiskers are plotted individually. Boxplots may include notches, which give a rough error estimate for the median of the distribution (we will learn more about error estimates for the *mean* in Chapter 8).

```
ggplot(chimps, aes(y = grey_score_avg)) +
  geom_boxplot(notch = TRUE)
```

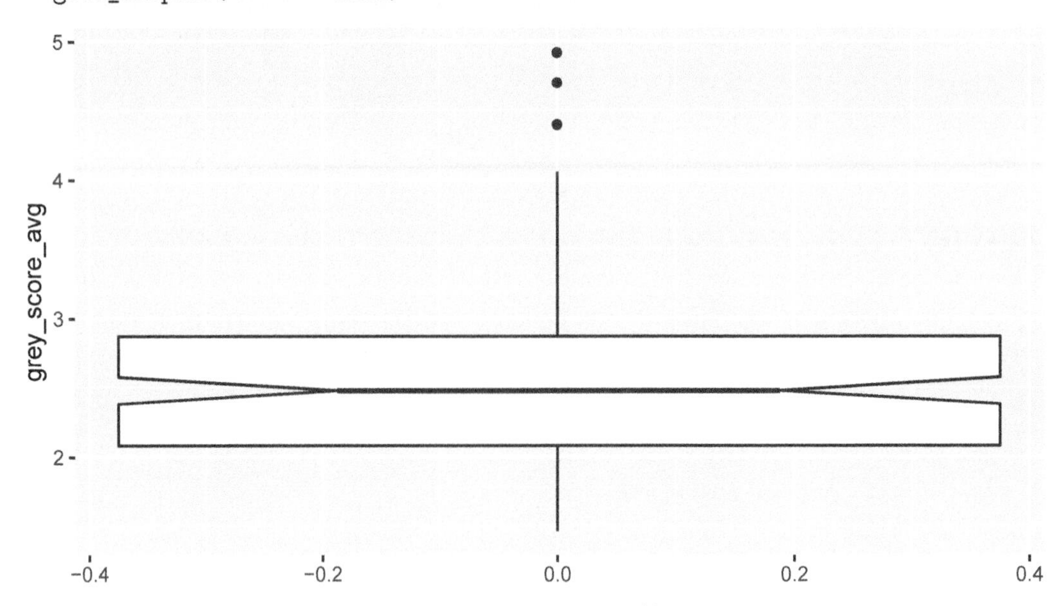


In this boxplot, the median is at 2.5, and the middle 50% of observations fall between about 2.1 and 2.9. The notches show that the true median for all chimps likely falls somewhere between 2.4 and 2.6. The top whisker is exactly 1.5 times as long as the box is tall, and the bottom whisker is truncated because there is no data point below 1.47. Three dots are displayed for observations that fall outside 1.5 IQR, identifying them as possible outliers.

Notice that the $x$-axis in our one-box boxplot had a meaningless scale. Typically, the x aesthetic in geom_boxplot is assigned to a factor variable to compare distributions across groups. Here are chimpanzee grey scores compared across three habitats:

```
ggplot(chimps, aes(x = population, y = grey_score_avg)) +
  geom_boxplot()
```

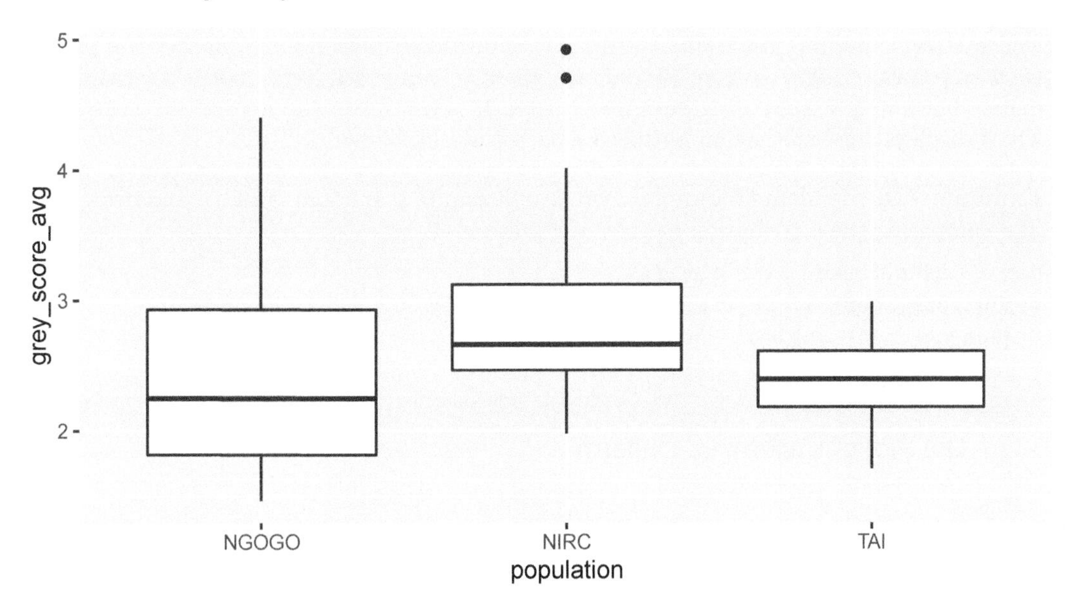


**Try It Yourself.**
Boxplots intentionally hide features of the data to provide a simple visualization. Compare `geom_boxplot` to `geom_violin`, which displays a smooth density of the same data.

### 7.2.5 QQ plots

Quantile-quantile plots, known as **qq plots**, are used to compare sample data to a known distribution. In this book, we have often done this by plotting a known distribution as a curve on top of a histogram or density estimation of the data. Overplotting the curve gives a large scale view but won't show small differences.

Unlike histograms or density plots, qq plots provide more precision and do not require binwidth/bandwidth selection. This book will use qq plots in Section 11.4 to investigate assumptions in regression.

A qq plot works by plotting the values of the data variable on the $y$-axis at positions spaced along the $x$-axis according to a theoretical distribution. Suppose you have a random sample of size $N$ from an unknown distribution. Sort the data points from smallest to largest, and rename them $y_{(1)}, \ldots, y_{(N)}$. Next, compute the expected values of $N$ points drawn from the theoretical distribution and sorted, $x_{(1)}, \ldots, x_{(N)}$. For example, to compare with a uniform distribution, the $x_{(i)}$ would be evenly spaced. To compare with a normal distribution, the $x_{(i)}$ are widely spaced near the tails and denser near the mean. The specific algorithm that R uses is a bit more sophisticated, but the complications only make a noticeable difference with small data sets.

The qq plot shows the points

$$\left(x_{(1)}, y_{(1)}\right), \ldots, \left(x_{(N)}, y_{(N)}\right)$$

If the sample data comes from a random variable that matches the theoretical distribution, then we would expect the ordered pairs to roughly fall on a straight line. If the sample does not match the theoretical distribution, the points will be far from a straight line.

We use the `geom_qq` geometry to create qq plots. `geom_qq` has one required aesthetic: `sample`, the variable containing the sample data. The known distribution for comparison is passed as an argument `distribution`. By default, `geom_qq` compares to the normal distribution `qnorm`, but any `q + root` distribution will work. It is helpful to add a `geom_qq_line` layer, which displays the line that an idealized data set would follow.

**Example 7.9.** Simulate 50 samples from a uniform $[0, 1]$ random variable and draw a qq plot comparing the distribution to that of a uniform $[0, 1]$.

```
dat <- data.frame(sam = runif(50, 0, 1))
ggplot(dat, aes(sample = sam)) +
  geom_qq(distribution = qunif) +
  geom_qq_line(distribution = qunif) +
  labs(title = "QQ Plot of Uniform vs Uniform")
```

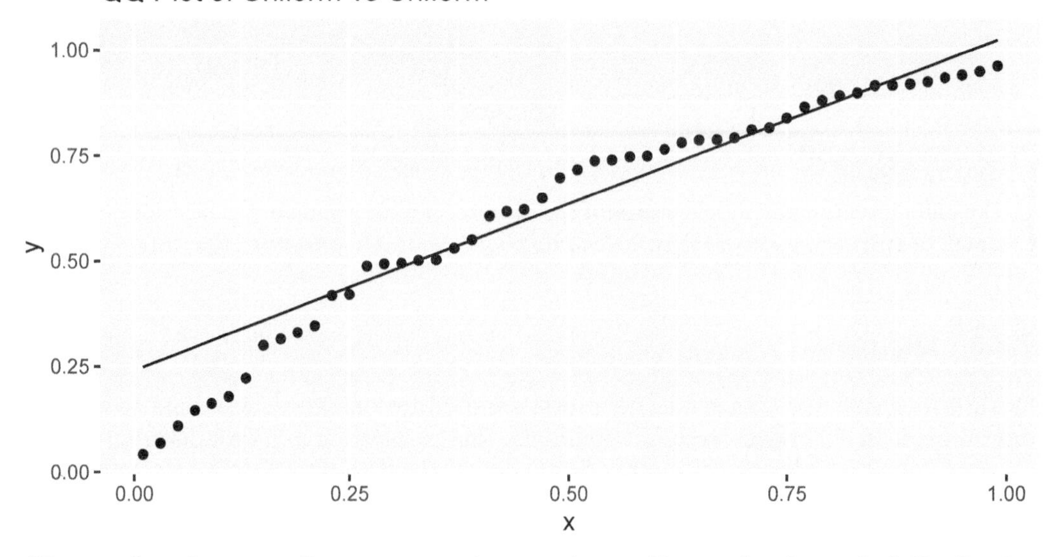


Observe that the $x$-coordinates are evenly spaced according to the theoretical distribution, while the $y$-coordinates are randomly spaced but overall evenly spread across the interval $[0, 1]$ leading to a scatterplot that closely follows a straight line.

We will primarily be using qq plots to plot a sample versus the quantiles of a normal distribution, which is the default for `geom_qq`. We will be looking to see whether the qq plot is U-shaped, S-shaped, or neither. Figure 7.3 illustrates how qq plots appear for data distributions with common shapes.

S-shaped qq plots indicate heavy or light tails relative to the normal distribution, depending on which direction the S is facing, and U-shaped qq plots indicate skewness.

**Example 7.10.** Consider the `budget_2013` variable in the `bechdel` data set. Create a qq plot versus a normal distribution and interpret.

```
ggplot(bechdel, aes(sample = budget_2013)) +
  geom_qq() +
  geom_qq_line()
```

The qq plot (Figure 7.4) has a very strong U shape, and the data is right-skewed.

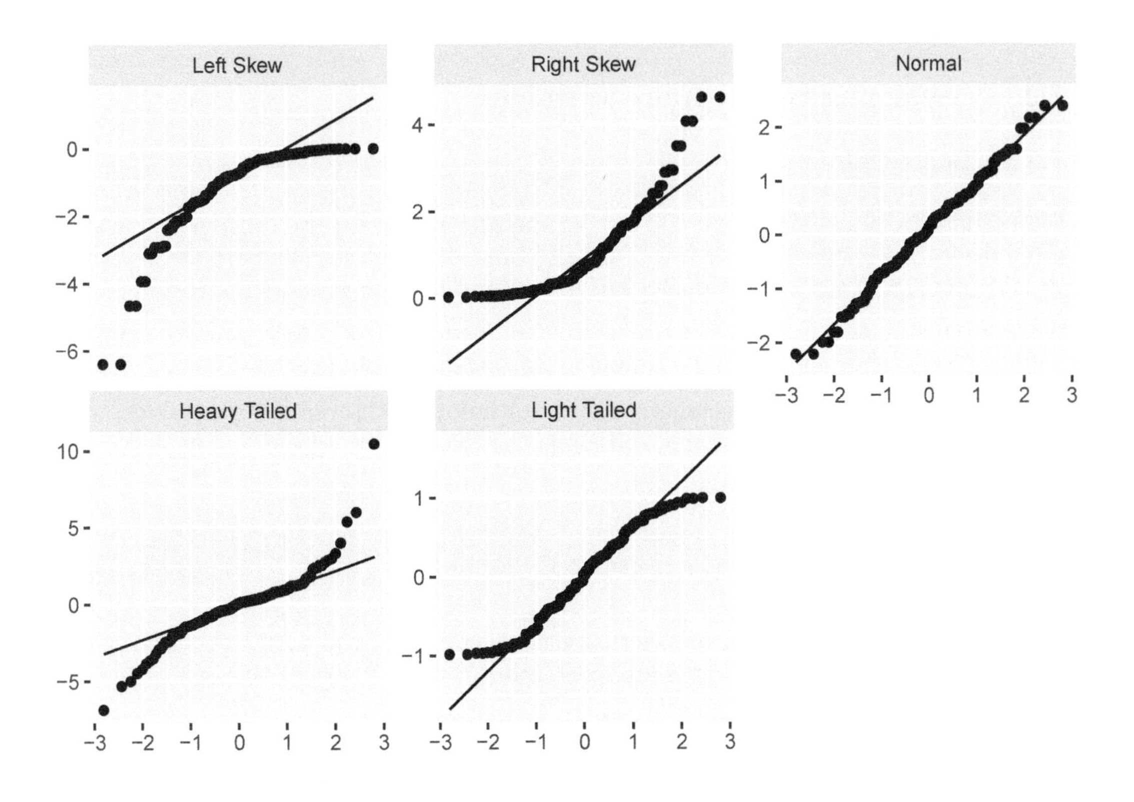


**FIGURE 7.3** Data distribution shapes with qq plots.

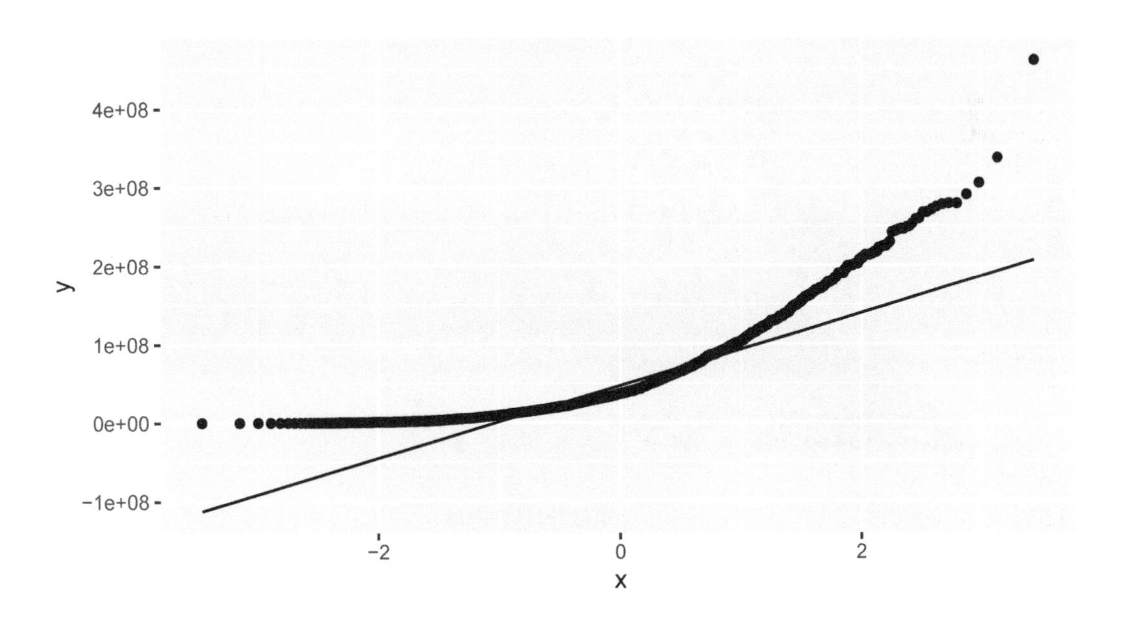


**FIGURE 7.4** A qq plot showing that the distribution of movie budgets is right-skewed.

## 7.3 Visualizing two or more variables

### 7.3.1 Scatterplots

Scatterplots are a natural way to display the interaction between two continuous variables, using the $x$- and $y$-axes as aesthetics. Typically, one reads a scatterplot with the $x$ variable as explaining the $y$ variable, so that $y$ values are dependent on $x$ values.

We have already seen examples of scatterplots in Section 7.1: a simple example using the built-in `CO2` data set and another example using `fosdata::housing`.

Scatterplots can be the base geometry for visualizations with even more variables, using aesthetics such as color, size, shape, and alpha that control the appearance of the plotted points.

**Example 7.11.** Explore the relationship between time at the charging station (`chargeTimeHrs`) and power usage (`kwhTotal`) for electric cars in the `fosdata::ecars` data set.

A first attempt at a scatterplot reveals that there is one data point with a car that charged for 55 hours. All other charge times are much shorter, so we filter out this one outlier and plot:

```
fosdata::ecars %>%
  filter(chargeTimeHrs < 24) %>%
  ggplot(aes(x = chargeTimeHrs, y = kwhTotal)) +
  geom_point()
```

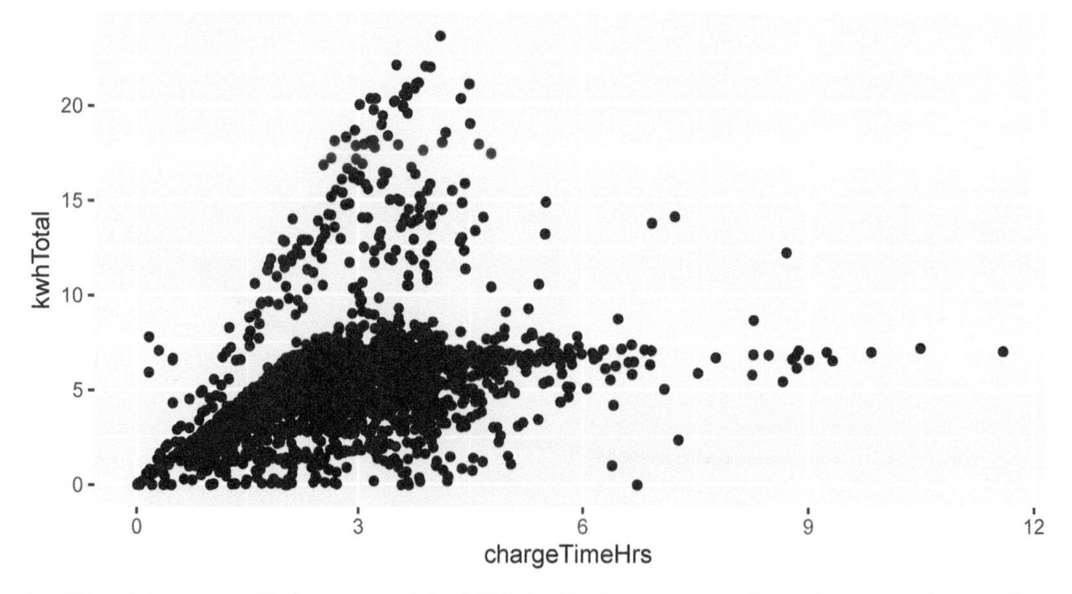


In this plot, we see that many points fall into the lower region bounded by a slanted line and horizontal line. This suggests cars charge at a standard speed, which accounts for the slant, and that after about 7.5 total kwh, most cars are fully charged. There is also a steeper slanted boundary running from the origin to the top of the chart, which suggests that some cars (or charging stations) have a higher maximum charge speed.

**Example 7.12.** Continuing with the `ecars` data set, let's explore the `distance` variable. Charging stations in this data set are generally workplaces, and the distance variable reports the distance to the drivers' home.

Only about 1/3 of the records in `ecars` contain valid distance data, so we filter for those. Then, we assign distance to the color aesthetic. The results look cluttered, so we also assign distance to the alpha aesthetic, which will make shorter distances partially transparent.

```
fosdata::ecars %>%
  filter(!is.na(distance)) %>%
  ggplot(aes(
    x = chargeTimeHrs, y = kwhTotal,
    color = distance, alpha = distance
  )) +
  geom_point() +
  labs(title = "Charge time, power usage, and distance from home")
```

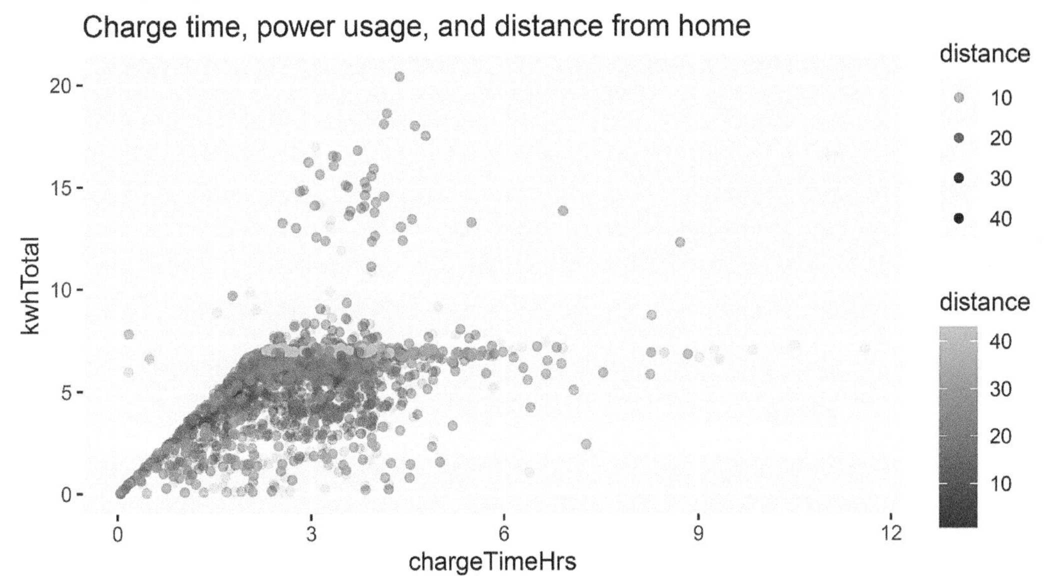


From this chart, we see a bright line at around 7.5 kwh total, which suggests that people who live further from work tend to need a full charge during the day.

> **Try It Yourself.**
> How does the plot in Example 7.12 look without filtering away the invalid data? How does it look without alpha transparency?

**Example 7.13.** Let's compare home sales prices between an urban and a suburban area of King County, Washington. We select two zip codes from the `fosdata::houses` data set. Zip code 98115 is a residential area in the city of Seattle, while 98038 is a suburb 25 miles southeast from Seattle's downtown.

In this example, multiple aesthetics are used to display a large amount of information in a single plot.

Sale price depends primarily on the size of the house, so we choose `sqft_living` as our $x$-axis and `price` as our $y$-axis. We distinguish the two zip codes by color, giving visual contrast even where the data overlaps. Because the data points do overlap quite a bit, we set the attribute `alpha` to 0.8, which makes all points 80% transparent.

The size of a house's property (lot) is also important to price, so this is displayed by mapping `sqft_lot` to the size aesthetic. Notice in the urban zip code there is little variation in lot size, while in the suburb there is a large variation in lot size and in particular larger lots sell for higher prices.

Finally, we map the `waterfront` variable to shape. There is only one waterfront property in this data set. Can you spot it?

```
fosdata::houses %>%
  filter(zipcode %in% c("98115", "98038")) %>%
  mutate(
    zipcode = factor(zipcode),
    waterfront = factor(waterfront)
  ) %>%
  ggplot(aes(
    x = sqft_living, y = price, color = zipcode,
    size = sqft_lot, shape = waterfront
  )) +
  geom_point(alpha = 0.8) +
  labs(title = "Housing in urban and suburban Seattle")
```

Housing in urban and suburban Seattle

price

2000000
1500000
1000000
500000

1000 2000 3000 4000 5000 6000

sqft_living

zipcode
98038
98115

waterfront
0
1

sqft_lot
2e+05
4e+05
6e+05

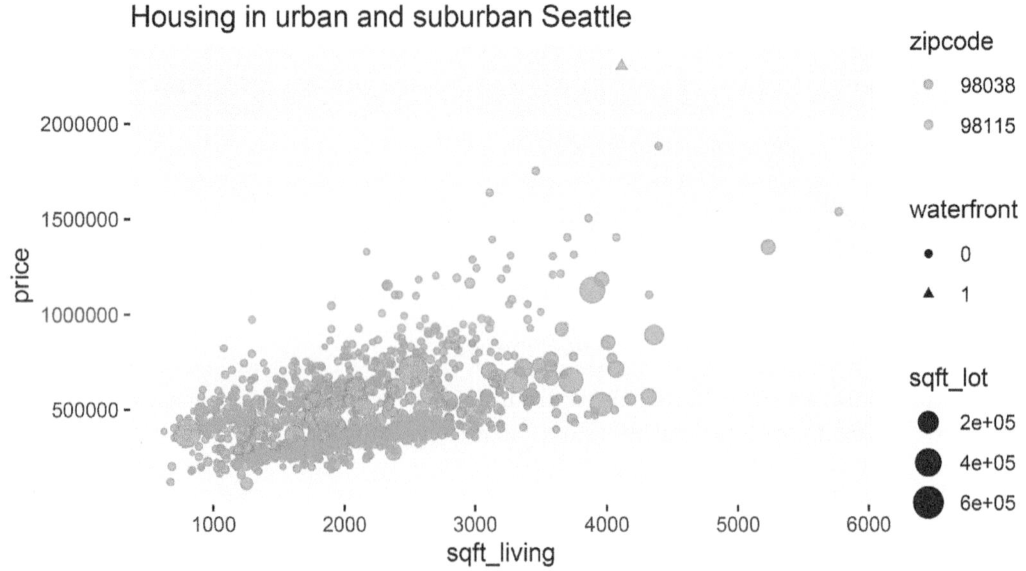

## 7.3.2 Line graphs and smoothing

When data contains time or some other sequential variable, a natural visualization is to put that variable on the $x$-axis. We can then display other variables over time using `geom_point` to produce dots, `geom_line` for a *line graph*, or `geom_smooth` for a smooth approximation.

**Example 7.14.** Create a line graph of the number of male babies named "Darrin" over time, using the data in the `babynames` data set. The time variable is `year` and the count of babies is in the variable `n`.

```
babynames <- babynames::babynames
babynames %>%
  filter(name == "Darrin", sex == "M") %>%
  ggplot(aes(x = year, y = n)) +
  geom_line()
```

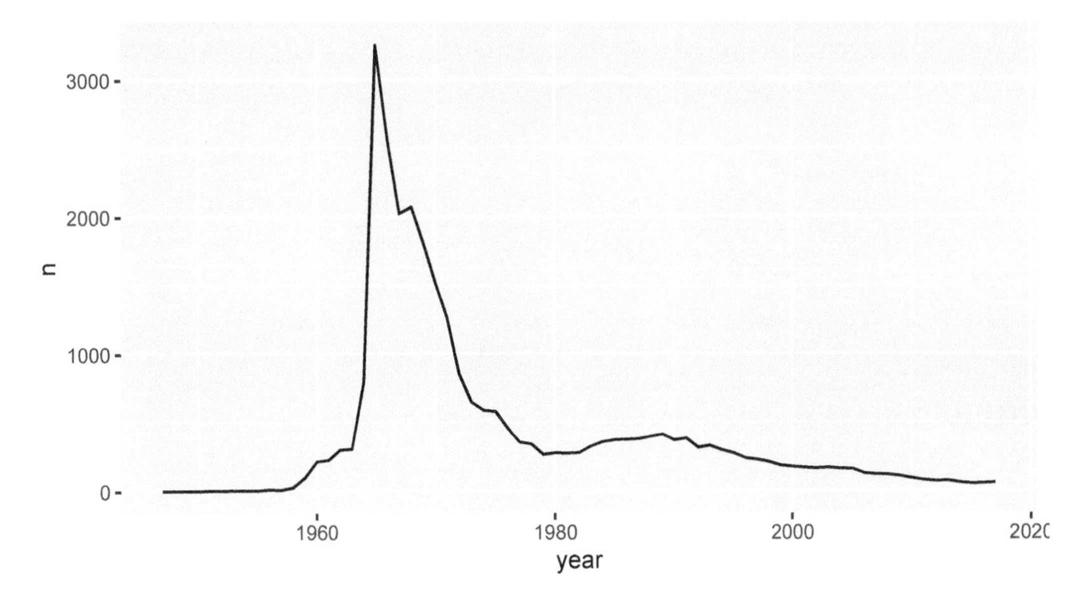


**FIGURE 7.5** Boys named Darrin, over time.

The plot, in Figure 7.5, shows that there was a peak of Darrins in 1965. If you meet someone named Darrin, you might be able to guess their age.

To focus attention on the general trend and remove some of the jagged year-to-year variation, plot a curve fitted to the data. The geometry `geom_smooth`[7] does this. We also show the data points to assess the fit.

```
babynames %>%
  filter(name == "Darrin", sex == "M") %>%
  ggplot(aes(x = year, y = n)) +
  geom_point() +
  geom_smooth()
```

```
## `geom_smooth()` using method = 'loess' and formula 'y ~ x'
```

The results, in Figure 7.6, aren't very good! Fitting a line to points like this is a bit like binning a histogram in that sometimes we need to play around with the fit until we get one that looks right. We can do that by changing the `span` value inside of `geom_smooth`. The `span` parameter is the percentage of points that `geom_smooth` considers when estimating a smooth curve to fit the data. The smaller the value of the `span`, the more the curve will go through the points.

[7]The geometry `geom_smooth` gives a helpful message as to how the line was fit to the scatterplot, which we will hide in the rest of the book.

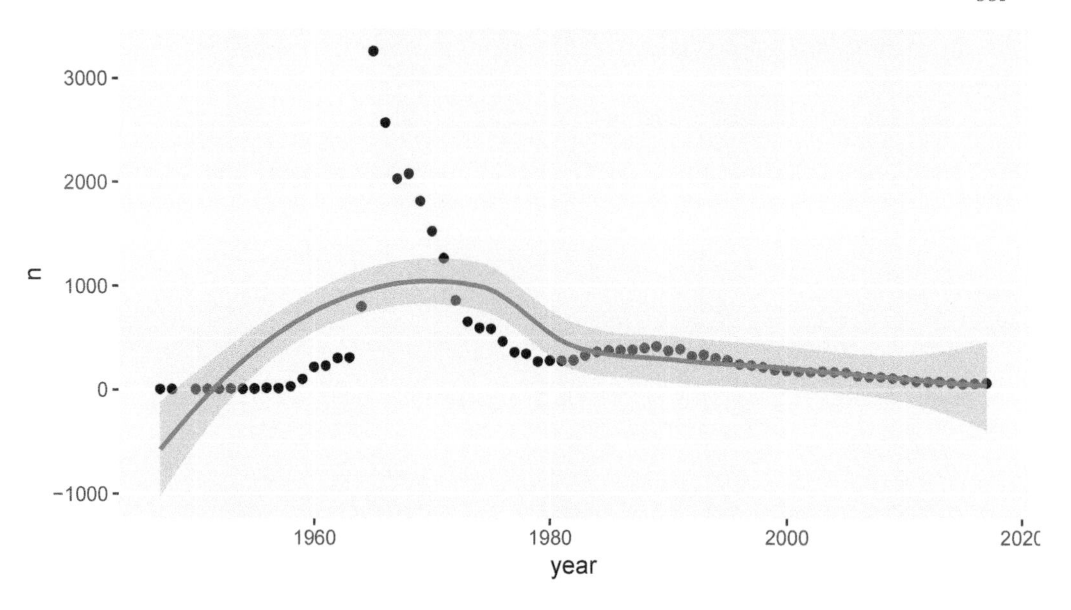


**FIGURE 7.6** Boys named Darrin, over time. A poorly fit curve.

> **Try It Yourself.**
> Add `geom_smooth(span = 1)` and `geom_smooth(span = .05)` to the above plot. You should see that with `span = .05`, the smoothed curve follows the points more closely.

After experimenting with some values of `span`, a value of 0.22 strikes a good compromise between smoothing out the irregularities of the points and following the curve. See Figure 7.7.

```
babynames %>%
  filter(name == "Darrin", sex == "M") %>%
  ggplot(aes(x = year, y = n)) +
  geom_point() +
  geom_smooth(span = .22)
```

We can also plot multiple lines on the same graph. For example, let's plot the number of babies named "Darrin," "Darren," or "Darin" over time. We map the `name` variable to the color aesthetic and then remove the error shadows from the fit curves to clean up the plot.

```
babynames %>%
  filter(name %in% c("Darrin", "Darren", "Darin"), sex == "M") %>%
  ggplot(aes(x = year, y = n, color = name)) +
  geom_point() +
  geom_smooth(span = .22, se = FALSE) +
  labs(x = "Birth year", y = "Number of male babies")
```

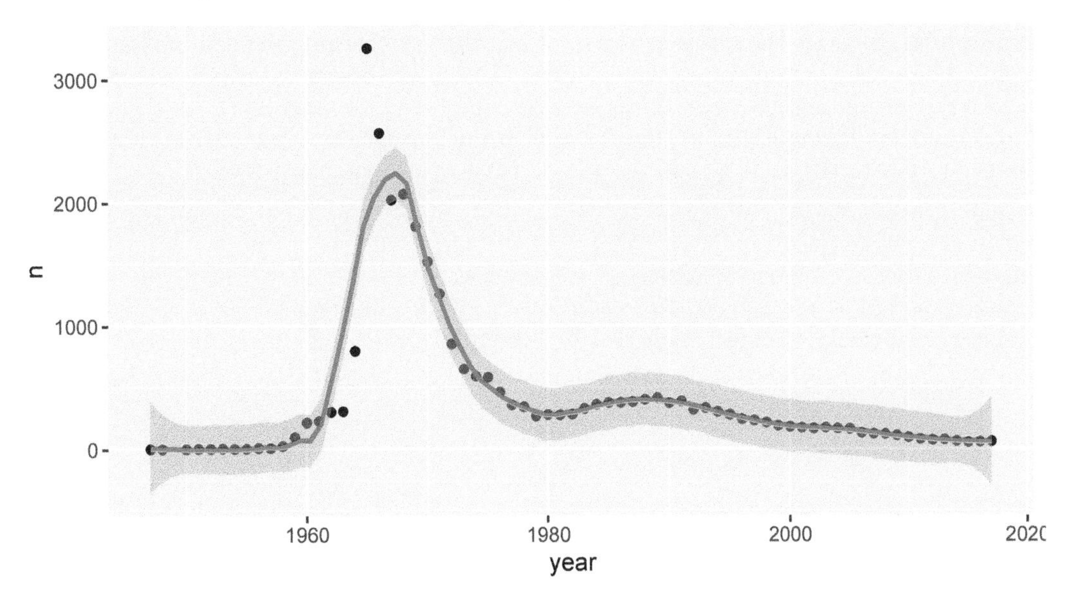


**FIGURE 7.7** Boys named Darrin, over time. Fitting the curve well by selecting a value for span.

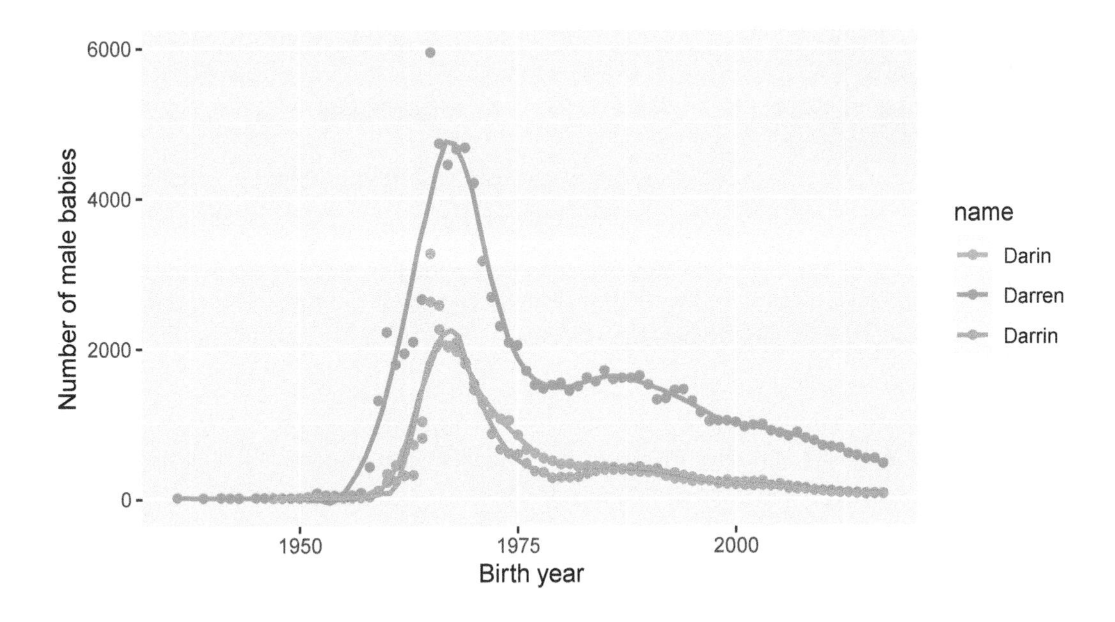


### 7.3.3 Faceting

An alternative way to add a categorical variable to a visualization is to **facet** on the categorical variable. Faceting packages many small, complete plots inside one larger visualization.

The `ggplot2` package provides `facet_wrap` and `facet_grid`, which create plots for each level of the faceted variables. `facet_wrap` places the multiple plots into a square grid, starting in the upper left and filling across and then down. It chooses the size of the grid based on the number of plots. `facet_grid` defines rows and columns from additional variables.

**Example 7.15.** Does the usage pattern for electric car charging depend on the type of facility where the charging station is located? We use the `ecars` data from `fosdata`, and remove the one outlier with a multi-day charge time. We plot total KWH as a function of charging time in four scatterplots, one for each of the four facility types. Notice that all four plots end up with the same $x$- and $y$-scales to make visual comparison valid.

```
fosdata::ecars %>%
  filter(chargeTimeHrs < 24) %>%
  ggplot(aes(x = chargeTimeHrs, y = kwhTotal)) +
  geom_point() +
  facet_wrap(vars(facilityType))
```

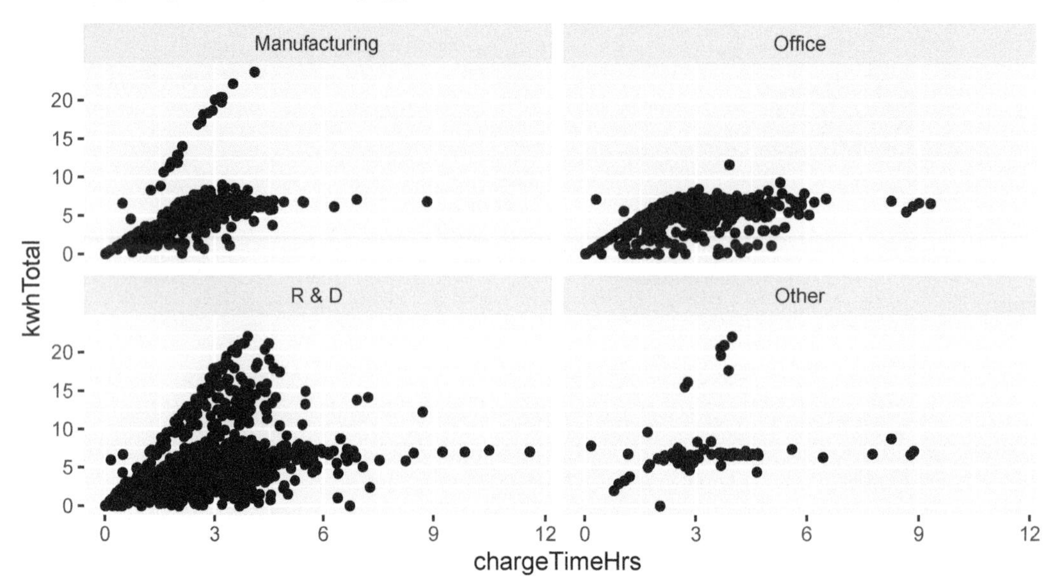


Looking at the plots, the steep line of cars that charge quickly is missing from the Office type of facility. This suggests that unlike Manufacturing and R&D facilities, the Office facilities in this study lacked high speed charging stations.

**Example 7.16.** Create stacked histograms that compare the number of words per sentence in the novels *Emma* and *Pride and Prejudice*. Here we use `facet_grid`, which takes as input a formula `y ~ x`, giving variables to use on the $y$- and $x$-axes respectively. Since we are not choosing to use an $x$ variable, we supply a `.` for `x`.

```
fosdata::austen %>%
  group_by(novel, sentence) %>%
  summarize(sentence_length = n()) %>%
  ggplot(aes(x = sentence_length)) +
  geom_histogram() +
  facet_grid(novel ~ .)
```

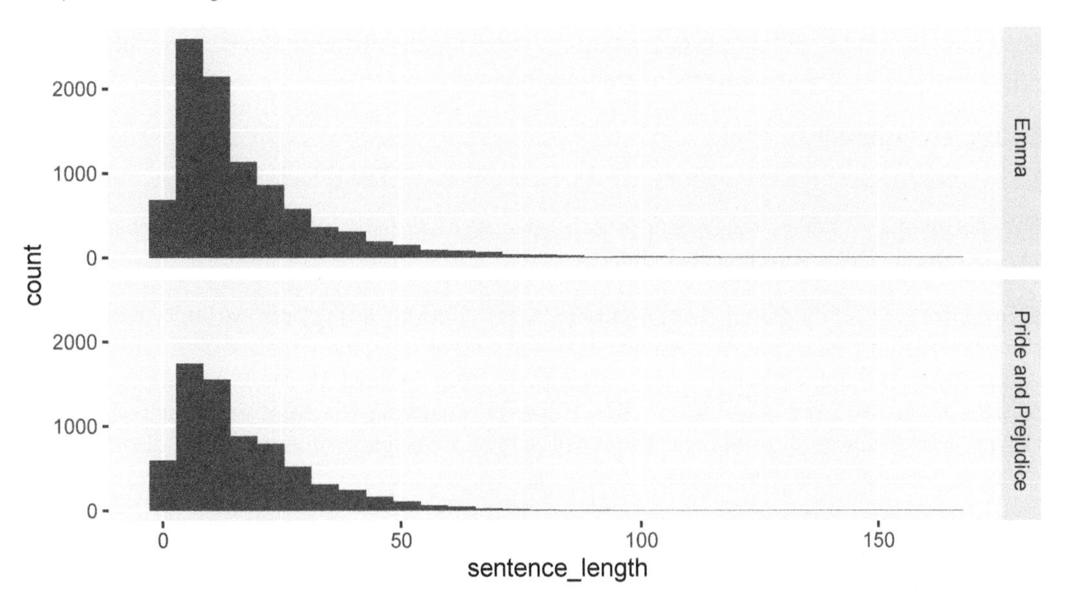


Since we probably wish to compare **proportions** of sentences of various lengths in the two novels, it might be a good idea to allow the $y$-axis to change for the various plots. This can be accomplished by adding `scales = "free_y"` inside of `facet_grid`. Other options include `scales = "free"` or `scales = "free_x"`.

> **Try It Yourself.**
> We chose a histogram because there were so many different sentence lengths. Recreate the above plot with `geom_bar` and see which one you like better.

## 7.4 Customizing

While `ggplot2` has good defaults for many types of plots, inevitably you will end up wanting to tweak things. In this section, we discuss customizing colors, adding themes, and adding annotations. For further customizations, we recommend consulting Wickham.[8]

### 7.4.1 Color

When you have two or especially when you have more than two variables that you want to visualize, then often it makes sense to use color as an aesthetic for one of the variables. Some ggplot geometries have two aesthetics, `color` and `fill`, that may be independently set to different colors. For example, the bars in a histogram have an outline set by `color` and the inside set by `fill`. Adjusting aesthetics is done through scales, which are added to the plot with `+`. The collection of colors used for a variable is called a *palette*.

[8]H Wickham, *ggplot2: Elegant Graphics for Data Analysis*, Use R! (Springer New York, 2009), https://books.google.com/books?id=bes-AAAAQBAJ.

For example, `scale_color_manual` sets the palette for the `color` aesthetic manually. Figure 7.1 in Section 7.1.2 demonstrated manual colors.

**Try It Yourself.**
R has 657 built-in color names. List them all with the `colors()` command.

For continuous variables, a simple approach is to use a *gradient* color palette, which smoothly interpolates between two colors.

**Example 7.17.** The `ecars` electric cars data set contains a `dollars` variable, which gives the price paid for charging. We can add this variable to our plot of power used and charging time. Using the default color scale (black to blue) it is hard to see the most interesting feature of the `dollars` variable (try it!). However, changing the color gradient reveals that the price of a charge depends (for the most part) on charging time only, and also that any charging session under four hours is free.

```
fosdata::ecars %>%
  filter(chargeTimeHrs < 24) %>%
  ggplot(aes(x = chargeTimeHrs, y = kwhTotal, color = (dollars))) +
  geom_point() +
  scale_color_gradient(low = "white", high = "red") +
  labs(title = "Cost of a charge depends only on charging time")
```

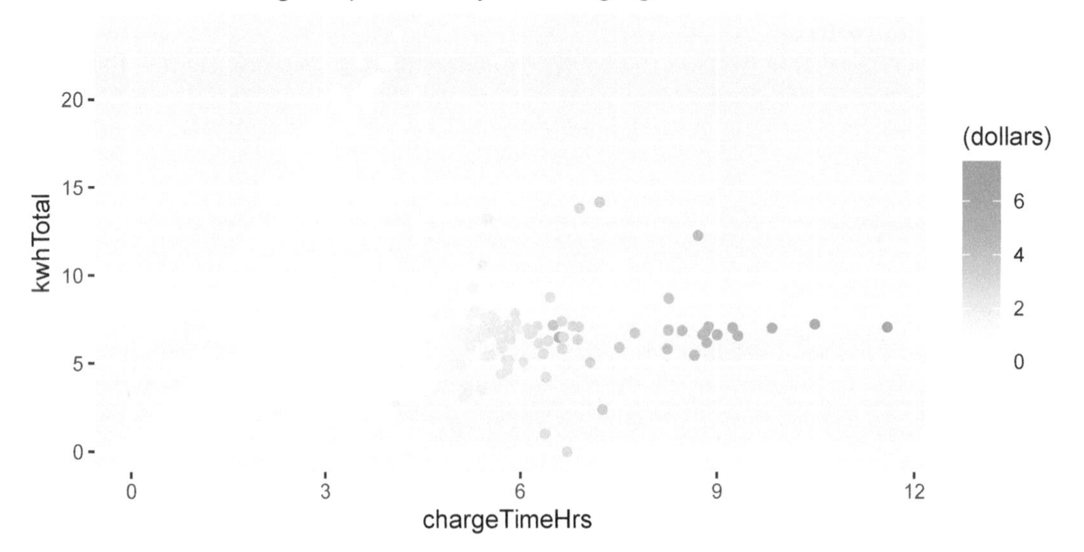


For better looking visualizations, it helps to choose colors from well-designed palettes. It is also good practice to choose colors that work well for people that have some form of colorblindness. Unless you are well versed in color design, you'll want a tool that can assist. The rest of this section will focus on the `RColorBrewer` package. You might also want to check out the `colorspace` package.

**Try It Yourself.**
See what color palettes are available in `RColorBrewer` by typing `RColorBrewer::display.brewer.all()`. To see just the colorblind friendly palettes, try `display.brewer.all(colorblindFriendly = T)`

The palette you choose should help to emphasize the important features of the variable mapped to the color aesthetic. Color palettes are often categorized into three types: categorical, sequential, or divergent.

| Type | Categorical | Sequential | Divergent |
|---|---|---|---|
| Use case | Distinguish between the levels of the variable. | Show the progression of the variable from small to large. | Neutral middle, stronger colors for extreme values of the variable. |
| RColorBrewer | `Set2` | `Blues` | `PiYG` |

The remainder of this section is a case study using the Bechdel data. Our goal is to visualize how the categorization of movies has evolved over time. A similar visualization was given in the *FiveThirtyEight* article, "The Dollar-And-Cents Case Against Hollywood's Exclusion of Women."[9]

The variable `clean_test` describes how a film is rated by the Bechdel test. It has levels `ok` and `dubious` for films that pass the test. Levels `nowomen`, `notalk`, and `men` describe how films fail the test: by having no women, no women who talk to each other, or women who talk only about men. As a first step, Figure 7.8 shows a barplot colored by the `clean_test` variable.

```
bechdel %>%
  ggplot(aes(x = "", fill = clean_test)) +
  geom_bar()
```

We would prefer `ok` and `dubious` to be next to each other, so we need to reorder the levels of the factor, as shown in Figure 7.9.

```
test_levels <- c("ok", "dubious", "men", "notalk", "nowomen")
bechdel$clean_test <- factor(bechdel$clean_test, levels = test_levels)
bechdel %>%
  ggplot(aes(x = "", fill = clean_test)) +
  geom_bar()
```

We want to assign a darkish blue to `ok`, a lighter blue to `dubious`, and shades of red to the others. We will be using `scale_fill_manual` to create the colors that we want to plot. (Recall: we use **scales** to tweak the mapping between variables and aesthetics.) We use `fill` since we are changing the color of the `fill` aesthetic. If we were changing the color of the `color` aesthetic, we would use `scale_color_manual`. The results are in Figure 7.10.

```
rbcolors <- c("darkblue", "blue", "red1", "red3", "darkred")
bechdel %>%
  ggplot(aes(x = "", fill = clean_test)) +
```

[9] Hickey, "The Dollar-and-Cents Case Against Hollywood's Exclusion of Women."

**FIGURE 7.8** Basic barplot colored by Bechdel test results.

**FIGURE 7.9** Bechdel test barplot with levels in correct order.

```
  geom_bar() +
  scale_fill_manual(values = rbcolors)
```

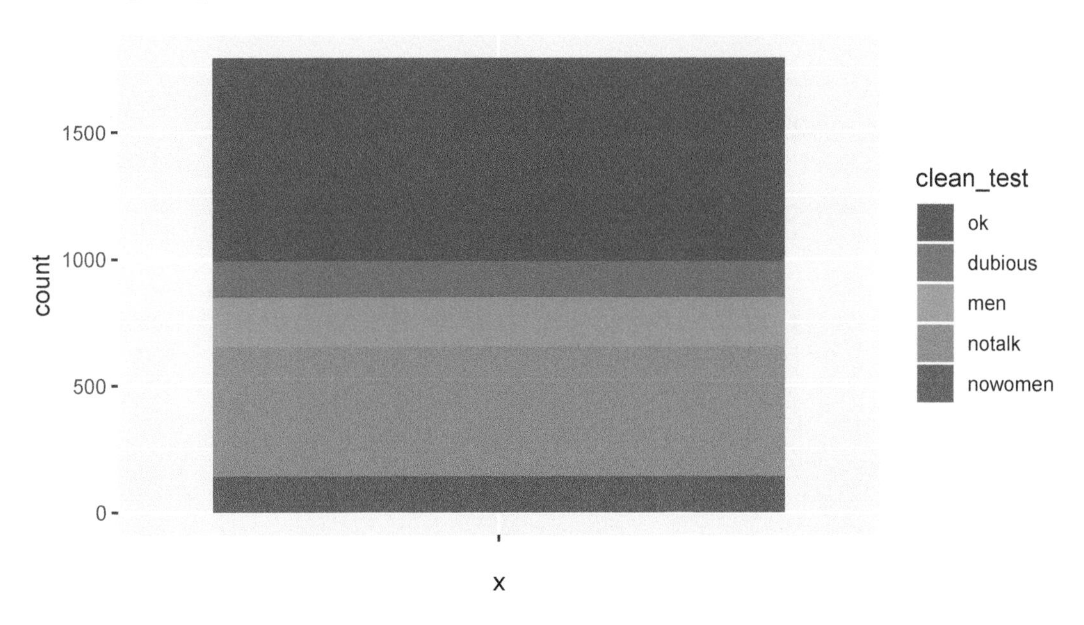


**FIGURE 7.10** Bechdel test barplot with custom colors.

While this is quick and easy, it is difficult to get nice looking colors. Instead, we use palettes from `RColorBrewer`. We want to use blues and reds, so the natural palettes to pick from are the `Blues` palette and the `Reds` palette.

```
library(RColorBrewer)
display.brewer.pal(n = 8, name = "Blues")
display.brewer.pal(n = 8, "Reds")
```

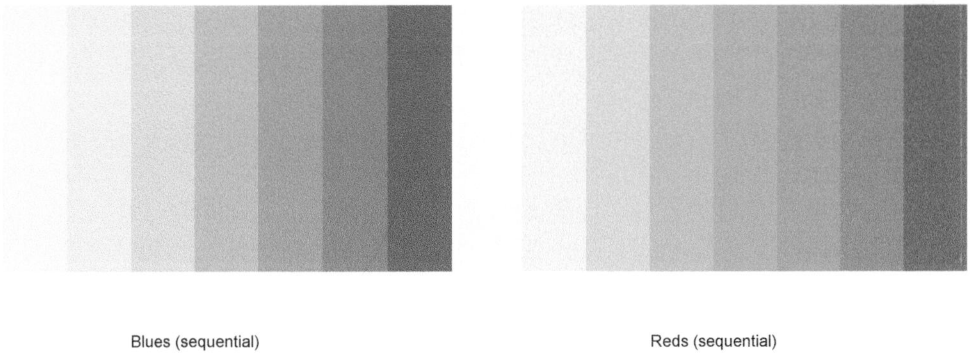

Blues (sequential)

Reds (sequential)

We want to have colors from the blue palette for two variables and colors from the red palette for the other three. A good choice might be colors 5 and 6 from the blues, and colors 4-6 from the reds. We construct our custom list of colors by indexing into the brewer palette vectors:

```
rbcolors <- c(
  brewer.pal(n = 8, "Blues")[6:5],
  brewer.pal(n = 8, "Reds")[4:6]
)
```

We can then supply the color codes to the values argument in scale_fill_manual.

```
bechdel %>%
  ggplot(aes(x = "", fill = clean_test)) +
  geom_bar() +
  scale_fill_manual(values = rbcolors)
```

**FIGURE 7.11** Bechdel test barplot with color brewer colors.

This (Figure 7.11) completes the color selection, but to finish the visualization we need to add a time component. It is possibly best to group the years into bins and create a similar plot as above for each year. The base R function cut will do this.

```
bechdel$year_discrete <- cut(bechdel$year, breaks = 8)
head(bechdel$year_discrete)
```

```
## [1] (2008,2013] (2008,2013] (2008,2013] (2008,2013] (2008,2013]
## [6] (2008,2013]
## 8 Levels: (1970,1975] (1975,1981] (1981,1986] ... (2008,2013]
```

Now, we let the x aesthetic be year_discrete, and produce Figure 7.12

```
bechdel %>%
  ggplot(aes(x = year_discrete, fill = clean_test)) +
  geom_bar() +
  scale_fill_manual(values = rbcolors)
```

Since there are many more movies in recent years, we might prefer to see the **proportion** of movies in each category. We do that with the position attribute of geom_bar, resulting in Figure 7.13.

```
bechdel %>%
  ggplot(aes(x = year_discrete, fill = clean_test)) +
```

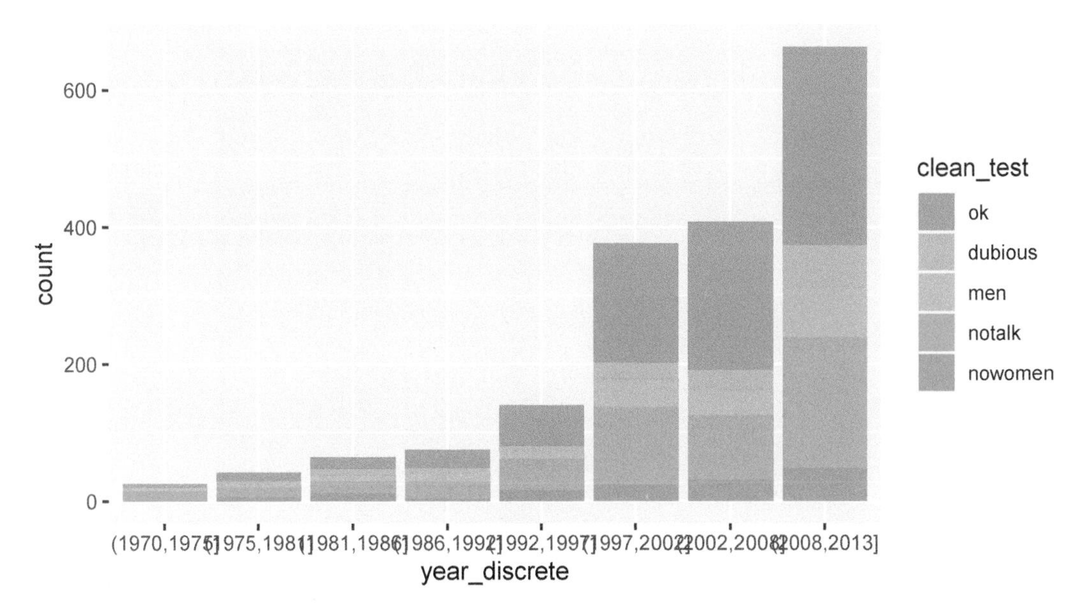


**FIGURE 7.12** Bechdel test barplot broken into year ranges.

```
  geom_bar(position = "fill") +
  scale_fill_manual(values = rbcolors)
```

In Section 7.4.2, we will make some final changes to this plot by adding appropriate labels, choosing a theme, and fixing the text overlap.

### 7.4.2 Labels and themes

We start by customizing the labels associated with a plot. We have added labels to a few of the plots above without providing a comprehensive explanation of what to do.

The function for adjusting labels is `labs`, which has arguments `title`, `subtitle`, `caption` and further arguments that are aesthetics in the plot. Let's return to the Bechdel data set visualization. We would like to have a meaningful title and possibly a subtitle for the plot. Some data visualization experts recommend that the title should not be a recap of the plot, but rather a summary of what the viewer should expect to get out of the plot. We choose rather to put that information in the subtitle. We also choose better names for the $x$-axis, the $y$-axis, and the legend title for the `fill`.

The $y$-axis is shown as a proportion, but would be easier to understand as a percentage. We make this change to the $y$ aesthetic with a scale. The $x$-axis labels came from the `cut` command. It would be a nice improvement to change them to say 1970-74, for example. That change would require either typing each label manually or heavy use of string manipulation. Instead, we simply rotate the labels by 20 degrees to make them fit a little better.

Finally, ggplot provides *themes* that control the overall look and feel of a plot. The default theme, with a gray background, is called `theme_gray`. Other useful themes are `theme_bw` for black-and-white plots and `theme_void` which shows nothing but the requested geometry. Add-on packages such as `ggthemes` and `ggthemr` provide more themes, or you can design your

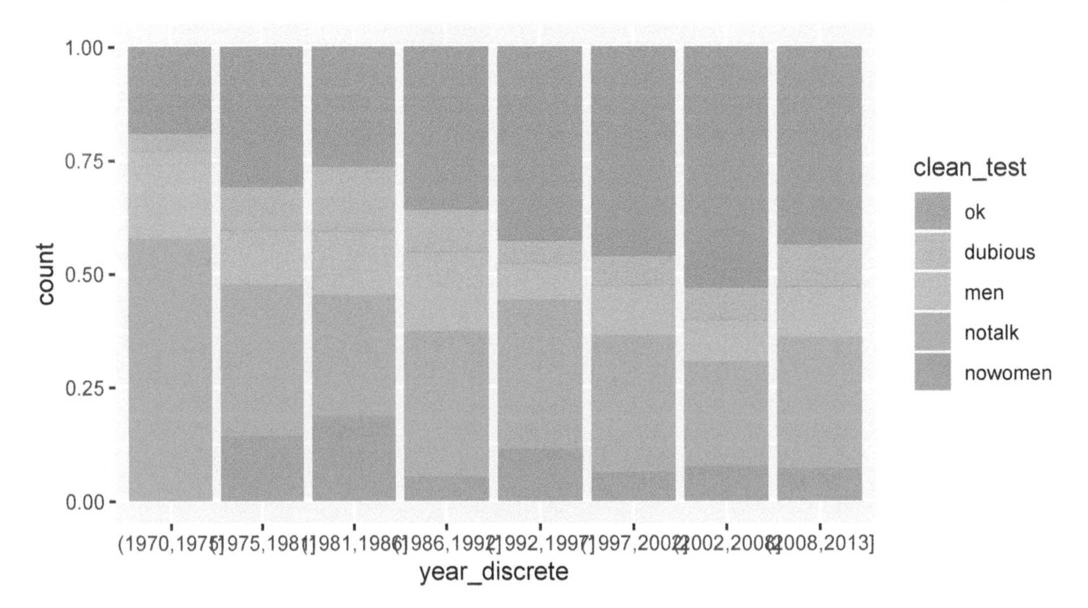


**FIGURE 7.13** Bechdel test barplot with proportions.

own theme so that your plots have your characteristic style. We will choose `theme_minimal` to complete our Bechdel test plot, see Figure 7.14.

```
bechdel %>%
  ggplot(aes(x = year_discrete, fill = clean_test)) +
  geom_bar(position = "fill") +
  scale_fill_manual(values = rbcolors) +
  scale_y_continuous(labels = scales::percent) +
  labs(
    title = "Percentages of Movies Passing Bechdel Test over Time",
    subtitle = "Not much progress since the 1990's",
    x = "Year",
    y = "Proportion",
    fill = "Result of Test"
  ) +
  theme_minimal() +
  theme(axis.text.x = element_text(angle = 20))
```

> **Try It Yourself.**
> Replace `theme_minimal()` with `theme_dark()` in the above code.
>
> Install `ggthemes` using `install.packages("ggthemes")` if necessary, and replace `theme_minimal()` with `ggthemes::theme_fivethirtyeight()`.

### 7.4.3 Text annotations

Many times, we will want to add text or other information to draw the viewers' attention to what is interesting about a visualization or to provide additional information. In this example, using the `babynames` data, we want to plot the number of female babies named

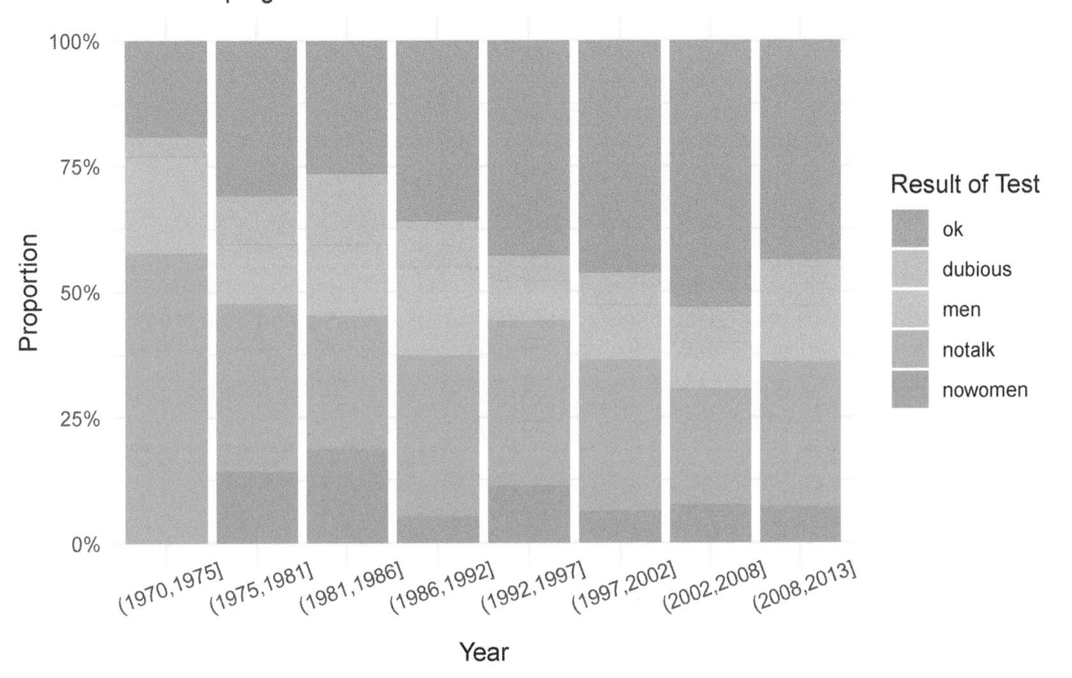


**FIGURE 7.14** The percentage of movies passing the Bechdel test over time has not seen much progress since the 1990's.

Peyton versus the number of male babies named Peyton from 1990 to 2010. In order to make it clear which data point is which, we will add text labels to some of the points.

We begin by filtering the data set so that it only contains information about Peyton from 1990 to 2010. The `babynames` data is in long format, with a row for each sex. It would be easier if it were in wide format, with both sex counts in a single row. So, we use `pivot_wider` in the `tidyr` package.

```
recentPeyton <- babynames::babynames %>%
  filter(year >= 1990, year <= 2010, name == "Peyton") %>%
  tidyr::pivot_wider(
    names_from = sex, values_from = n,
    id_cols = c(name, year)
  )
head(recentPeyton)
```

```
## # A tibble: 6 x 4
##   name    year     F     M
##   <chr>  <dbl> <int> <int>
## 1 Peyton  1990    62   121
## 2 Peyton  1991    67   126
## 3 Peyton  1992   399   187
## 4 Peyton  1993   617   242
## 5 Peyton  1994   585   357
## 6 Peyton  1995   588   453
```

A first attempt at the plot gives a sense that the number of male and female Peyton babies are roughly equal, since the dots are mostly near the diagonal line of slope 1. However, from the visualization in Figure 7.15 we cannot tell which year is which.

```
recentPeyton %>%
  ggplot(aes(x = M, y = F)) +
  geom_point() +
  geom_abline(slope = 1, linetype = "dotted")
```

To add the year information to the plot, we add text labels to each point using `geom_text`. Along with `x` and `y` aesthetics, the `geom_text` geometry requires the aesthetic `label`. The `label` aesthetic tells `geom_text` what text to use on each point. See Figure 7.16.

```
recentPeyton %>%
  ggplot(aes(x = M, y = F, label = year)) +
  geom_point() +
  geom_abline(slope = 1, linetype = "dotted") +
  geom_text()
```

By default, `geom_text` puts the text exactly at the $(x, y)$-coordinate given in the data. In this case, that is not what we want! We would like the text to be nudged a little bit so that it doesn't overlap the scatterplot. We can do this by hand using `nudge_x` and `nudge_y`, but it is easier to use the `ggrepel` package.

```
recentPeyton %>%
  ggplot(aes(x = M, y = F, label = year)) +
  geom_point() +
  geom_abline(slope = 1, linetype = "dotted") +
  ggrepel::geom_text_repel() +
```

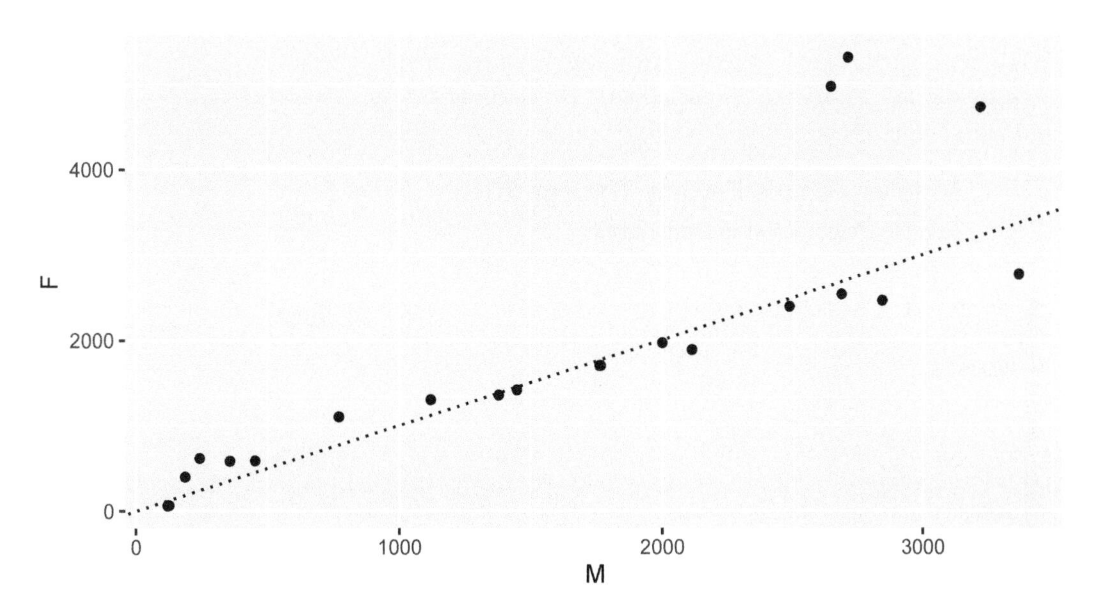


**FIGURE 7.15** Male and female Peyton babies, one point per year.

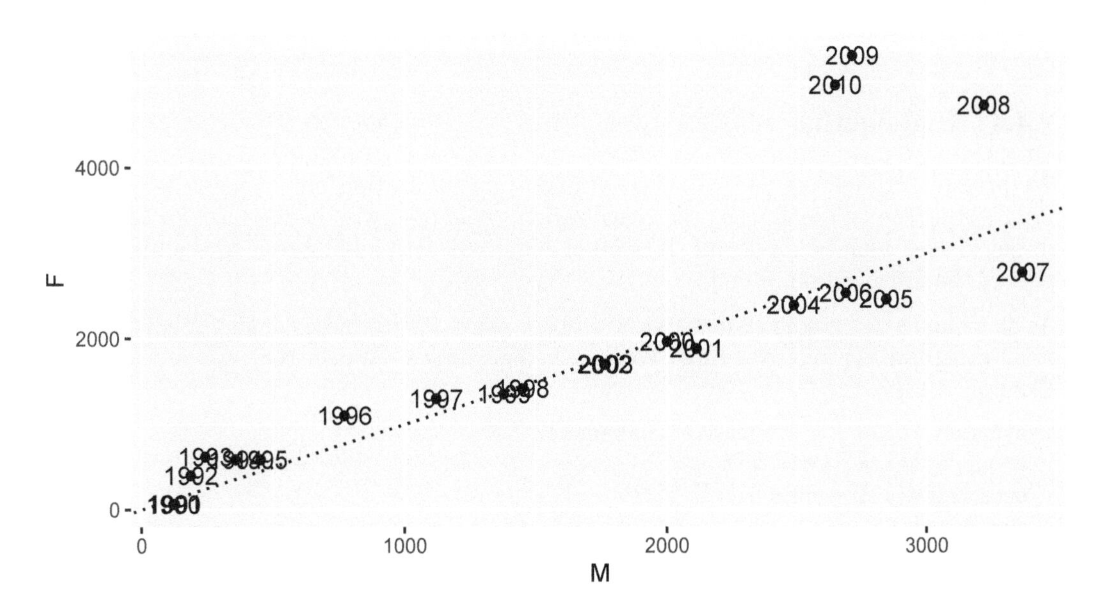


**FIGURE 7.16** Male and female Peyton babies, years labeled.

```
labs(
  x = "Male babies named Peyton",
  y = "Female babies named Peyton",
  title = "Male vs. Female Peytons Through 2010",
  subtitle = "Female Peytons Set to Dominate?"
)
```

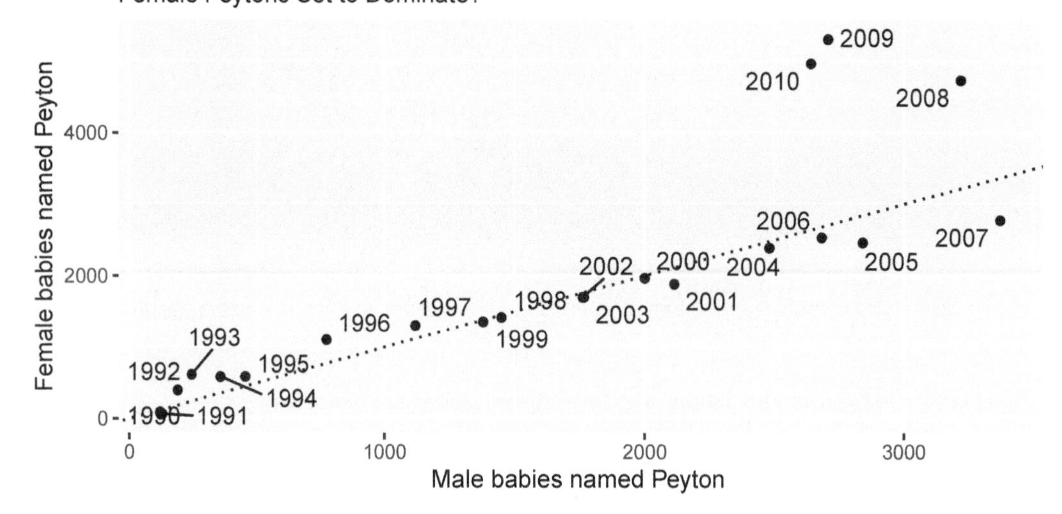


**FIGURE 7.17** Male and female Peyton babies, finished visualization.

Our finished visualization in Figure 7.17 also gets better labels with the `labs` command.

### 7.4.4 Highlighting

We present highlighting as one last example of customization out of many possibilities. Highlighting can be useful when the visualization would otherwise be too busy for the viewer to quickly see what is important. It is often combined with adding text to the highlighted part of the visualization.

As an example[10], suppose that we wish to observe how the 20 most commonly given names of all-time have evolved over time. We use the `babynames` data set that was introduced in Example 6.14.

```
babynames_top20 <- babynames::babynames %>%
  group_by(name, sex) %>%
  summarize(tot = sum(n)) %>%
  ungroup() %>%
  slice_max(n = 20, tot) %>%
  select(name, sex) %>%
  left_join(babynames::babynames)
```

[10]This example is similar to one found at the R Graph Gallery, https://www.r-graph-gallery.com/, which is maintained by Yan Holtz. That site has many interesting data visualization techniques and tricks.

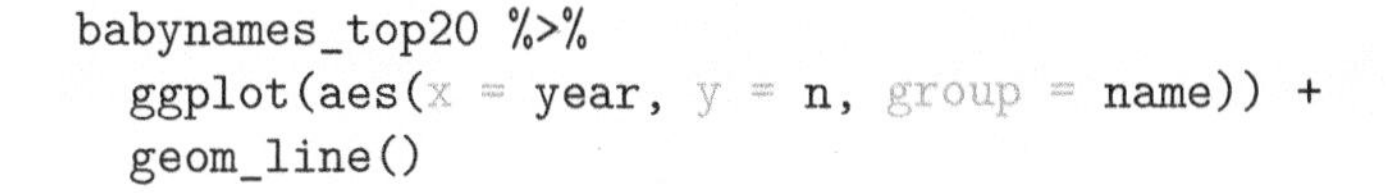

```
babynames_top20 %>%
  ggplot(aes(x = year, y = n, group = name)) +
  geom_line()
```

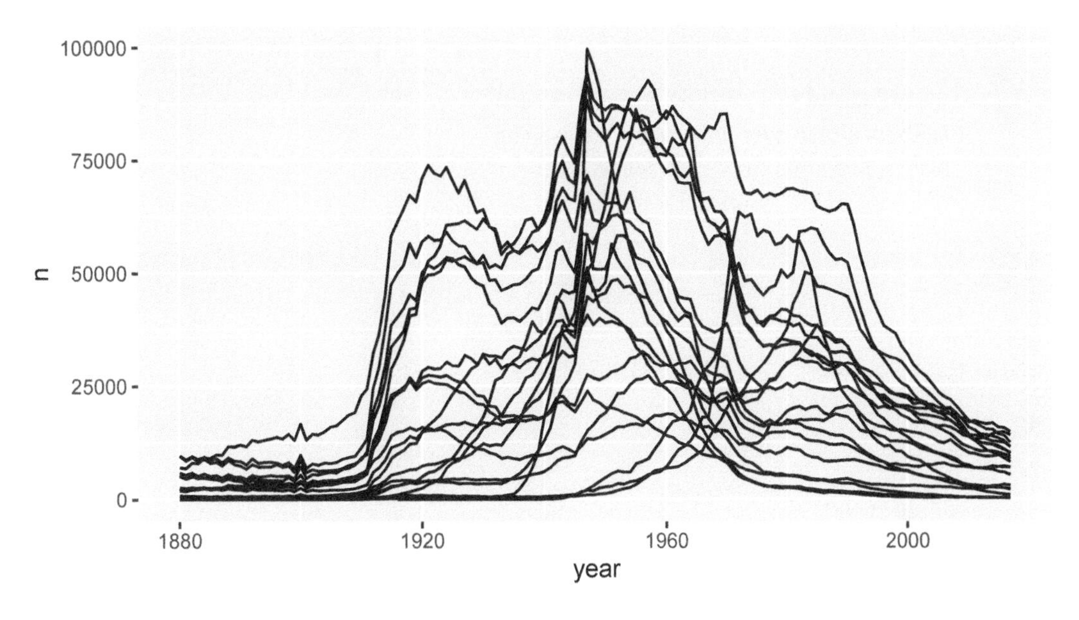


**FIGURE 7.18** The 20 most popular baby names over time.

As you can see, Figure 7.18 is a bit of a mess. It does give us good information about the general patterns that the most commonly given names of all-time have taken, but it is hard to pick up any more information than that.

**Try It Yourself.**

1. Change `group = name` to `color = name`. Does that help?
2. Change `y = n` to `y = prop` to get a feel for how the **proportions** of babies with the 20 most commonly given names has evolved over time.
3. Think about why this code doesn't work.

```
babynames::babynames %>%
  group_by(name, sex) %>%
  mutate(tot = sum(n)) %>%
  ungroup() %>%
  slice_max(n = 20, tot) %>%
  ggplot(aes(x = year, y = n, group = name)) +
  geom_line()
```

To customize Figure 7.18, we might decide to illustrate one or more of the names. The R package `gghighlight` provides an easy way to highlight one or more data groupings in a plot. We simply add `+ gghighlight()` with the condition on the variable(s) we want to highlight as an argument. Figure 7.19 shows the result of highlighting Elizabeth and Mary.

```
babynames_top20 %>%
  ggplot(aes(x = year, y = prop, color = name)) +
  geom_line() +
  gghighlight::gghighlight(name == "Elizabeth" | name == "Mary") +
```

```
  labs(
    title = "Twenty most commonly given names",
    y = "proportion of babies",
    subtitle = "Elizabeth stable over time; Mary decreasing"
  )
```

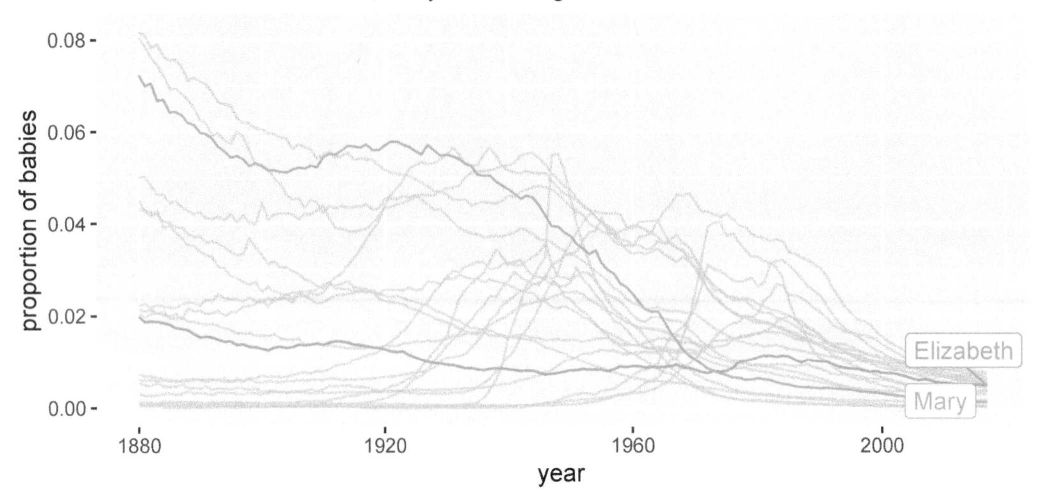


**FIGURE 7.19** Highlighting data groupings with gghighlight.

## Vignette: Choropleth maps

There are many ways of visualizing data. One very useful way of visualizing is via **choropleth maps**. A *choropleth map* is a type of map where regions on the map are shaded according to some variable that has been observed that is related to that region. One of the most common choropleth maps in the US is the national election map, colored by state. Other examples include weather maps and radar images.

These types of maps can seem intimidating to create. However, R has tools to make basic choropleth maps of the US colored along counties or states easier than it would otherwise be. The hardest part, as usual, is getting the data in the format that is required by the mapping tools.

Let's look at the 2012 US presidential election results in Alabama, and visualize them. We start by loading the data, and computing the percentage of people who voted for Romney from among those who voted either for Romney or Obama.

```
pres_election <- fosdata::pres_election
pres_election <- filter(
  pres_election, year == 2012,
  stringr::str_detect(candidate, "Obama|Romney")
```

```
)
pres_election <- pres_election %>%
  filter(state == "Alabama") %>%
  arrange(state, county, party) %>%
  group_by(FIPS, state, county) %>%
  summarize(
    percent_romney =
      last(candidatevotes) / sum(candidatevotes, na.rm = T)
  ) %>%
  ungroup()
head(pres_election)

## # A tibble: 6 x 4
##    FIPS state   county  percent_romney
##   <int> <chr>   <chr>            <dbl>
## 1  1001 Alabama Autauga          0.732
## 2  1003 Alabama Baldwin          0.782
## 3  1005 Alabama Barbour          0.484
## 4  1007 Alabama Bibb             0.736
## 5  1009 Alabama Blount           0.875
## 6  1011 Alabama Bullock          0.236
```

To plot this, we need the map data.

```
library(maps)
alabama_counties <- map_data("county") %>%
  filter(region == "alabama")
head(alabama_counties)

##        long      lat group order  region subregion
## 1 -86.50517 32.34920     1     1 alabama   autauga
## 2 -86.53382 32.35493     1     2 alabama   autauga
## 3 -86.54527 32.36639     1     3 alabama   autauga
## 4 -86.55673 32.37785     1     4 alabama   autauga
## 5 -86.57966 32.38357     1     5 alabama   autauga
## 6 -86.59111 32.37785     1     6 alabama   autauga
```

Notice that the counties in this data set aren't capitalized, while in our other data set, they are. There are likely other differences in county names, so we really need to get this map data in terms of FIPS. The map data needs to contain three variables, and their names are important. It will have a variable called `id` that matches to the id variable from the fill data, a variable `x` for longitude and `y`, for latitude.

```
fips_data <- maps::county.fips
fips_data <- filter(
  fips_data,
  stringr::str_detect(polyname, "alabama")
) %>%
  mutate(county = stringr::str_remove(polyname, "alabama,"))
alabama_counties <- left_join(alabama_counties, fips_data,
  by = c("subregion" = "county")
)
alabama_counties <- rename(alabama_counties, FIPS = fips, x = long, y = lat)
```

```
alabama_counties <- select(alabama_counties, FIPS, x, y, group, order) %>%
  rename(id = FIPS)
```

And this is how we want the data. We want the value that we are plotting `percent_romney` repeated multiple times with `long` and `lat` coordinates. The rest is easy.

```
ggplot(pres_election, aes(fill = percent_romney)) +
  geom_map(
    mapping = aes(map_id = FIPS),
    map = data.frame(alabama_counties)
  ) +
  expand_limits(alabama_counties) +
  scale_fill_gradient2(
    low = "blue", high = "red",
    mid = "white", midpoint = .50
  ) +
  coord_map()
```

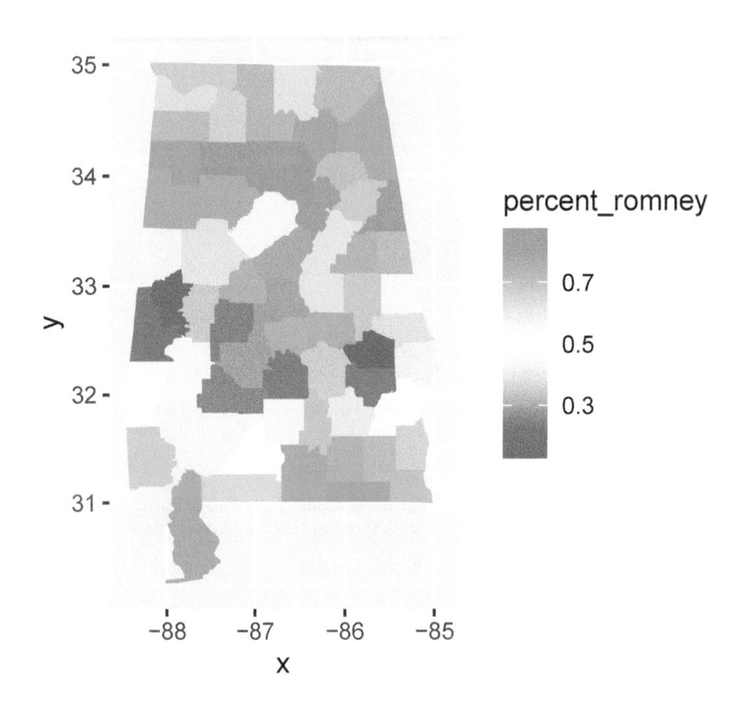


**FIGURE 7.20** Choropleth map for the 2012 presidential election in Alabama.

In Figure 7.20 you can see that Alabama has a belt across its middle that voted in favor of Obama, while the rest of the state voted for Romney.

## Vignette: COVID-19

Possibly the most visualized data of all time comes from the COVID-19 pandemic. We apply some of the techniques from this chapter to create the types of visualizations that were commonly used to track the spread of the disease and to compare its prevalence in different geographical regions.

This vignette will use the `covid` data included the `fosdata` package. This data was sourced on September 13, 2021 from the GitHub repository[11] run by *The New York Times.* We encourage you to download the most recent data set from that site and rework through these examples, or to download the larger data set that gives information by county.

```
covid <- fosdata::covid
```

Let's plot the cumulative number of cases in New York by date.

```
filter(covid, state == "New York") %>%
  ggplot(aes(x = date, y = cases)) +
  geom_point()
```

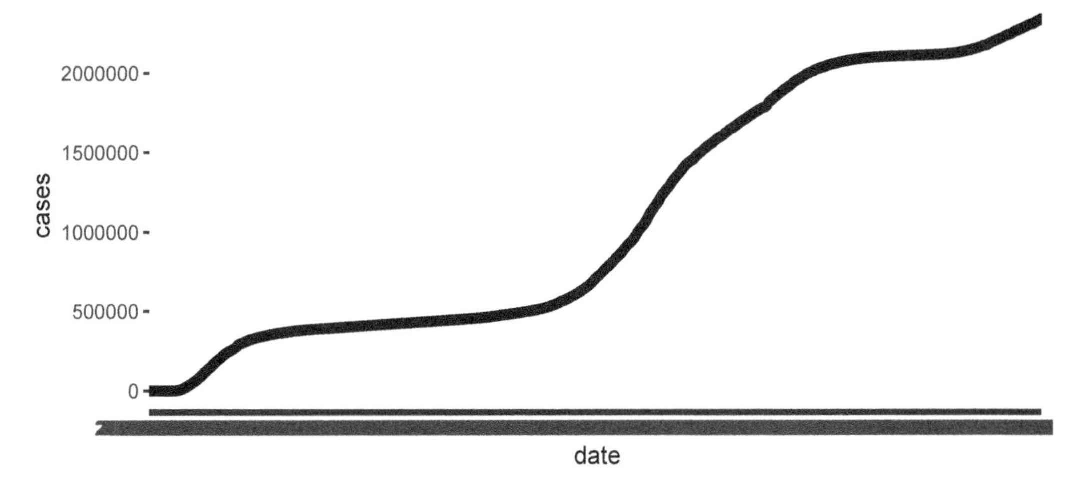


In this plot, the thick black stripe along the $x$-axis shows each date, but they are unreadable. The solution is to change the type of the date variable. It is a `chr` variable, but it should be of type `Date`. Base R can work with dates, but it is much easier with the `lubridate` package. In order to convert a character variable into a date, you need to specify the format that the date is in from among `ymd`, `ydm`, `mdy`, `myd`, `dym`, or `dmy`. If there are times involved, you add an underscore and then the correct combination of `h` `m` and `s`.

The bad dates prevented our first graph from using `geom_line` (try it!), but now with correct typing `geom_line` works.

```
covid <- mutate(covid, date = lubridate::ymd(date))
```

```
filter(covid, state == "New York") %>%
  ggplot(aes(x = date, y = cases)) +
```

[11]https://github.com/nytimes/covid-19-data

```
  geom_line() +
  labs(title = "Total COVID-19 cases in New York")
```

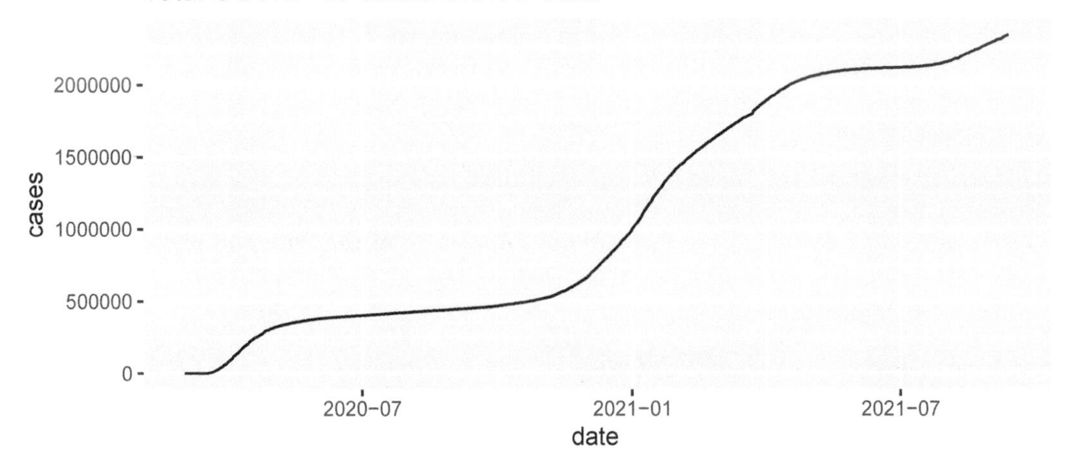


Next, let's consider the **daily new cases**. We can compute the number of new cases on a given day by taking the total number of cases on the given day and subtracting the total number of cases on the previous day. The R command for doing this is `diff`. For example,

```
diff(c(1, 4, 6, 2))
```

```
## [1]  3  2 -4
```

Notice that the length of the output of `diff` is one less than the length of the input, so in our setting we will want to add a leading zero. Let's see how to do it.

```
covid <- covid %>%
  group_by(state) %>%
  mutate(daily_cases = diff(c(0, cases))) %>%
  ungroup()
```

One issue that arises is that states sometimes make mistakes and need to correct them later. For example, if a state is systematically double counting some cases, and they realize it a month later, then they may subtract all of the double counted cases from the total cases on the same day. This can lead to **negative** daily cases, which does not make sense.

> **Try It Yourself.**
> Find the date and state which has the largest (in absolute value) negative value for daily cases.

The best way to handle negative daily cases is not clear, but for data visualization purposes it is better to set all negative values of daily cases to zero.

```
covid <- covid %>%
  mutate(daily_cases = pmax(daily_cases, 0))
```

Now, we can plot the daily cases by date in New York as follows.

```
covid %>%
  filter(state == "New York") %>%
  ggplot(aes(x = date, y = daily_cases)) +
  geom_point()
```

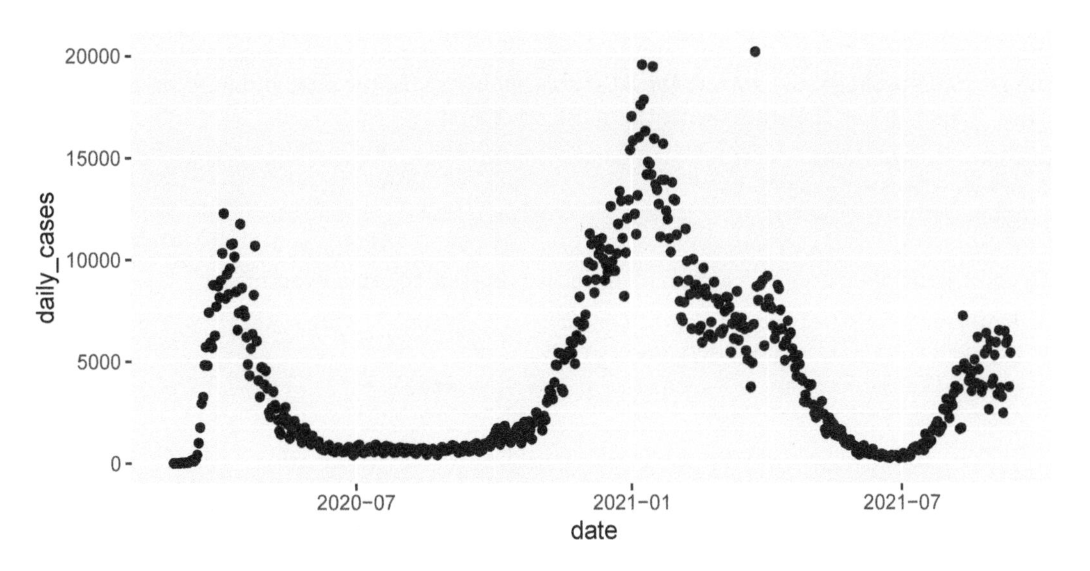


**FIGURE 7.21** Basic plot of daily COVID-19 cases in New York State.

The results, in Figure 7.21, lack visual appeal. For the more finished plot in Figure 7.22 we add a title, resize the points, and add a smoothed line with the span chosen carefully. A larger span gives a smoother curve but no longer follows the peaks of the data well.

```
covid %>%
  filter(state == "New York") %>%
  ggplot(aes(x = date, y = daily_cases)) +
  geom_point(size = 0.5) +
  geom_smooth(span = 0.1, method = "loess") +
  labs(title = "Daily COVID-19 cases in New York", y = "Daily new cases")
```

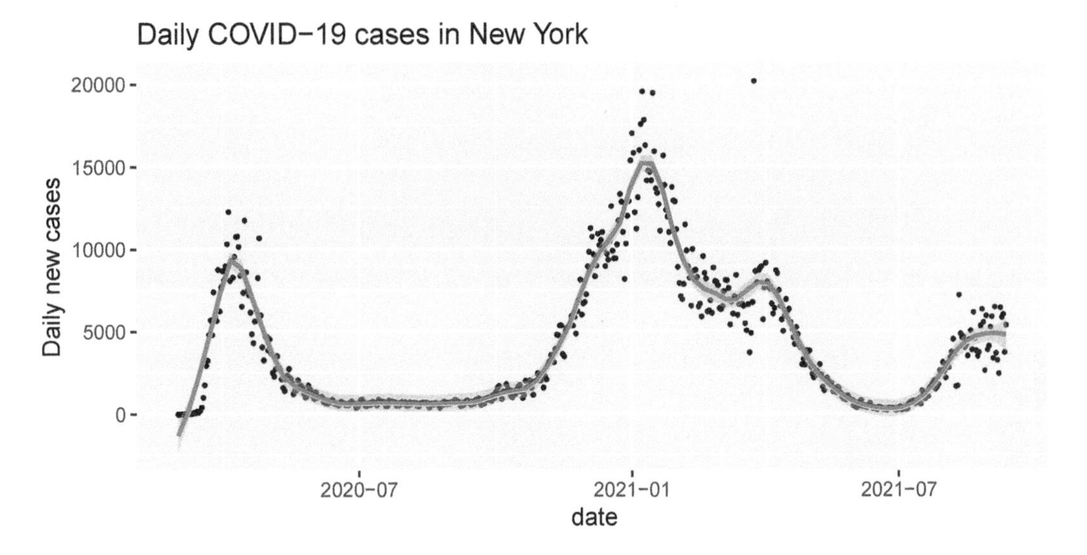


**FIGURE 7.22** Finished plot of daily COVID-19 cases in New York State.

### State-by-state comparisons

When comparing states, we want our chart legends to list states in a reasonable order. Using the `forcats` package, we convert the `state` variable to a factor after ordering by number of cases. This will list the states with the most cases first.

```
covid <- covid %>%
  arrange(desc(cases)) %>%
  mutate(state = forcats::as_factor(state))
```

To show multiple states on the same graph, we use the `color` aesthetic.

```
bigstates <- c("California", "Florida", "Illinois", "New York", "Texas")
filter(covid, state %in% bigstates) %>%
  ggplot(aes(x = date, y = cases, color = state)) +
  geom_line()
```

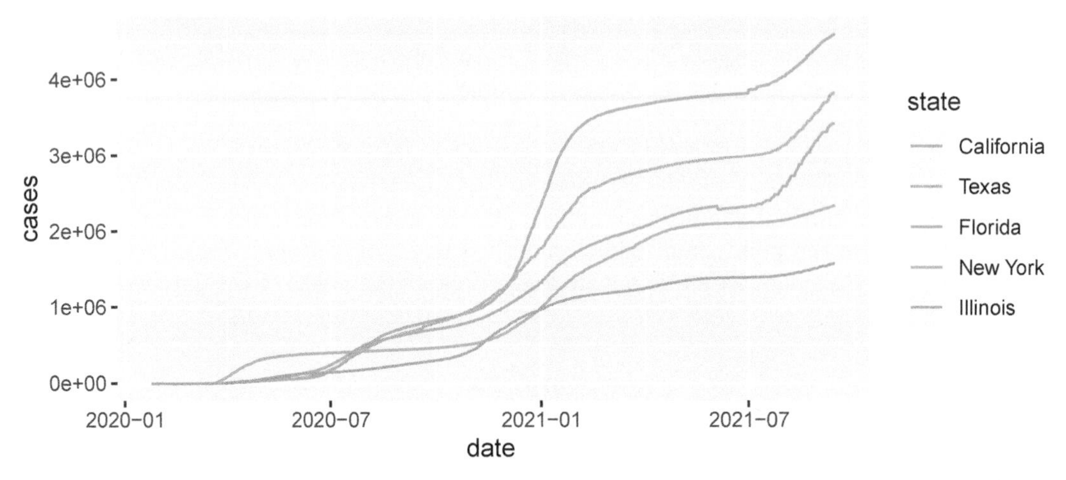


When tracking a disease, the rate of growth is important. It is visible as the slope of the graph when plotted with the $y$-axis on a logarithmic scale.

```
filter(covid, state %in% bigstates) %>%
  ggplot(aes(x = date, y = cases, color = state)) +
  geom_line() +
  scale_y_log10(labels = scales::comma)
```

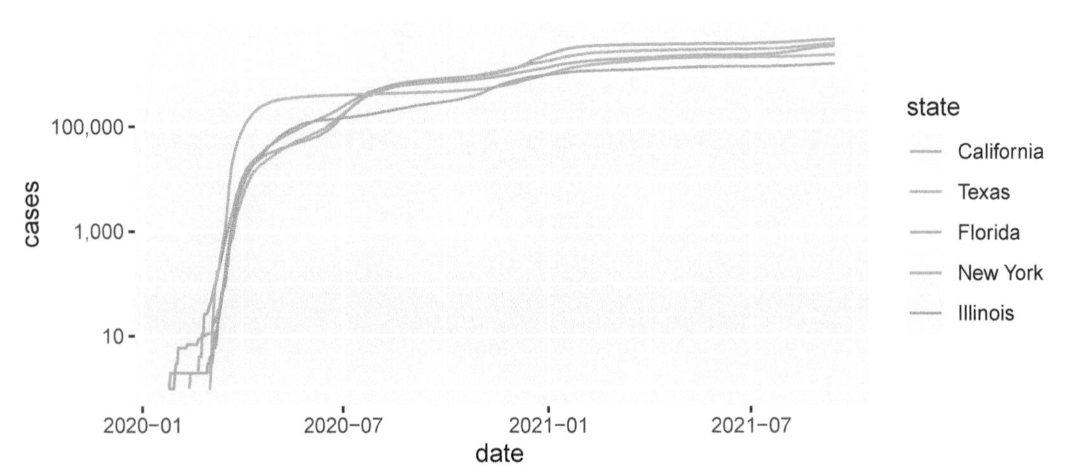

Unless you are used to reading graphs with log scale on the $y$-axis, the plot can be misleading. It is not obvious that Texas has had more than twice the total number of cases as Illinois as of September 2021.

Finally, we use highlighting to show the five big states' daily new cases with the other states shown in gray for comparison. We add custom colors and some theme elements. The finished plot is Figure 7.23.

```
custom_colors <- RColorBrewer::brewer.pal(5, name = "Set1")
covid %>%
  ggplot(aes(x = date, y = daily_cases, color = state)) +
  geom_smooth(span = 0.11, se = FALSE, size = .6) +
  gghighlight::gghighlight(state %in% bigstates) +
  labs(
    title = "States with most cases",
    subtitle = "compared to other U.S. states",
    x = "Date", y = "Daily new cases"
  ) +
  scale_color_manual(values = custom_colors) +
  theme_minimal() +
  theme(plot.title = element_text(face = "bold"))
```

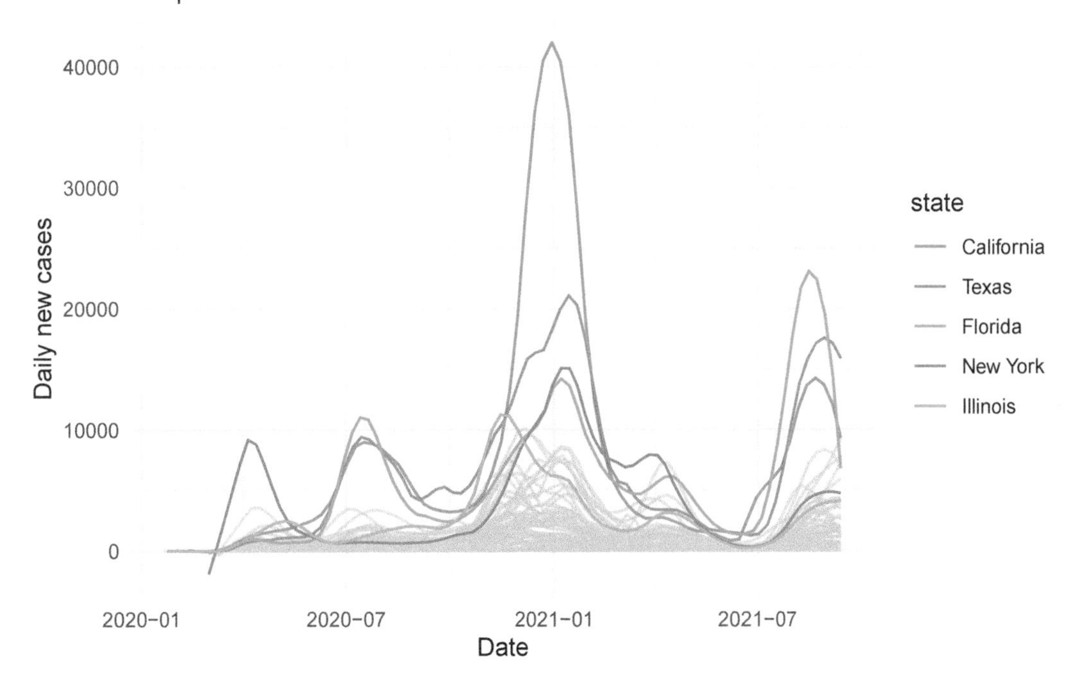


**FIGURE 7.23** Finished visualization of a multiple state comparison.

> **Try It Yourself.**
> Why didn't we use `geom_line` for the daily cases? Try it and see which plot you prefer.

## Exercises

---

Exercises 7.1 – 7.7 require material in Sections 7.1 and 7.2.

**7.1.** The built-in R data set `quakes` gives the locations of earthquakes off of Fiji in the 1960's. Create a plot of the locations of these earthquakes, showing depth with color and magnitude with size.

**7.2.** Create a boxplot of highway mileage for each different cylinder in `mtcars`, and display on one plot with highway mileage on the $y$-axis and cylinder on the $x$-axis. You will need to change `cyl` to a factor.

**7.3.** Consider the `brake` data set in the `fosdata` package. The variable `p1_p2` gives the total time that it takes a participant to press the "brake," realize they have accidentally pressed the accelerator, and release it after seeing a red light. Create a density plot of the variable `p1_p2` and use it to answer the following questions.

a. Does the data appear to be symmetric or skewed?
b. What appears to be approximately the most likely amount of time to release the accelerator?

**7.4.** Consider the `brake` data set in the `fosdata` package. The variable `latency_p1` gives the length of time (in ms) that a participant needed to step on the brake pedal after experiencing a stimulus. Create a qq plot of `latency_p1` and comment on whether the variable appears to be normal.

**7.5.** The `mpg` data from the `ggplot2` package contains fuel economy data for popular cars. Create a scatterplot of highway mileage versus city mileage colored by the number of cylinders. Use `geom_jitter` so the points don't overlap.

The `cyl` variable in `mpg` is an integer. Change it to a factor and recreate the plot. Which do you like better?

**7.6.** The data set `storms` is included in the `dplyr` package. It contains information about 425 tropical storms in the Atlantic.

a. Produce a histogram of the `wind` speeds in this data set. Fill your bars using the `category` variable so you can see the bands of color corresponding to the different storm categories.
b. Repeat part (a) but make a histogram of the `pressure` variable. You should observe that high category storms have low pressure.
c. Describe the general shape of these two distributions.
d. What type is the `category` variable in this data set? How did that affect the plots?

**7.7.** Continue using the `storms` data described in Exercise 7.6. Produce a plot showing the position track of each storm from 2014 (use `long` for x and `lat` for y). Color your points by the name of the storm so you can distinguish the seven storm tracks. Which storm in 2014 made it the furthest North?

---

Exercises 7.8 – 7.27 require material through Section 7.3. Exercises which come from the same data set (such as `austen`) may not all require Section 7.3, but at least one exercise from that data set will.

---

Exercises 7.8 – 7.10 use the `austen` data set from the `fosdata` package. This data set contains the full text of *Emma* and of *Pride and Prejudice*, two novels by Jane Austen. Each observation is a word, together with the sentence, chapter and novel it appears in, as well as the word length and sentiment score of the word.

**7.8.** Create a barplot of the word lengths of the words in the data set, faceted by novel.

**7.9.** In *Emma*, restrict to words that have non-zero sentiment score. Create a scatterplot of the percentage of words that have a positive sentiment score versus chapter. Add a line using `geom_line` or `geom_smooth` and explain your choice.

**7.10.** Use `geom_col()` to create a barplot of the mean sentiment score per chapter for *Emma*.

**7.11.** Consider the `juul` data set in the `ISwR` package. This data set is 1339 observations of, among other things, the `igf1` (insulin-like growth factor) levels in micrograms/liter, the `age` in years, and the `tanner` puberty level of participants.

a. Create a scatterplot of `igf1` level versus `age`.
b. Add categorical coloring based on `tanner`.
c. Is `geom_smooth`, `geom_line` or neither appropriate here?

**7.12.** Consider the `combined_data` presidential election data set that was created in Section 6.6.2. For the 2000 election, create a scatterplot of the percent of votes cast for George W. Bush versus the unemployment rate, where each point represents the results of one county.

**7.13.** Consider the Bechdel test data, `fosdata::bechdel`.

a. Create a scatterplot of the percentage of movies that pass the Bechdel test in a given year versus the year that the movie was released.
b. The cost to earnings ratio is the earnings divided by the budget. Create boxplots of the cost to earnings ratio (in 2013 dollars) of movies made in 2000 or later, split by whether they pass the Bechdel test.
c. Re-do part (b), except make the boxplot of the **log** of the cost to earnings ratio. Explain the pros and cons of making this transformation.

**7.14.** Make a scatterplot showing $CO_2$ uptake as a function of concentration level for the built-in data set `CO2`. Include a smoothed fit line and color by Type. Facet your plot to one plot for each `Plant`.

**7.15.** The `fosdata::pres_election` data set gives voting results from the 2000-2016 U.S. presidential elections. Produce five bar charts, one for each election, that show the total number of votes received by each political party. Use `facet_wrap` to put all five charts into the same visualization.

**7.16.** The `ecars` data set from `fosdata` was introduced in this chapter. Create a visualization showing scatterplots with the `chargeTimeHrs` variable on the $x$-axis and the `kwhTotal` variable on the $y$-axis. Facet your visualization with one plot per day of week and platform. Remove the `web` platform cars, so you have 14 facets in two rows and seven columns. Be sure your weekdays display in a reasonable order.

---

Exercises 7.17 – 7.19 all use the `Batting` data set from the `Lahman` package. This gives the batting statistics of every player who has played baseball from 1871 through the present day.

**7.17.** a. Create a scatterplot of the number of doubles hit in each year of baseball history.
b. Create a scatterplot of the number of doubles hit in each year, in each league. Show only the leagues 'NL' and 'AL,' and color the NL blue and the AL red.

**7.18.** Create boxplots for total runs scored per year in the AL and the NL from 1969 to the present.

**7.19.** a. Create a histogram of lifetime batting averages (H/AB) for all players who have at least 1000 career AB's.
b. In your histogram from (d), color the NL blue and the AL red. (If a player played in both the AL and NL, count their batting average in each league if they had more than 1000 AB's in that league.)

**7.20.** Use the `People` data set from the `Lahman` package.

a. Create a barplot of the birth months of all players. (Hint: `geom_bar()`)
b. Create a barplot of the birth months of the players born in the USA after 1970.

---

Exercises 7.21 – 7.27 use the `babynames` data set from the `babynames` package, which was introduced in Section 6.6.1.

**7.21.** Make a line graph of the total number of babies of each sex versus year.

**7.22.** Make a line graph of the number of different names used for each sex versus year.

**7.23.** Make a line graph of the total number of babies with your name versus year. If your name doesn't appear in the data, use the name "Alexa."

**7.24.** Make a line graph comparing the number of boys named "Bryan" and the number of boys named "Brian" from 1920 to the present.

**7.25.** I wish that I had Jessie's girl, or maybe Jessie's guy? On one graph, plot the number of male and female babies named "Jessie" over time.

Three time periods show up in the history of Jessie:

- More male than female Jessie.
- More female than male Jessie.
- About the same male and female Jessie.

Approximately what range of years does each time period span?

**7.26.** Using the data from `fosdata::movies`, create a scatterplot of the ratings of *Twister (1996)* versus the date of review, and add a trend line using `geom_smooth`. It may be useful to use `lubridate::as_datetime(timestamp)`.

**7.27.** Consider the `frogs` data set in the `fosdata` package. This data was used to argue that a new species of frog had been found in a densely populated area of Bangladesh. Create a scatterplot of head length distance from tip of snout to back of mandible versus forearm length distance from corner of elbow to proximal end of outer palmar metacarpal tubercle, colored by species. Explain whether this plot is visual evidence that the physical characteristics of the dhaka frog are different than the other frogs.

---

Exercises 7.28 – 7.33 require material through Section 7.4.

**7.28.** Consider the `scotland_births` data set in the `fosdata` package. This data set contains the number of births in Scotland by age of the mother for each year from 1945-2019. In Exercise 7.28, you converted this data set into long format.

a. Create a line plot of births by year from 1945-2019 for each age group represented in the data.
b. Highlight and color ages 20 and 30, and provide meaningful labels and titles.

**7.29.** The data set `Arbuthnot` from the `HistData` library gives information about birth and death in London from 1629-1710.

a. Make a plot of Mortality showing a point for each Year. Set the size and color of your points to the `Plague` variable.
b. Hide the legend for size. Adjust the color to be a gradient from black to red. Add text annotations to show the years of the major plague events.

**7.30.** In a paper[12] by Hallmann et al., a plot similar to Figure 7.24 was created to visualize the change in biomass over time with the following explanation: "Seasonal distribution of insect biomass showing that highest insect biomass catches in mid summer show most severe declines. Color gradient ... ranges from 1989 (blue) to 2016 (orange)."

Use the data in `fosdata::biomass` to recreate the plot.

**FIGURE 7.24** Visualization of seasonal insect biomass.

**7.31.** The `msleep` data set is part of the `ggplot2` package. It contains information about the sleep patterns of 83 mammal species.

Make a plot showing the log of brain weight on the $x$-axis, total sleep on the $y$-axis, and color by `vore`. Label points with the names of the animals, but only label the species with brain weight bigger than 1 or that sleep more than 17 hours a day.

[12]Caspar A Hallmann et al., "More Than 75 Percent Decline over 27 Years in Total Flying Insect Biomass in Protected Areas," *PLOS One* 12, no. 10 (October 2017): 1–21, https://doi.org/10.1371/journal.pone.0185809.

**7.32.** The data `flint` from `fosdata` gives the results of tap water lead testing during the Flint, Michigan water crisis in 2015. Figure 7.25 is a graph showing lead levels at first draw (`Pb1`) for Flint's eight geographical areas, called "Wards." The red horizontal line represents the EPA's "action level" for lead in water, at 15 ppb.

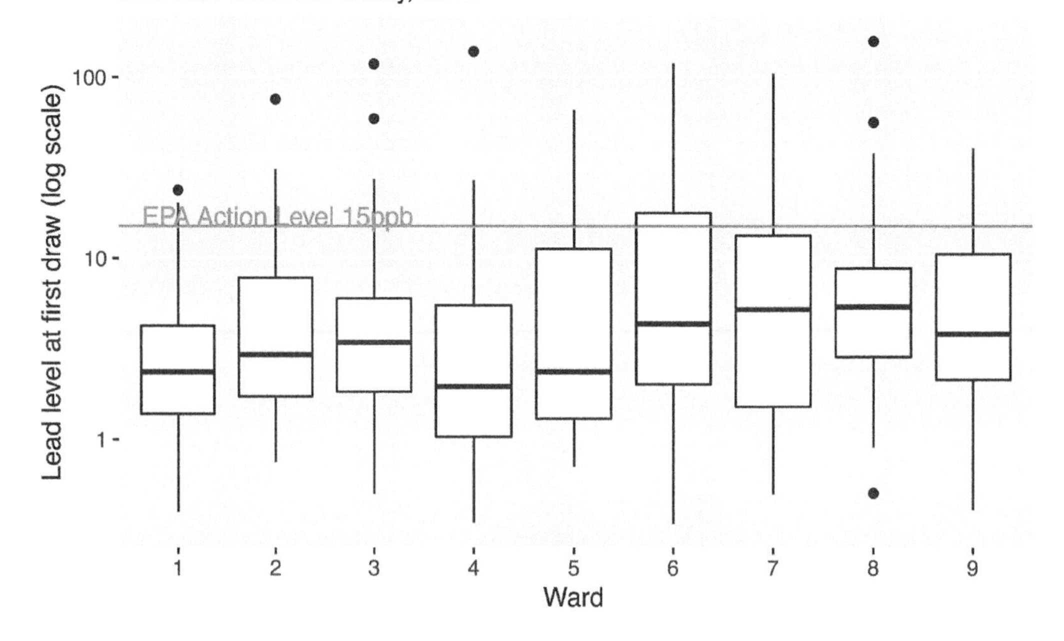


**FIGURE 7.25** Visualization of lead levels in Flint, Michigan.

Reproduce this graph as well as you can. The $y$-axis scale is logarithmic, which you can accomplish with `scale_y_log10()`. Note that there is no Ward 0 in Flint.

**7.33.** Data scientist Kieran Healy created a widely circulated figure similar to the one shown in Figure 7.26.

Recreate this plot as well as you can, using Healy's GitHub data. Read Healy's blog post and his follow-up post. Do you think this figure is a reasonable representation of gun violence in the United States?

Data:  Post:  Follow-up: 

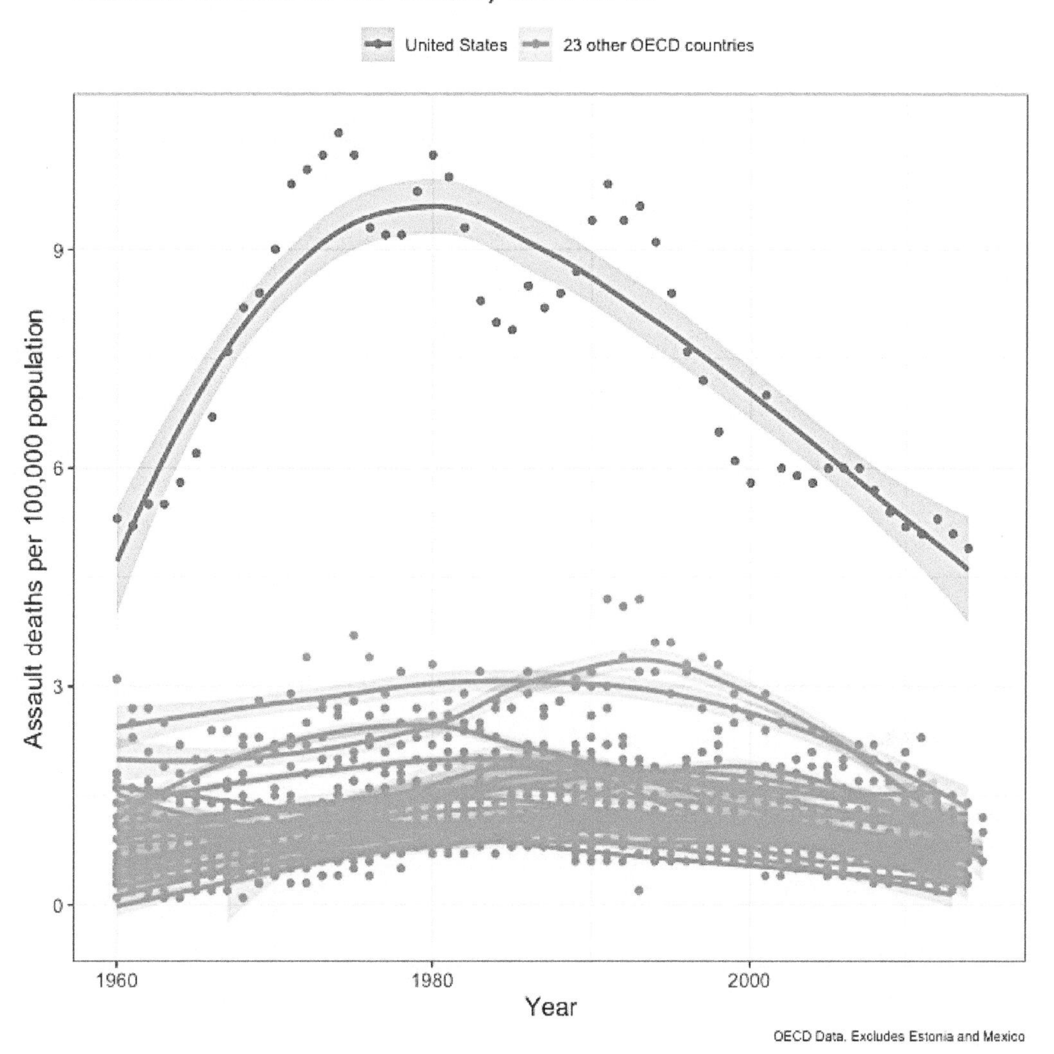


**FIGURE 7.26** Visualization of gun violence in the United States.

# 8

# *Inference on the Mean*

Car accidents are sometimes caused by unintended acceleration, when the driver fails to realize that the car is accelerating in time to apply the brakes. This type of accident seems to happen more with older drivers. Hasegawa, Kimura, and Takeda[1] developed an experiment to test whether older drivers are less adept at correcting unintended acceleration.

The set-up of the experiment was as follows. Older (65+) drivers and younger (18-32) drivers were recruited. The drivers were put in a simulation and told to apply the brake when a light turned red. Ninety percent of the time, this would work as desired, but ten percent of the time, the brake would instead **speed up** the car. When the participants became aware of the unintended acceleration, they were to release the center pedal and push the left pedal to stop the acceleration.

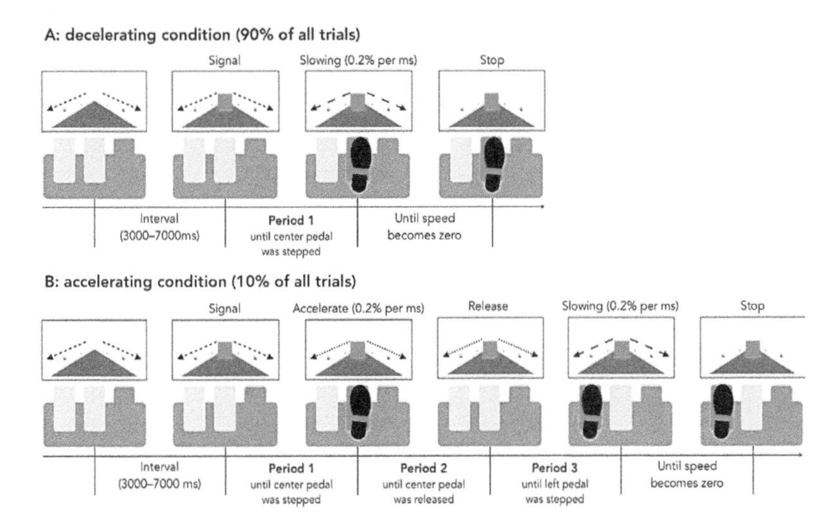


**FIGURE 8.1** Figure from Hasegawa, et al.

The data from this experiment is available in the `brake` data set from the `fosdata` package.

```
brake <- fosdata::brake
```

The authors collected a lot more data than we will be using in this chapter! Let's focus on the latency (length of time) variables. The experimenters measured three things in the case when the brake pedal activated acceleration:

- `latency_p1`: The length of time before the initial application of the brake.
- `latency_p2`: The length of time before the brake was released.

[1]K Hasegawa, M Kimura, and Y Takeda, "Age-Related Differences in Correction Behavior for Unintended Acceleration," *PLOS One* 15, no. 7 (2020): e0236053, https://doi.org/10.1371/journal.pone.0236053.

DOI: 10.1201/9781003004899-8

- `latency_p3`: The length of time before the pedal to the left of the brake was pressed, which would then stop the car.

The experimenters desired to determine whether there is a difference in latency between older and younger drivers in the population of all drivers. The experiment measured a sample of drivers and produced the data shown in the accompanying figure.

```
brake %>%
  tidyr::pivot_longer(
    cols = matches("^lat"),
    names_to = "type", values_to = "time"
  ) %>%
  ggplot(aes(x = type, y = time, color = age_group)) +
  geom_boxplot()
```

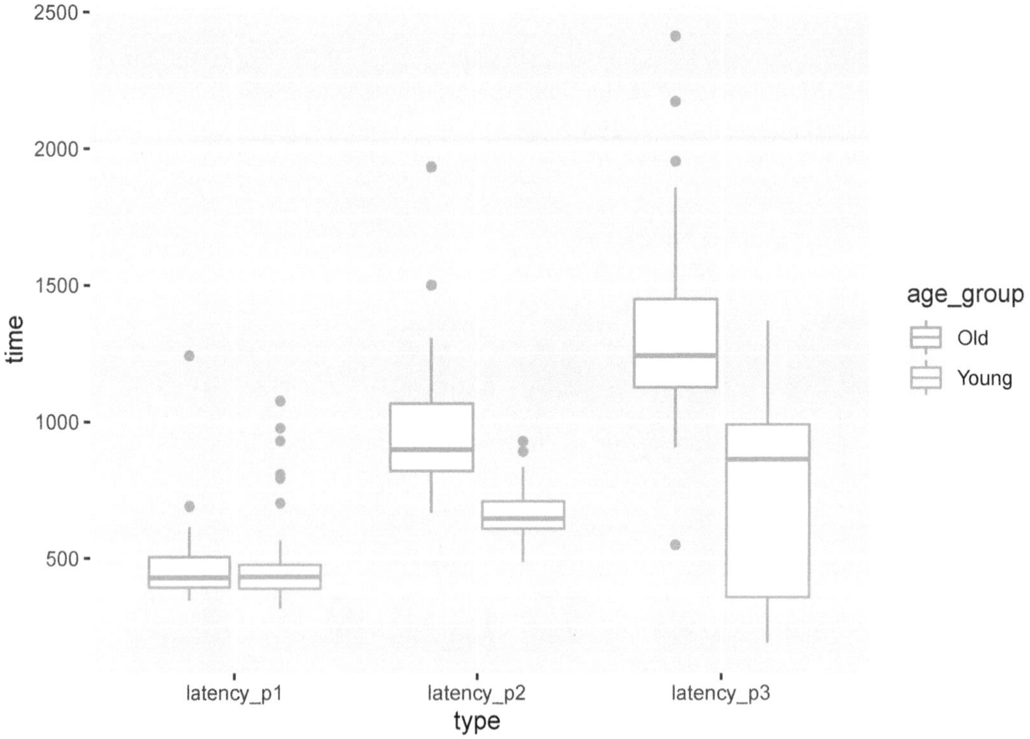

At a first glance, it seems like there may not be a difference in time to react to the red light between old and young, but there may well be a difference in the time to realize a mistake has been made. However, there is considerable overlap between the distributions for old and young. Is it possible that the apparent difference is due to the random nature of the sample, rather than a true difference in the population? The purpose of this chapter is to be able to systematically answer such questions through a process known as *statistical inference*.

Once the theory is developed, we return to this experiment in Example 8.14 and test whether the mean length of time for an older driver to release the pedal after it resulted in an unintended acceleration is the same as the mean length of time for a younger driver.

## 8.1 Sampling distribution of the sample mean

In Section 5.5 we discussed various sampling distributions for statistics related to a random sample from a population. The $t$ distribution first described in Section 5.5.2 will play an essential role in this chapter, so we review and expand upon that discussion here. Theorem 5.5 states that if $X_1, \ldots, X_n$ are iid normal random variables with mean $\mu$ and standard deviation $\sigma$, then

$$\frac{\overline{X} - \mu}{S/\sqrt{n}}$$

has a $t$ distribution with $n-1$ degrees of freedom, where $S$ is the sample standard deviation and $\overline{X}$ is the sample mean. We write this as

$$\frac{\overline{X} - \mu}{S/\sqrt{n}} \sim t_{n-1}.$$

A random variable that has a $t$ distribution with $n-1$ degrees of freedom is said to be a $t$ random variable with $n-1$ degrees of freedom.

**Try It Yourself.**
Simulate the pdf of $\frac{\overline{X}-\mu}{S/\sqrt{n}}$ when $n = 3$, $\mu = 5$, and $\sigma = 1$, and compare to the pdf of a $t$ random variable with 2 degrees of freedom.

Figure 8.2 compares the pdf of $t$ random variables with various degrees of freedom.

We see from the graphs that $t$ distributions are bump-shaped and symmetric about 0. For all $t$ random variables, 0 is the most likely outcome, and the interval $[-k, k]$ has the highest probability of occurring of any interval of length $2k$.

As the number of degrees of freedom increases, the tails of the $t$ distribution become lighter. This means that

$$\lim_{M\to\infty} \frac{P(|T_\nu| > M)}{P(|T_\tau| > M)} = \infty$$

for $T_\nu$ and $T_\tau$ $t$ random variables with degrees of freedom $\nu < \tau$.

Thinking of the $t$ distribution as the sampling distribution of $\frac{\overline{X}-\mu}{S/\sqrt{n}}$, a smaller value of $n$ means that $S$ gives a worse approximation to $\sigma$, resulting in a distribution that is more spread out. As the sample size goes to $\infty$, we become more and more sure of our estimate of $\sigma$, until in the limit the $t$ distribution is standard normal.

**Theorem 8.1.** *For $n = 1, 2, \ldots$ let $T_n$ be a $t$ random variable with $n$ degrees of freedom. Let $Z$ be a standard normal random variable. Then,*

$$T_n \to Z$$

*in the sense that for every $a < b$,*

$$\lim_{n\to\infty} P(a < X_n < b) = P(a < Z < b)$$

The difference between the pdfs of a $t$ random variable and a standard normal random variable may not appear large in the picture, but it can make a substantial difference in the computation of **tail** probabilities, as 8.1 explores. The results of statistical inference are especially sensitive to these tail probabilities.

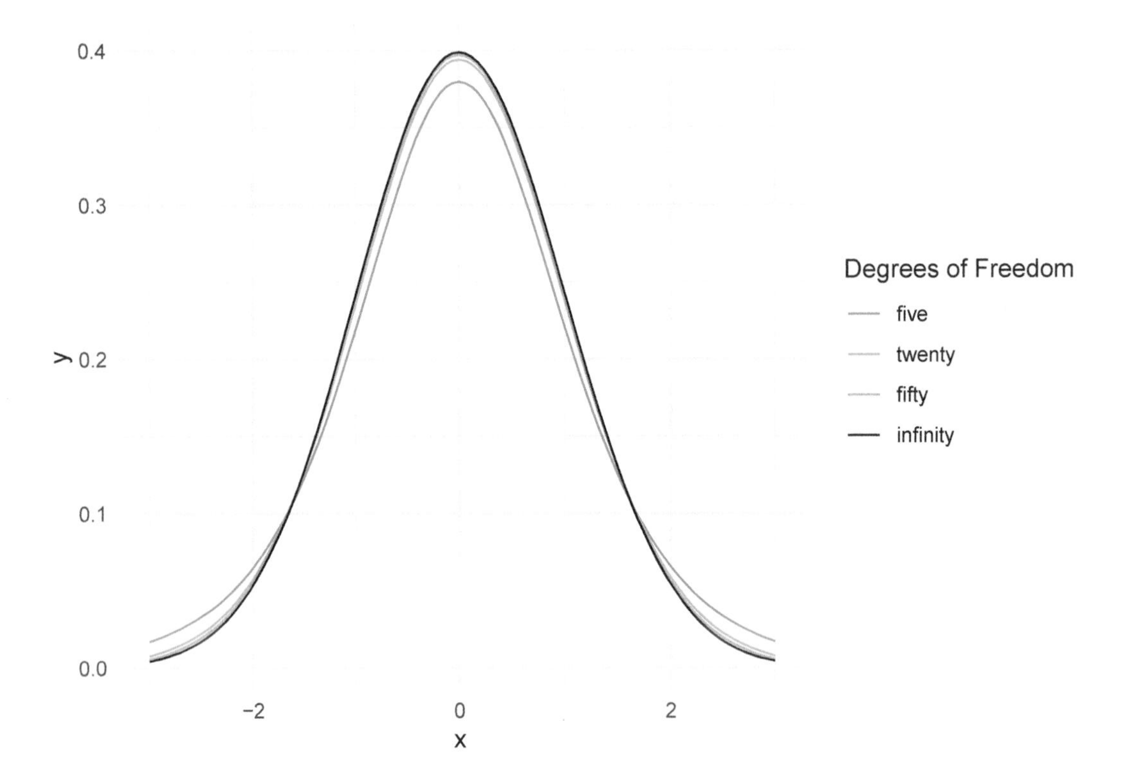


**FIGURE 8.2** Probability density functions of standard normal and $t$ random variables.

**Example 8.1.** Let $T$ be a $t$ random variable with 3 degrees of freedom, and let $Z$ be a standard normal random variable. Compare $P(T < -2.5)$ with $P(Z < -2.5)$.

On one hand, if we compute the absolute difference between the two probabilities, we get $|P(Z < -2.5) - P(T < -2.5)| \approx 0.038$, which may seem small.

```
pt(-2.5, df = 3) - pnorm(-2.5)
```

```
## [1] 0.03764366
```

On the other hand, if we compute the relative magnitude of the probabilities, we see that $P(T < -2.5)$ is approximately **7 times** larger than $P(Z < -2.5)$, which is quite a difference!

```
pt(-2.5, df = 3) / pnorm(-2.5)
```

```
## [1] 7.062107
```

> **Try It Yourself.**
> About how many degrees of freedom for a $t$ random variable $T$ do you need before $P(T < -2.5)$ is only 50% larger than $P(Z < -2.5)$?

Now suppose that $X_1, \ldots, X_n$ represent data which is randomly sampled from a population which is normally distributed with mean $\mu$ and standard deviation $\sigma$. We do not know the values of $\mu$ and $\sigma$, but we do know that $\frac{\overline{X}-\mu}{S/\sqrt{n}}$ has the $t$ distribution with $n-1$ degrees of freedom.

Let $t_\alpha$ be the unique real number such that $P(T > t_\alpha) = \alpha$, where $T$ is a $t$ random variable with $n-1$ degrees of freedom. Note that $t_\alpha$ depends on $n$, because the $t$ distribution has more area in its tails when $n$ is small.

Then, we know that

$$P\left(t_{1-\alpha/2} < \frac{\overline{X}-\mu}{S/\sqrt{n}} < t_{\alpha/2}\right) = 1 - \alpha.$$

Because the $t$ distribution is symmetric, $t_{1-\alpha/2} = -t_{\alpha/2}$, so we can rewrite the equation symmetrically as:

$$P\left(-t_{\alpha/2} < \frac{\overline{X}-\mu}{S/\sqrt{n}} < t_{\alpha/2}\right) = 1 - \alpha.$$

Now with probability $1-\alpha$,

$$-t_{\alpha/2} < \frac{\overline{X}-\mu}{S/\sqrt{n}} < t_{\alpha/2}$$

so

$$-t_{\alpha/2}S/\sqrt{n} < \overline{X} - \mu < t_{\alpha/2}S/\sqrt{n}.$$

Multiplying through by $-1$ and then reversing the order of terms,

$$-t_{\alpha/2}S/\sqrt{n} < \mu - \overline{X} < t_{\alpha/2}S/\sqrt{n}$$

with probability $1-\alpha$. Finally, add $\overline{X}$ to all terms to conclude

$$P\left(\overline{X} - t_{\alpha/2}S/\sqrt{n} < \mu < \overline{X} + t_{\alpha/2}S/\sqrt{n}\right) = 1 - \alpha.$$

Let's think about what this means. Take $n = 10$ and $\alpha = 0.10$ just to be specific. This means that if we were to repeatedly take 10 samples from a normal distribution with fixed mean $\mu$

and any $\sigma$, then 90% of the time, $\mu$ would lie between $\overline{X} + t_{\alpha/2}S/\sqrt{n}$ and $\overline{X} - t_{\alpha/2}S/\sqrt{n}$. That sounds like something that we need to double check using a simulation!

```
t05 <- qt(.05, 9, lower.tail = FALSE)
# Take samples from a normal population with mean 5 and sd 8
mu <- 5
sigma <- 8
mean(replicate(10000, {
  X <- rnorm(10, mu, sigma) # sample of size 10
  Xbar <- mean(X) # sample mean
  SE <- sd(X) / sqrt(10) # standard error
  (Xbar - t05 * SE < mu) & (mu < Xbar + t05 * SE)
}))
```

```
## [1] 0.9002
```

The first line computes $t_{\alpha/2}$ using `qt`, the quantile function for the $t$ distribution, here with 9 degrees of freedom. Then we check the inequality given above. It works; the interval contains the value $\mu = 5$ about 90% of the time.

> **Try It Yourself.**
> Repeat the above simulation with $n = 20$, $\mu = 7$, and $\alpha = .05$.

## 8.2 Confidence intervals for the mean

In the previous section, we saw that if $X_1, \ldots, X_n$ is a random sample from a normal population with mean $\mu$, then

$$P\left(\overline{X} - t_{\alpha/2}S/\sqrt{n} < \mu < \overline{X} + t_{\alpha/2}S/\sqrt{n}\right) = 1 - \alpha.$$

This leads naturally to the following definition.

**Definition 8.1.** Given a sample of size $n$ from iid normal random variables, the interval

$$\left(\overline{X} - t_{\alpha/2}S/\sqrt{n},\ \ \overline{X} + t_{\alpha/2}S/\sqrt{n}\right)$$

is a $100(1-\alpha)\%$ *confidence interval* for $\mu$.

Confidence intervals are random because they depend on the random sample. Sometimes a sample will produce a confidence interval that contains $\mu$, and sometimes a sample will produce a confidence interval that does not contain $\mu$. Since $\mu$ is typically not known when performing experiments, there is typically no way to determine whether a particular confidence interval succeeded in capturing the population mean. However, the procedure of taking a sample and computing a confidence interval will succeed in capturing the population mean $100(1-\alpha)\%$ of the time. This is what "confidence" means.

Over all scientific experiments that use confidence intervals to describe population means, we accept the fact that some proportion ($\alpha$, specifically) will fail to contain the true mean, and we will not know which ones.

**Example 8.2.** Consider the heart rate and body temperature data `normtemp`.[2]

You can read `normtemp` as follows.

```
normtemp <- fosdata::normtemp
```

What kind of data is this?

```
str(normtemp)
```

```
## 'data.frame':	130 obs. of  3 variables:
##  $ temp  : num  96.3 96.7 96.9 97 97.1 97.1 97.1 97.2 97.3 97.4 ...
##  $ gender: int  1 1 1 1 1 1 1 1 1 1 ...
##  $ bpm   : int  70 71 74 80 73 75 82 64 69 70 ...
```

```
head(normtemp)
```

```
##   temp gender bpm
## 1 96.3      1  70
## 2 96.7      1  71
## 3 96.9      1  74
## 4 97.0      1  80
## 5 97.1      1  73
## 6 97.1      1  75
```

We see that the data set is 130 observations of body temperature and heart rate. There is also gender information attached, that we will ignore for the purposes of this example.

Let's visualize the data through a boxplot (turned sideways with `coord_flip`).

```
ggplot(normtemp, aes(x = "", y = temp)) +
  geom_boxplot(notch = TRUE) +
  coord_flip()
```

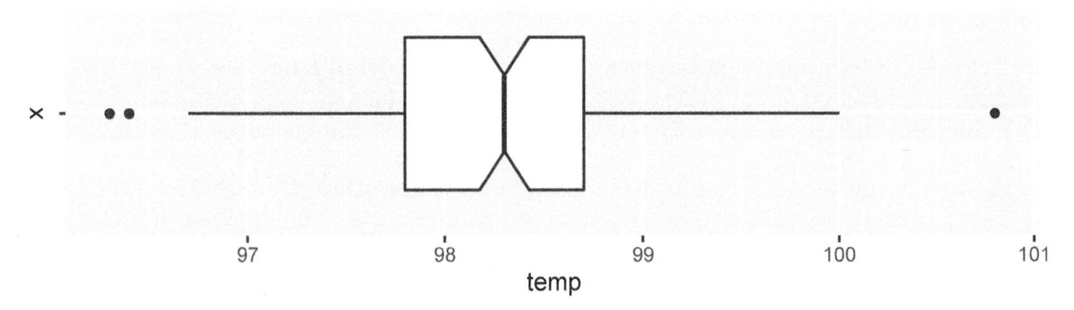


The data looks symmetric without concerning outliers. According to the plot's notches, the true median should be about between 98.2 and 98.4.

Let's find a 98% confidence interval for the mean body temperature of healthy adults. In order for this to be valid, we are assuming that we have a *random sample* from all healthy adults (highly unlikely to be formally true), and that the body temperature of healthy adults

[2]This data is derived from a data set presented in P Mackowiak, S Wasserman, and M Levine, "A Critical Appraisal of 98.6 Degrees F, the Upper Limit of the Normal Body Temperature, and Other Legacies of Carl Reinhold August Wunderlich." *JAMA* 268 12 (1992): 1578–80. Data was constructed to match as closely as possible the histograms and summary statistics presented in that article. Additional information about this data set can be found in L Shoemaker Allen, "What's Normal? – Temperature, Gender, and Heart Rate," *Journal of Statistics Education* 4, no. 2 (1996), https://doi.org/10.1080/10691898.1996.11910512.

is normally distributed (also unlikely to be formally true). Later, we will discuss deviations from our assumptions and when we should be concerned.

Using Definition 8.1, we have

```
xbar <- mean(normtemp$temp)
s <- sd(normtemp$temp)
lower_ci <- xbar - s / sqrt(130) * qt(.99, df = 129)
upper_ci <- xbar + s / sqrt(130) * qt(.99, df = 129)

lower_ci
```

```
## [1] 98.09776
```

```
upper_ci
```

```
## [1] 98.40071
```

The 98% confidence interval is $(98.1, 98.4)$. The R command that finds a confidence interval for the mean in one easy step is `t.test`.

```
t.test(normtemp$temp, conf.level = .98)
```

```
##
##  One Sample t-test
##
## data:  normtemp$temp
## t = 1527.9, df = 129, p-value < 2.2e-16
## alternative hypothesis: true mean is not equal to 0
## 98 percent confidence interval:
##  98.09776 98.40071
## sample estimates:
## mean of x
##  98.24923
```

We get a lot of information that we will discuss in more detail later. But we see that the confidence interval computed this way agrees with the values we computed above. We are 98% confident that the mean body temperature of healthy adults is between 98.1 and 98.4.

We are not saying anything about the mean body temperature of the people in the study. That is exactly 98.24923, and we don't need to do anything fancy. We are making an inference about the mean body temperature of *all* healthy adults. To be more precise, we might want to say it is the mean body temperature of all healthy adults as would be measured by this experimenter. Other experimenters could have a different technique for measuring body temperatures which could lead to a different answer.

Once we have done the experiment and have a particular interval, it doesn't make sense to talk about the *probability* that $\mu$ (a constant) is in that fixed interval. There is nothing random remaining, and either $\mu$ is in the interval or it is not. Since $\mu$ is typically an unknown value, it is impossible to know if the specific confidence interval produced actually does contain $\mu$.

**Assumptions 8.1 (For producing a confidence interval).**

1. *The population is normal, or the sample size is large enough for the Central Limit Theorem to reasonably apply.*
2. *The sample values are independent.*

Assumption 1, normality, can be tested by examining histograms, boxplots, or qq plots of the data. We might measure skewness. We may also run simulations to determine how large of a sample size is necessary for the Central Limit Theorem to make up for departures from normality.

Assumption 2 is harder to verify, because there are several ways that a sample can be dependent. Here are a few things to look out for. First, if there is a time component to the data, you should be wary that the observations might depend on time. For example, if the data you are collecting is outdoor temperature, then the recorded temperatures may very well depend on time. Second, if you are repeating measurements, then the data is not independent. For example, if you have ratings of movies, then it is likely that each person has their own distribution of ratings that they give. So, if you have 10 ratings from each of 5 people, you would not expect the 50 ratings to be independent. Third, if there is a change in the experimental setup, then you should not assume independence.

We systematically study the robustness to departures from the assumptions in Section 8.5 using simulations.

Although we focus on confidence intervals for the mean, the concept of a confidence interval applies to estimating any population parameter:

**Definition 8.2.** Suppose that $X_1, \ldots, X_n$ is a random sample from a population with a parameter $\theta$. A $100(1-\alpha)\%$ *confidence interval* for a parameter $\theta$ is an interval $(L, U)$, where $L$ and $U$ are functions of $X_1, \ldots, X_n$, such that the probability that $\theta \in (L, U)$ is $(1-\alpha)$ for each value of $\theta$.

With this more general definition of confidence interval, we see that there is also a more subtle issue with thinking of confidence as a probability; a procedure which will contain the true parameter value 95% of the time does not necessarily always create intervals that are about 95% "likely" (whatever meaning you attach to that) to contain the parameter value. As an extreme example, consider the following procedure for estimating the mean. It ignores the data and 95% of the time says that the mean is between $-\infty$ and $\infty$. The other 5% of the time, the procedure says the mean is in the empty interval. This absurd procedure will contain the true parameter 95% of the time, but the intervals obtained are either 100% or 0% likely to contain the true mean, and when we see which interval we have, we **know** what the true probability is of it containing the mean.[3] Nonetheless, it is widely accepted to say that we are $100(1-\alpha)\%$ *confident* that the mean is in the interval, which is what we will also do.

## 8.3 Hypothesis tests of the mean

The most fundamental technique of statistical inference is the hypothesis test. There are many types of hypothesis tests but all follow the same logical structure, so we begin with hypothesis testing of a population mean. The terminology and methods are developed with a single example that runs throughout this section: human body temperature.

[3]This example and a thought provoking examination of related issues can be found in Richard D Morey et al., "The Fallacy of Placing Confidence in Confidence Intervals," *Psychonomic Bulletin & Review* 23, no. 1 (2016): 103–23, https://doi.org/10.3758/s13423-015-0947-8.

In 1868, Carl Wunderlich[4] claimed to have "got together a material which comprises many thousand complete cases of thermometric observations of disease, and millions of separate readings of temperature." Based on the measurements he made, Wunderlich claimed that the mean temperature of healthy adults is 37 degrees Celsius, or 98.6 degrees Fahrenheit. His values are still the temperatures most people consider to be normal, today.

Now suppose that we wish to revisit the findings of Wunderlich. Perhaps measuring devices have become more accurate in the last 150 years, or perhaps the mean body temperature of healthy adults has changed for some other reason. We wish to **test** whether the true mean body temperature of healthy adults is 98.6 degrees or whether it is not. This is an example of a hypothesis test.

Hypothesis testing begins with a *null hypothesis* and an *alternative hypothesis*. Both the null and the alternative hypotheses are statements about a population. In this chapter, that statement will be a statement about the mean $\mu$ of the population.

For a test of mean body temperature, the null hypothesis is $H_0 : \mu = 98.6$ and the alternative hypothesis is $H_a : \mu \neq 98.6$, where $\mu$ is the true mean body temperature of healthy adults.

Typically, the null hypothesis is the default assumption, or the conventional wisdom about a population. Often it is exactly the thing that a researcher is trying to show is false. The alternative hypothesis states that the null hypothesis is false, sometimes in a particular way.

The purpose of hypothesis testing is to either **reject** or **fail to reject** the null hypothesis. A researcher would collect data relating to the population being studied and use a hypothesis test to determine whether the evidence against the null hypothesis (if any) is strong enough to reject the null hypothesis in favor of the alternative hypothesis. In terms of body temperature, we might collect 130 body temperatures of healthy adults. If the mean body temperature of the sample is far enough away from 98.6 relative to the standard deviation of the sample, then that would be sufficient evidence to reject the null hypothesis. Of course, we still need to quantify **far enough away** in the previous sentence, and it will also depend on the sample size.

For a hypothesis test of the mean, we will construct a test statistic

$$T = \frac{\overline{X} - \mu_0}{S/\sqrt{n}}$$

where $\mu_0$ is the value of the mean under the null hypothesis.

Under the same Assumptions 8.1 as for confidence intervals, $T$ will have the $t$ distribution with $n - 1$ degrees of freedom. That is, we require the population to be normal or a large sample, and the samples must be independent.

**Definition 8.3.** The *rejection region* of a hypothesis test is the set of possible values of the test statistic which will lead to a rejection of the null hypothesis.

The $\alpha$ level of a hypothesis test is the probability that a test statistic will have a value in the rejection region given that the null hypothesis is true.

**Example 8.3.** Suppose in the body temperature example, we have a rejection region of $|T| > 2$. The $\alpha$ level of the test is the probability that $|T| > 2$ when $\mu = 98.6$. Note that when $\mu = 98.6$, $T \sim t_{129}$, so we can compute $\alpha$ via

[4]C A Wunderlich and W B Woodman, *On the Temperature in Diseases: A Manual of Medical Thermometry*, New Sydenham Society Publications (New Sydenham Society, 1871).

```
pt(-2, 129) + pt(2, 129, lower.tail = FALSE)
```

```
## [1] 0.04760043
```

The level of the test in this example would be $\alpha = .0476$.

In practice, we construct the rejection region to have a pre-specified level $\alpha$.

**Definition 8.4.** In a one sample hypothesis test of the mean at the level $\alpha$, the rejection region is $|T| > t_{\alpha/2}$, where $t_{\alpha/2}$ = `qt(alpha/2, n - 1, lower.tail = FALSE)` is the unique value such that $P(T > t_{\alpha/2}) = \alpha/2$.

**Example 8.4.** Consider again the `temp` variable in the `normtemp` data set in the `fosdata` package. To construct a rejection region for a hypothesis test at the $\alpha = .02$ level, we compute

```
qt(.02 / 2, 130 - 1, lower.tail = FALSE)
```

```
## [1] 2.355602
```

So, our rejection region is $|T| > 2.355602$. Our actual value of $|T|$ in this example is

```
abs((mean(normtemp$temp) - 98.6) / (sd(normtemp$temp) / sqrt(130)))
```

```
## [1] 5.454823
```

Since this is **in** the rejection region, we would reject the null hypothesis at the $\alpha = .02$ level that the true mean body temperature of healthy adults is 98.6 degrees Fahrenheit.

Note that in the previous example, the observed value of the test statistic was considerably larger than the smallest value for which we would have rejected the null hypothesis. Indeed, consider Table 8.1, which provides rejection regions corresponding to various levels $\alpha$.

We see that we would have rejected the null hypothesis for $\alpha = 0.000001$ but not when $\alpha = 0.0000001$. The **smallest** value of $\alpha$ for which we would still reject $H_0$ is called the $p$-value associated with the hypothesis test.

**Definition 8.5.** Let $X_1, \ldots, X_n$ be a random sample from a normal population with unknown mean $\mu$ and standard deviation. When testing $H_0 : \mu = \mu_0$ versus $H_a : \mu \neq \mu_0$, the *p-value* associated with the test is $2 * P(T > |t|)$, where $t = \frac{\overline{x} - \mu_0}{s/\sqrt{n}}$ is the observed value of the test statistic, and $T$ is a $t$ random variable with $n - 1$ degrees of freedom. The $p$-value is also the smallest level $\alpha$ for which we would reject $H_0$.

When conducting a hypothesis test at the level $\alpha$, if the $p$-value is less than $\alpha$, then we reject $H_0$. If the $p$-value is greater than $\alpha$, then we fail to reject $H_0$.

**TABLE 8.1** Rejection regions when n = 130.

| Alpha | Rejection Region \|T\| > |
|---|---|
| 0.01 | 2.614479 |
| 0.001 | 3.367546 |
| 0.0001 | 4.015720 |
| 0.00001 | 4.599057 |
| 0.000001 | 5.138220 |
| 0.0000001 | 5.645460 |

**Definition 8.6.** In a hypothesis test, if $p < \alpha$, we say the results are *statistically significant.*

Throughout scientific literature, the term "significance" is used to mean that $H_0$ was rejected by an experiment resulting in a $p$-value less than $\alpha$. More often than not, the value of $\alpha$ is unstated, and implicitly $\alpha = 0.05$ so that "significant" means that $p < 0.05$.

**Example 8.5.** Continuing with the `normtemp` data from above, we can compute the $p$-value using R in multiple ways. First, we recall that under the assumption that the underlying population is normal, $T$ has a $t$ distribution with 129 degrees of freedom. The observed test statistic in our case is $|t| = 5.454823$. The smallest rejection region which would still lead to rejecting $H_0$ is $|T| \geq 5.454823$, which has probability given by

```
pt(-5.454823, df = 129) + pt(5.454823, df = 129, lower.tail = F)
```

```
## [1] 2.410635e-07
```

This is the $p$-value associated with the test.

As one would expect, R also has automated this procedure in the function `t.test`. In this case, one would use

```
t.test(normtemp$temp, mu = 98.6)
```

```
##
##  One Sample t-test
##
## data:  normtemp$temp
## t = -5.4548, df = 129, p-value = 2.411e-07
## alternative hypothesis: true mean is not equal to 98.6
## 95 percent confidence interval:
##  98.12200 98.37646
## sample estimates:
## mean of x
##  98.24923
```

Either way, we get a $p$-value of about 0.0000002, which is the probability of obtaining data at least this compelling against $H_0$ when $H_0$ is true. In other words, it is very unlikely that data with a test statistic as large as 5.45 would be collected if the true mean temp of healthy adults is 98.6. (Again, we are assuming a random sample of healthy adults and normal distribution of temperatures of healthy adults.)

**Example 8.6.** Let's consider the `Cavendish` data set in the `HistData` package. Henry Cavendish carried out experiments in 1798 to determine the mean density of the earth. He collected 29 measurements, but changed the exact set-up of his experiment after the 6th measurement.

The current accepted value for the mean density of the earth is $5.515\, g/cm^3$. Let's examine Cavendish's data from that modern perspective and test $H_0 : \mu = 5.515$ versus $H_a : \mu \neq 5.515$. The null hypothesis is that Cavendish's experimental measurements were unbiased, and that the true mean was the correct value. The alternative is that his setup had some inherent bias.

A first question that comes up is: what are we going to do with the first six observations? Since the experimental setup was changed after the 6th measurement, it seems likely that the first six observations are **not** independent of the rest of the observations. Hence, we will proceed by throwing out the first six observations. The variable `density3` contains the measurements with the first 6 values removed.

Let's look at a boxplot of the data.

```
cavendish <- HistData::Cavendish
ggplot(cavendish, aes(x = "", y = density3)) +
  geom_boxplot(outlier.size = -1) +
  geom_jitter(height = 0, width = 0.2) +
  coord_flip()
```

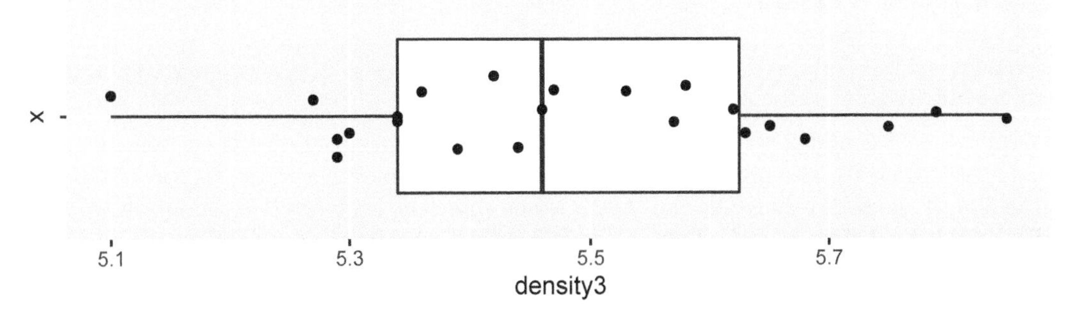


The data appears plausibly to be from a normal distribution, and there are no apparent outliers. Since the data was collected sequentially over time, we may wonder whether there is any time dependence. A full examination of that is beyond the scope of this text, but we can examine a scatterplot of the values against observation number to see if there are any obvious patterns. We would also want to confirm with the experimenter that the observation order given in the data set is the order in which the observations were made.

```
cavendish %>%
  mutate(observation = 1:29) %>%
  ggplot(aes(x = observation, y = density3)) +
  geom_point()
```

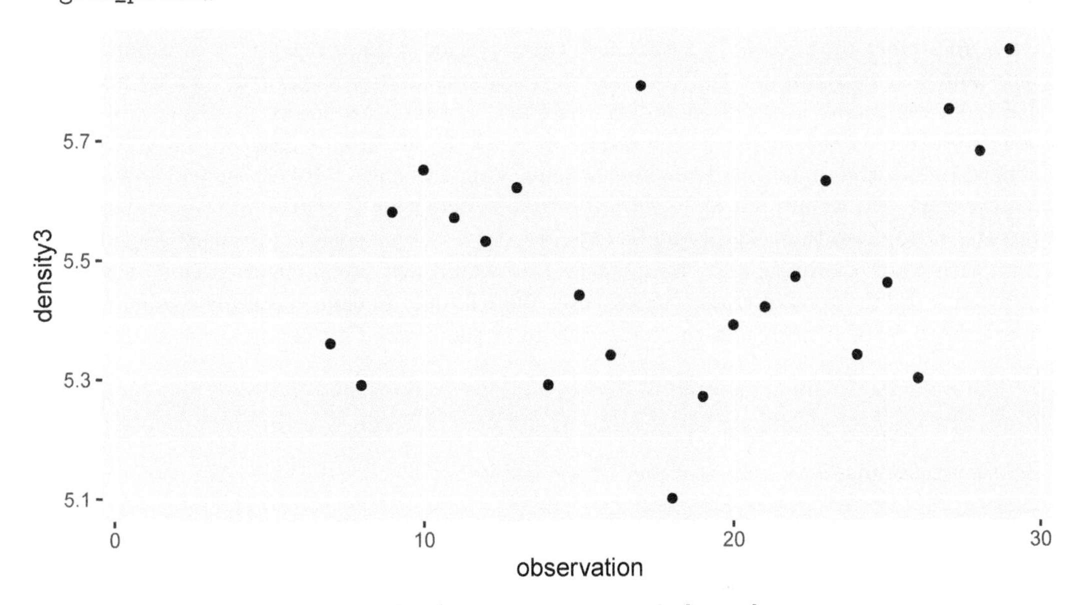


There is no obvious pattern to the data, so we assume independence.

```
t.test(cavendish$density3, mu = 5.5155)

##
##  One Sample t-test
##
## data:  cavendish$density3
## t = -0.80648, df = 22, p-value = 0.4286
## alternative hypothesis: true mean is not equal to 5.5155
## 95 percent confidence interval:
##  5.401134 5.565822
## sample estimates:
## mean of x
##  5.483478
```

We get a 95% confidence interval of $(5.40, 5.57)$, which does contain the current accepted value of 5.5155. We also obtain a $p$-value of .4286, so we would not reject $H_0 : \mu = 5.5155$ at the $\alpha = .05$ level. We conclude that there is not sufficient evidence that the mean value of Cavendish's experiment differed from the currently accepted value.

## 8.4 One-sided confidence intervals and hypothesis tests

In this short section, we introduce one-sided confidence intervals and hypothesis tests through an example.

Consider the `cows_small` data set from the `fosdata` package. If you are interested in the full, much more complicated data set, that can be found in `fosdata::cows`.

```
cows_small <- fosdata::cows_small
```

On hot summer days, cows in Davis, CA were sprayed with water for three minutes, and the change in temperature as measured at the shoulder was measured.[5] There are three different treatments of cows, including the control, which we ignore in this example. For now, we are only interested in the nozzle type `tk_0_75`. In this case, we have reason to believe before the experiment begins that spraying the cows with water will **lower** their temperature. So, we are not as much interested in detecting a *change* in temperature, as a *decrease* in temperature. Additionally, should it turn out by some oddity that spraying the cows with water increased their temperature, we would **not** recommend spraying cows with water in order to increase their temperature. We will only make a recommendation of using nozzle `tk_0_75` if it decreases the temperature of the cows. This changes our analysis.

The same data cannot be used to decide whether to do a one-sided test and for the test itself. Doing so will inflate the $p$-value by a factor of 2 and is highly inappropriate.

Before proceeding, let's look at a plot of the data.

[5]Jennifer Van Os, Karin Schutz, and Cassandra Tucker, "Cooling Cows Efficiently with Sprinklers: Physiological Responses to Water Spray," *Journal of Dairy Science* 98 (July 2015), https://doi.org/10.3168/jds.2015-9434.

```
ggplot(cows_small, aes(x = "", y = tk_0_75)) +
  geom_boxplot(outlier.size = -1) +
  geom_jitter(height = 0, width = 0.2) +
  coord_flip()
```

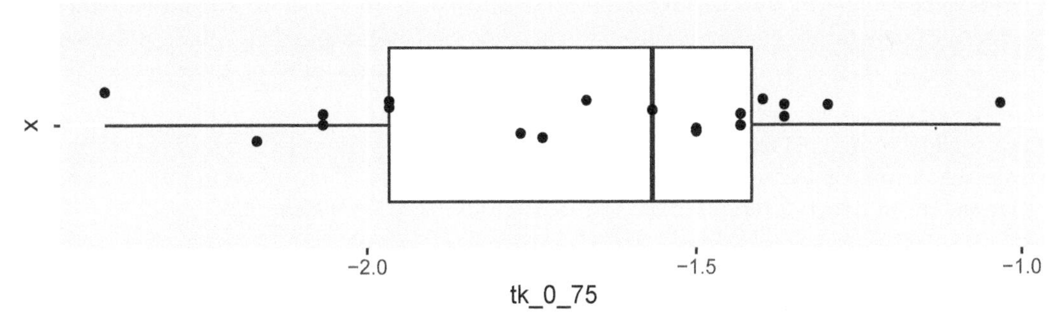


The data seems to plausibly come from a normal distribution, there don't seem to be any outliers, and we assume that the observations are independent.

Let $\mu$ denote the true (unknown) mean change in temperature as measured at the shoulder of cows sprayed with the `tk_0_75` nozzle for three minutes. We will construct a 98% confidence interval for $\mu$ of the form $(-\infty, u)$. If $u = -1$ for example, this would mean that we are 98% confident that the **drop** in temperature is at least one degree Celsius.

We will also perform the following hypothesis test:

- $H_0 : \mu \geq 0$ vs.
- $H_a : \mu < 0$

If we reject $H_0$ in favor of $H_a$, then we are concluding that the temperature of cows sprayed with `tk_0_75` will drop on average.

The outcomes that are most likely under $H_a$ relative to $H_0$ are of the form $\overline{X} < u$, so the rejection region for this test is

$$\frac{\overline{X} - \mu_0}{S/\sqrt{n}} < t_{.98}$$

where $t_{.98}$ = `qt(.98, 18, lower.tail = FALSE)` = `-2.213703`. After collecting the data, we will reject the null hypothesis if $t = \frac{\overline{x} - 0}{s/\sqrt{n}} < -2.213703$, and the $p$-value is `pt(t, 18)`.

R computes this for us using `t.test` with `alternative = "less"`:

```
t.test(cows_small$tk_0_75, mu = 0, alternative = "less", conf.level = .98)
```

```
##
##  One Sample t-test
##
## data:  cows_small$tk_0_75
## t = -20.509, df = 18, p-value = 3.119e-14
## alternative hypothesis: true mean is less than 0
## 98 percent confidence interval:
##       -Inf -1.488337
## sample estimates:
## mean of x
## -1.668421
```

Note that the $p$-value is given as we discussed above via:

```
pt(-20.509, 18)

## [1] 3.119495e-14
```

The $p$-value associated with the hypothesis test is $p = 3.1 \times 10^{-14}$, so we would reject $H_0$ that the mean temperature change is zero. We conclude that use of nozzle TK-0.75 is associated with a mean drop in temperature after three minutes as measured at the shoulder of the cows.

The confidence interval for the change in temperature associated with nozzle TK-0.75 is read from the output of `t.test` as $(-\infty, -1.49)$ degrees Celsius. That is, we are 98% confident that the mean **drop** in temperature is at least 1.49 degrees Celsius.

## 8.5 Assessing robustness via simulation

In the chapter up to this point, we have assumed that the underlying data is a random sample from a normal distribution. Let's look at how various types of departures from those assumptions affect the validity of our statistical procedures. For the purposes of this discussion, we will restrict to hypothesis testing at a specified $\alpha$ level, but our results will hold for confidence intervals and reporting $p$-values just as well.

Our point of view is that if we design our hypothesis test at the $\alpha = 0.1$ level, say, then we want to incorrectly reject the null hypothesis 10% of the time. That is how our test is designed. In order to talk about this, we introduce some new terminology.

**Definition 8.7.** For a hypothesis test, a *type I error* is rejecting the null hypothesis when it is true. The *effective type I error rate* of a hypothesis test is the probability that a type I error will occur.

When all of the assumptions of a hypothesis test at the $\alpha$ level are met, the effective type I error rate is exactly $\alpha$. If one or more of the assumptions are violated, then we want to estimate the effective type I error rate and compare it to $\alpha$. If the effective error rate is far from $\alpha$, then the test is not performing as designed. We will note whether the test makes too many type I errors or too few, but the main objective is to see whether the test is performing as designed.

Our first observation is that when the sample size is large, then the Central Limit Theorem tells us that $\overline{X}$ is approximately normal. Since the test statistic $T$ is also approximately normal when $n$ is large, this gives us some measure of protection against departures from normality in the underlying population.

In the sequel, it will be very useful to pull out the $p$-value of the `t.test`. Let's see how we can do that. We begin by doing a t.test on "garbage data" just to see what `t.test` returns.

```
str(t.test(c(1, 2, 3)))

## List of 10
##  $ statistic  : Named num 3.46
##   ..- attr(*, "names")= chr "t"
##  $ parameter  : Named num 2
```

```
##    ..- attr(*, "names")= chr "df"
##  $ p.value    : num 0.0742
##  $ conf.int   : num [1:2] -0.484 4.484
##    ..- attr(*, "conf.level")= num 0.95
##  $ estimate   : Named num 2
##    ..- attr(*, "names")= chr "mean of x"
##  $ null.value : Named num 0
##    ..- attr(*, "names")= chr "mean"
##  $ stderr     : num 0.577
##  $ alternative: chr "two.sided"
##  $ method     : chr "One Sample t-test"
##  $ data.name  : chr "c(1, 2, 3)"
##  - attr(*, "class")= chr "htest"
```

It's a list of 10 things, and we can pull out the $p$-value by using `t.test(c(1,2,3))$p.value`.

### 8.5.1 Symmetric, light tails

As a model for a symmetric, light-tailed population, we choose a uniform random variable.

**Example 8.7.** Estimate the effective type I error rate in a $t$-test of $H_0 : \mu = 0$ versus $H_a : \mu \neq 0$ when the underlying population is uniform on the interval $(-1, 1)$. Take a sample size of $n = 10$ and test at the $\alpha = 0.1$ level.

In this scenario, $H_0$ is true. So, we are going to estimate the probability of rejecting $H_0$. Let's build up our replicate function step by step.

```
runif(10, -1, 1) # Sample of size 10 from uniform rv
```

```
##  [1] -0.4689827 -0.2557522  0.1457067  0.8164156 -0.5966361  0.7967794
##  [7]  0.8893505  0.3215956  0.2582281 -0.8764275
```

```
t.test(runif(10, -1, 1))$p.value # p-value of test
```

```
## [1] 0.5089686
```

```
t.test(runif(10, -1, 1))$p.value < .1 # true if we reject
```

```
## [1] FALSE
```

```
mean(replicate(10000, t.test(runif(10, -1, 1))$p.value < .1))
```

```
## [1] 0.1007
```

Our result is 10.07%. That is really close to 10%, which is what the test is designed to give. The $t$-test is working well in this situation.

**Remark.** How do we know we have run enough replications to get a reasonable estimate for the probability of a type I error? Well, we don't. So we repeat our experiment a couple of times:

```
mean(replicate(10000, t.test(runif(10, -1, 1))$p.value < .1))
```

```
## [1] 0.0968
```

```
mean(replicate(10000, t.test(runif(10, -1, 1))$p.value < .1))
```

```
## [1] 0.0993
```

The results are pretty similar each time. That's reason enough to believe the true type I error rate is relatively close to what we computed.

> **Try It Yourself.**
> Estimate the effective type I error rate as in Example 8.7 for $\alpha = .05$ and $\alpha = .01$, and for $n = 10$ and $n = 20$. How close are the effective error rates to the designed values?

### 8.5.2 Skew

As a model for a skewed population, we use an exponential random variable.

**Example 8.8.** Estimate the effective type I error rate in a $t$-test of $H_0 : \mu = 1$ versus $H_a : \mu \neq 1$ when the underlying population is exponential with rate 1. Use a sample of size $n = 20$ and test at the $\alpha = 0.05$ level.

We have to specify the null hypothesis `mu = 1` in the $t$-test so that $H_0$ will be true.

```
mean(replicate(10000, t.test(rexp(20, 1), mu = 1)$p.value < .05))
```

```
## [1] 0.0824
```

Here we get an effective type I error rate (rejecting $H_0$ when $H_0$ is true) of 8.24%. Since the test is designed to have a 5% error rate, we made 1.648 times as many type I errors as we should have. Most people would argue that the test is not working as designed.

> **Try It Yourself.**
> Repeat Example 8.8 with sample size of $n = 50$ for $\alpha = .1$, $\alpha = .01$ and $\alpha = .001$. We recommend increasing the number of replications as $\alpha$ gets smaller. Don't go overboard though, because this can take a while to run.

You should see that at $n = 50$, the effective type I error rate gets worse in relative terms as $\alpha$ gets smaller. The estimate at the $\alpha = .001$ level is an effective error rate more than 6 times what is designed! That is unfortunate, because small $\alpha$ are exactly what we are interested in. That's not to say that $t$-tests are useless in this context, but $p$-values should be interpreted with some level of caution when the underlying population is skewed. In particular, it would be misleading to report a $p$-value with 7 significant digits, as R does by default.

We simulate the effective type I error rates for $H_0 : \mu = 1$ versus $H_a : \mu \neq 1$ at $\alpha = .1, .05, .01$ and .001 levels for sample sizes $N = 10, 20, 50, 100$ and 200.

```
effectiveError <- sapply(c(.1, .05, .01, .001), function(y) {
  sapply(c(10, 20, 50, 100, 200), function(x) {
    mean(replicate(20000, t.test(rexp(x, 1), mu = 1)$p.value < y))
  })
})
colnames(effectiveError) <- c(".1", ".05", ".01", ".001")
rownames(effectiveError) <- c("10", "20", "50", "100", "200")
effectiveError
```

```
##          0.1    0.05    0.01   0.001
```

```
## 10  0.14310 0.10015 0.04600 0.01460
## 20  0.12730 0.08085 0.03240 0.01120
## 50  0.11475 0.06850 0.02310 0.00605
## 100 0.10945 0.05800 0.01755 0.00365
## 200 0.10135 0.05345 0.01490 0.00235
```

Figure 8.3 plots these results, illustrating how the test improves as sample size increases.

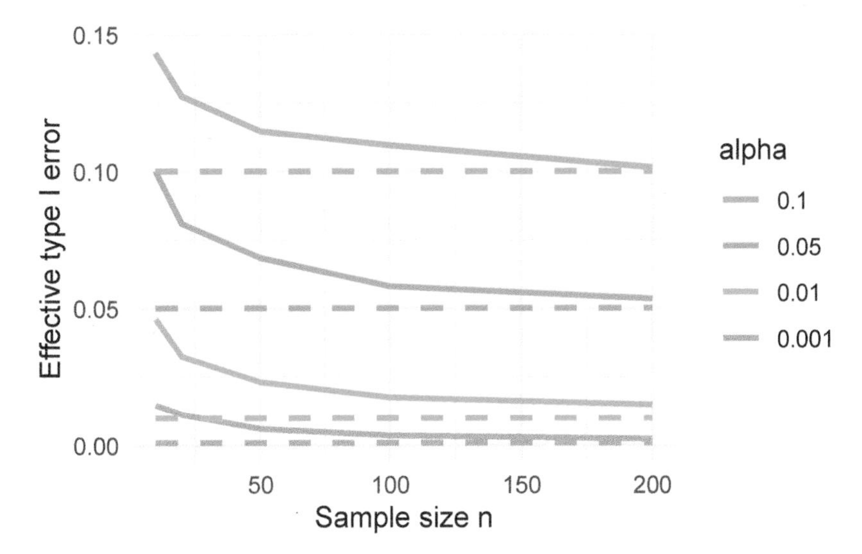


**FIGURE 8.3** Effective type I errors when applying a $t$-test to exponential data. Dashed lines show the type I error a correctly behaving test should have.

### 8.5.3 Heavy tails and outliers

If the results of the preceding section didn't convince you that you need to understand your underlying population before applying a $t$-test, then this section will. A typical "heavy tail" distribution is the $t$ random variable. Indeed, the $t$ random variable with 1 or 2 degrees of freedom doesn't even have a finite standard deviation!

**Example 8.9.** Estimate the effective type I error rate in a $t$-test of $H_0 : \mu = 0$ versus $H_a : \mu \neq 0$ when the underlying population is $t$ with 3 degrees of freedom. Take a sample of size $n = 30$ and test at the $\alpha = 0.1$ level.

```
mean(replicate(20000, t.test(rt(30, 3), mu = 0)$p.value < .1))
```

```
## [1] 0.09535
```

Not too bad. It seems that the $t$-test has an effective error rate *less* than how the test was designed.

**Example 8.10.** Let's model data that contains an outlier. Assume a population is normal with mean 3 and standard deviation 1. However, there is an error in measurement, and one of the values is multiplied by 100 due to a missing decimal point. Estimate the effective type I error rate in a $t$-test of $H_0 : \mu = 3$ versus $H_a : \mu \neq 3$ at the $\alpha = 0.1$ level with sample size $n = 100$.

```
mean(replicate(10000, {
  dat <- rnorm(100, 3, 1)
  dat[1] <- dat[1] * 100
  t.test(dat, mu = 3)$p.value < .1
}))
```

```
## [1] 2e-04
```

The mean is 0.0002 which means that the test only rejected the null hypothesis 2 of the 10,000 times we ran it. This is terrible, because the test is designed to reject $H_0$ 1,000 times out of 10,000!

Both examples in this section resulted in fewer type I errors than expected. That may seem like a good thing. However, one way to avoid type I errors is to simply *never* reject $H_0$, and that's what is happening here. Using a $t$-test on heavy-tailed data or data with outliers results in a test with very low *power*, a concept we will explore with one example here and in greater detail in Section 8.7.

**Example 8.11.** Use the same population as in Example 8.10, a normal population with mean 3 and standard deviation 1, with an outlier caused by multiplying one of the values by 100. Change the null hypothesis to $H_0 : \mu = 7$, so it is no longer true.

Estimate the percentage of time you would *correctly* reject $H_0$ at the $\alpha = 0.1$ level with sample size $n = 100$.

We have a null hypothesis for $\mu$ that is 4 standard deviations away from the true value of $\mu$. That should be easy to spot. Let's see.

```
mean(replicate(20000, {
  dat <- c(rnorm(99, 3, 1), 100 * rnorm(1, 3, 1))
  t.test(dat, mu = 7)$p.value < .1
}))
```

```
## [1] 0.0706
```

We have estimated the probability of correctly rejecting the null hypothesis as 0.0706. The value 0.0706 is the power of the test. There is no "right" value for the power, but 0.0706 is very low. We only detect a difference of four standard deviations (which is huge!) about 7% of the time that we run the test. The test is hardly worth doing.

### 8.5.4 Independence

The $t$-test assumes that data values are independent of each other. What happens when that assumption is violated? We investigate via simulation, looking at two different types of dependence that frequently occur in real experiments. One is when the sample consists of repeated measures on a common observational unit, and the other is when there is dependence between observations when ordered by another variable such as time.

**Example 8.12.** Suppose that an experimental design consists of multiple measurements from a single unit. For example, we might measure the temperature of 20 people five times each in order to determine the mean temperature of healthy adults. In this case, it would be a mistake to assume that the observations are independent.

Assume that the true mean temperature of healthy adults is 98.6 with a standard deviation of 0.7. We model repeated samples as follows. First, choose a mean temperature for each of the 20 people. Then, assume that each person's own temperatures are centered at their personal mean temperature with standard deviation 0.5.

```
# choose a mean temperature for each person
people_means <- rnorm(20, 98.6, .7)
# take five readings of each person
temps <- rnorm(100, mean = rep(people_means, 5), sd = 0.5)
head(temps)
```

```
## [1]  98.65737 100.25115  98.23785  97.59733  97.89514  98.82903
```

We now test $H_0 : \mu = 98.6$ using all 100 data points, and then replicate this test to estimate the type I error rate:

```
sim_data <- replicate(1000, {
  people_means <- rnorm(20, 98.6, .7)
  temps <- rnorm(100, mean = rep(people_means, 5), sd = 0.5)
  t.test(temps, mu = 98.6)$p.value
})
mean(sim_data < .05)
```

```
## [1] 0.344
```

The effective type I error rate in this case is much larger than 5%! This is strong evidence that one should not ignore this type of dependence in data.

> **Try It Yourself.**
> 1. Change the population standard deviation from 0.7 to 2. Does the effective type I error rate increase or decrease?
> 2. Change the personal standard deviation from 0.5 to 2 (while returning the population standard deviation to 0.7). Does the effective type I error rate increase or decrease?
> 3. Return the standard deviations to 0.7 and 0.5, and now take 10 readings from 10 different people. What happens to the effective type I error rate?

**Example 8.13.** This example explores dependence between observations when ordered by a second variable. A common special case is when the observations are ordered via time. For example, if you wish to determine whether the high temperatures in a specific city in 2020 are higher than they were in 2000, you could subtract the high temperatures in 2000 from those in 2020. The resulting data would likely be dependent on time: if it were warmer in 2020 on February 1, then it will usually be warmer in 2020 on February 2.

In order to simulate this, we will create 100 data points $x_1, \ldots, x_{100}$ such that $x_1$ is standard normal, and $x_n = 0.5x_{n-1} + z_n$, where $z_n$ is again standard normal. We see that the $n^{\text{th}}$ data point will depend on the previous data point relatively strongly, and on ones before that to a smaller extent.

We show two ways to simulate data of this type. The first uses a `while` loop, which we only covered briefly in the *Loops in R* vignette at the end of Chapter 3. Readers who do not wish to see this can safely bypass to the built-in R function which samples from this distribution.

```
x <- rep(NA, 100) # create a vector of length 100 that we will fill in below
x[1] <- rnorm(1)
i <- 2
while (i < 101) {
```

```
  x[i] <- .5 * x[i - 1] + rnorm(1)
  i <- i + 1
}
plot(x, type = "l")
```

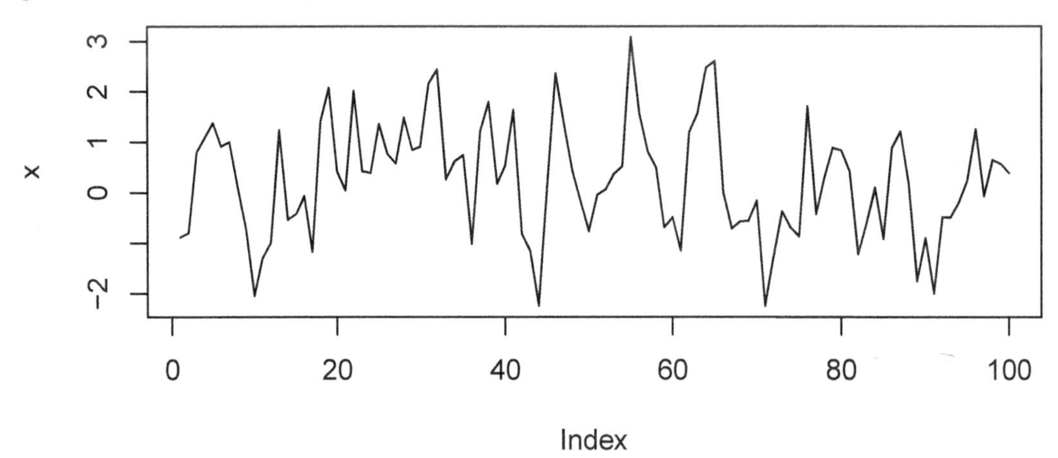


This is a common enough process that there is a built-in R function `arima.sim` which does essentially this and much, much more. For our purposes, we only care about the `ar` parameter in the model list. That is the value $c$ such that $x_n = cx_{n-1} + z_n$.

```
x <- arima.sim(model = list(ar = .5), n = 100)
```

Now we suppose that we have data of this type, and we estimate the effective type I error rate when testing $H_0 : \mu = 0$ versus $H_a : \mu \neq 0$.

```
sim_data <- replicate(10000, {
  x <- arima.sim(model = list(ar = .5), n = 100)
  t.test(x, mu = 0)$p.value
})
mean(sim_data < .05)
```

```
## [1] 0.2619
```

The effective type I error rate is much higher than the desired rate of $\alpha = .05$. It would be inappropriate to use a one sample $t$-test on data of this type.

> **Try It Yourself.**
> Change the `ar` parameter to .1 and to .9 and estimate the effective type I error rate. Which one is performing closest to the desired rate of $\alpha = .05$?

### 8.5.5 Summary

When doing $t$-tests on populations that are not normal, the following types of departures from design were noted:

1. When the underlying distribution is skewed, the exact $p$-values should not be reported to many significant digits (especially the smaller they get). If a $p$-value close to the $\alpha$ level of a test is obtained, then we would want to note the fact that we are unsure of our $p$-values due to skewness, and fail to reject.

2. When the underlying distribution has outliers, the *power* of the test is severely compromised. A single outlier of sufficient magnitude will force a $t$-test to *never* reject the null hypothesis. If your population has outliers, you do not want to use a $t$-test.
3. When the sample is not independent, then the source and type of dependence will need to be understood and adjusted for. It is not typically appropriate to use `t.test` without making these adjustments.

Boxplots created in base R and in `ggplot2` show outliers in the data. Outliers that show up in boxplots are **not necessarily** problematic for `t.test`.

**Try It Yourself.**
Generate a sample of size 50 from a normal population, and plot it with a boxplot. Do you see outliers? Repeat this a few times to get a sense of how often apparent outliers show up in perfectly normal data.

For real experiments with apparent outliers, run simulations with normal data and compare. If the number and especially the magnitude of outliers in your data are worse than what you see in simulations, then you are right to be concerned.

## 8.6 Two sample hypothesis tests

This section presents hypothesis tests for comparing the means of two populations. In this situation, we have two populations that we are sampling from, and we ask whether the mean of population one differs from the mean of population two.

For tests of the mean, we must assume the underlying populations are normal, or that we have enough samples so that the Central Limit Theorem takes hold. We are also *as always* assuming that there are no outliers. Remember, outliers can kill $t$-tests. Both the one sample and the two sample variety.

For two sample tests, the (unknown) means of the two populations are $\mu_1$ and $\mu_2$. Our hypotheses are:

- $H_0 : \mu_1 = \mu_2$ versus
- $H_a : \mu_1 \neq \mu_2$.

As with one sample tests, the alternate hypothesis may be one-sided if justified by the experiment. One-sided tests have $H_a : \mu_1 \leq \mu_2$ or $H_a : \mu_1 \geq \mu_2$. One-sided tests should be used judiciously, and the same warnings apply as were presented in Section 8.4.

The most straightforward experiment that results in two independent samples is a study comparing the same measurement on two different populations. In this case, a sample is chosen from population one, a second sample is chosen independently from population two, and both samples are measured. Another common setting is a *controlled experiment.* In a controlled experiment, the two populations are a group that receives some treatment and a control group that does not[6]. Subjects are randomly assigned to either the control or treatment group. In both of these designs, the *two independent samples t-test* is appropriate.

[6]With human subjects, the control group usually receives a *placebo* treatment so that the subjects and experimenters are unaware (blind) of which group they are in.

The *two sample paired t-test* is used when data comes in pairs. Pairs may be two different subjects matched for some common traits, or they may come from two measurements on a single subject.

For example, suppose you wish to determine whether a new fertilizer increases the mean number of apples that apple trees produce. Here are three reasonable experimental designs:

**Design 1**
: Randomly select two samples of trees. Apply fertilizer to one sample and leave the other sample alone as a control group. At the end of the growing season, count the apples on each tree in each group.

**Design 2**
: Choose one sample of trees for the experiment. Count the number of apples on each tree after one growing season. The next season, apply the fertilizer and count the number of apples again.

**Design 3**
: Carefully choose trees in pairs, so that each pair of trees is of similar maturity and will grow under similar conditions (soil quality, water, and sunlight). Randomly assign one tree of each pair to receive fertilizer and the other to act as control. Count the number of apples on the fertilized trees and the unfertilized trees.

*Image credit: George Chernilevsky.*

In design 1, the two independent samples $t$-test is appropriate. Data from this experiment will likely be in long form, with a two-valued grouping variable in one column and a measurement variable giving apple counts in a second column.

Designs 2 and 3 result in *paired data*. The measures in the two populations are dependent. You expect there to be some relationship between the number of apples you count on each pair. For example, big trees would have more apples each season than small trees. In general, it is a hard problem to deal with dependent data, but paired data is a dependency we can handle with the two sample paired $t$-test.

In designs 2 and 3, the data will likely be in wide form for analysis: each row representing a pair, and two columns with apple counts. In design 2, the rows are individual trees and the columns are the two measurements, before and after. In design 3, the rows are pairs of trees and the columns are measurements for the fertilized and unfertilized member of each pair. Table 8.2 shows the typical structure of data for each of these designs.

**TABLE 8.2** Data from three experimental designs: independent samples, before/after, and matched pairs.

| group | apples |
|---|---|
| fertilized | 660 |
| control | 533 |
| fertilized | 418 |
| fertilized | 549 |
| control | 574 |
| fertilized | 558 |

| before | after |
|---|---|
| 469 | 496 |
| 651 | 498 |
| 539 | 594 |
| 438 | 582 |
| 279 | 559 |
| 612 | 592 |

| fertilized | control |
|---|---|
| 578 | 353 |
| 507 | 452 |
| 301 | 542 |
| 562 | 636 |
| 494 | 490 |
| 484 | 539 |

### 8.6.1 Two independent samples $t$-test

For the two independent samples $t$-test, we require data $x_{1,1}, \ldots, x_{1,n_1}$ drawn from one population and $x_{2,1}, \ldots, x_{2,n_2}$ drawn independently from a second population.

**Assumptions 8.2.**

- *The samples are independent.*
- *The underlying populations are normal, or the sample sizes are large.*
- *There are no outliers.*

We calculate the means of the two samples $\overline{x}_1$ and $\overline{x}_2$ and their variances $s_1^2$ and $s_2^2$. The test statistic is given by

$$t = \frac{\overline{x}_1 - \overline{x}_2}{\sqrt{s_1^2/n_1 + s_2^2/n_2}}.$$

If the assumptions are met, then the test statistic has a $t$ distribution with a complicated (and non-integer) df. In the past, when computations like this were made by hand, it was common to add an assumption of "equal variances" and use **both** populations to estimate the variance, in which case the expression for $t$ simplifies somewhat. In general, it is a more conservative approach not to assume the variances of the two populations are equal, and we generally recommend not to assume equal variance unless you have a good reason for doing so.

In R, the two independent samples $t$-test is performed with the syntax `t.test(value ~ group, data = dataset)`. Here `value` is the name of the measurement variable, `group` should be a variable that takes on exactly two values, and `dataset` is the name of the data set. The `value ~ group` expression is known as a *formula*, and should be read as "value which depends on group." The `~` is a tilde character. We will see this syntax used frequently in Chapter 11 when performing regression analysis.

**Example 8.14.** We return to the `brake` data in the `fosdata` package. Recall that we were interested in whether there is a difference in the mean time for an older person and the mean time for a younger person to release the brake pedal after it unexpectedly causes the car to speed up.

Let $\mu_1$ be the mean time to release the pedal for the older group, and let $\mu_2$ denote the mean time of the younger group in milliseconds.

$$H_0 : \mu_1 = \mu_2$$

versus

$$H_a : \mu_1 \neq \mu_2$$

It might be reasonable to do a one-sided test here if you have data or other evidence suggesting that older drivers are going to have slower reaction times, but we will stick with the two-sided tests.

Let's think about the data that we have for a minute. Do we expect the reaction times to be approximately normal? How many samples do we have? We compute the mean, sd, skewness, and number of observations in each group and also plot the data.

```
brake <- fosdata::brake
brake %>%
  group_by(age_group) %>%
  summarize(
    mean = mean(latency_p2),
    sd = sd(latency_p2),
    skew = e1071::skewness(latency_p2),
    N = n()
  )
```

```
## # A tibble: 2 x 5
##   age_group  mean    sd  skew     N
##   <chr>     <dbl> <dbl> <dbl> <int>
## 1 Old        956.  242.  1.89    40
## 2 Young      664.  95.9 0.780    40
```

```
ggplot(brake, aes(x = age_group, y = latency_p2)) +
  geom_boxplot(outlier.size = -1) +
  geom_jitter(height = 0, width = .2)
```

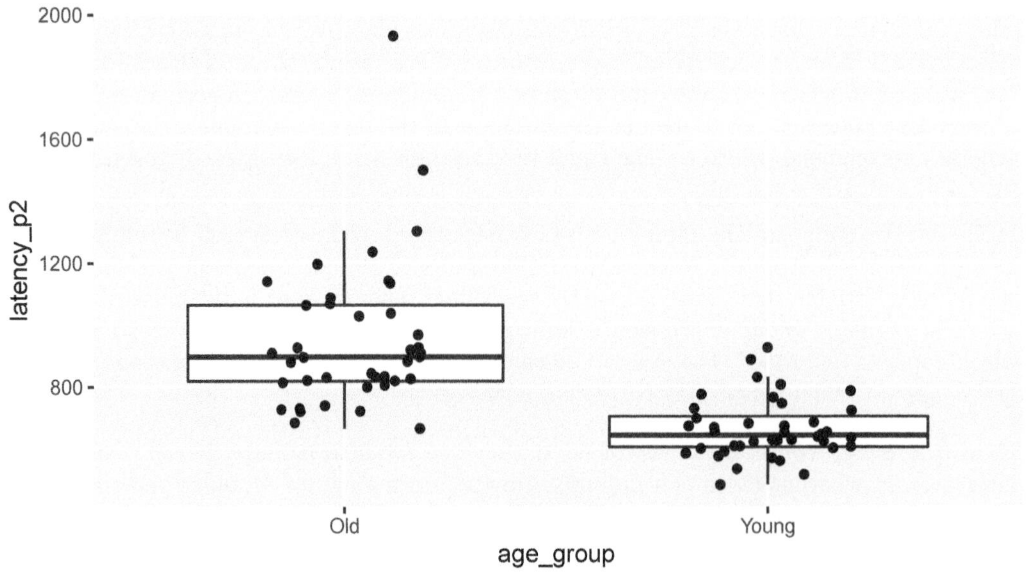

The older group in particular is right skewed, and there is one extreme[7] outlier among the older drivers. These are concerns, but with 80 observations our assumptions are met well

[7]We say the outlier is extreme because if we run `boxplot(rnorm(40, 956, 242))`, it is extremely rare to see an outlier that is so far away from the mean.

enough. We tell `t.test` to break up `latency_p2` by `age_group` and perform a two sample $t$-test.

```
t.test(latency_p2 ~ age_group, data = brake)
```

```
##
##  Welch Two Sample t-test
##
## data:  latency_p2 by age_group
## t = 7.0845, df = 50.97, p-value = 4.014e-09
## alternative hypothesis: true difference in means is not equal to 0
## 95 percent confidence interval:
##  208.7430 373.8345
## sample estimates:
##   mean in group Old mean in group Young
##            955.7747            664.4859
```

We reject the null hypothesis that the two means are the same ($p = 4.0 \times 10^{-9}$). A 95% confidence interval for the difference in mean times is $(209, 374)$ ms. That is, we are 95% confident that the mean time for older drivers is between 209 and 374 ms slower than the mean time for younger drivers.

In Exercise 8.40, you are asked to redo this analysis after removing the largest older driver outlier. When doing so, you will see that the skewness is improved, the $p$-value gets smaller, and the confidence interval gets more narrow. Therefore, we have no reason to doubt the validity of the work we did with the outlier included.

> **Try It Yourself.**
> Is there a significant difference between young and old drivers in the mean time to press the brake after seeing the red light (`latency_p1`)?

### 8.6.2 Paired two sample $t$-test

For the paired two sample $t$-test, we require data $x_{1,1}, \ldots, x_{1,n}$ drawn from one population and $x_{2,1}, \ldots, x_{2,n}$ drawn from a second population. The data come in pairs, so that $x_{1,i}$ and $x_{2,i}$ are not independent.

**Assumptions 8.3.**

- *The samples are paired.*
- *The underlying populations are normal, or the sample size is large.*
- *There are no outliers.*

If we compute the difference $d_i = x_{1,i} - x_{2,i}$ for each pair, then the resulting differences will be independent and contain information about which population is larger. Then we simply perform a one sample $t$-test to test the differences against zero. The mean and variance of $d$ are $\overline{d} = \overline{x}_1 - \overline{x}_2$ and $s_d^2$. So the test statistic is

$$T = \frac{\overline{x}_1 - \overline{x}_2}{s_d/\sqrt{n}}.$$

When the test assumptions are met, $T$ will have the $t$ distribution with $n - 1$ degrees of freedom.

In R, the syntax for a paired $t$-test is `t.test(dataset$var1, dataset$var2, paired = TRUE)`.

It is possible to perform a paired $t$-test on long format data with formula notation `value ~ group`. This requires you to be sure that the data pairs are in order. More often than not, long format data is unpaired, and using formula notation to perform a paired test is a mistake.

**Example 8.15.** Return to the `cows_small` data set in the `fosdata` package. We wish to determine whether there is a difference between nozzle TK-0.75 and nozzle TK-12 in terms of the mean amount that they cool down cows in a three-minute period.

Let $\mu_1$ denote the mean change in temperature of cows sprayed with TK-0.75 and let $\mu_2$ denote the mean change in temperature of cows sprayed with TK-12. Our null hypothesis is $H_0 : \mu_1 = \mu_2$ and our alternative is $H_a : \mu_1 \neq \mu_2$. According to the manufacturer of the TK-0.75, it sprays 0.4 liters per minute at 40 PSI, while the TK-12 sprays 4.5 liters per minute. Since the TK-12 sprays considerably more water, it would be worth considering a one-sided test in this case. However, we don't know any other characteristics of the nozzles which might mitigate the flow advantage of the TK-12, so we stick with a two-sided test.

The data is in wide format, with one column for each nozzle type. To check assumptions, we compute and plot the difference variable. Here we chose to use a qq plot but a histogram or boxplot would have also been appropriate.

```
cows_small <- fosdata::cows_small
ggplot(cows_small, aes(sample = tk_0_75 - tk_12)) +
  geom_qq() +
  geom_qq_line()
```

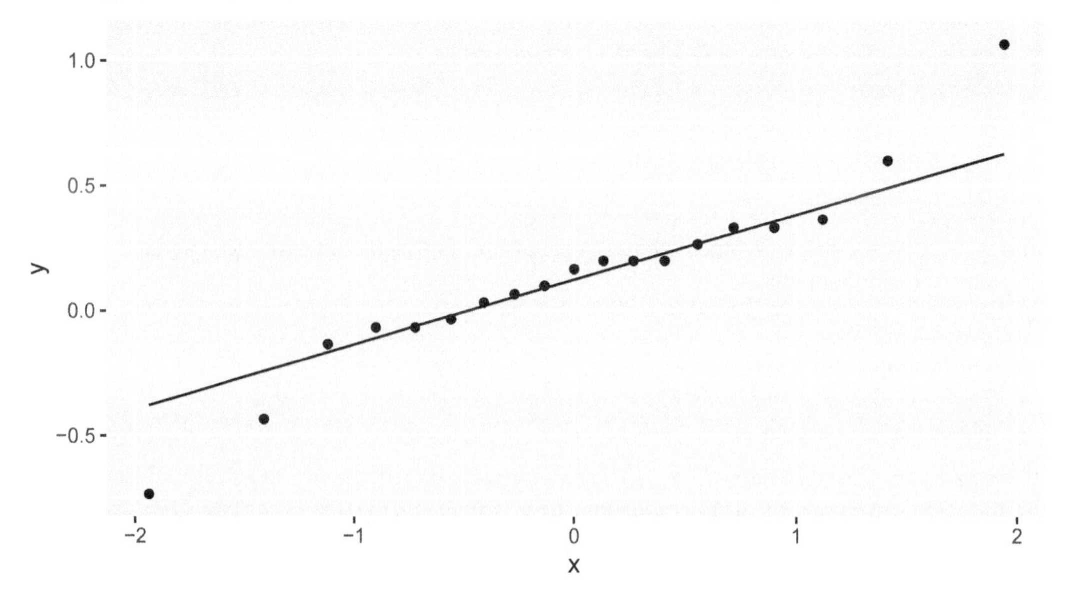


The distribution is somewhat heavy-tailed but good enough to proceed. We perform the paired two sample $t$-test.

```
t.test(cows_small$tk_12, cows_small$tk_0_75, paired = TRUE)
```

```
##
##  Paired t-test
```

```
##
## data:  cows_small$tk_12 and cows_small$tk_0_75
## t = -1.5118, df = 18, p-value = 0.1479
## alternative hypothesis: true difference in means is not equal to 0
## 95 percent confidence interval:
##  -0.31023781  0.05058868
## sample estimates:
## mean of the differences
##              -0.1298246
```

We find a 95% confidence interval for the difference in the means to be $(-0.31, 0.05)$, and we do not reject $H_0$ at the $\alpha = .05$ level ($p = .1479$). We cannot conclude that there is a difference in the mean temperature change associated with the two nozzles.

> **Try It Yourself.**
> Is there a significant difference in the mean temperature change between the TK-0.75 nozzle and the control?

## 8.7 Type II errors and power

Up to this point, we have been concerned about the probability of incorrectly rejecting the null hypothesis when the null hypothesis is true. There is another type of error that can occur, however, and that is failing to reject the null hypothesis when the null hypothesis is false.

**Definition 8.8.** In a hypothesis test,

- A *type I error* is rejecting the null when it is true.
- A *type II error* is failing to reject the null when it is false.

**Definition 8.9.** The *power* of a test is defined to be one minus the probability of a type II error. That is, *power* is the probability of correctly rejecting the null hypothesis.

One issue that arises is that the null hypothesis is (usually) a *simple hypothesis*; that is, it consists of a complete specification of parameters, while the alternative hypothesis is often a *composite hypothesis*; that is, it consists of parameters in a range of values. Therefore, when we say the null hypothesis is false, we are not completely specifying the distribution of the population. For this reason, the type II error is a function that depends on the value of the parameter(s) in the alternative hypothesis.

The examples in this section imagine we are designing an experiment to collect data like `fosdata::normtemp`, which measured body temperature and heart rate of 130 healthy adults.

**Example 8.16.** Suppose we wish to conduct an experiment to determine if the mean heart rate of healthy adults is 80 beats per minute. Let $\mu$ be the population mean heart rate. Then the hypotheses are:

- $H_0 : \mu = 80$.
- $H_a : \mu \neq 80$.

With a sample size of $n = 130$ and significance level $\alpha = 0.01$, what is the probability of a type II error?

Our plan is to simulate the experiment. However, for a type II error to happen, the alternate hypothesis must be true. The alternate hypothesis does not tell us the population distribution, so we must make assumptions about it.

We start by assuming the population heart rate is normally distributed. This is reasonable, as many physiological measurements are normal.

Next, assume the true population mean is 78. Of course, there is no point in performing the experiment if we *know* that $\mu = 78$, but the hypothetical computation will still be useful. It will tell us the probability we fail to detect a difference of 2 bpm with our experimental setup, or equivalently the power our test will have to detect such a difference.

Finally, we need the population standard deviation. The American Heart Association says that 60-100 bpm is a range for normal heart rate. We interpret to mean that about 99% of healthy adult heart rates will fall in this range, so we will estimate the standard deviation to be (100 - 60)/6 = 6.67 bpm.[8]

Now that the alternate hypothesis is specified, we can perform a simulation. Start by taking a sample of size 130 from a normal distribution with mean 78 and standard deviation 6.67. Then compute the $t$ statistic and determine whether $H_0$ is rejected at the $\alpha = 0.01$ level.

```
dat <- rnorm(130, 78, 6.67)
t_stat <- (mean(dat) - 80) / (sd(dat) / sqrt(130))
t_stat
```

```
## [1] -3.843193
```

```
qt(.005, 129) # critical value
```

```
## [1] -2.614479
```

Since `t_stat` is less than `qt(.005, 129)` we would (correctly) reject the null hypothesis. It is even easier to use the built-in R function `t.test`:

```
dat <- rnorm(130, 78, 6.67)
t.test(dat, mu = 80)$p.value > .01
```

```
## [1] FALSE
```

Next, repeatedly sample and then compute the percentage of times that the $p$-value of the test is greater than .01. That will be our estimate for the probability of a type II error.

```
simdata <- replicate(10000, {
  dat <- rnorm(130, 78, 6.67)
  t.test(dat, mu = 80)$p.value > .01
})
mean(simdata)
```

```
## [1] 0.2136
```

We get about a 0.2136 chance for a type II error in this case, or a power of 0.7864.

This experimental setup will be able to detect a 2 bpm difference in heart rate about 78.6% of the time.

[8]When the standard deviation of a population is unknown, it is often easier to estimate the range of possible values and divide by 4 or 6 to get a rough estimate of the standard deviation.

For many experimental designs, simulation is the best way to compute power. However, the $t$-test on a normal population is simple enough mathematically that power computations are possible to do exactly. The built-in function `power.t.test` does this. It has the following arguments:

`n` the number of samples

`delta`
the difference between the null and alternative mean

`sd` the standard deviation of the population

`sig.level`
the $\alpha$ significance level of the test

`power`
one minus the probability of a type II error

`type`
either `one.sample`, `two.sample` or `paired`, default is two

`alternative`
`two.sided` or `one.sided`, default is two

To use `power.t.test`, specify any four of the first five parameters. The function then solves for the unspecified parameter.

**Example 8.17.** With the experimental setup of Example 8.16, compute the power exactly.

The assumptions made in Example 8.16 for the simulation are exactly the inputs needed for `power.t.test`. The value of `delta` is 2, which is the difference between our null hypothesis $\mu = 80$ and our assumed reality of $\mu = 78$.

```
power.t.test(
  n = 130, delta = 2, sd = 6.67,
  sig.level = .01, type = "one"
)
```

```
##
##      One-sample t test power calculation
##
##               n = 130
##           delta = 2
##              sd = 6.67
##       sig.level = 0.01
##           power = 0.7878322
##     alternative = two.sided
```

The power is 0.7878322, so the probability of a type II error is one minus that, or

```
1 - .7878322
```

```
## [1] 0.2121678
```

which is close to what we computed in the simulation.

A traditional approach in designing experiments is to specify the significance level and power, and use a simulation or `power.t.test` to determine what sample size is necessary to achieve that power. In order to do so, we must estimate the standard deviation of the underlying population and we must specify the *effect size* that we are interested in.

Sometimes prior work or a pilot study could be used to determine a reasonable effect size. Other times, we can pick a size that would be clinically significant (as opposed to statistically significant). Think about it like this: if the true mean body temperature of healthy adults is 98.59, would that make a difference in treatment or diagnosis? No. What about if it were 100? Then yes, that would make a difference in diagnoses! It is subjective and can be challenging to specify what `delta` to provide, but we should provide one that would make an important difference in some context.

**Example 8.18.** How large a sample is needed to detect a clinically significant difference in body temperature from 98.6 degrees with a power of 0.8?

The null hypothesis is that the mean temperature of healthy adults is 98.6 degrees. We estimate the standard deviation of the underlying population to be 0.3 degrees. After consulting with our subject area expert, we decide that a clinically significant difference from 98.6 degrees would be 0.2 degrees. That is, if the true mean is 98.4 or 98.8, then we want to detect it with power 0.8. The sample size needed is

```
power.t.test(
  delta = 0.2, sd = .3,
  sig.level = .01, power = 0.8, type = "one.sample"
)
```

```
##
##      One-sample t test power calculation
##
##               n = 29.64538
##           delta = 0.2
##              sd = 0.3
##       sig.level = 0.01
##           power = 0.8
##     alternative = two.sided
```

We would need to collect 30 observations in order to have an appropriately powered test.

### 8.7.1 Effect size

We used the term *effect size* in Section 8.7 without really saying what we mean. A common measure of the effect size is given by Cohen's $d$.

**Definition 8.10.** *Cohen's d is given by*

$$d = \frac{\overline{x} - \mu_0}{s},$$

the difference between the sample mean and the mean from the null hypothesis, divided by the sample standard deviation.

The parametrization of `power.t.test` is redundant in the sense that the power of a test only relies on `delta` and `sd` through the effect size $d$. As an example, we note that when `delta = .1` and `sd = .5` the effect size is 0.2, as it also is when `delta = .3` and `sd = 1.5`. Both of these give the same power:

```
power.t.test(n = 30, delta = .3, sd = 1.5, sig.level = .05, type = "one")
```

```
##
##      One-sample t test power calculation
##
```

```
##                n = 30
##            delta = 0.3
##               sd = 1.5
##        sig.level = 0.05
##            power = 0.1839206
##      alternative = two.sided
```

```
power.t.test(n = 30, delta = .1, sd = .5, sig.level = .05, type = "one")
```

```
##
##      One-sample t test power calculation
##
##                n = 30
##            delta = 0.1
##               sd = 0.5
##        sig.level = 0.05
##            power = 0.1839206
##      alternative = two.sided
```

It is common to report the effect size, especially when there is statistical significance. Effect sizes are often reported along with adjectives such as "small," "medium," or "large." These adjectives are domain dependent, and what is a very small effect in one discipline could be a very large effect in another.

You do not want your test to be underpowered for two reasons. First, you likely will not reject the null hypothesis in instances where you would like to be able to do so. Second, when effect sizes are estimated using underpowered tests, they tend to be overestimated. This leads to a reproducibility problem, because when people design further studies to corroborate your study, they are using an effect size that is too high. Let's run some simulations to verify. We assume that we have an experiment with effect size $d = 0.2$.

```
power.t.test(
  delta = 0.2, sd = 1, sig.level = .05,
  power = 0.25, type = "one"
)
```

```
##
##      One-sample t test power calculation
##
##                n = 43.24862
##            delta = 0.2
##               sd = 1
##        sig.level = 0.05
##            power = 0.25
##      alternative = two.sided
```

This says when $d = 0.2$, we need a sample size of 43 for the power to be 0.25. Remember, power this low is **not** recommended! We use simulation to estimate the percentage of times that the null hypothesis is rejected. We assume we are testing $H_0 : \mu = 0$ versus $H_a : \mu \neq 0$.

```
simdata <- replicate(10000, {
  dat <- rnorm(43, 0.2, 1)
  t.test(dat, mu = 0)$p.value
})
mean(simdata < .05)
```

```
## [1] 0.2416
```

And we see that indeed, we get about 25%, which is what `power.t.test` predicted. Note that the true effect size is 0.2. The estimated effect size from tests which are significant is given by:

```
simdata <- replicate(10000, {
  dat <- rnorm(43, 0.2, 1)
  ifelse(t.test(dat, mu = 0)$p.value < .05,
    abs(mean(dat)) / sd(dat),
    NA
  )
})
mean(simdata, na.rm = TRUE)
```

```
## [1] 0.4066708
```

We see that the mean estimated effect size is double the actual effect size, so we have a *biased* estimate for effect size. If we have an appropriately powered test, then this bias is much less.

```
power.t.test(
  delta = 0.2, sd = 1, sig.level = .05,
  power = 0.8, type = "one"
)
```

```
##
##      One-sample t test power calculation
##
##               n = 198.1513
##           delta = 0.2
##              sd = 1
##       sig.level = 0.05
##           power = 0.8
##     alternative = two.sided
```

```
simdata <- replicate(10000, {
  dat <- rnorm(199, 0.2, 1)
  ifelse(t.test(dat, mu = 0)$p.value < .05,
    abs(mean(dat)) / sd(dat),
    NA
  )
})
mean(simdata, na.rm = TRUE)
```

```
## [1] 0.2252131
```

With power 0.8, the estimated effect size is only a little over the true value of 0.2.

## 8.8 Resampling

Resampling is a modern technique for doing statistical analyses. Commonly used techniques include the bootstrap, permutation tests and cross validation. This section covers using the bootstrap and permutation tests to perform inference on the mean when sampling from one or two populations. Cross validation is covered in the context of linear regression in Section 11.7. For more information on these techniques, see Efron and Tibshirani.[9]

### 8.8.1 Bootstrapping

The key fact we used for inference on the mean in this chapter is that

$$\frac{\overline{X} - \mu_0}{S/\sqrt{n}}$$

is a $t$ random variable with $n-1$ degrees of freedom when $X_1, \ldots, X_n$ is a random sample from a normal population with mean $\mu_0$.

Bootstrapping provides an alternative approach for performing inference on the mean, where we estimate the sampling distribution of $\overline{X}$ directly from the random sample that we obtain as data. The idea is to repeatedly resample with replacement from the data that we obtained, and compute the value $\overline{X}$ for each resample. The obtained sample is an estimate for the distribution of $\overline{X}$. In this section, we will illustrate the basic ideas of the bootstrap through examples.

**Example 8.19.** Consider the `brake` data set in the `fosdata` package. We will construct the sampling distribution for $\overline{X}$. Let's start by plotting a histogram of the data.

```
brake_old <- fosdata::brake %>%
  filter(age_group == "Old")
ggplot(brake_old, aes(x = latency_p1)) +
  geom_histogram() +
  scale_x_continuous(limits = c(300, 1300))
```

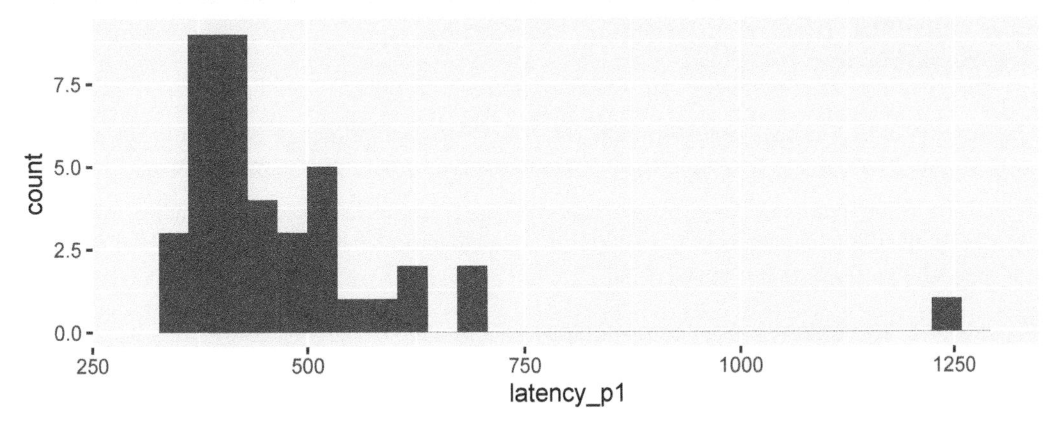


This latency data has mean 474.202 and does not appear to be normally distributed.

[9]Bradley Efron and Robert J Tibshirani, *An Introduction to the Bootstrap*, Monographs on Statistics and Applied Probability 57 (New York: Chapman & Hall, 1993).

To bootstrap, we resample from the data **with replacement**. If we do this one time, here is what we get.

```
latency_p1_resample <- sample(brake_old$latency_p1, replace = TRUE)
data.frame(latency_p1_resample) %>% ggplot(aes(x = latency_p1_resample)) +
  geom_histogram() +
  scale_x_continuous(limits = c(300, 1300)) +
  geom_vline(xintercept = mean(latency_p1_resample), color = "red")
```

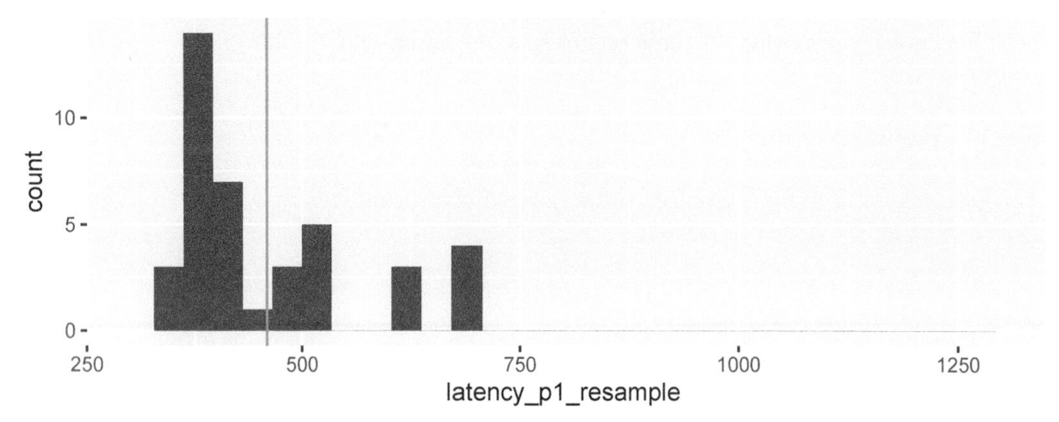


We see that the characteristics of this sample are similar to, but different from, the characteristics of the original data. In particular, the largest value in the original sample was not chosen in the bootstrap resample. If we repeat this, the largest value could be chosen once, twice or more times. The mean of the resample is 459.1, shown in red on the plot.

To estimate the sampling distribution of $\overline{X}$, we resample many times and get many different possible values for the sample mean. Each sample mean represents one possible result of an experiment that was not run. Together, these resamples provide an estimate of the distribution of $\overline{X}$. The histogram in Figure 8.4 is our estimate for the sampling distribution of $\overline{X}$.

```
Xbar <- replicate(10000, {
  brake_lat <- sample(brake_old$latency_p1, replace = TRUE)
  mean(brake_lat)
})
data.frame(Xbar) %>% ggplot(aes(x = Xbar)) +
  geom_histogram()
```

**Definition 8.11.** Using bootstrapping, the $100(1-\alpha)\%$ confidence interval is the interval from the $\alpha/2$ to the $1-\alpha/2$ quantiles of the bootstrap estimated sampling distribution.

The R function `quantile` computes quantiles of a numeric vector. Quantiles are often multiplied by 100 and referred to as percentiles, so that the 50th percentile is the 0.5 quantile and also the median of the data.

**Example 8.20.** Using the `brake` data, construct a 95% bootstrap confidence interval for $\mu$, the mean latency in older drivers until they press the brake.

To compute the 95% confidence interval for the mean latency in older drivers, find the 0.025 and 0.975 quantiles of the $\overline{X}$ distribution that was simulated in Example 8.19.

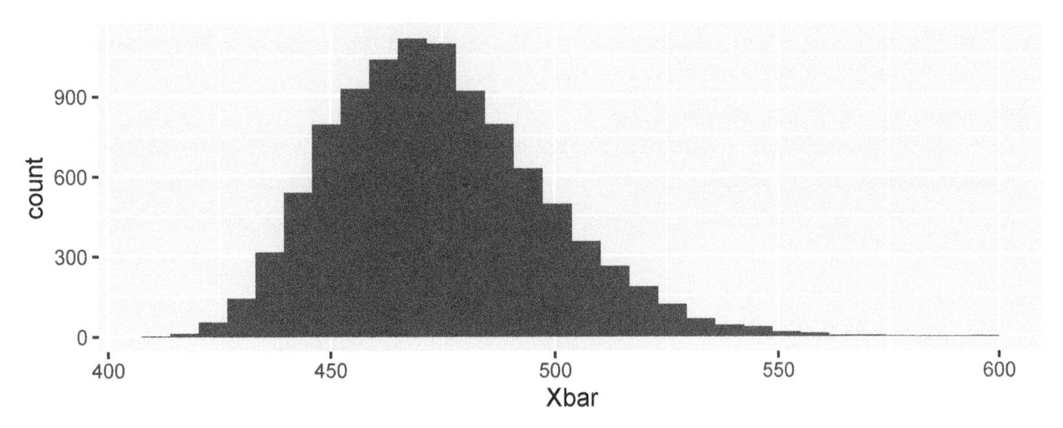


**FIGURE 8.4** Bootstrap resampling estimate for the distribution of $\overline{X}$.

```
quantile(Xbar, c(.025, .975))
```

```
##     2.5%    97.5%
## 434.0751 525.7794
```

The 95% confidence interval is 434 – 526. Compare this to the confidence interval computed using `t.test`:

```
t.test(brake_old$latency_p1)$conf.int
```

```
## [1] 425.4956 522.9084
## attr(,"conf.level")
## [1] 0.95
```

The bootstrap interval is narrower and both endpoints are higher, which one could attribute to the right skewness of this data.

The bootstrap method also works when we wish to create confidence intervals for other test statistics. For example, the *trimmed mean* removes a fraction of the most extreme values of a vector and then computes the mean. The trimmed mean is sometimes used in the presence of outliers, when we wish to understand the mean of the "typical" data from the population. The trimmed mean is implemented in R via `mean` with the argument `trim`, which is "the fraction (0 to 0.5) of observations to be trimmed from each end of [the vector] before the mean is computed."

**Example 8.21.** Find a 95% bootstrap confidence interval for the mean after trimming by 10% on each side of the `p1_latency` for older drivers.

We first compute the trimmed mean. Note that this is the same as the mean after removing the smallest 4 and largest 4 values.

```
mean(brake_old$latency_p1, trim = 0.1)
```

```
## [1] 447.3073
```

```
mean(sort(brake_old$latency_p1)[-c(1:4, 37:40)])
```

```
## [1] 447.3073
```

We then follow the same procedure as above, sampling with replacement from the population and computing the trimmed mean each time.

```
Xbar_trimmed <- replicate(10000, {
  latency_p1_resample <- sample(brake_old$latency_p1, replace = TRUE)
  mean(latency_p1_resample, trim = 0.1)
})
quantile(Xbar_trimmed, c(.025, .975))
```

```
##     2.5%    97.5%
## 419.5230 483.2291
```

Figure 8.5 shows the histogram estimate for the sampling distribution of the trimmed mean. The distribution is more symmetric than the $\overline{X}$ distribution was, and the distribution for the trimmed mean has lighter tails. The 95% confidence interval for the trimmed mean is considerably narrower and recentered lower than the interval for $\overline{X}$.

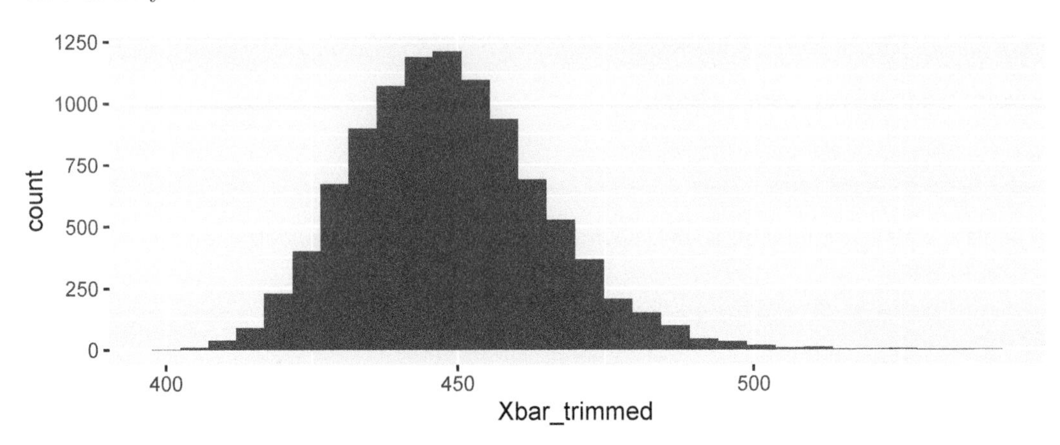


**FIGURE 8.5** Bootstrap resampling estimate for the distribution of the trimmed mean.

Hypothesis testing is also possible with bootstrap methods, although we will only discuss the one sample test.

Given data $x_1, \ldots, x_n$, we want to test $H_0 : \mu = \mu_0$ against $H_a : \mu \neq \mu_0$. We wish to estimate the sampling distribution of $\overline{X}$ under the null hypothesis. One problem with using our data to do so is that the mean of our data is not $\mu_0$! The solution is to transform the data by adding $\mu_0 - \overline{x}$ to each data point so that the mean of our data is $\mu_0$:

$$y_i = x_i + \mu_0 - \overline{x}.$$

By resampling $y_i$, we get an estimate of the $\overline{X}$ distribution under the null hypothesis. The $p$-value is then the proportion of the estimated distribution which is as extreme as the observed value $\overline{x}$.

**Example 8.22.** Consider the `weight_estimate` data set in the `fosdata` package.

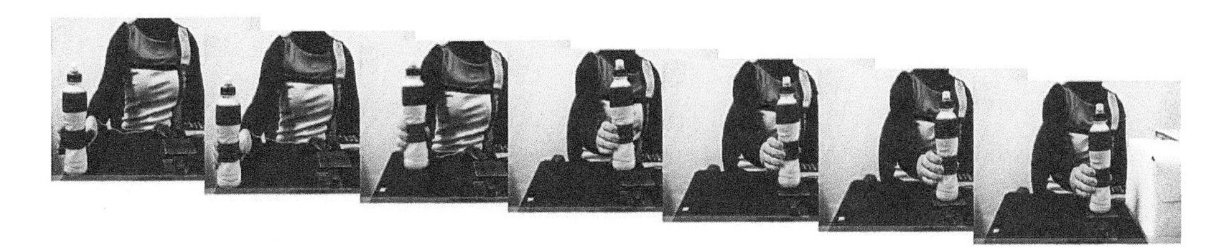

*Image credit: Sciutti, et al..*

In this experiment,[10] participants watched actors pick up objects of various weights, and then estimated the weight of the object. Let $\mu$ be the mean value that a six-year-old child would estimate for the 200-gram object. We wish to test $H_0 : \mu = 200$ versus $H_a : \mu \neq 200$. We start by examining the data. The histogram of `mean200` is shown in Figure 8.6 with the hypothesized mean in blue and the sample mean in red.

```
weight <- fosdata::weight_estimate %>%
  filter(age == "6")
mean(weight$mean200)
```

```
## [1] 241.5789
```

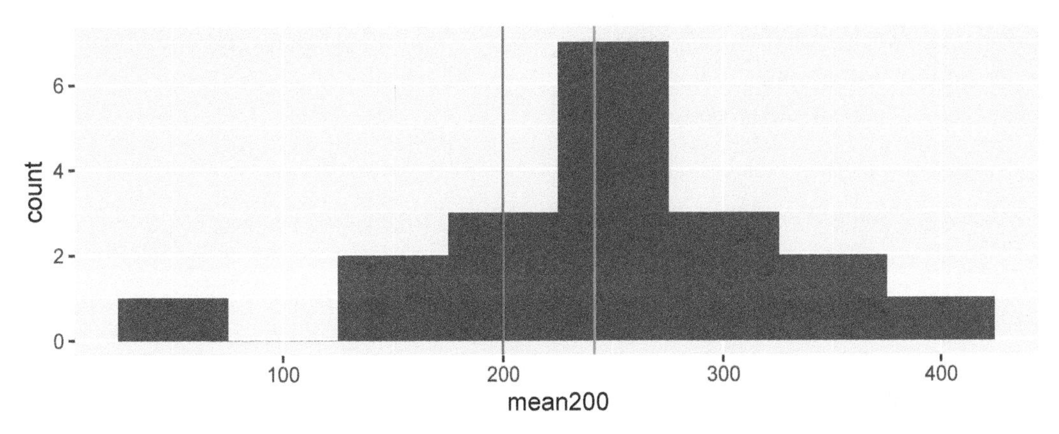


**FIGURE 8.6** Six-year-old child estimates of the weight of a 200-gram object. Sample mean in red, hypothesized mean in blue.

The sample mean is 241.6 grams, which is higher than the value under $H_0$. We now shift the data so it has mean 200 to match the null hypothesis.

```
weight_forboot <- weight$mean200 - mean(weight$mean200) + 200
mean(weight_forboot)
```

```
## [1] 200
```

We resample from this data and see how often we get a sample mean that is 241.6 or larger.

```
Xbar <- replicate(10000, {
  boot_weight <- sample(weight_forboot, replace = TRUE)
```

[10] Alessandra Sciutti, Laura Patanè, and Giulio Sandini, "Development of Visual Perception of Others' Actions: Children's Judgment of Lifted Weight," *PLOS One* 14, no. 11 (November 2019): e0224979.

```
  mean(boot_weight)
})
```

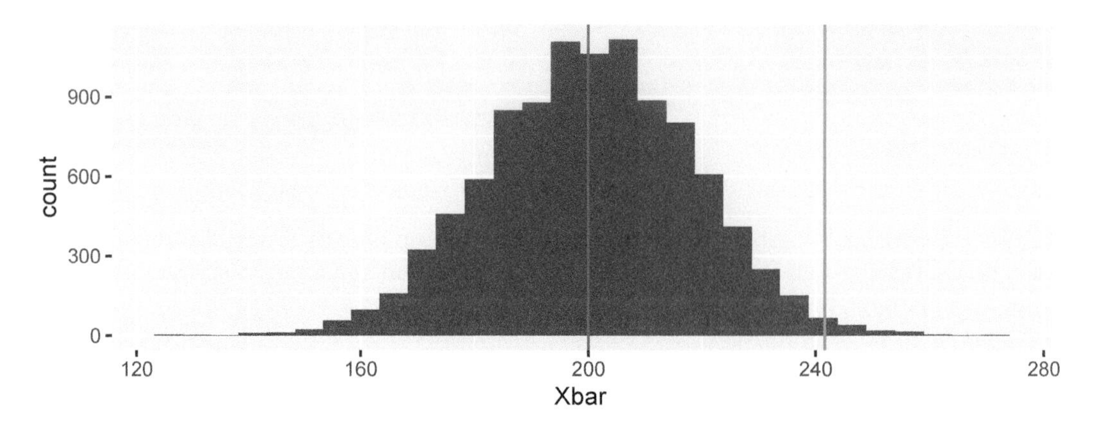


**FIGURE 8.7** Bootstrap resampled distribution of mean weight estimates. Sample mean in red, hypothesized mean in blue.

Figure 8.7 shows the bootstrap resampled estimate for the distribution of $\overline{X}$. It is unlikely, but not impossible, that the sample mean is 241.6. We can estimate the probability that $\overline{X} \geq 241.6$ by computing[11]

```
mean(Xbar >= mean(weight$mean200))

## [1] 0.0111
```

Since we are doing a two-sided test, we multiply by 2 to get the $p$-value.

```
2 * mean(Xbar >= mean(weight$mean200))

## [1] 0.0222
```

Since the $p$-value is less than 0.05, we reject the null hypothesis that the true mean weight estimate of six-year-olds is 200 grams. We get a similar result if we use `t.test`, though the $p$-value in `t.test` is somewhat higher.

```
t.test(weight$mean200, mu = 200)

##
##  One Sample t-test
##
## data:  weight$mean200
## t = 2.2456, df = 18, p-value = 0.03753
## alternative hypothesis: true mean is not equal to 200
## 95 percent confidence interval:
##  202.6782 280.4797
## sample estimates:
## mean of x
##  241.5789
```

[11]It can be argued that we should count the data that we obtained as one of the bootstrap samples and estimate the $p$-value via `2 * (sum(Xbar >= mean(weight$mean200)) + 1)/10001`. One advantage of this is that it prevents an estimated $p$-value of exactly 0.

### 8.8.2 Permutation tests

Another popular resampling technique is the **permutation test**. If we imagine that there was no difference between two populations, then the assignments of the grouping variables can be considered random. A permutation test performs random permutations of the grouping variables, and re-computes the test statistic. Under $H_0$ : "the response is independent of the grouping," we should relatively frequently get values as extreme as the value obtained in the sample. We reject $H_0$ if the proportion of times we get something as extreme as in the sample is less than $\alpha$.

Let's consider the `masks` data set in the `fosdata` package. The `nasal_swab` variable gives the viral load measured via a nasal swab for two types of influenza. We are interested in whether the mean viral loads are different in the two populations. Load the data with:

```
masks <- fosdata::masks
```

Next, we compute the mean viral load for virus type "A" minus the mean viral load for virus type "B."

```
xbar_A <- masks %>%
  filter(pcr_type == "A") %>%
  pull(nasal_swab) %>%
  mean()
xbar_B <- masks %>%
  filter(pcr_type == "B") %>%
  pull(nasal_swab) %>%
  mean()
xbar_diff <- xbar_A - xbar_B
xbar_diff
```

```
## [1] -2484975
```

Type B had a larger viral load on average. Let's look at a plot.

```
ggplot(masks, aes(x = pcr_type, y = nasal_swab)) +
  geom_boxplot(outlier.size = -1) +
  geom_jitter(height = 0, width = 0.2)
```

The distribution in Figure 8.8 is visibly skewed, so there should be at least some hesitation to use a $t$-test. Instead, we permute the age groups many times and recompute the test statistic.

```
mu_diff <- replicate(10000, {
  # generate new randomly permuted age groups
  masks$permuted_type <- sample(masks$pcr_type)

  # recompute the difference in means
  mu_A <- masks %>%
    filter(permuted_type == "A") %>%
    pull(nasal_swab) %>%
    mean()
  mu_B <- masks %>%
    filter(permuted_type == "B") %>%
    pull(nasal_swab) %>%
    mean()
```

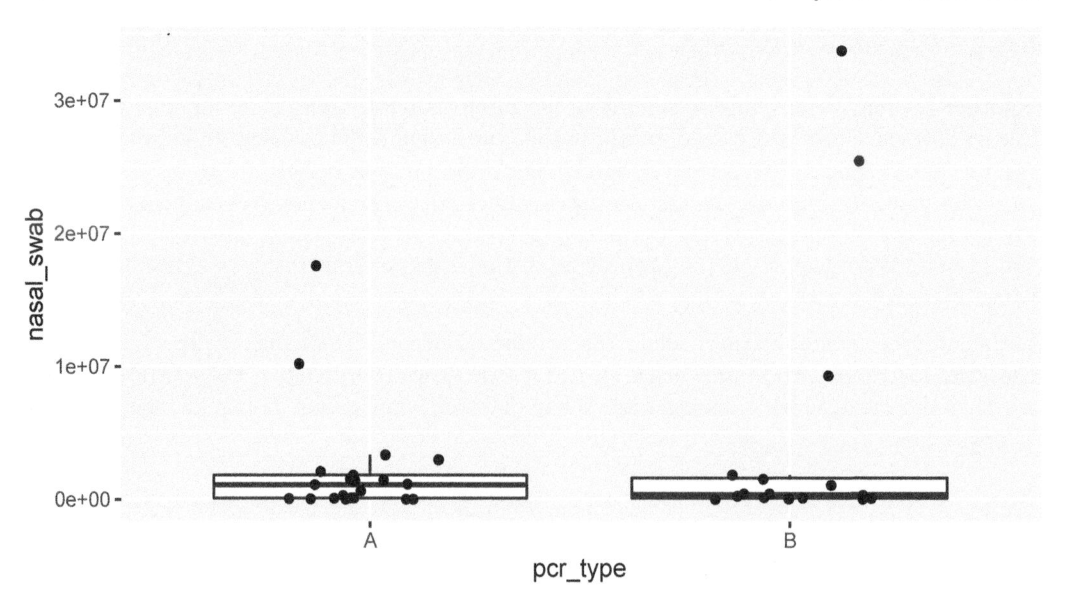


**FIGURE 8.8** Viral load as measured by a nasal swab for two types of influenza.

```
  mu_A - mu_B
})

data.frame(mu_diff) %>% ggplot(aes(x = mu_diff)) +
  geom_histogram() +
  geom_vline(xintercept = xbar_diff, color = "red")
```

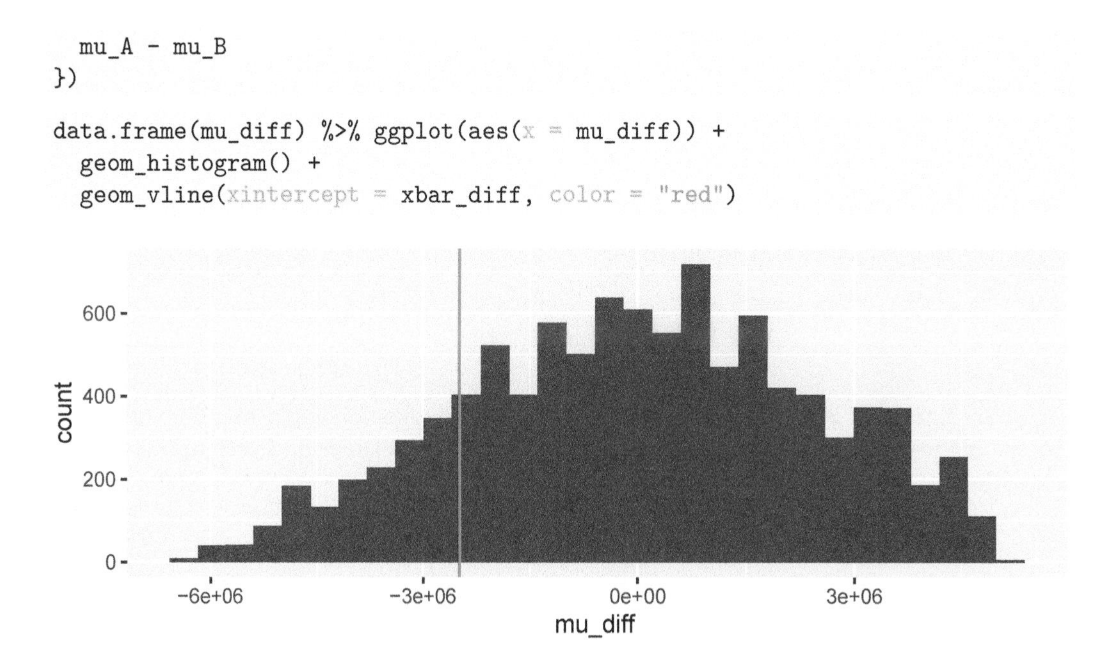


**FIGURE 8.9** Difference in means for 10,000 permutations of the A and B groups. Observed value shown in red.

We see from the histogram (Figure 8.9) that it is not that unusual to get a value of -2484975. To get a $p$-value, we compute the percentage of resamples to the left (in Figure 8.9) of the observed value and double it.

```
2 * mean(mu_diff < xbar_diff)

## [1] 0.3328
```

We have a $p$-value of 0.3328, and we would not reject the null hypothesis. In the $p$-value computation, we only test for `mu_diff < xbar_diff` because `xbar_diff` is negative. For a positive observation, the inequality would reverse. We double it because we are performing a two-sided test.

## Exercises

Exercises 8.1 – 8.5 require material through Section 8.1.

**8.1.** Let $X_1, \ldots, X_{12}$ be independent normal random variables with mean 1 and standard deviation 3. Simulate 10000 values of

$$T = \frac{\overline{X} - 1}{S/\sqrt{12}}$$

and plot the density function of $T$. On your plot, add a curve in blue for $t$ with 11 degrees of freedom. Also add a curve in red for the standard normal distribution. Confirm that the distribution of $T$ is $t$ with 11 df.

**8.2.** This exercise illustrates Theorem 8.1 in a specific instance. Let $a = 0$ and $b = 1$. Find the smallest value of $n$ such that $|P(a < X_n < b) - P(a < Z < b)| < .01$, where $X_n$ is a $t$ random variable with $n$ degrees of freedom and $Z$ is a standard normal random variable.

**8.3.** Let $X$ be a $t$ random variable with 6 degrees of freedom. Find a constant $k > 0$ such that

$$P(-k < X < k) = .8$$

**8.4.** Compute $P(Z > 2)$, where $Z$ is a standard normal random variable. Compute $P(T > 2)$ for $T$ having a $t$ distribution with each of 40, 20, 10, and 5 degrees of freedom. How does the degrees of freedom affect the area in the tail of the $t$ distribution?

**8.5.** Let $X_1, \ldots, X_{20}$ be independent normal random variables with mean 7 and unknown standard deviation. Find the value of $k$ such that

$$P(\overline{X} - k(S/\sqrt{n}) < \mu < \overline{X} + k(S/\sqrt{n})) = 0.99,$$

where $S$ as usual is the sample standard deviation.

Exercises 8.6 – 8.13 require material through Section 8.2.

**8.6.** The data set `fosdata::plastics` gives measurements of plastic microfibers found in snow. Give the 99% confidence interval for the mean diameter of plastic microfibers.

**8.7.** The data set `morley` is built into R. The `Speed` variable contains 100 measurements of the speed of light, conducted by A. A. Michelson in 1879. Measurements have 299000 km/s subtracted from them.

Compute the 95% confidence interval for the speed of light. Does your confidence interval contain the modern accepted speed of light of 299792 km/s?

**8.8.** Consider the `fosdata::brake` data set discussed in the chapter. Recall that `latency_p2` is the time it takes subjects to release the pedal once they have pressed it and the car has started to accelerate rather than decelerate. Find a 95% confidence interval for the P2 latency for young drivers based on the data.

**8.9.** The data set `fosdata::cows_small` gives the change in shoulder temperature of cows standing in the heat.

a. Give the 95% confidence interval for the shoulder temperature change after spraying with the `tk_12` nozzle.
b. Give the 95% confidence interval for the shoulder temperature change after no spraying (`control`).
c. Which of the two intervals is wider?
d. What is the standard deviation of temperature change for the `tk_12` condition? For the `control` condition?
e. How are your answers to parts (c) and (d) related?

**8.10.** Suppose that $X_1, \ldots, X_5$ is a random sample from a normal distribution with unknown mean and **known** standard deviation 2. If $\overline{X} = 3$, then find a 95% confidence interval for the mean using the fact that $\frac{\overline{X}-\mu}{\sigma/\sqrt{n}}$ is a standard normal random variable.

**8.11.** Some data randomly sampled from a normal population has a sample mean of 30. Match each interval on the left with one of the descriptions on the right:

| Interval | Description |
|---|---|
| a. (28.5, 31.5) | I. 90% confidence interval. |
| b. (28.0, 32.0) | II. 95% confidence interval. |
| c. (28.7, 31.3) | III. 99% confidence interval. |

**8.12.** Suppose you collect a random sample of size 20 from a normal population with unknown mean and standard deviation. Suppose $\overline{x} = 2.3$ and $s = 1.2$.

a. Find a 98% confidence interval for the mean.
b. The interval $(2.0, 2.6)$ is a confidence interval for the mean. What percent confidence interval is it?

**8.13.** Suppose a population has an exponential distribution with $\lambda = 0.5$. We can simulate drawing a sample of size 10 from this population with `rexp(10,0.5)` and compute a 95% confidence interval with `t.test(rexp(10,0.5),mu=2)$conf.int`.

a. What is the population mean $\mu$?
b. Write code to simulate 10000 confidence intervals and determine what percent of the time $\mu$ is in the 95% confidence interval.
c. Why is your answer different from 95%?

---

Exercises 8.14 – 8.19 require material through Section 8.3.

**8.14.** Consider the `react` data set from `ISwR`. Find a 98% confidence interval for the mean difference in measurements of the tuberculin reaction. Does the mean differ significantly from zero according to a $t$-test? Interpret.

**8.15.** The data `fosdata::weight_estimate` gives children's estimates of weights lifted by actors (see Example 8.22). State and carry out a hypothesis test that the mean of the children's estimates of the 100 g object differ from the actual weight, 100 g. Repeat this test for the 200, 300, and 400 g objects. Which of the estimates were significantly different from the correct weights?

**8.16.** Consider the `bp.obese` data set from `ISwR`. It contains data from a random sample of Mexican American adults in a small California town. Consider only the blood pressure data for this problem. Normal diastolic blood pressure is 120. Is there evidence to suggest that the mean blood pressure of the population differs from 120? What is the population in this example?

**8.17.** For the `ISwR::bp.obese` data set, consider the `obese` variable. What is the natural null hypothesis? Is there evidence to suggest that the obesity level of the population differs from the null hypothesis?

**8.18.** Suppose you collect a random sample of size 20 from a normal population with unknown mean and standard deviation. You wish to test $H_0 : \mu = 2$ versus $H_a : \mu \neq 2$.

a. The region $|T| > 1.6$ is a rejection region for this hypothesis test. What is the $\alpha$ level of the rejection region?
b. Find a rejection region that corresponds to $\alpha = 0.005$.

**8.19.** Suppose we wish to test $H_0 : \mu = 1$ versus $H_a : \mu \neq 1$. Let $X_1, \ldots, X_5$ be a random sample from a normal population with known standard deviation $\sigma = 2$. We construct the test statistic $Z = \frac{\overline{X}-1}{\sigma/\sqrt{n}}$.

a. If the rejection region is $|Z| > 2.2$, then what is the significance level of the test?
b. What would a rejection region be that corresponds to a significance level of $\alpha = 0.01$?

---

Exercise 8.20 requires material through Section 8.4.

**8.20.** Suppose that a **dishonest** statistician is doing a $t$-test of $H_0 : \mu = 0$ at the $\alpha = 0.05$ level. The statistician waits until they get the data to specify the alternative hypothesis. If $\overline{X} > 0$, then they choose $H_a : \mu > 0$ and if $\overline{X} < 0$, they choose $H_a : \mu < 0$. Suppose the statistician collects 20 independent samples and the underlying population is standard normal. Use simulation to confirm that the null hypothesis is rejected 10% of the time.

---

Exercises 8.21 – 8.24 require material through Section 8.5.

**8.21.** This problem illustrates that `t.test` performs as expected when the underlying population is normal.

a. Choose 10 random values of $X$ having the Normal(4, 1) distribution. Use `t.test` to compute the 95% confidence interval for the mean. Is 4 in your confidence interval?
b. Replicate the experiment in part (a) 10,000 times and compute the percentage of times the population mean 4 was included in the confidence interval.

**8.22.** Mackowiak et al.[12] studied the mean normal body temperature of healthy adults. One hundred forty-eight subjects' body temperatures were measured one to four times daily for 3 consecutive days for a total of 700 temperature readings. Let $\mu$ denote the true mean

[12]Mackowiak, Wasserman, and Levine, "A Critical Appraisal of 98.6 Degrees F, the Upper Limit of the Normal Body Temperature, and Other Legacies of Carl Reinhold August Wunderlich."

normal body temperature of healthy adults and $X_1, \ldots, X_{700}$ the temperature readings from the study. Explain why

$$\frac{\overline{X} - \mu}{S/\sqrt{n}}$$

is not a $t$ random variable with 699 degrees of freedom.

**8.23.** This problem explores the accuracy of `t.test` when the underlying population is skewed.

a. Choose 10 random values of $X$ having an exponential distribution with rate 1/4. Use `t.test` to compute the 95% confidence interval for the mean. Is the true mean, 4, in your confidence interval?
b. Replicate the experiment in part (a) 10,000 times and compute the percentage of times that the population mean 4 was included in the confidence interval. Explain why this number is not 95%.
c. Repeat part (b) using 100 values of $X$ to make each confidence interval. What percentage of times did the confidence interval contain the population mean? Why is it closer to 95%?

**8.24.** This problem explores the accuracy of `t.test` when the data is not independent. Suppose $X_1, \ldots, X_{20}$ is a random sample from normal population with mean 0 and standard deviation 1. Let $Y_1 = X_1 + X_2$, $Y_2 = X_2 + X_3, \ldots,$ and $Y_{19} = X_{19} + X_{20}$. Note that $Y_1, \ldots, Y_{19}$ are **not** independent, but they **do** have mean 0.

Estimate the effective type I error rate when testing $H_0 : \mu = 0$ versus $H_a : \mu \neq 0$ at the $\alpha = .05$ level using $Y_1, \ldots, Y_{19}$.

(Hint: `Y <- X[1:19]+X[2:20]` computes $Y$ as needed for this simulation.)

---

Exercises 8.25 – 8.40 require material through Section 8.6.

**8.25.** Each of these studies has won the prestigious Ig Nobel prize.[13] For each, state the null hypothesis.

a. Marina de Tommaso, Michele Sardaro, and Paolo Livrea, for measuring the relative pain people suffer while looking at an ugly painting, rather than a pretty painting, while being shot (in the hand) by a powerful laser beam.
b. Atsuki Higashiyama and Kohei Adachi, for investigating whether things look different when you bend over and view them between your legs.
c. Patricia Yang, David Hu, Jonathan Pham, and Jerome Choo, for testing the biological principle that nearly all mammals empty their bladders in about 21 seconds.

**8.26.** The didgeridoo is an Indigenous Australian musical instrument. In a 2006 study[14] that recently won the Ig Nobel Peace Prize, researchers investigated didgeridoo playing as a treatment for sleep apnoea, a breathing disorder that interferes with sleep. The researchers separated 25 patients into a treatment group that received didgeridoo lessons and a control group that did not. From the paper:

> Participants in the didgeridoo group practiced an average of 5.9 days a week (SD 0.86) for 25.3 minutes (SD 3.4). Compared with the control group in the didgeridoo group daytime sleepiness (difference -3.0, 95% confidence interval -5.7 to -0.3, $P = 0.03$)

[13] See https://www.improbable.com/ig/

[14] Milo A Puhan et al., "Didgeridoo Playing as Alternative Treatment for Obstructive Sleep Apnoea Syndrome: Randomised Controlled Trial," *BMJ: British Medical Journal* 332, no. 7536 (2006): 266–68.

and apnoea-hypopnoea index (difference -6.2, -12.3 to -0.1, $P = 0.05$) improved significantly and partners reported less sleep disturbance (difference -2.8, -4.7 to -0.9, $P < 0.01$). There was no effect on the quality of sleep (difference -0.7, -2.1 to 0.6, $P = 0.27$). The combined analysis of sleep related outcomes showed a moderate to large effect of didgeridoo playing.

a. The paper reports four measurements. All four differences are reported as negative numbers. What does that tell you?
b. Only one of their reported measures did not show a significant improvement for the didgeridoo group. Which one? How do you know it was not significant?
c. Which of their reported measures was the most significant evidence in favor of didgeridoo playing?

**8.27.** The data set `fosdata::chimps` was introduced in Example 7.8. The variable `grey_score_avg` is a measurement of the grey hair in chimpanzees in the study. Two subspecies of chimps are in this study: *verus* and *schweinfurthii*. Is there evidence at the $\alpha = 0.05$ level to suggest that the mean grey score differs between the two subspecies?

**8.28.** Experimenters measured the leg strength of adults using two methods: a stationary isometric dynamometer and a Wii Balance Board. The strength (in kg) of each adult is reported in `fosdata::leg_strength`.

Test for a difference in the mean leg strength as measured with these two devices.

**8.29.** The data set `cigs` from the `fosdata` package contains measurements of cigarette brands that were tested by the U.S. Federal Trade Commission for chemical content. Check with a plot that the nicotine (`nic`) and `tar` variables are reasonably bump shaped and symmetric, without major outliers.

a. Test for a difference in nicotine content between filtered and unfiltered cigarette brands. State your result with a $p$-value.
b. Test for a difference in tar content between filtered and unfiltered cigarette brands. State your result with a $p$-value.
c. Test for a difference in nicotine content between menthol and non-menthol cigarette brands. State your result with a $p$-value.
d. Test for a difference in tar content between menthol and non-menthol cigarette brands. State your result with a $p$-value.

**8.30.** For the `ISwR::bp.obese` data set from Exercise 8.16, is there evidence to suggest that the mean blood pressure of Mexican American men is different from that of Mexican American women in this small California town? Would it be appropriate to generalize your answer to the mean blood pressure of men and women in the USA?

**8.31.** Consider the `ex0112` data in the `Sleuth3` package. The researchers randomly assigned 7 people to receive fish oil and 7 people to receive regular oil, and measured the decrease in their diastolic blood pressure. Is there sufficient evidence to conclude that the mean decrease in blood pressure of patients taking fish oil is different from the mean decrease in blood pressure of those taking regular oil?

**8.32.** The `barnacles` data in `fosdata` has measurements of barnacle density on various coral reefs. Some reefs are considered deep and some shallow, as given by the `deep` variable. For this exercise, only consider the reefs with location `FGB`, the Flower Garden Banks in the Gulf of Mexico.

a. Check that `barnacle_density` is right-skewed, but that the log of `barnacle_density` is close to normally distributed.

b. Perform a two sample $t$-test to determine if the log of `barnacle_density` is different for deep and shallow reefs. Report your results with a $p$-value.

**8.33.** Consider the `ex0333` data in the `Sleuth3` package. This is an observational study of brain size and litter size in mammals. Create a boxplot of the brain size for large litter and small litter mammals. Does the data look normal? Create a histogram or a qq plot. Repeat after taking the logs of the brain sizes. Is there evidence to suggest that there is a difference in mean brain sizes of large litter animals and small litter animals?

**8.34.** Consider the `case0202` data in the `Sleuth3` package. This data is from a study of twins, where one twin was schizophrenic and the other was not. The researchers used magnetic resonance imaging to measure the volumes (in $cm^3$) of several regions and subregions of the twins' brains. State and carry out a hypothesis test that there is a difference in the volume of these brain regions between the Affected and the Unaffected twins.

**8.35.** Consider the `cows_small` data set in the `fosdata` package that was discussed in this chapter.

a. Is there sufficient evidence to conclude that the mean temperature change associated with the TK-12 nozzle is less than the mean temperature change associated with not spraying the cows at all; that is, with the control?

b. Is there sufficient evidence to conclude that the mean temperature change associated with the TK-12 nozzle is different than the mean temperature change associated with the TK-0.75 nozzle?

---

The next exercises use the data set `fosdata::child_tasks`, which comes from a study by Chan and Morgan[15] that studies the length of time needed for children to complete various tasks.

**8.36.** Consider the `child_tasks` data set in the `fosdata` package. Let $\mu_1$ be the mean time for girls to finish the day night task, and let $\mu_2$ be the mean time for boys to finish the same task. Perform a two-sided hypothesis test of $H_0 : \mu_1 = \mu_2$ at the $\alpha = .01$ level.

**8.37.** Consider the `child_tasks` data set in the `fosdata` package. It seems likely that age of participant is an important factor in the length of time the children take to perform tasks. Since the sampling method for obtaining participants might not be a true random sample, we might wonder whether the mean age of boys and girls via the sampling method for this experiment are the same. Let $\mu_1$ be the mean age for girls and let $\mu_2$ be the mean age for boys. Perform a two-sided hypothesis test of $H_0 : \mu_1 = \mu_2$ at the $\alpha = .05$ level.

**8.38.** Consider the `child_tasks` data set in the `fosdata` package. Let $\mu_1$ be the mean time for children in the 7 year old group to finish the day night task, and let $\mu_2$ be the mean time for children in the 6 year old group to finish the same task. Perform a two-sided hypothesis test of $H_0 : \mu_1 = \mu_2$ at the $\alpha = .01$ level.

**8.39.** Consider the `child_tasks` data set in the `fosdata` package. The "card sort" task asks children to sort cards either by the shape pictured on the card or the color of that shape. They first do one task, then switch to the other task. Children's times at these tasks are recorded in the `card_sort_preswitch_time_secs` and `card_sort_postswitch_time_secs` variables.

Let $\mu_1$ be the mean time for the pre-switch task and $\mu_2$ be the mean time for the post-switch

[15] Amy Y C Chan and Sarah-Jane Morgan, "Assessing Children's Cognitive Flexibility with the Shape Trail Test." *PLOS One* 13, no. 5 (2018): e0198254.

task. Perform a two-sided hypothesis test of $H_0 : \mu_1 = \mu_2$ at the $\alpha = .01$ level. Are the children faster pre- or post-switch?

**8.40.** Example 8.14 tested the difference in `latency_p2` times between young and older subjects, with data from `fosdata::brake`. Repeat the analysis with the largest older outlier removed. Does removing the largest older outlier change the conclusion of your test at the $\alpha = .05$ level?

---

Exercises 8.41 – 8.50 require material through Section 8.7.

**8.41.** In a jury trial, the null hypothesis is that the defendant is innocent ("presumed innocent until proven guilty"). The alternate hypothesis is that the defendant is guilty.

a. If the defendant actually committed the crime but is found not guilty, what type of error is that?
b. If the defendant is innocent but is found guilty, what type of error is that?
c. In your opinion, which type of error is more serious, and why?

**8.42.** In a drug trial, patients are given a drug to test if it will cure their illness.

a. What is the null hypothesis for a drug trial?
b. What would a type I error be for a drug trial?
c. What would a type II error be for a drug trial?
d. What are the implications for each type of error for ill patients?
e. What are the implications for each type of error for drug manufacturers?

**8.43.** Suppose you wish to test the side effects of a new vaccine by forming two groups of subjects, one that gets the vaccine and another that gets a placebo. You would like to detect a $1°F$ change in body temperature at the $\alpha = 0.05$ level of significance with a power of 99%. Data on human body temperature (such as `fosdata::normtemp`) suggests that the standard deviation of body temperature is around $0.73°F$. How many subjects will you need in each group?

**8.44.** The compressive strength of concrete is the pressure (N/mm$^2$) at which it fails. The compressive strength of a particular type of concrete follows a Norm(65, 6) distribution. When concrete is poured at a construction site, five cubes are poured and taken to a lab for testing. Suppose the actual concrete pour has a mean compressive strength of only 60. What power does this test have to detect the difference in means? Use a 0.05 significance level.

**8.45.** Suppose a population has mean 0 and standard deviation $\sqrt{3}$. We want to test the (false) null hypothesis $H_0 : \mu = 1$ with a sample size of $n = 40$ and $\alpha = 0.05$.

a. Suppose the population is Norm$(0, \sqrt{3})$. What is the power of a $t$-test to detect this difference in means?
b. Suppose the population is $t$ with 3 degrees of freedom, which has standard deviation $\sqrt{3}$. Perform a simulation to determine the power of the $t$-test to detect the difference in means.

**8.46.** The purpose of this exercise is to examine the effect of sample size, effect size, and significance level on the power of a test.

a. Fix an effect size of $d = 0.2$ and a significance level of $\alpha = 0.05$, and compute the power of a two-sided one sample $t$-test for $n = 200$ and $n = 400$. What happens to the power?
b. Fix a significance level of $\alpha = 0.05$ and $n = 200$, and compute the power of a two-sided one sample $t$-test for $d = 0.2$ and $d = 0.4$. What happens to the power?

c. Fix $n = 200$ and an effect size of $d = 0.2$, and compute the power of a two-sided one sample $t$-test for $\alpha = 0.05$ and $\alpha = 0.1$. What happens to the power?

---

The next two exercises use the data set `fosdata::wrist`, which comes from a study by Raittio et al.[16] that considers two different methods for applying casts to wrist fractures.

**8.47.** The authors performed a power analysis in order to determine sample size. They wanted a power of 80% at the $\alpha = .05$ level, and they assumed that 30% of their patients would drop out. Previous studies had shown that the standard deviation of the `prwe12m` variable would be about 14. They concluded that they needed 40 participants in each group. What difference in means did they use in their power computation? (You can check your answer by looking up their paper.)

**8.48.** The authors used the `fosdata::wrist` data set to find a 95% confidence interval for the difference in the mean of `prwe12m` between the functional cast position and the volar-flexion and ulnar deviation cast position. They **assumed** equal variances in the two populations to find the 95% confidence interval.

a. Find the 95% confidence interval of the mean difference assuming equal variances.
b. Compare your answer to the result in the published paper.
c. Was assuming equal variance justified?

**8.49.** This exercise explores how the $t$-test changes when data values are transformed.

a. Suppose you wish to test $H_0 : \mu = 0$ versus $H_a : \mu \neq 0$ using a $t$-test. You collect data $x_1 = -1, x_2 = 2, x_3 = -3, x_4 = -4$ and $x_5 = 5$. What is the $p$-value?
b. Now, suppose you multiply everything by 2: $H_0 : \mu = 0$ versus $H_a : \mu \neq 0$ and your data is $x_1 = -2$, $x_2 = 4$, $x_3 = -6$, $x_4 = -8$, and $x_5 = 10$. What happens to the $p$-value? (Try to answer this without using R. Check your answer using R, if you must.)
c. Now, suppose you square the magnitudes of everything. $H_0 : \mu = 0$ versus $H_a : \mu \neq 0$, and your data is $x_1 = -1$, $x_2 = 4$, $x_3 = -9$, $x_4 = -16$, and $x_5 = 25$. What happens to the $p$-value?

**8.50.** This problem explores how the $t$-test behaves in the presence of an outlier.

a. Create a data set of 20 random values $x_1, \ldots, x_{20}$ with a normal distribution with mean 10 and sd 1. Replace $x_{20}$ with the number 1000. Perform a $t$-test with $H_0 : \mu = 0$, and observe the value of $t$. It should be close to 1. Is the $t$-test able to find a significant difference between the mean of this data and 0?

   The next parts of this problem ask you to prove that the $t$-test statistic is always close to 1 in the presence of a large outlier.

b. Assume that $x_1, \ldots, x_{n-1}$ are not changing, but $x_n$ varies. Let $\overline{x} = \frac{1}{n}\sum_{i=1}^{n} x_i$ as usual. Show that
$$\lim_{x_n \to \infty} \frac{\overline{x}}{x_n} = \frac{1}{n}.$$
c. Let the sample variance be $s^2 = \frac{1}{n-1}\sum_{i=1}^{n}(x_i - \overline{x})^2$ as usual. Show that
$$\lim_{x_n \to \infty} \frac{s^2}{x_n^2} = \frac{1}{n}.$$

[16] Lauri Raittio et al., "Two Casting Methods Compared in Patients with Colles' Fracture: A Pragmatic, Randomized Controlled Trial," *PLOS One* 15, no. 5 (May 2020): e0232153.

d. Finally show that

$$\lim_{x_n \to \infty} \frac{\overline{x} - \mu_0}{s/\sqrt{n}} = 1,$$

where $\mu_0$ is any real number. (Hint: divide top and bottom by $x_n$, then use parts (b) and (c)).

e. What does this say about the ability of a $t$-test to reject $H_0 : \mu = \mu_0$ at the $\alpha = .05$ level as $x_n \to \infty$?

---

Exercises 8.51 – 8.55 require material through Section 8.8.

**8.51.** Consider the `weight_estimate` data from the `fosdata` package. Construct a bootstrap 95% confidence interval for the true mean weight that adults estimate for the 400-gram object.

**8.52.** Consider the `masks` data from the `fosdata` package. Thirty-seven patients with confirmed influenza were given nasal swabs and the viral load was computed in the variable `nasal_swab`. Our goal is to use this data to make a 95% confidence interval for the mean viral load in influenza patients.

a. Is it appropriate to use a the $t$ statistic and `t.test` to make a 95% confidence interval? Explain.
b. Create a bootstrap 95% confidence interval for the true mean viral load for influenza patients.

**8.53.** Consider the `brake` data from the `fosdata` package. Construct a bootstrap 95% confidence interval for the **median** of the `latency_p1` variable for older drivers.

**8.54.** Consider the `react` data from the `ISwR` package. Construct a bootstrap 98% confidence interval for the true mean difference in estimates of reaction size.

**8.55.** The data set `morley` is built into R. The `Speed` variable contains 100 measurements of the speed of light, conducted by A. A. Michelson in 1879. Measurements have 299000 km/s subtracted from them.

The currently accepted speed of light is 299792 kilometers/second. Perform a bootstrap hypothesis test at the $\alpha = .05$ level of $H_0 : \mu = 792$ versus $H_a : \mu \neq 792$.

# 9

# *Rank Based Tests*

In a 2013 paper, D.K. Milton et al.[1] studied the impact of wearing a surgical facemask on exhaled aerosol droplets for patients with influenza. Each subject performed two 30-minute trials, exhaling into a collection device (Figure 9.1). The subjects were tested once while wearing a facemask and once without, and the researchers counted the number of copies of the influenza virus in their fine particle droplets.

**FIGURE 9.1** Inlet cone for the human exhaled breath air sampler used to measure influenza virus, from Milton et al.

Virus counts are in the variables `mask_fine` and `no_mask_fine` within the data set `fosdata::masks`. The difference `no_mask_fine - mask_fine` will be positive if the subject exhaled less influenza virus while wearing a mask.

```
masks <- fosdata::masks %>%
  mutate(virus_difference = no_mask_fine - mask_fine)
summary(masks$virus_difference)
```

```
##    Min. 1st Qu.  Median    Mean 3rd Qu.    Max.
##    -531       2      75    3809     417  102348
```

The `virus_difference` variable has two large positive outliers (26422 and 102348) that

[1] Donald K Milton et al., "Influenza Virus Aerosols in Human Exhaled Breath: Particle Size, Culturability, and Effect of Surgical Masks," *PLOS Pathogens* 9, no. 3 (March 2013): 1–7, https://doi.org/10.1371/journal.ppat.1003205.

DOI: 10.1201/9781003004899-9

make visualization challenging. Figure 9.2 shows a dotplot[2] of the variable with those two values removed.

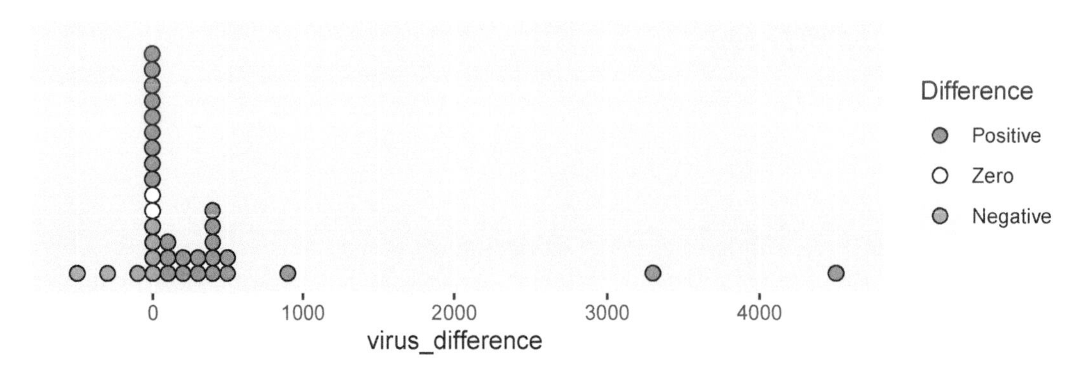


**FIGURE 9.2** Each dot represents one patient and gives the difference in fine particle virus count between their masked and maskless trials. A positive value means the exhaled virus count was larger when not wearing a mask. Two large positive points are omitted.

Even without the largest values, it is obvious from the plot that most patients shed fewer virus particles while wearing a mask. The seven patients who shed more particles while wearing a mask had smaller differences in viral count than the 28 who shed fewer particles while wearing a mask. However, the presence of outliers (the two shown in the figure and the two even larger ones not shown) renders the $t$-test powerless to detect a difference between masked and unmasked patients. A paired two-sample $t$-test on this data results in a $p$-value of 0.19, not significant.

Wilcoxon rank based tests take the data and sort it from smallest to largest, replacing the actual measurements with their ranks. In this example, the ranks range from 1 to 37, for the 37 subjects of the study. Using ranks instead of the actual data values makes the test resistant to outliers while still accounting for the preponderance of positive values and the fact that those values tend to be larger.

```
wilcox.test(masks$mask_fine, masks$no_mask_fine, paired = TRUE)
```

```
##
##  Wilcoxon signed rank test with continuity correction
##
## data:  masks$mask_fine and masks$no_mask_fine
## V = 88.5, p-value = 0.000214
## alternative hypothesis: true location shift is not equal to 0
```

We interpret this result as saying that the median difference in viral count is positive, with $p = 0.0002$. The low $p$-value suggests that masks reduce exhaled viral particles.

In this chapter we discuss rank based tests that can be used, and are still effective, on a population with outliers. Rank based tests may also be used when the measured quantity is **ordinal** rather than **numeric**.

[2]A *dotplot* is similar to a histogram, with each data point shown as an individual dot. This one was made with `geom_dotplot`.

## 9.1 One sample Wilcoxon signed rank test

The *Wilcoxon signed rank test* tests $H_0 : m = m_0$ versus $H_a : m \neq m_0$, where $m$ is the median of the underlying population. We will assume that the data is **centered symmetrically** around the median (so that the median is also the mean). There can be outliers, bimodality or any kind of tails.

Let's look at how the test works by hand by examining a simple, made up data set. Suppose you wish to test

$$H_0 : m = 6 \qquad \text{vs} \qquad H_a : m \neq 6$$

You collect the data $x_1 = 15, x_2 = 7, x_3 = 3, x_4 = 10$ and $x_5 = 13$. The test works as follows:

1. Compute $y_i = x_i - m_0$ for each $i$. Here $m_0 = 6$ and we get $y_1 = 9, y_2 = 1, y_3 = -3, y_4 = 4$ and $y_5 = 7$.
2. Let $R_i$ be the rank of the absolute value of $y_i$. That is, $R_1 = 5, R_2 = 1, R_3 = 2, R_4 = 3, R_5 = 4$ since $|y_1|$ is largest, $|y_2|$ is smallest, etc.
3. Let $r_i$ be the *signed* rank of $R_i$; i.e., $r_i = R_i \times \text{sign}(y_i)$, so $r_1 = 5, r_2 = 1, r_3 = -2, r_4 = 3, r_5 = 4$.
4. Add all of the positive ranks. We get $r_1 + r_2 + r_4 + r_5 = 13$. That is the test statistic for this test, and it is traditionally called $V$.
5. Compute the $p$-value for the test, which is the probability that we get a test statistic $V$ which is this, or more, extreme relative to the expected value under the assumption of $H_0$.

In order to perform the last step, we need to understand the sampling distribution of $V$, under the assumption of the null hypothesis, $H_0$. The ranks of our five data points will always be the numbers 1, 2, 3, 4, and 5. When $H_0$ is true, each data point is equally likely to be positive or negative, and its corresponding rank will be included in the sum in item 4 above half of the time. So, the expected value is

$$E(V) = \frac{1}{2} \cdot 1 + \frac{1}{2} \cdot 2 + \frac{1}{2} \cdot 3 + \frac{1}{2} \cdot 4 + \frac{1}{2} \cdot 5 = \frac{15}{2} = 7.5$$

In our example, $V = 13$, which is 5.5 away from the expected value of 7.5.

For the probability distribution of $V$, Table 9.1 lists all $2^5 = 32$ possibilities of how the ranks could be signed. Since we have assumed that the distribution is centered about its mean, each of the possibilities is equally likely. Therefore, we can compute the proportion of rows in the table that lead to a test statistic at least 5.5 away from the expected value of 7.5.

The last column in Table 9.1 is "Yes" if the sum of positive ranks is greater than or equal to 13 or less than or equal to 2. As you can see, we have that 6 of the 32 possibilities are at least as far away from the test statistic $V = 13$ as our data, so the $p$-value would be $\frac{6}{32} = .1875$.

Let's check it with the built-in R command.

```
wilcox.test(c(15, 7, 3, 10, 13), mu = 6)
```

**TABLE 9.1** All possible sums of positive signed ranks.

| r1 | r2 | r3 | r4 | r5 | Sum of positive ranks | Far from 7.5? |
|---|---|---|---|---|---|---|
| -1 | -2 | -3 | -4 | -5 | 0 | Yes |
| -1 | -2 | -3 | -4 | 5 | 5 | |
| -1 | -2 | -3 | 4 | -5 | 4 | |
| -1 | -2 | -3 | 4 | 5 | 9 | |
| -1 | -2 | 3 | -4 | -5 | 3 | |
| -1 | -2 | 3 | -4 | 5 | 8 | |
| -1 | -2 | 3 | 4 | -5 | 7 | |
| -1 | -2 | 3 | 4 | 5 | 12 | |
| -1 | 2 | -3 | -4 | -5 | 2 | Yes |
| -1 | 2 | -3 | -4 | 5 | 7 | |
| -1 | 2 | -3 | 4 | -5 | 6 | |
| -1 | 2 | -3 | 4 | 5 | 11 | |
| -1 | 2 | 3 | -4 | -5 | 5 | |
| -1 | 2 | 3 | -4 | 5 | 10 | |
| -1 | 2 | 3 | 4 | -5 | 9 | |
| -1 | 2 | 3 | 4 | 5 | 14 | Yes |
| 1 | -2 | -3 | -4 | -5 | 1 | Yes |
| 1 | -2 | -3 | -4 | 5 | 6 | |
| 1 | -2 | -3 | 4 | -5 | 5 | |
| 1 | -2 | -3 | 4 | 5 | 10 | |
| 1 | -2 | 3 | -4 | -5 | 4 | |
| 1 | -2 | 3 | -4 | 5 | 9 | |
| 1 | -2 | 3 | 4 | -5 | 8 | |
| 1 | -2 | 3 | 4 | 5 | 13 | Yes |
| 1 | 2 | -3 | -4 | -5 | 3 | |
| 1 | 2 | -3 | -4 | 5 | 8 | |
| 1 | 2 | -3 | 4 | -5 | 7 | |
| 1 | 2 | -3 | 4 | 5 | 12 | |
| 1 | 2 | 3 | -4 | -5 | 6 | |
| 1 | 2 | 3 | -4 | 5 | 11 | |
| 1 | 2 | 3 | 4 | -5 | 10 | |
| 1 | 2 | 3 | 4 | 5 | 15 | Yes |

```
##
##  Wilcoxon signed rank exact test
##
## data:  c(15, 7, 3, 10, 13)
## V = 13, p-value = 0.1875
## alternative hypothesis: true location is not equal to 6
```

We see that our test statistic $V = 13$ and the $p$-value is 0.1875, just as we calculated.

In general, if you have $n$ data points, the expected value of $V$ is $E(V) = \frac{n(n+1)}{4}$ (Exercise 9.3). To deal with ties, give each data point the average rank of the tied values.

The sampling distribution of the test statistic $V$ under the null hypothesis is a built-in R distribution with root `signrank`. As usual, this function has prefixes `d`, `p`, `q`, and `r`, which correspond to the pmf, cdf, quantile function, and random generator, respectively. So, in the above example, we also could have computed the $p$-value as the probability that $V$ is in $\{0, 1, 2, 13, 14, 15\}$ as

```
sum(dsignrank(c(0:2, 13:15), 4))
```

```
## [1] 0.1875
```

When the sample size $n$ is large, the sampling distribution of the test statistic $V$ is approximately normal with mean $\frac{n(n+1)}{4}$ and variance $\frac{n(n+1)(2n+1)}{24}$. Figure 9.3 shows the distribution of $V$ with $n = 10$ with a superimposed normal curve that has mean 27.5 and standard deviation 9.81.

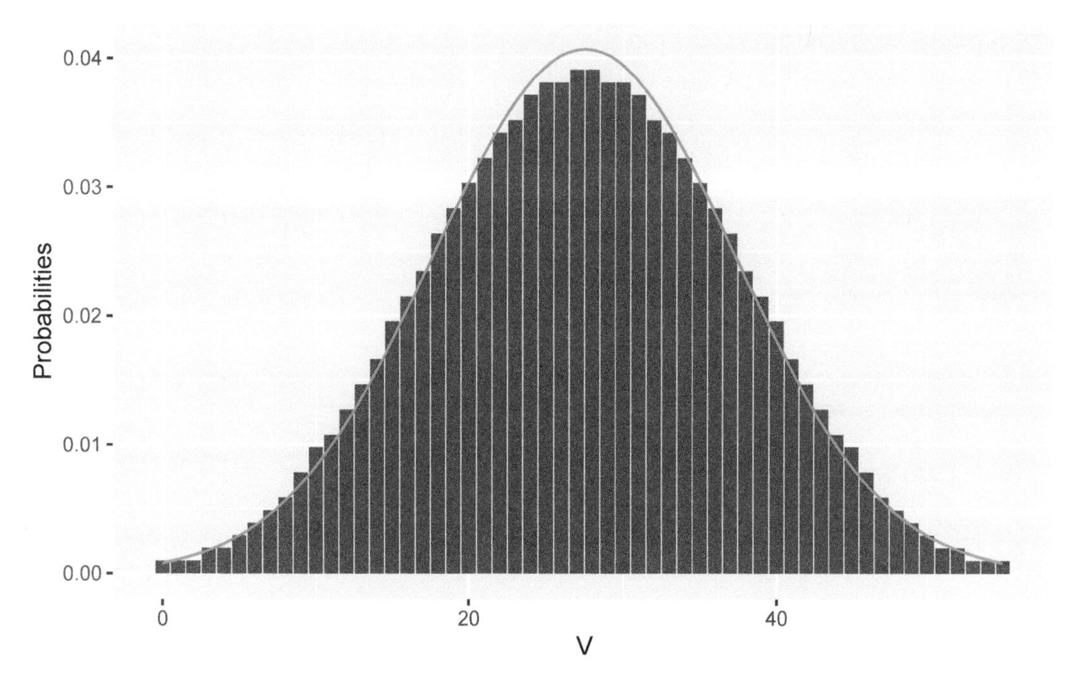


**FIGURE 9.3** pmf of $V$ when $n = 10$ with superimposed normal distribution.

**Example 9.1.** Consider the `normtemp` data set in the `fosdata` package. Recall that this data is the gender, body temperature and heart rate of 130 patients. We are interested in whether the mean or median body temperature is 98.6. A $t$-test looks like this:

```
normtemp <- fosdata::normtemp
t.test(normtemp$temp, mu = 98.6)
```

```
##
##  One Sample t-test
##
## data:  normtemp$temp
## t = -5.4548, df = 129, p-value = 2.411e-07
## alternative hypothesis: true mean is not equal to 98.6
## 95 percent confidence interval:
##  98.12200 98.37646
## sample estimates:
## mean of x
##  98.24923
```

resulting in a $p$-value of $2 \times 10^{-7}$. If we instead perform the Wilcoxon signed rank test, we get

```
wilcox.test(normtemp$temp, mu = 98.6)
```

```
##
##  Wilcoxon signed rank test with continuity correction
##
## data:  normtemp$temp
## V = 1774.5, p-value = 1.174e-06
## alternative hypothesis: true location is not equal to 98.6
```

The $p$-value still leads to the same conclusion at most reasonable $\alpha$ levels, but it is about 5 times as large as the $t$-test.

Both the one sample $t$-test and the Wilcoxon signed rank test are estimating the probability of obtaining data "this unlikely" or "more unlikely," given that the null hypothesis is true. However, the two tests are summarizing what it means to be "this unlikely" in different ways. We will explore the pros and cons of the Wilcoxon signed rank test versus the one sample $t$-test in Sections 9.3 and 9.4.

## 9.2 Two sample Wilcoxon tests

In this section, we will explain how to use `wilcox.test` to compare two populations. We will split the discussion into the cases when the data is **paired** and when it is **independent**.

### 9.2.1 Paired two sample test

In many studies, observations are naturally paired. Each observation consists of two measurements coming from two different populations. For example, the two measurements might be test scores before and after studying, opinions of a married couple, or pairs of agricultural fields matched for similar soil and weather characteristics. Any setting from Section 8.6 where a paired $t$-test is appropriate may also be studied with a paired Wilcoxon test.

The paired two sample Wilcoxon test has the following hypotheses and assumptions:

- $H_0$ : The two populations have the same distribution.
- $H_a$ : The two populations have different distributions.

**Assumptions 9.1 (for paired Wilcoxon test).**

1. *Subtracting measured values is possible and meaningful. We will need to assume that a difference of 2 is less than a difference of 3, for example.*
2. *The only dependence is between pairs of data points, one from each population. Unpaired observations are independent.*

Under these assumptions, the **differences** in the paired values in the two populations will be a symmetric distribution under the null hypothesis. Therefore, assuming the null hypothesis is true, we can use the Wilcoxon signed rank test on the differences of the values between the two populations.

**Example 9.2.** The `flint` data set in the `fosdata` package gives the results of tap water lead testing during the Flint, Michigan water crisis in 2015.[3] In the study, households filled three sample collection bottles from their faucets. One bottle was filled at "first draw," after the faucet had been off for 6 hours. The second and third bottles were filled after the faucet had been running for 45 seconds, and after it had been running for 2 minutes. All bottles were sent to a lab at Virginia Tech for lead testing.

Does the length of time that a faucet is left on influence the measurement of lead in the water coming from the faucets?

We restrict to two populations: the amount of lead in the water on first draw (`Pb1`) and the amount of lead in the water two minutes later (`Pb3`). This is paired data. It would be wrong to assume independence between these two measurements within one household in this experiment. A faucet that has high levels of lead at first draw is also likely to have higher levels of lead two minutes later.

A careful look at the data reveals two houses that have notes saying they were tested twice. It is unreasonable to expect those pairs of data points to be independent of each other, so we remove the observations that correspond to those houses. This is a reasonable thing to do, especially if the houses weren't tested twice based on the outcome of the first test.

```
flint <- fosdata::flint
flint <- flint %>%
  filter(Notes == "")
```

The null hypothesis is that the distribution of `Pb1` (lead draw in first sample) and `Pb3` (lead draw after 2 minutes) are the same. Under the null hypothesis, the difference between values in `Pb1` and `Pb3` will be symmetric, so the assumptions of the Wilcoxon signed rank test are met on the differences.

We start by providing a histogram of the differences.

```
flint %>% ggplot(aes(x = Pb1 - Pb3)) +
  geom_histogram() +
  labs(title = "Difference in lead levels: first draw and after 2 minutes")
```

[3] Rebekah Martin et al., "Lead Results from Tap Water Sampling in Flint, MI," 2015, http://flintwaterstudy.org/2015/12/complete-dataset-lead-results-in-tap-water-for-271-flint-samples/.

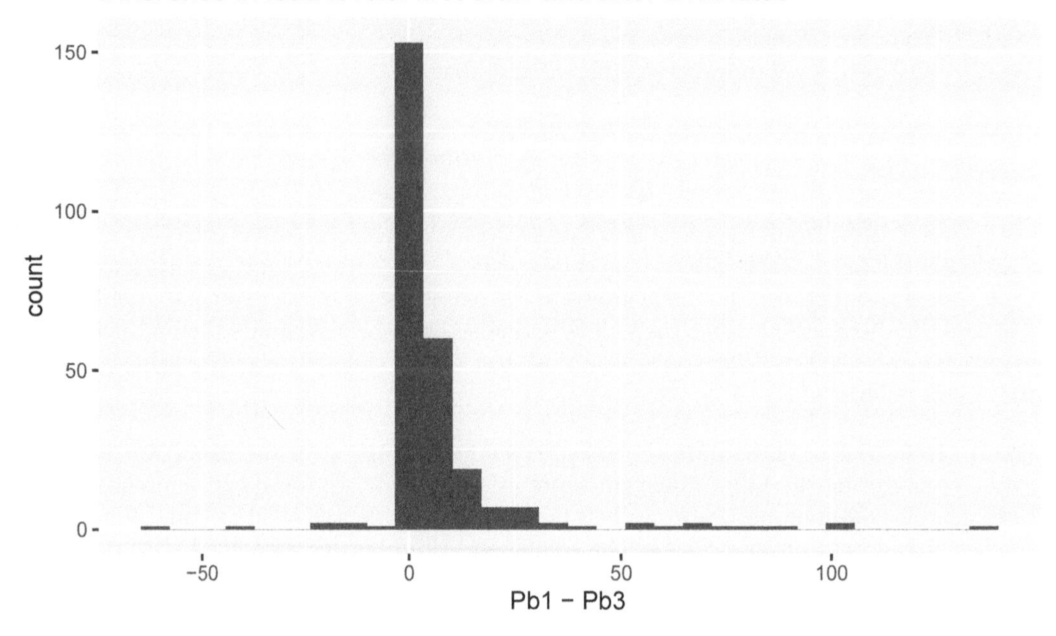


The histogram indicates that the difference in lead levels is positive and smallish, with a heavy right tail and outliers both positive and negative. The heavy tail and outliers make this data unsuitable for a $t$-test.

We perform the paired two sample Wilcoxon test as follows:

```
wilcox.test(flint$Pb1, flint$Pb3, paired = TRUE)
```

```
##
##  Wilcoxon signed rank test with continuity correction
##
## data:  flint$Pb1 and flint$Pb3
## V = 32931, p-value < 2.2e-16
## alternative hypothesis: true location shift is not equal to 0
```

The $p$-value is very low, so we reject $H_0$ and conclude that running the water does make a significant difference to the lead level.

### 9.2.2 Independent two sample test

This section uses `wilcox.test` to compare independent samples from two populations, referred to as the *Wilcoxon rank sum test*. We no longer need to assume that the population distributions are symmetric, but we must assume that all of the observations are independent.

The null hypothesis and alternative hypotheses can be stated as follows.

- $H_0$: the distribution of population one is the same as the distribution of population two.
- $H_a$: the distribution of population one is different than the distribution of population two.

To perform the test, we choose a random sample of size $m$ from population one and independently choose a random sample of size $n$ from population two. For each data point $x$

in group one, count the number of data points in group two that are less than $x$. The total number of "wins" for group one (counting ties as 0.5) is the test statistic $W$. Under the null hypothesis, the expected value of the test statistic is

$$E[W] = \frac{mn}{2},$$

which is exactly half of the possible pairings between the two groups.

Let's look at a simple example for concreteness. Suppose that we obtain the following data, which has two observations in group one and three observations in group two.

| group | value |
|---|---|
| 1 | 5 |
| 1 | 11 |
| 2 | 0 |
| 2 | 3 |
| 2 | 10 |

The group one value $x = 5$ is bigger than two observations in group two, and the group one value $x = 11$ is larger than three observations in group two. So, the value of the test statistic is $W = 2 + 3 = 5$. The expected value of the test statistic is $2 \times 3/2 = 3$. The $p$-value of the test is the probability (under the null) of obtaining a test statistic either 5 or larger, or obtaining one that is 1 or smaller, i.e., $P(|W - 3| \geq 2)$. To compute this probability, we imagine that we have sorted the 5 values from smallest to largest. Under the null hypothesis, each possible arrangement of group one and group two within the sorted values would be equally likely. There are $\binom{5}{3} = 10$ possible permutations of the values, shown as the ten rows of Table 9.2.

Under the assumption that the distribution of population one is the same as the distribution of population two, each outcome is equally likely. There are four rows of Table 9.2 with $|W - 3| \geq 2$, so the $p$-value is $P(|W - 3| \geq 2) = 4/10$.

The sampling distribution of the test statistic $W$ is in R with root name `wilcox`, where as always, prefixes `d`, `p`, `q`, and `r` correspond to the pmf, cdf, quantile function, and random generator, respectively.

For example, if we wish to compute $P(|W - 3| \geq 2)$ when the two sample sizes are 2 and 3 as above, we could do the following.

**TABLE 9.2** Possible permutations of three 2's and two 1's, with associated test statistic.

| V1 | V2 | V3 | V4 | V5 | W |
|---|---|---|---|---|---|
| 2 | 2 | 2 | 1 | 1 | 6 |
| 2 | 2 | 1 | 2 | 1 | 5 |
| 2 | 1 | 2 | 2 | 1 | 4 |
| 1 | 2 | 2 | 2 | 1 | 3 |
| 2 | 2 | 1 | 1 | 2 | 4 |
| 2 | 1 | 2 | 1 | 2 | 3 |
| 1 | 2 | 2 | 1 | 2 | 2 |
| 2 | 1 | 1 | 2 | 2 | 2 |
| 1 | 2 | 1 | 2 | 2 | 1 |
| 1 | 1 | 2 | 2 | 2 | 0 |

```
sum(dwilcox(c(0, 1, 5, 6), m = 2, n = 3))
```

```
## [1] 0.4
```

Figure 9.4 shows the distribution of the test statistic $W$ with group sizes $m = 5$ and $n = 10$. The sampling distribution of the test statistic $W$ is approximately normal when the groups are even modestly large.

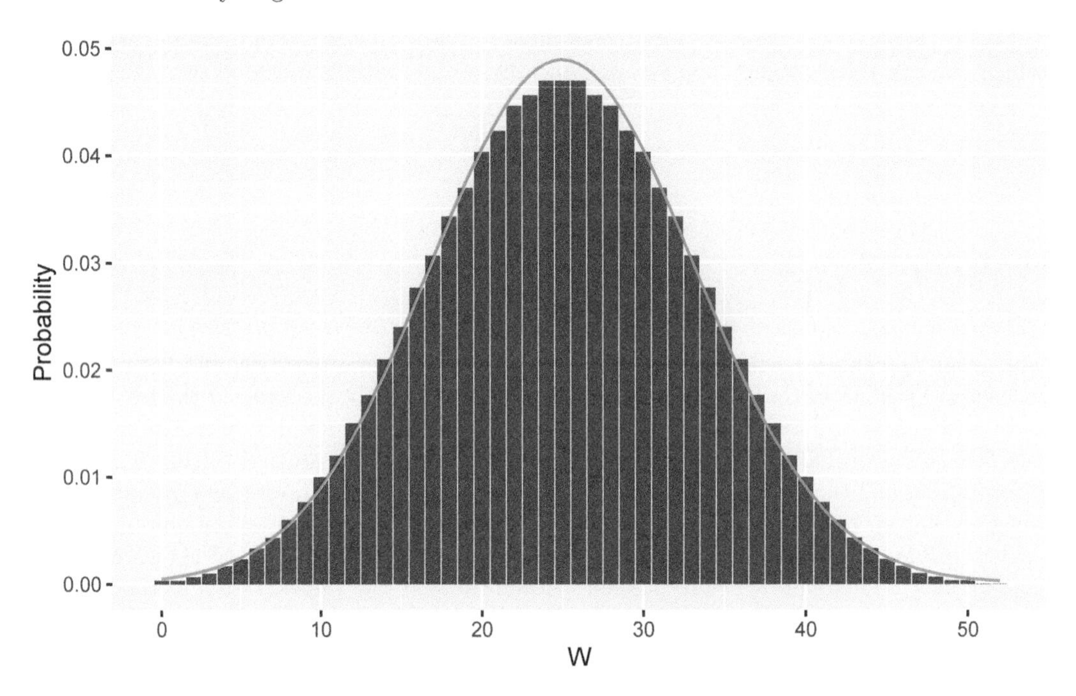


**FIGURE 9.4** pmf of $W$ when $m = 5$, $n = 10$, with superimposed normal distribution.

The function `wilcox.test` computes the test statistic $W$ and the $p$-values in one simple step.

```
wilcox.test(c(5, 11), c(0, 3, 10))
```

```
##
##  Wilcoxon rank sum exact test
##
## data:  c(5, 11) and c(0, 3, 10)
## W = 5, p-value = 0.4
## alternative hypothesis: true location shift is not equal to 0
```

**Example 9.3.** Hillier et al.[4] studied whether academic papers on climate change got more or fewer citations based on their narrative style. They assessed various aspects of the narrative style of the papers and also counted the number of citations. The data set `climate` in the `fosdata` package summarizes their data for articles published in three well-respected journals.

Let's load the data and do some exploring.

[4]Ann Hillier, Ryan P Kelly, and Terrie Klinger, "Narrative Style Influences Citation Frequency in Climate Change Science," *PLOS One* 11, no. 12 (2016): e0167983.

```
climate <- fosdata::climate
```

For the purposes of this example, we are going to look at whether the distribution of articles with an appeal in the abstract are associated with different citation rates than those without an appeal. An appeal being "Yes" indicates that the author made an explicit appeal to the reader or a clear call for action. The `normalized_citation` variable gives the number of citations per year since publication. Let's restrict to those two variables to clean things up a bit.

```
climate <- climate %>%
  select(binary_appeal, normalized_citations, abstract_number)
```

```
climate <- climate %>%
  mutate(climate,
    binary_appeal =
      factor(binary_appeal, levels = 0:1, labels = c("No", "Yes"))
  )
ggplot(climate, aes(x = binary_appeal, y = normalized_citations)) +
  geom_boxplot(outlier.size = -1) +
  geom_jitter(height = 0, width = 0.2)
```

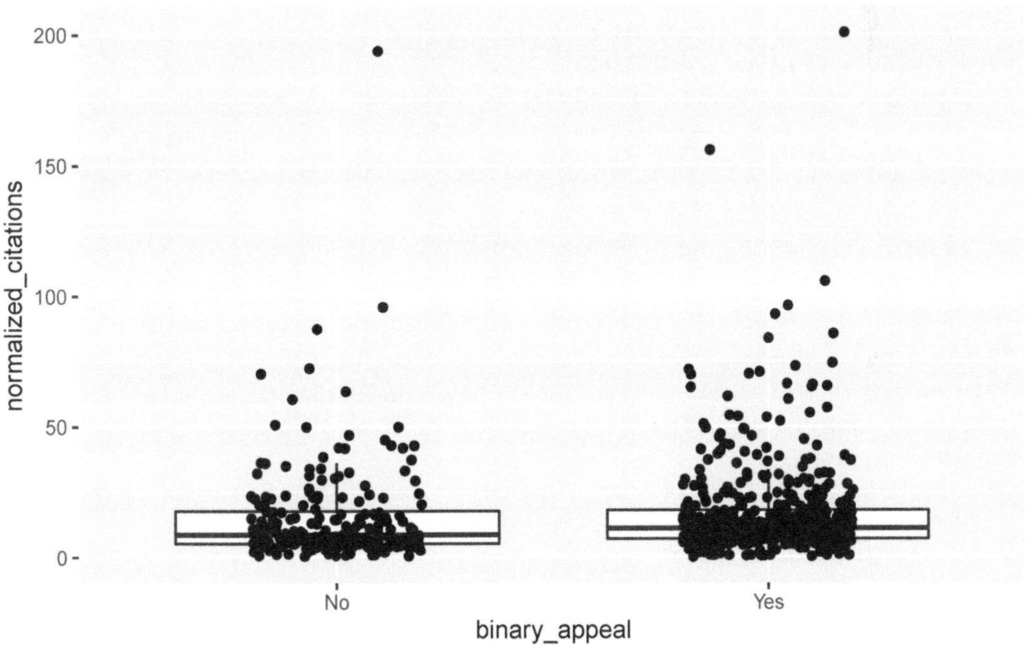

It's a bit hard to see what is going on. Let's compute some summary statistics.

```
climate %>%
  group_by(binary_appeal) %>%
  summarize(
    mean = mean(normalized_citations),
    sd = sd(normalized_citations),
    skew = e1071::skewness(normalized_citations),
    N = n()
  )
```

```
## # A tibble: 2 x 5
##   binary_appeal  mean    sd  skew     N
##   <fct>         <dbl> <dbl> <dbl> <int>
## 1 No             15.5  19.6  4.54   213
## 2 Yes            16.7  18.4  4.17   519
```

One thing that we notice is that there is quite a bit of skewness in these variables. The authors took the `log` of the citation variable, which lowered the skewness and made the data look more normal. A $t$-test without transforming the data would be problematic even with 200-500 samples because of the level of skewness of the data, as well as the outliers. Instead, we will use a Wilcoxon rank sum test with hypotheses:

- $H_0$: the distribution of papers with an appeal is the same as the distribution of papers without an appeal.
- $H_a$: the distribution of papers with an appeal is different than the distribution of papers without an appeal.

```
wilcox.test(normalized_citations ~ binary_appeal, data = climate)
```

```
##
##  Wilcoxon rank sum test with continuity correction
##
## data:  normalized_citations by binary_appeal
## W = 47608, p-value = 0.003182
## alternative hypothesis: true location shift is not equal to 0
```

Note the use of the formula notation ~ in `wilcox.test`; this is similar to its use in `t.test`, where we are telling R how we want to group the data. We reject the null hypothesis that the two distributions are the same at the $\alpha = .05$ level.

Recall that the expected value of the test statistic is

```
213 * 519 / 2
```

```
## [1] 55273.5
```

The test statistic that we obtained is 47608, which is lower than the expected value. This indicates that the "no appeal" papers are cited less frequently than would be expected under the null hypothesis.

### 9.2.3 Ordinal data

Another common use of `wilcox.test` is when the data is ordinal rather than numeric.

**Definition 9.1.** Data is *ordinal* when, given data points $p_1$ and $p_2$, we can determine whether $p_1$ is bigger than $p_2$, less than $p_2$, or equal to $p_2$.

For example, airplane passengers may fly first class, business class, or coach. The seat type of an airline passenger is ordinal data because the three types of seats are ordered by quality.

Ordinal data is different from numeric data because there is no natural way to assign numbers to ordinal data. It is quite common to assign numbers to ordinal categories, but that doesn't necessarily mean that it is a correct thing to do.

**Example 9.4.** Surveys commonly ask for responses in the form of *Likert-scale* data. The

Likert scale is an ordinal five-point scale with choices reflecting agreement with a statement. Often, those options are given numeric values, as in this example:

**Circle the response that most accurately reflects your opinion**

| Question | Strongly Agree | Agree | Neutral | Disagree | Strongly Disagree |
|---|---|---|---|---|---|
| *Pride and Prejudice* is a great book | 1 | 2 | 3 | 4 | 5 |
| `wilcox.test` is a useful R command | 1 | 2 | 3 | 4 | 5 |

Although the options are listed as numbers, it does not necessarily make sense to treat the responses numerically. For example, if one respondent Strongly Disagrees that `wilcox.test` is useful, and another Strongly Agrees, is the mean of those responses Neutral? What is the mean of Strongly Agree and Agree? If the responses are numeric, what units do they have? So, it may not (and often does not) make sense to treat the responses to the survey as numeric data, even though it is encoded as numbers. For compelling examples of how numbering Likert scales can lead you astray, see the somewhat whimsical booklet, "Do not use averages with Likert scales."[5]

What we **can** do with Likert scale data is to order the responses in terms of amount of agreement. It is clear that Strongly Agree > Agree > Disagree > Strongly Disagree. Neutral is omitted because Neutral responses open a whole new can of worms. For example, we don't know whether someone who put Neutral for Question (1) did so because they have never read *Pride and Prejudice,* or because they read it and don't have an opinion on whether it is a great book. In one case, the response might better be treated as missing data, while in the other it would be natural to include the result between Agree and Disagree.

**Example 9.5.** Consider the MovieLens data set `fosdata::movies` from Chapter 6. Each movie review has a star rating, which is a subjective opinion of how much enjoyment the viewer had. The number of stars that a movie receives can definitely be **ordered** from 5 stars being the best, down to 1/2 star being the worst. It is not as clear that the number of stars should be treated numerically. To perform inference with the star ratings, we prefer the Wilcoxon test, which depends only on the ordinal nature of the data.

Let's compare the movies *Toy Story* and *Toy Story 2*. Recall that the data that we have in our MovieLens data set is a random sample from all users. Is there sufficient evidence to suggest that there is a difference in the ratings of *Toy Story* and *Toy Story 2* in the full data set?

The `movies` data set consists of observations pulled from 610 distinct users. If one user rates both *Toy Story* and *Toy Story 2*, those two ratings are not independent of each other. So, let's recast the question as follows. Among those people who have only rated one or the other movie, is there evidence of a difference in the ratings of *Toy Story* and *Toy Story 2* in the full data set?

We should also decide whether we want to do a one-sided or a two-sided test. Some people have strong opinions about which *Toy Story* movie is better, but that is not sufficient evidence or reason to do a one-sided test. So, let $X$ be the rating of *Toy Story* for a randomly

[5]Dwight Barry, "Do Not Use Averages with Likert Scale Data," 2017, https://bookdown.org/Rmadillo/likert/.

sampled person in the large data set who only rated one of the two movies. Let $Y$ be the rating of *Toy Story 2* for a randomly sampled person who only rated one of the two movies. Our null and alternative hypotheses are:

- $H_0$: $X$ and $Y$ have the same distribution.
- $H_a$: $X$ and $Y$ do not have the same distribution.

After loading the data, we create a data frame that only has the ratings of the first two *Toy Story* movies (by excluding *Toy Story 3* which came out years later).

```
toy_story <- fosdata::movies %>%
  filter(stringr::str_detect(title, "Toy Story [^3]"))
```

Next, we filter out only those ratings from people who rated just one of the two movies.

```
toy_story_once <- toy_story %>%
  group_by(userId) %>%
  mutate(N = n()) %>%
  filter(N == 1)
```

Finally, we perform the Wilcoxon test:

```
wilcox.test(rating ~ title, data = toy_story_once)
```

```
##
##  Wilcoxon rank sum test with continuity correction
##
## data:  rating by title
## W = 1060.5, p-value = 0.9448
## alternative hypothesis: true location shift is not equal to 0
```

There is not a statistically significant difference ($p = 0.94$) between the ratings of *Toy Story* and *Toy Story 2*. We do not reject the null hypothesis.

## 9.3 Power and sample size

In this section, we compare the power of a $t$-test with that of a one sample Wilcoxon test. We do so with three populations: normal, heavy tails, and normal with an outlier.

We begin by comparing the power of Wilcoxon rank sum test with $t$-test when the underlying data is normal. We assume we are testing $H_0 : \mu = 1$ versus $H_a : \mu \neq 1$ at the $\alpha = .05$ level, when the underlying population is truly normal with mean 0 and standard deviation 1 with a sample size of 10. Let's first estimate the percentage of time that a $t$-test correctly rejects the null-hypothesis.

```
mean(replicate(10000, t.test(rnorm(10, 0, 1), mu = 1)$p.value < .05))
```

```
## [1] 0.8011
```

We see that we correctly reject $H_0$ 80.1% of the time in this simulation. Let's see what happens with the Wilcoxon test.

```
mean(replicate(10000, wilcox.test(rnorm(10, 0, 1), mu = 1)$p.value < .05))
```

```
## [1] 0.7864
```

Here, we see that we correctly reject $H_0$ 78.6% of the time in this simulation. If you repeat the simulations, you will see that indeed a $t$-test correctly rejects $H_0$ more often than the Wilcoxon test does, on average.

However, if there is a departure from normality in the underlying data, Wilcoxon can outperform `t.test`. Let's repeat the above simulations with a heavy tailed distribution. We sample from a $t$ distribution with 3 degrees of freedom.

```
mean(replicate(10000, t.test(rt(10, 3), mu = 1)$p.value < .05))
```

```
## [1] 0.5378
```

```
mean(replicate(10000, wilcox.test(rt(10, 3), mu = 1)$p.value < .05))
```

```
## [1] 0.544
```

Here, we see that Wilcoxon outperforms $t$, but not by a tremendous amount.

Finally, we look at the third example, which is normal with a single outlier. Our model for an outlier will be that we multiply one of the values by 100. This does not change the mean of the distribution (since it is still zero), but it will often introduce relatively large outliers. We can simulate data of this type as follows:

```
dat <- rnorm(10, 0, 1)
dat[10] <- dat[10] * 100
dat
```

```
##  [1]  -0.07719331   1.84950460  -0.08484561   0.09717147  -0.59400957
##  [6]  -0.60122548   0.55263885  -1.04205402  -0.09278696 -57.35763723
```

If you run the above code multiple times, you will see that the last value usually does not appear to come from a standard normal distribution, though sometimes it can appear to do so.

We again estimate the power of a $t$-test and a one sample Wilcoxon test on this type of data, again with $H_0 : \mu = 1$. Since the simulations are a bit more complicated, we write out the simulations in a longer format.

```
p_vals <- replicate(10000, {
  dat <- rnorm(10, 0, 1)
  dat[10] <- dat[10] * 100
  t.test(dat, mu = 1)$p.value
})
mean(p_vals < .05)
```

```
## [1] 0.033
```

```
p_vals <- replicate(10000, {
  dat <- rnorm(10, 0, 1)
  dat[10] <- dat[10] * 100
  wilcox.test(dat, mu = 1)$p.value
})
mean(p_vals < .05)
```

```
## [1] 0.4275
```

Now we see a substantial difference in the power of the two tests, with the one sample Wilcoxon test being much more powerful.

If your data is symmetric with outliers, the one sample Wilcoxon test will be much more powerful than the one sample $t$-test.

To summarize the results of this section:

1. If your population is normal, then a $t$-test will be a little bit more powerful than a one sample Wilcoxon test.
2. If your population has moderately heavy tails, then a one sample Wilcoxon test will likely be a little bit more powerful than a $t$-test.
3. If your population has outliers, you should use a one sample Wilcoxon test rather than a $t$-test.

### 9.3.1 Sample size

When performing rank-based tests, we recommend using simulation to estimate power and sample sizes. With power computations, we will set the significance level $\alpha$ we wish to test, and we will need some estimate of the effect size we wish to detect. This can be challenging to do well. We illustrate the technique with an example.

**Example 9.6.** Suppose we are designing an experiment in which students are randomly assigned to either a traditional classroom or a flipped classroom. At the end of the semester, students will be given a statement and asked whether they Strongly Agree, Agree, Disagree or Strongly Disagree with the statement. We wish to determine whether there is a statistically significant difference in the responses between those in the traditional classroom and those in the flipped classroom. For an experiment with significance level $\alpha = 0.05$ and power of 80%, what number of students would need to be assigned to each group?

To perform the simulation, we need to estimate the specific distribution of responses in the two groups under the alternate hypothesis. The power computation will tell us how large a sample we need to take to be able to detect the difference between our estimated distributions. Without knowledge of what sort of effect size matters to educational decisions, this is impossible. So imagine we read literature about educational studies, and decide that a reasonable effect would be probability distributions given in the table below.

We recommend estimating the power for various sample sizes and plotting a smoothed version of the resulting curve. When using this technique, we don't need to estimate each probability as accurately as we would if we were only doing it once.

```
p_trad <- c(.4, .3, .1, .2)
p_flip <- c(.6, .3, .1, 0)
```

| | Traditional | Flipped |
|---|---|---|
| Strongly Agree | 0.4 | 0.6 |
| Agree | 0.3 | 0.3 |
| Disagree | 0.1 | 0.1 |
| Strongly Disagree | 0.2 | 0.0 |

```
sample_sizes <- seq(10, 150, length.out = 20)
powers <- sapply(sample_sizes, function(x) {
  tt <- replicate(500, {
    d1 <- sample(1:4, x, T, prob = p_trad)
    d2 <- sample(1:4, x, T, prob = p_flip)
    suppressWarnings(wilcox.test(d1, d2))$p.value
  })
  mean(tt < .05)
})
for_plot <- data.frame(
  sample_size = sample_sizes,
  power = powers
)

ggplot(for_plot, aes(x = sample_size, y = power)) +
  geom_smooth(se = F) +
  geom_point()
```

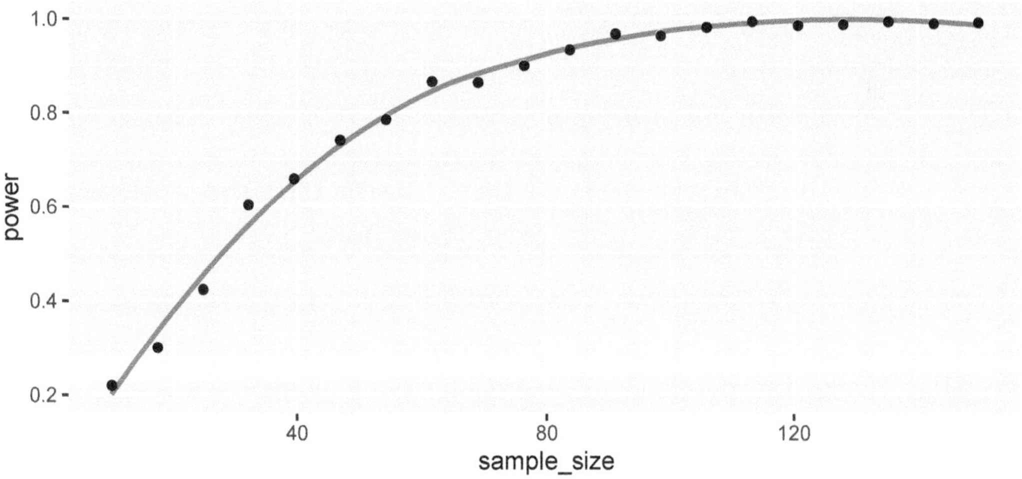

We read off of the plot that in order to have a power of about 0.8, we would need about 58 samples in each group.

## 9.4 Effect size and consistency

### 9.4.1 Effect size

When communicating statistical results to others, it may be useful to include a *common language effect size.* The goal of a common language effect size statistic is to present results in a way that is easy to understand without advanced statistical training. We recommend the following for the two sample Wilcoxon rank sum test.

**Definition 9.2.** Given two populations, *Vargha and Delaney's A* is the probability that a sample from one population will be larger than a sample from the other population plus one-half the probability that they will be equal.

Intuitively, Vargha and Delaney's $A$ is the probability that a sample from one population will be larger than a sample from the other population, where we are assuming that ties are broken 50-50.

The effect size $A$ is directly related to the test statistic for the two sample Wilcoxon rank sum test. The Wilcoxon test statistic $W$ counts the number of times that samples from population one are larger than those in population two, plus one half the number of ties. To compute $A$, divide $W$ by the number of comparisons that are made.

**Example 9.7.** Let's make sense of the difference between *Sense and Sensibility* (1995) and *The Sixth Sense* (1999), the two most popular movies with "sense" in the title. The corresponding `movieId`'s for those movies are 17 and 2762. Which movie is preferred, is there a significant difference, and what is the effect size?

To maintain independence of the samples, restrict to those raters who only rated one of the two movies:

```
sensible_movies <- fosdata::movies %>%
  filter(movieId %in% c(17, 2762)) %>%
  group_by(userId) %>%
  mutate(N = n()) %>%
  filter(N == 1)
```

Now test whether the distribution of ratings are the same for these two sensible movies:

```
wilcox.test(rating ~ title, data = sensible_movies)
```

```
##
##  Wilcoxon rank sum test with continuity correction
##
## data:  rating by title
## W = 3089.5, p-value = 0.65
## alternative hypothesis: true location shift is not equal to 0
```

The $p$-value is 0.65, so there is not a significant difference in the ratings. For the common language effect size $A$, we need to know the number of comparisons. How many ratings of each sensible movie were there?

```
table(sensible_movies$title)
```

```
##
## Sense and Sensibility (1995)      Sixth Sense, The (1999)
##                           42                          154
```

Since each pair of ratings is compared, there are $42 \times 154 = 6468$ comparisons. Then $A = W/6468 = 3089.5/6468 \approx 47.8\%$. Since *Sense and Sensibility* comes first in the ordering of titles, *Sense and Sensibility* is preferred by about 47.8% of raters, while *The Sixth Sense* is preferred by about 53.2% of raters. (This is only formally true if we imagine that for ties we flip a coin.)

Without using the $W$ statistic, we can compute $A$ directly:

```
s_and_s_ratings <- filter(sensible_movies, movieId == 17) %>%
  pull(rating)
```

```
t_s_s_ratings <- filter(sensible_movies, movieId == 2762) %>%
  pull(rating)
sum(sapply(s_and_s_ratings, function(x) {
  sum(x > t_s_s_ratings) + 1 / 2 * sum(x == t_s_s_ratings)
})) / (length(s_and_s_ratings) * length(t_s_s_ratings))
```

```
## [1] 0.4776592
```

There is a package, **effsize**, that has a function to compute Vargha and Delaney's $A$ from the data with no manipulation:

```
effsize::VD.A(rating ~ title, data = sensible_movies)
```

```
##
## Vargha and Delaney A
##
## A estimate: 0.4776592 (negligible)
```

It's still 47.8% in favor of *Sense and Sensibility*, described as a "negligible" effect. You should take the adjectives assigned to effect sizes by the **effsize** package with a grain of salt, as effect sizes are domain dependent. As with Cohen's $d$, what is considered a negligible effect in one field might be considered a large effect in a different field.

### 9.4.2 Consistency

In this section, we discuss the consistency of the one sample `t.test` and the two independent sample `wilcox.test`.

**Definition 9.3.** A hypothesis test is *consistent* if the probability of rejecting the null hypothesis given that the null hypothesis is false converges to 1 as the sample size goes to infinity.

Consistency is a desirable property for a hypothesis test because it means that a sufficiently large sample size will always have the power to detect a false $H_0$. We will not provide proofs of consistency, but instead examine it via simulations.

**Example 9.8.** Consider the `t.test` with one population, where we are testing $H_0 : \mu = 0$ versus $H_a : \mu \neq 0$. We assume that the underlying population has mean $\mu_a \neq 0$, so $H_0$ is false. Also assume the population is normally distributed with standard deviation $\sigma = 1$, and choose significance level $\alpha = 0.01$.

We start by computing the percentage of times $H_0$ will be rejected when $\mu_a = 1$ and the sample size is $n = 20$.

```
sigma <- 1
nsize <- 20
mua <- 1
pvals <- replicate(10000, {
  dat <- rnorm(nsize, mua, sigma)
  t.test(dat, mu = 0)$p.value
})
mean(pvals < .01)
```

```
## [1] 0.9338
```

We see that with 20 samples, we already reject $H_0$ about 93% of the time. By repeating this simulation with various values of $n$ and $\mu_a$, we create the plot in Figure 9.5. It appears that the probability of rejecting $H_0$ is getting larger as the sample size increases, and it also appears to be converging to 1. We would need to take more samples to see this for $\mu_a = 0.02$, but 0.02 is a very small difference between the null and alternative hypothesis.

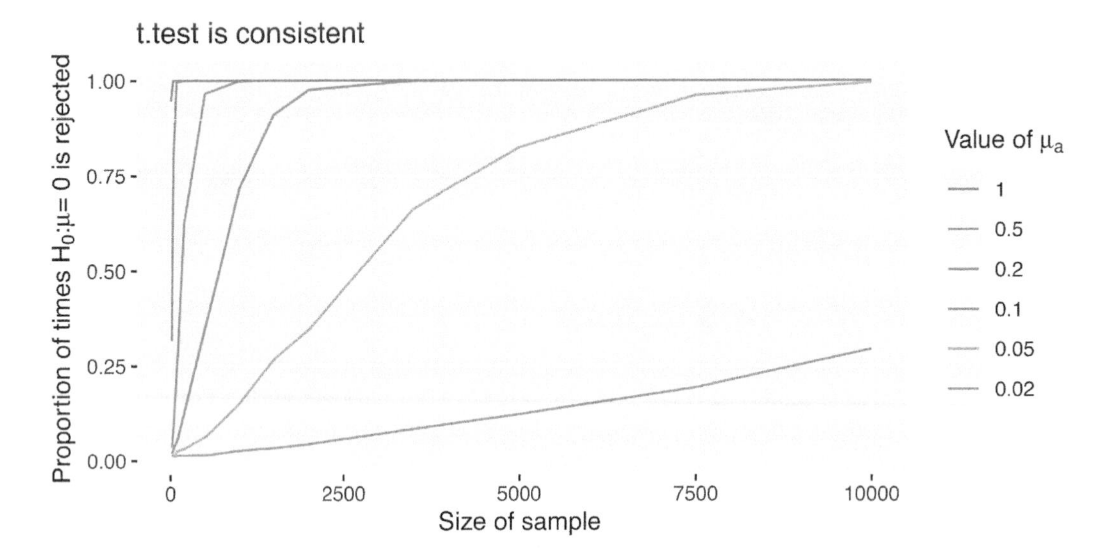


**FIGURE 9.5** Each curve shows the probability that a $t$-test rejects $H_0$ as sample size grows. The different curves correspond to different population means $\mu_a$. As $\mu_a$ gets closer to the hypothesized $\mu = 0$, larger samples are needed to detect the difference.

The story for the Wilcoxon rank sum test is a bit more involved. We begin with an example where the null hypothesis is false, but we see that the test is not consistent.

**Example 9.9.** Let population 1 consist of numbers uniformly distributed on $[-1, 1]$. Let population 2 consist of numbers uniformly distributed on $[-2, -1]$ with probability 1/2, and numbers uniformly distributed on $[1, 2]$ with probability 1/2, as shown.

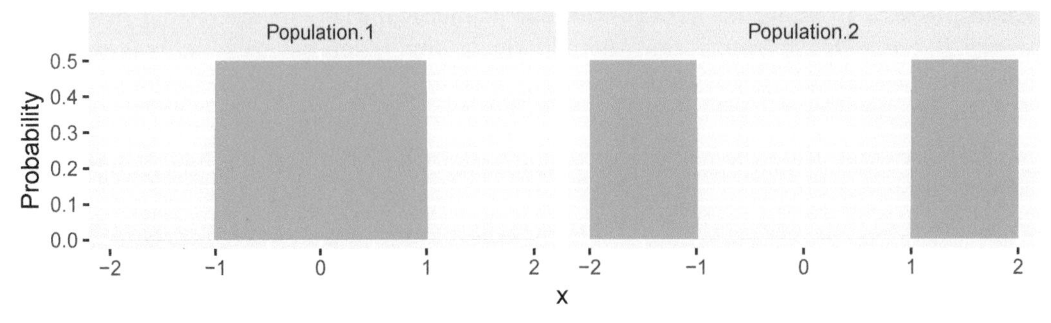


Fixing $\alpha = 0.05$, let's see what percentage of times, on average, the Wilcoxon test rejects $H_0$ with a sample size of $n = 100$.

```
N <- 100
sim_data <- replicate(10000, {
  d1 <- runif(N, -1, 1)
  d2 <- sample(c(-1, 1), N, replace = TRUE) * runif(N, 1, 2)
```

```
  wilcox.test(d1, d2)$p.value
})
mean(sim_data < .05)
```

```
## [1] 0.0889
```

We see that the test rejects $H_0$ more than with probability $\alpha = .05$, but the test still has very low power. In other words, it does not do a good job of distinguishing population 1 from population 2. Repeating the computation with different sample sizes $n$ produces Table 9.4.

From Table 9.4 we see that as $n$ grows, the power of the test is not approaching 1. It can be shown that when $n \to \infty$, the power converges to about 0.1095. With these populations, the Wilcoxon rank sum test is not consistent.

However, we do have the following theorem:

**Theorem 9.1.** *Suppose $X$ and $Y$ are random variables with finite variance and*

$$P(X > Y) \neq P(X < Y).$$

*Then the Wilcoxon rank sum test of $X$ against $Y$ is consistent. That is, as the number of samples goes to infinity, the power converges to 1.*

The condition in Theorem 9.1 is equivalent to a statement about Vargha and Delaney's $A$. If $A \neq 0.5$, then the two sample Wilcoxon test is consistent.

We recommend using the Wilcoxon rank sum test only when we suspect, or wish to detect, a difference in populations that is of the type $P(X > Y) \neq P(X < Y)$. In this case, the test is consistent.

To finish off this section, we provide an example of two populations such that $P(X > Y) \neq P(X < Y)$, and give evidence with simulation that the Wilcoxon rank sum test is consistent.

**Example 9.10.** Let $X$ and $Y$ be random variables, with $X \sim \text{Unif}(1, 2)$ and $Y$ taking values 0 and 3 with probabilities $p < 1/2$ and $1 - p$. Then $P(X > Y) = p$ and $P(X < Y) = 1 - p$. Fix significance level $\alpha = 0.01$.

Here is a single computation when $p = 0.4$ and the sample size $n = 100$.

```
p <- 0.4
N <- 100
```

**TABLE 9.4** Power of Wilcoxon test.

| n | power |
|---|---|
| 10 | 0.105 |
| 20 | 0.098 |
| 50 | 0.125 |
| 100 | 0.092 |
| 200 | 0.107 |
| 500 | 0.101 |
| 1000 | 0.106 |

```
pvals <- replicate(1000, {
  x <- runif(N, 1, 2)
  y <- sample(c(0, 3), size = N, prob = c(p, 1 - p), replace = TRUE)
  suppressWarnings(wilcox.test(x, y))$p.value
})
mean(pvals < 0.01)
```

```
## [1] 0.416
```

The probability of correctly rejecting $H_0$ in this case is 0.416. Repeating the simulation for different values of $p$ and $n$ produces the plot shown in Figure 9.6. From the figure, it appears that the percentage of times $H_0$ is rejected converges to 1 as the sample size goes to infinity, so the test is consistent. Values of $A$ closer to 1/2 require larger sample sizes.

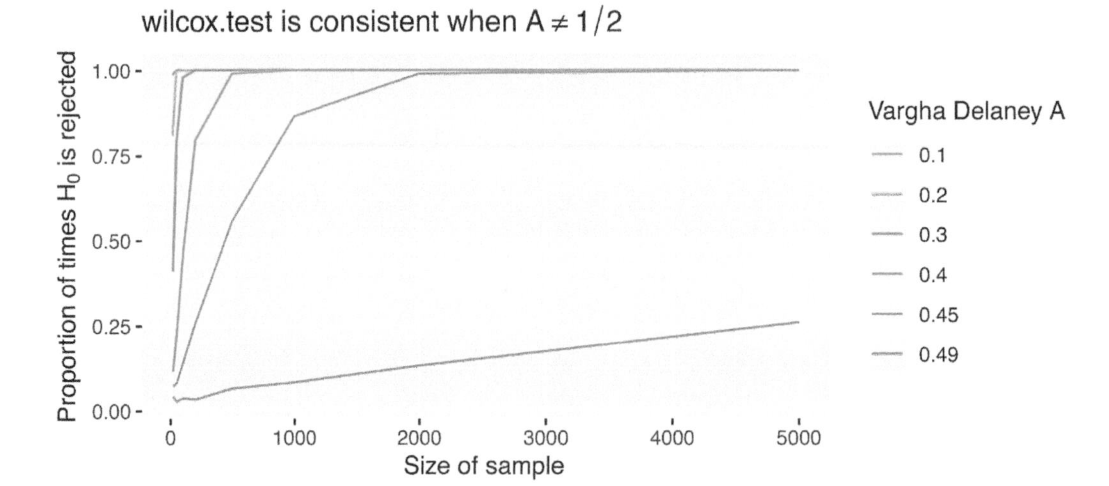


**FIGURE 9.6** Each curve shows the probability that a Wilcoxon test rejects $H_0$ as sample size grows. The curves correspond to fixed population values of $A = P(X < Y)$. As $A$ gets closer to 1/2, larger samples are needed to detect the difference between $X$ and $Y$.

## 9.5 Summary

In this chapter, we provided an alternative to `t.test`. In the one sample case, we can use the Wilcoxon signed rank test for a random sample from a symmetric population. The Wilcoxon signed rank test is preferred over `t.test` when either the data is ordinal and means cannot be taken, or when in the presence of outliers. For a population that is normally distributed, use `t.test` because it has slightly larger power. When the population distribution is skewed or unknown (but with no outliers), use `t.test` with the caveats regarding sample size presented in Chapter 8.

In the independent two sample case, we can use the Wilcoxon rank sum test as an alternative to `t.test`. Use the Wilcoxon rank sum test when the data is ordinal, when there may be outliers, or when the sample sizes are small enough that the Central Limit Theorem may not apply. Recall that the sample size required for the Central Limit Theorem to apply depends on the distribution. Use `t.test` when the populations are normal or when the sample size is large enough that the Central Limit Theorem applies. If the data is ordinal, but can be transferred meaningfully to a numeric scale, then you can use `t.test` on the transferred data. However, it can be difficult to re-interpret your result in the original scale when you do this.

For the two sample Wilcoxon rank sum test, the null hypothesis is that the populations are the same and the alternative is that they are different. The power of a Wilcoxon rank sum test goes to 1 as the sample size goes to infinity in the case that $P(X > Y) \neq P(X < Y)$. We recommend only using the Wilcoxon rank sum test when you wish to detect differences between the populations of the type $P(X > Y) \neq P(X < Y)$. If the null hypothesis is rejected, then we recommend reporting Vargha and Delaney's $A$ as an estimate of effect size, and as an estimate of $P(X > Y)$. If the populations appear to be different for reasons other than $P(X > Y) \neq P(X < Y)$, such as having different variances, then the Wilcoxon rank sum test will typically have poor power, and should not be used. For a more advanced theoretical treatment of this topic, we recommend the text by Lehmann.[6]

In the paired two sample case, we can use the Wilcoxon signed rank test on the differences as an alternative to a $t$-test on the differences. For the Wilcoxon signed rank test to be valid, the *differences* in values must be meaningful, and in particular, must satisfy the conditions of the Wilcoxon signed rank test. Use Wilcoxon signed rank when the data is ordinal with meaningful differences or when there are outliers.

## Vignette: ROC curves and the Wilcoxon rank sum statistic

Suppose you wish to classify an object into one of two groups. It would be helpful if there were a variable $X$ and a value $x_0$ such that whenever $X < x_0$ we would classify the object into group 1, and whenever $X \geq x_0$ we would classify the object into group 2. A *receiver operating characteristic (ROC) curve* is a graphical measurement of how well a variable discriminates between two alternatives.

As an example, let's consider `palmerpenguins::penguins` and suppose that we are trying to distinguish between Adelie and chinstrap penguins, based solely on bill length.

```
penguins <- palmerpenguins::penguins %>%
  filter(species %in% c("Adelie", "Chinstrap")) %>%
  mutate(species = droplevels(species))
ggplot(
  penguins,
  aes(x = bill_length_mm, y = flipper_length_mm, color = species)
) +
  geom_point() +
  geom_vline(xintercept = c(40.9, 46), linetype = "dashed")
```

[6] E L Lehmann, *Nonparametrics: Statistical Methods Based on Ranks*, Holden-Day Series in Probability and Statistics (Holden-Day, Inc., San Francisco, Calif.; McGraw-Hill International Book Co., New York-Düsseldorf, 1975).

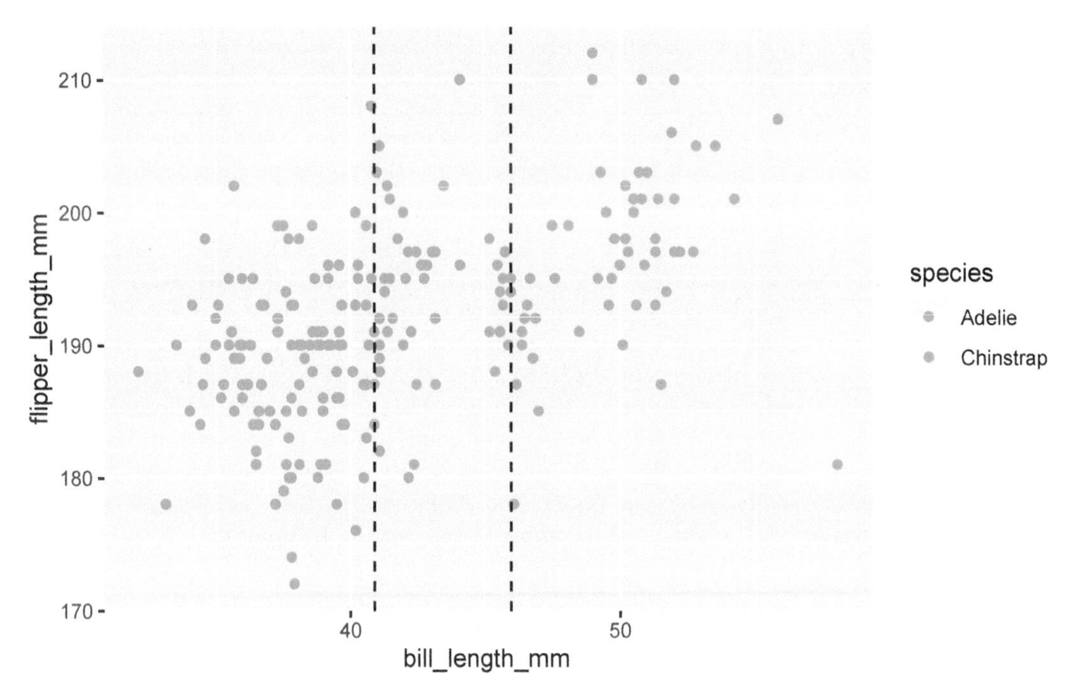


Based on this graph, we can see that there is not a value of bill length that completely separates the two variables. Whatever $x_0$ we pick so that we assign any penguin with bill length less than $x_0$ to be Adelie, we will be misclassifying some of the penguins. The ROC curve is computed as follows. Start at the smallest bill length and imagine that is your split value; all penguins with bill length less than that are classified as Adelie. The $y$-coordinate is the percentage of penguins correctly classified as Adelie with that split value, and the $x$-coordinate is the percentage of penguins incorrectly classified as Adelie with that split value.

An ideal ROC curve would look like this:

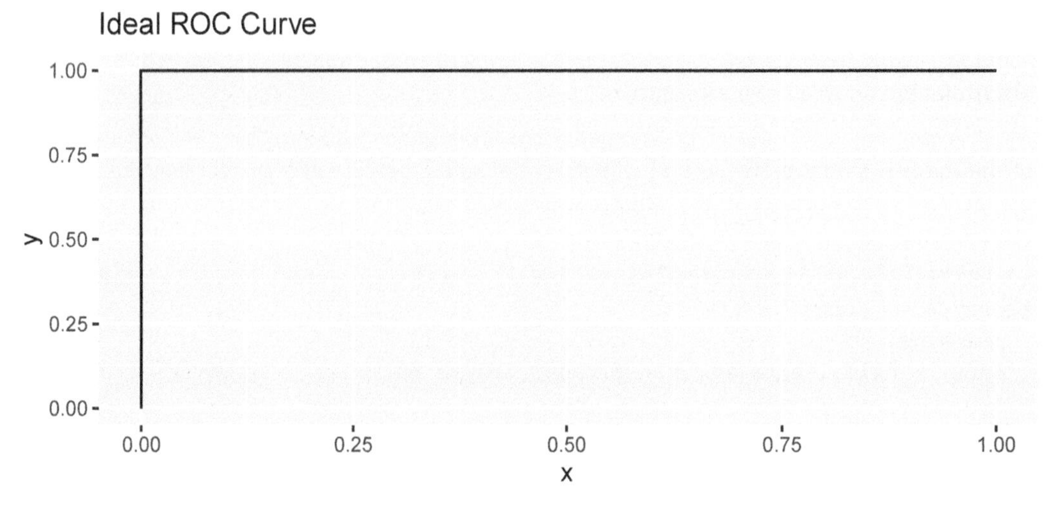


This means that there is at least one $x_0$ such that all of the objects in group 1 have values less than $x_0$, while all of the objects in group 2 have values at least $x_0$. Let's look at the ROC curve for the penguins using the `ROCR` package.

```
penguins <- penguins %>% select(species, bill_length_mm)
penguins <- penguins[complete.cases(penguins), ]
pred <- ROCR::prediction(penguins$bill_length_mm, penguins$species)
perf <- ROCR::performance(pred, "tpr", "fpr")
plot(perf, colorize = TRUE, main = "ROC curve for two penguin species")
```

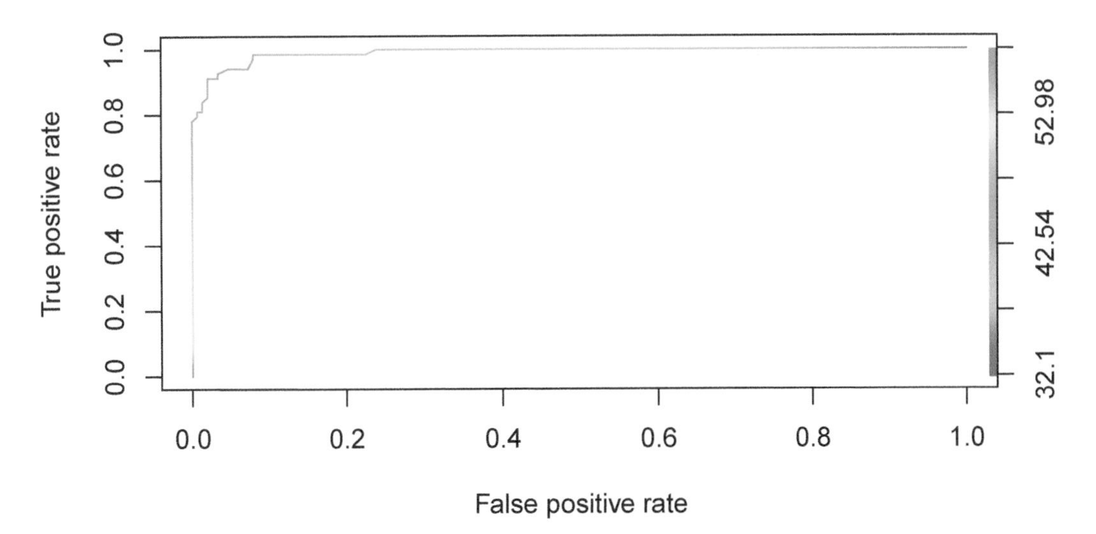


Note that the curve looks similar to the idealized version, except that it cuts off the corner at the top left. The curve there is green, which corresponds to bill lengths of around 45 mm. Those values of bills are exactly where we are unsure how the penguin should be classified.

The area underneath an ideal ROC curve is 1. For the penguin ROC curve, a little bit of area is missing in the upper left corner. The area under the ROC curve is abbreviated *AUC* and measures the separation between the two groups. It is used as a performance measure for machine learning algorithms.

**Proposition 9.1.** *The area under the ROC curve (AUC) is equal to Vargha and Delaney's A.*

Proposition 9.1 allows another interpretation of the AUC: it measures the probability that a randomly selected member of group 2 has a higher value of the variable under question than a randomly selected member of group 1. We will not prove Proposition 9.1, but the following code verifies that three ways of computing the AUC are the same in this context.

```
penguins <- penguins %>%
  mutate(species = factor(species, levels = c("Chinstrap", "Adelie")))
wilcox.test(bill_length_mm ~ species, data = penguins)$statistic / (151 * 68)
```

```
##         W
## 0.9901636
```

```
effsize::VD.A(bill_length_mm ~ species, data = penguins)
```

```
##
## Vargha and Delaney A
##
```

```
## A estimate: 0.9901636 (large)
```

```
perf <- ROCR::performance(pred, "auc")
perf@y.values[[1]]
```

```
## [1] 0.9901636
```

## Exercises

Exercises 9.1 – 9.6 require material through Section 9.1.

**9.1.** Suppose you wish to test $H_0 : \mu = 2$ versus $H_a : \mu \neq 2$ using a Wilcoxon signed rank test. If you collect 30 samples, what is the expected value of the test statistic $V$ under the null hypothesis?

**9.2.** Suppose you are testing $H_0 : \mu = 0$ versus $H_a : \mu \neq 0$. You collect 40 data points and compute $V = 601$. Use the built-in R function `dsignrank` or `psignrank` to compute the $p$-value associated with the test.

**9.3.** Suppose $x_1, \ldots, x_n$ is a random sample from a symmetric distribution with mean 0.

a. Show that the expected value of the Wilcoxon test statistic $V$ is $E(V) = \dfrac{n(n+1)}{4}$.
b. Find $\lim_{x_n \to \infty} E[V]$, which is the scenario of a single, arbitrarily large outlier.

**9.4.** Suppose you wish to test $H_0 : \mu = 3$ versus $H_a : \mu \neq 3$. You collect the data points $x_1 = -4$, $x_2 = 0$, $x_3 = 2$ and $x_4 = 8$. Go through all of the steps of a Wilcoxon signed rank test and determine a $p$-value. Check your answer using R.

**9.5.** In this example, we explore what happens when applying the Wilcoxon signed rank test to skewed data. Suppose you take a random sample of size 20 from an exponential random variable with rate 1. The **mean** of the distribution is 1, and the **median** is log(2). All tests below are to be conducted at the $\alpha = .05$ level.

a. Estimate the effective type I error when testing $H_0 : \mu = 1$ versus $H_a : \mu \neq 1$ in this setting.
b. Estimate the effective type I error when testing $H_0 : m = \log(2)$ versus $H_a : m \neq \log(2)$ in this setting.
c. Even though the test is not working correctly for either of those values, there **is** a value that makes the test work approximately correctly. The value is called a *pseudo-median*, and for exponential random variables with rate 1 it is approximately 0.84. Confirm that the effective type I error of testing $H_0 : m = .84$ versus $H_a : m \neq .84$ is approximately 0.05. (It is, based on our simulations, slightly larger than .05.)

**9.6.** Consider the `weight_estimate` data set in the `fosdata` package. Children, and some adults, were asked to estimate the weight of an object while watching a professional actor pick it up. For the purposes of this problem, consider only the `mean200` variable.

a. Plot a boxplot of `mean200`. Does it appear to be reasonable symmetric?
b. Conduct a Wilcoxon signed rank test of $H_0 : \mu = 200$ versus $H_a : \mu \neq 200$ at the $\alpha = .05$ level and interpret.

a. Explain why it is very likely that each observation of `Pb2` is dependent on the observation of `Pb3`.
b. There are two houses that are sampled twice. What does this imply about the independence of `Pb2` and `Pb3` **between** observations? Since this only affects 4 of the 271 observations, we will ignore this.
c. Plot the data. Would it be appropriate to use a $t$-test directly on this data?
d. If appropriate, test using a Wilcoxon signed rank test at the $\alpha = .01$ level.
e. State your conclusions, including either a point estimate or a confidence interval for the pseudo-median (see Exercise 9.5 on the difference in lead levels).

---

Exercises 9.17 – 9.21 require material through Section 9.3.

**9.17.** Compare the effective power of `t.test` versus `wilcox.test` in the case of testing $H_0 : \mu = 0$ versus $H_a : \mu \neq 0$ when the underlying population is uniform on the interval $[-0.5, 1]$, and the sample size is 30.

**9.18.** This problem explores the effective type I error rate for a one sample Wilcoxon and $t$-tests. Choose a sample of 21 values where $X_1, \ldots, X_{20}$ are iid normal with mean 0 and sd 1, and $X_{21}$ is 10 or -10 with equal probability. Test $H_0 : m = 0$ versus $H_a : m \neq 0$ at the $\alpha = .05$ level. How often does the Wilcoxon test reject $H_0$? Compare with the effective type I error rate for a $t$-test of the same data. Which test is performing closer to how it was designed?

**9.19.** How well can hypothesis tests detect a small change? Suppose the population is normal with mean $\mu = 0.1$ and standard deviation $\sigma = 1$. We test the hypothesis $H_0 : \mu = 0$ versus $H_a : \mu \neq 0$.

a. When $n = 100$, what percent of the time does a $t$-test correctly reject $H_0$?
b. When $n = 100$, what percent of the time does a Wilcoxon test correctly reject $H_0$?
c. Repeat parts (a) and (b) with $n = 1000$.
d. The assumptions for both tests are satisfied, since the population is normal. Which test would you recommend?

**9.20.** In this problem, we estimate the probability of a type II error. Suppose you wish to test $H_0 : \mu = 1$ versus $H_a : \mu \neq 1$ at the $\alpha = .05$ level.

a. Suppose the true underlying population is $t$ with 3 degrees of freedom, and you take a sample of size 20.
   i. What is the true mean of the underlying population?
   ii. What type of error would be possible to make in this context, type I or type II? In the problems below, if the error is impossible to make in this context, the probability would be zero.
b. Approximate the probability of a type I error if you use a $t$-test.
c. Approximate the probability of a type I error if you use a Wilcoxon test.
d. Approximate the probability of a type II error if you use a $t$-test.
e. Approximate the probability of a type II error if you use a Wilcoxon test.
f. Note that $t$ random variables with small degrees of freedom have heavy tails and contain data points that look like outliers. Does it appear that Wilcoxon or $t$-test is more powerful with this type of population?

**9.21.** Suppose that you are planning an experiment, and you wish to use a two sample Wilcoxon test on your data. You wish your experiment to have power of 80% when the test is performed at the $\alpha = .05$ level and when population one is normal with mean 0 and standard deviation 1, and population 2 is normal with mean 0.4 and standard deviation 1. Assume that you will take equal sample sizes from each population.

Create a power versus sample size plot for 5 to 10 sample sizes between 20 and 200, and indicate how many samples from each population you would recommend.

---

Exercises 9.22 – 9.27 require material through Section 9.4.

**9.22.** Consider the MovieLens data `fosdata::movies`.

a. Is there sufficient evidence at the $\alpha = .05$ level to conclude that among people who have only seen one of the movies, those people have a preference between *Sleepless in Seattle* and *While You Were Sleeping?*
b. Report Vargha and Delaney's $A$ effect size for this data, and interpret.

**9.23.** Consider the `sharks` data in the `fosdata` package.[7] Participants either watched a video or listened to an audio documentary about sharks while different types of music played. Participants were then asked to rate sharks on various dimensions such as gracefulness and viciousness. For this problem, consider **only** those participants who watched a video and heard either ominous or uplifting music.

a. Is there a difference in those participants' responses to how vicious sharks are? Test at the $\alpha = .05$ level.
b. If there is a difference, provide the common language effect size and explain.

**9.24.** Consider the `plastics` data set in the `fosdata` package. Snow was filtered in both the Arctic and Europe to find microfibers, which are microplastics that are in fibrous shape.

a. Perform a two sample Wilcoxon rank sum test to determine whether the length of the microfibers is the same in the Arctic as in Europe.
b. The authors in the paper[8] reported a test statistic of 13723, which is not what R reports. Confirm that the authors used the following method to obtain their test statistic. They computed the test statistic $W$ from `wilcox.test` and then added the total number of comparisons between lengths of fibers from the Arctic with other lengths of fibers from the Arctic.
c. Provide an effect size in terms of Vargha and Delaney's $A$ and interpret.

**9.25.** Provide an estimate for the effect size in Exercise 9.16 and interpret.

**9.26.** Use simulation to show that if population 1 is standard normal, and population 2 is normal with mean 1 and standard deviation 1, then the Wilcoxon rank sum test is consistent at the $\alpha = .05$ level.

**9.27.** Suppose population 1 is standard normal, and population 2 is normal with mean 0 and standard deviation 5.

a. Use simulation to show that $P(X > Y) = .5$, where $X$ is a random sample from population 1 and $Y$ is a random sample from population 2.
b. Use sample sizes of $n = 10, 100, 1000$ and 5000 for each population, and estimate the probability that the null hypothesis will be rejected at the $\alpha = .05$ level. Use 1000 replications rather than 10,000 so that it runs in a reasonable amount of time.
c. Does the Wilcoxon rank sum test appear to be consistent in this context?

---

[7]Andrew P Nosal et al., "The Effect of Background Music in Shark Documentaries on Viewers' Perceptions of Sharks," *PLOS One* 11, no. 8 (August 2016): 1–15, https://doi.org/10.1371/journal.pone.0159279.

[8]Melanie Bergmann et al., "White and Wonderful? Microplastics Prevail in Snow from the Alps to the Arctic," *Science Advances* 5, no. 8 (2019): eaax1157, https://doi.org/10.1126/sciadv.aax1157.

# 10

# Tabular Data

Tabular data is data on entities that has been aggregated in some way. A typical example would be to count the number of successes and failures in an experiment, and to report those aggregate numbers rather than the outcomes of the individual trials. Another way that tabular data arises is via binning, where we count the number of outcomes of an experiment that fall in certain groups, and report those numbers.

Inference on categorical variables has traditionally been performed by approximating counts with continuous variables and performing *parametric methods* such as the $z$-tests of proportions and the $\chi^2$-tests. With modern computing power, it is possible to calculate the probability of each experimental outcome exactly, leading to *exact methods* that do not rely on continuous approximation. These include the binomial test and the multinomial test. A third approach is to use *Monte Carlo methods*, where the computer performs simulations to estimate the probability of events under the null hypothesis.

## 10.1 Tables and plots

For categorical (factor) variables, the most basic information of interest is the count of observations in each value of the variable. Often, the data is better presented as proportions, which are the count divided by the total number of observations. For visual display, categorical variables are naturally shown as barplots or pie charts.

In this section, we demonstrate with two data sets. The first is `fosdata::wrist`, from a study of wrist fractures that recorded the fracture side and handedness of 104 elderly patients. The `wrist` data was used by Raittio et al.[1] to evaluate the effectiveness of two types of casts for treating a common type of wrist fracture. The second is `fosdata::snails` which records features of snail shells collected in England. The `snails` data was collected in 1950 by Cain and Sheppard[2] as an investigation into natural selection. They explored the relationship between the appearance of snails and the environment in which snails live.

Let's begin with the `wrist` data set. Each row in the wrist data is an individual patient. Here we only pay attention to two variables, both coded as 1 for "right" and 2 for "left":

```
wrist <- fosdata::wrist
wrist %>%
  select(handed_side, fracture_side) %>%
  head()
```

[1] Raittio et al., "Two Casting Methods Compared in Patients with Colles' Fracture."

[2] A J Cain and P M Sheppard, "Selection in the Polymorphic Land Snail Cepæa Nemoralis," *Heredity* 4, no. 3 (1950): 275–94.

DOI: 10.1201/9781003004899-10

```
## # A tibble: 6 x 2
##   handed_side fracture_side
##         <dbl>         <dbl>
## 1           1             2
## 2           1             1
## 3           1             1
## 4           1             2
## 5           1             1
## 6           1             2
```

For ease of interpretation, let's change the variables into factors, which is really what they are.

```
wrist <- wrist %>%
  mutate(
    handed_side = factor(handed_side, labels = c("right", "left")),
    fracture_side = factor(fracture_side, labels = c("right", "left"))
  )
```

The built-in command `table` can count the number of rows that take each value.

```
table(wrist$handed_side)
```

```
##
## right  left
##    97     7
```

This table shows that there were 97 right-handed patients and 7 left-handed patients. The `proportions`[3] function converts the table of counts to a table of proportions:

```
table(wrist$handed_side) %>% proportions()
```

```
##
##      right       left
## 0.93269231 0.06730769
```

So only 6.7% of patients in this study were left-handed. Does that sound reasonable for a random sample? We will investigate this question in the next section.

Passing two variables to `table` will produce a matrix of counts for each pair of values, but a better tool for the job is the `xtabs` function. The `xtabs` function builds a table, called a *contingency table* or *cross table*. The first argument to `xtabs` is a formula, with the factor variables to be tabulated on the right of the ~ (tilde).

```
xtabs(~ handed_side + fracture_side, data = wrist)
```

```
##            fracture_side
## handed_side right left
##       right    41   56
##       left      3    4
```

One could ask if people are more likely to fracture their wrist on their non-dominant side, since more right-handed patients fractured their left hand (56) than their right hand (41).

Categorical data is often given as counts, rather than individual observations in rows. The

[3]The R function `proportions` is new to R 4.0.1 and is recommended as a drop-in replacement for the unfortunately named `prop.table`.

snails data gives a count for each combination of Location, Color, and Banding. It does not have a row for each individual snail.

```
snails <- fosdata::snails
head(snails)
```

```
## # A tibble: 6 x 5
##   Location  Habitat        Color  Banding Count
##   <chr>     <fct>          <fct>  <fct>   <int>
## 1 Hackpen   Downland Beech Yellow X00000     15
## 2 Hackpen   Downland Beech Yellow X00300     24
## 3 Hackpen   Downland Beech Yellow X12345      0
## 4 Hackpen   Downland Beech Yellow Others      0
## 5 Rockley 1 Downland Beech Yellow X00000     17
## 6 Rockley 1 Downland Beech Yellow X00300     25
```

To make a table of Color vs. Banding for snails, use `xtabs` and give the Count for each group on the left side of the formula:

```
xtabs(Count ~ Banding + Color, data = snails)
```

```
##         Color
## Banding  Brown Pink Yellow
##   X00000   339  433    126
##   X00300    48  421    222
##   X12345    16  395    352
##   Others    23  373    156
```

Frequently when creating tables of this type, we will want to know the row and column sums as well. These are generated by the function `addmargins`.

```
xtabs(Count ~ Banding + Color, data = snails) %>%
  addmargins()
```

```
##         Color
## Banding  Brown Pink Yellow  Sum
##   X00000   339  433    126  898
##   X00300    48  421    222  691
##   X12345    16  395    352  763
##   Others    23  373    156  552
##   Sum      426 1622    856 2904
```

Other times, we are interested in the proportions that are in each cell. The `proportions` function could convert these counts to overall proportions, but more interesting here is to ask what the color distribution was for each type of banding. This is called a *marginal distribution*, and `proportions` will compute it with the margin option:

```
xtabs(Count ~ Banding + Color, data = snails) %>%
  proportions(margin = 1) %>%
  round(digits = 2)
```

```
##         Color
## Banding  Brown Pink Yellow
##   X00000  0.38 0.48   0.14
##   X00300  0.07 0.61   0.32
##   X12345  0.02 0.52   0.46
```

```
## Others 0.04 0.68 0.28
```

The sum of proportions is 1 across each row. We see that 38% of unbanded (X0000) snails were brown, but only 2% of five-banded (X12345) snails were brown. The comparison of different banding types is easier to see with a plot. Tables produced by `xtabs` are not tidy, and therefore not suitable for sending to ggplot. Converting the table to a data frame with `as.data.frame` works, but instead we compute the counts with dplyr:

```
snails %>%
  group_by(Color, Banding) %>%
  summarize(Count = sum(Count)) %>%
  ggplot(aes(x = Banding, y = Count, fill = Color)) +
  geom_bar(stat = "identity") +
  scale_fill_manual(values = c("brown", "pink", "yellow")) +
  coord_flip()
```

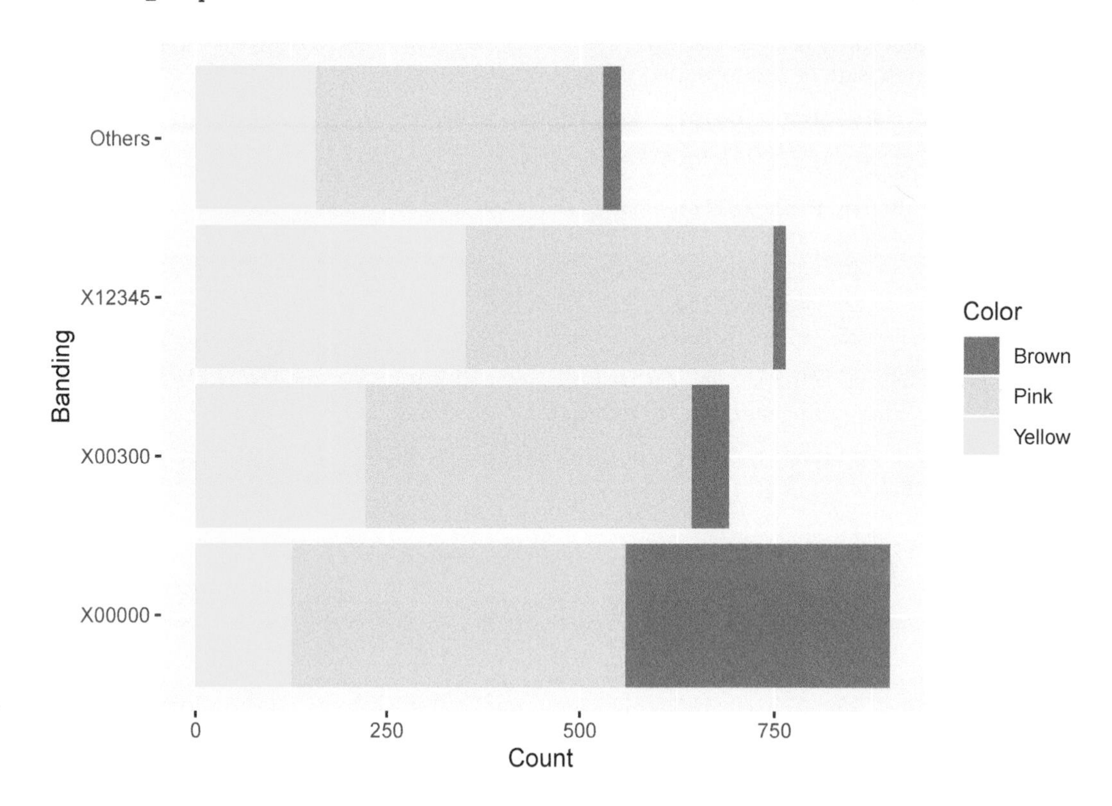


**FIGURE 10.1** Snail color and banding.

A common approach to visualizing categorical variables is with a pie chart. Pie charts are out of favor among data scientists because colors are hard to distinguish and the sizes of wedges are difficult to compare visually. In fact, ggplot2 does not include a built-in pie chart geometry. Instead, one applies polar coordinates to a barplot. Here is an example showing the proportions of snails found in each habitat:

```
snails %>%
  group_by(Habitat) %>%
  summarize(Count = sum(Count)) %>%
```

```
ggplot(aes(x = NA, y = Count, fill = Habitat)) +
geom_bar(stat = "identity") +
coord_polar("y") +
theme_void()
```

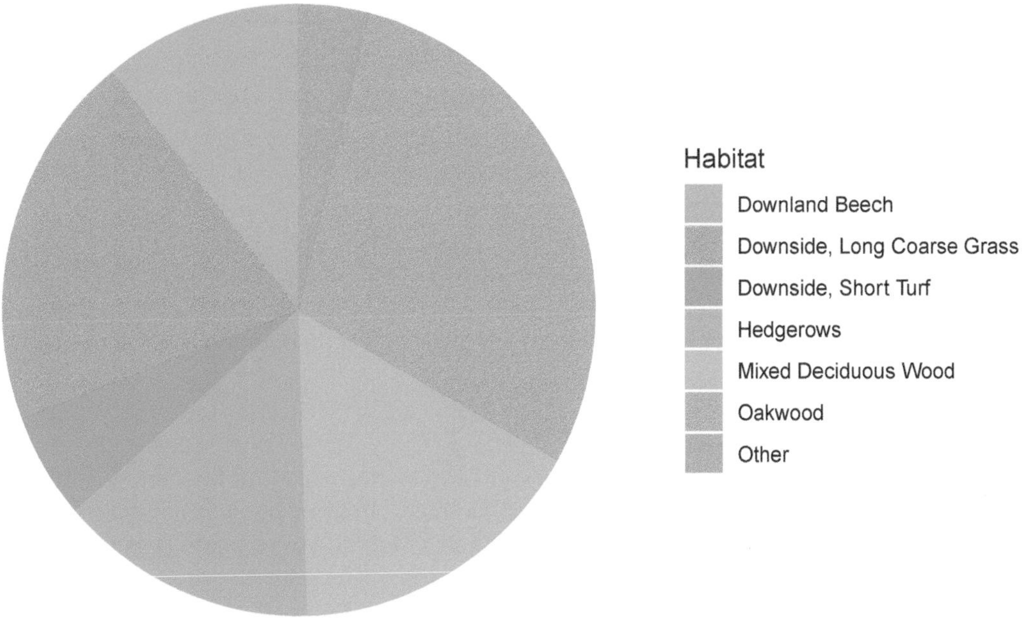

Can you tell whether there were more snails in the Hedgerows or in the Mixed Deciduous Wood? If you have colorblindness or happen to be reading a black and white copy of this text, you probably cannot even tell which wedge is which.

## 10.2 Inference on a proportion

The simplest setting for categorical data is that of a single binomial random variable. Here $X \sim \text{Binom}(n, p)$ is a count of successes on $n$ independent identical trials with probability of success $p$. A typical experiment would fix a value of $n$, perform $n$ Bernoulli trials, and produce a value of $X$. From this single value of $X$, we are interested in learning about the unknown population parameter $p$. For example, we may want to test whether the true proportion of times a die shows a 6 when rolled is actually $1/6$. We might choose to toss the die $n = 1000$ times and count the number of times $X$ that a 6 occurs.

Polling is an important application. Before an election, a polling organization will sample likely voters and ask them whether they prefer a particular candidate. The results of the poll should give an estimate for the true proportion of voters $p$ who prefer that candidate. The case of a voter poll is not formally a Bernoulli trial unless you allow the possibility of asking the same person twice; however, if the population is large then polling approximates a Bernoulli trial well enough to use these methods.

If $X$ is the number of successes on $n$ trials, the point estimate for the true proportion $p$ is given by

$$\hat{p} = \frac{X}{n}$$

Recall that $E[\hat{p}] = \frac{1}{n}E[X] = \frac{1}{n}np = p$, so $\hat{p}$ is an unbiased estimate of $p$. The standard deviation $\sigma(\hat{p})$ is $\sqrt{p(1-p)/n}$, so that a larger sample size $n$ will lead to less variation in $\hat{p}$ and therefore a better estimate of $p$.

Our goal is to use the sample statistic $\hat{p}$ to calculate confidence intervals and perform hypothesis testing with regards to $p$.

This section introduces *one sample tests of proportions*. Here, we present the theory associated with performing *exact* binomial hypothesis tests using `binom.test`, as well as `prop.test`, which uses the normal approximation.

A one sample test of proportions requires a hypothesized value of $p_0$. Often $p_0 = 0.5$, meaning we expect success and failure to be equally likely outcomes of the Bernoulli trial. Or, $p_0$ may come from historic values or a known larger population. The hypotheses are:

$$H_0 : p = p_0; \qquad H_a : p \neq p_0$$

You run $n$ trials and obtain $x$ successes, so your estimate for $p$ is given by $\hat{p} = x/n$. (We are thinking of $x$ as **data** rather than as a random variable.) Presumably, $\hat{p}$ is not exactly equal to $p_0$, and you wish to determine the probability of obtaining an estimate that unlikely or more unlikely, assuming $H_0$ is true.

### 10.2.1 Exact binomial test

Our first approach is the *binomial test*, which is an exact test in that it calculates a $p$-value using probabilities coming from the binomial distribution.

For the $p$-value, we are going to add the probabilities of all outcomes that are no more likely than the outcome that was obtained, since if we are going to reject when we obtain $x$ successes, we would also reject if we obtain a number of successes that was even less likely to occur. Formally, the $p$-value for the exact binomial test is given by:

$$\sum_{y:\ P(X=y) \leq P(X=x)} P(X = y)$$

**Example 10.1.** Consider the `wrist` data. Approximately 10.6% of the world's population is left-handed[4]. Is this sample of elderly Finns consistent with the proportion of left-handers in the world? In this binomial random variable, we choose left-handedness as success. Then $p$ is the true proportion of elderly Finns who are left-handed and $p_0 = 0.106$. Our hypotheses are:

$$H_0 : p = 0.106; \qquad H_a : p \neq 0.106$$

The sample contains 104 observations and has 7 left-handed patients, giving $\hat{p} = 7/104 \approx 0.067$, which is lower than $p_0$. The probability of getting exactly 7 successes under $H_0$ is

[4] M Papadatou-Pastou et al., "Human Handedness: A Meta-Analysis." *Psychological Bulletin* 146, no. 6 (2020): 481–524, https://doi.org/10.1037/bul0000229.

`dbinom(7, 104, 0.106)`, or 0.061. Anything less than 7 successes is less likely under the null hypothesis, so we would add all of those to get part of the $p$-value. To determine which values we add for successes *greater than* 7, we look for all outcomes that have probability of occurring (under the null hypothesis) less than 0.061. That is all outcomes 15 through 104, since $X = 14$ is more likely than $X = 7$ (`dbinom(14, 104, 0.106)` $= 0.075 > 0.061$) while $X = 15$ is less likely than $X = 7$ (`dbinom(15, 104, 0.106)` $= 0.053 < 0.061$).

The calculation is illustrated in Figure 10.2, where the dashed red line indicates the probability of observing exactly 7 successes. We sum all of the probabilities that are at or below the dashed red line.

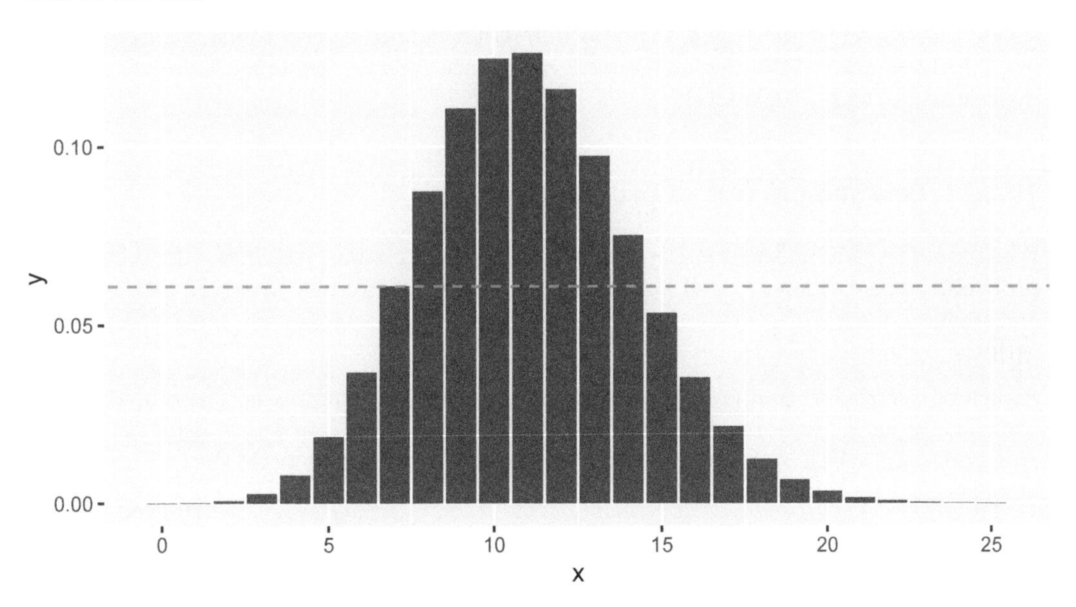


**FIGURE 10.2** The pmf for $X \sim \text{Binom}(104, 0.106)$, with a line at $P(X = 7)$. $X$ values past 25 are negligible and not shown.

The $p$-value is

$$P(X \le 7) + P(X \ge 15)$$

where $X \sim \text{Binom}(n = 104, p = 0.106)$.

```
sum(dbinom(x = 0:7, size = 104, prob = 0.106)) +
  sum(dbinom(x = 15:104, size = 104, prob = 0.106))
```

```
## [1] 0.2628259
```

R will make these computations for us, naturally, in the following way.

```
binom.test(x = 7, n = 104, p = 0.106)
```

```
##
##  Exact binomial test
##
## data:  7 and 104
## number of successes = 7, number of trials = 104, p-value =
## 0.2628
## alternative hypothesis: true probability of success is not equal to 0.106
```

```
## 95 percent confidence interval:
##  0.02748752 0.13377135
## sample estimates:
## probability of success
##             0.06730769
```

With a $p$-value of 0.26, we fail to reject the null hypothesis. There is not sufficient evidence to conclude that elderly Finns have a different proportion of left-handers than the world's proportion of lefties.

The `binom.test` function also produces the 95% confidence interval for $p$. In this example, we are 95% confident that the true proportion of left-handed elderly Finns is in the interval $[0.027, 0.134]$. Since 0.106 lies in the 95% confidence interval, we failed to reject the null hypothesis at the $\alpha = 0.05$ level.

### 10.2.2 One sample test of proportions

When $n$ is large and $p$ isn't too close to 0 or 1, binomial random variables with $n$ trials and probability of success $p$ are well approximated by normal random variables with mean $np$ and standard deviation $\sqrt{np(1-p)}$. This can be used to get approximate $p$-values associated with the hypothesis test $H_0 : p = p_0$ versus $H_a : p \neq p_0$.

As before, we need to compute the probability under $H_0$ of obtaining an outcome that is as likely or less likely than obtaining $x$ successes. However, in this case we are using the normal approximation, which is *symmetric* about its mean. The $p$-value is twice the area of the tail outside of $x$.

The `prop.test` function performs this calculation, and has identical syntax to `binom.test`.

**Example 10.2.** We return to the `wrist` example, testing $H_0 : p = 0.106$ versus $H_a : p \neq 0.106$. Let $X$ be a binomial random variable with $n = 104$ and $p = 0.106$. $X$ is approximated by a normal variable $Y$ with

$$\mu(Y) = np = 104 \cdot 0.106 = 11.024 \tag{10.1}$$

$$\sigma(Y) = \sqrt{np(1-p)} = \sqrt{104 \cdot 0.106 \cdot 0.894} = 3.13934. \tag{10.2}$$

Figure 10.3 is a plot of the pmf of $X$ with its normal approximation $Y$ overlaid. The shaded area corresponds to $Y \leq 7$. The $p$-value is twice that area, which we compute with `pnorm`:

```
2 * pnorm(7, 11.024, 3.13934)
```

```
## [1] 0.1999135
```

For a better approximation, we perform a *continuity correction* (see also Example 4.20). The basic idea is that $Y$ is a continuous rv, so when $Y = 7.3$, for example, we need to decide what *integer* value that should be associated with. Rounding suggests that $Y = 7.3$ should correspond to $X = 7$ and be included in the shaded area. The continuity correction includes values from 7 to 7.5 in the $p$-value, resulting in a corrected $p$-value of:

```
2 * pnorm(7.5, 11.024, 3.13934)
```

```
## [1] 0.2616376
```

The continuity correction gives a more accurate approximation to the underlying binomial rv, but not necessarily a closer approximation to the exact binomial test.

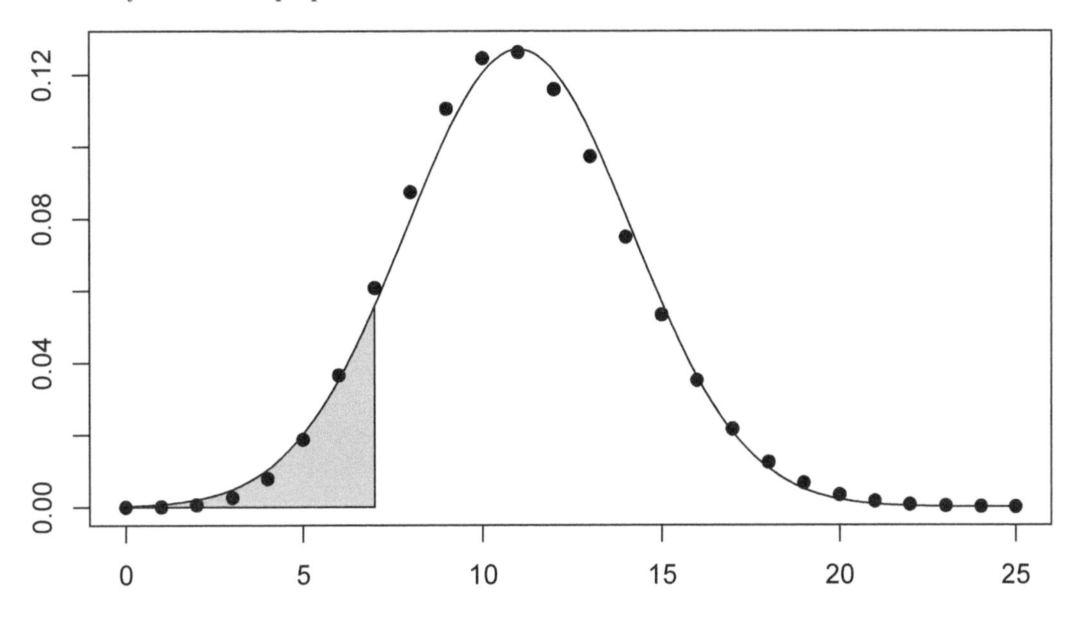


**FIGURE 10.3** Binomial rv with normal approximation overlaid.

The built-in R function for the one sample test of proportions is `prop.test`:

```
prop.test(x = 7, n = 104, p = 0.106)$p.value
```

```
## [1] 0.2616376
```

The `prop.test` function performs continuity correction by default. The $p$-value here is almost identical to the result of `binom.test`, and as before we fail to reject $H_0$. The confidence interval produced is also quite similar to the exact binomial test.

> **Try It Yourself.**
> Look at the full output of `prop.test(x = 7, n = 104, p = 0.106)`, and observe that the test statistic is given as a $\chi^2$ random variable with 1 degree of freedom. Confirm that the test statistic is $c = \left((\tilde{x} - np_0)/\sqrt{np_0(1-p_0)}\right)^2$, where $\tilde{x}$ is the number of successes after a continuity correction (in this case, $\tilde{x} = 7.5$).
>
> Use `pchisq(c, 1, lower.tail = FALSE)` to recompute the $p$-value using this test statistic. You should get the same answer $p = 0.2616377$.

**Example 10.3.** The Economist/YouGov Poll leading up to the 2016 presidential election sampled 3669 likely voters and found that 1798 intended to vote for Clinton. Assuming that this is a random sample from all likely voters, find a 99% confidence interval for $p$, the true proportion of likely voters who intended to vote for Clinton at the time of the poll.

```
prop.test(x = 1798, n = 3699, conf.level = 0.99)$conf.int
```

```
## [1] 0.4648186 0.5073862
## attr(,"conf.level")
## [1] 0.99
```

We are 99% confident that the true proportion of likely Clinton voters was between .465

and .507. In fact, 48.2% of voters did vote for Clinton, and the true value does fall in the 99% confidence interval range.

Most polls do not report a confidence interval. Typically, they report the point estimator $\hat{p}$ and the *margin of error*, which is half the width of the 95% confidence interval. For this poll, $\hat{p} \approx 0.486$ and the 95% confidence interval is $[0.470, 0.502]$ so the pollsters would report that they found 48.6% in favor of Clinton with a margin of error of 1.6%.

## 10.3 $\chi^2$ tests

The $\chi^2$ test is a general approach to testing the hypothesis that tabular data follows a given distribution. It relies on the Central Limit Theorem, in that the various counts in the tabular data are assumed to be approximately normally distributed.

The setting for $\chi^2$ testing requires tabular data. For each cell in the table, the count of observations that fall in that cell is a random variable. We denote the observed counts in the $k$ cells by $X_1, \ldots, X_k$. The null hypothesis requires an *expected count* for each cell, $E[X_i]$. The test statistic is the $\chi^2$ statistic.

**Definition 10.1.** If $X_1, \ldots, X_k$ are the observed counts of cells in tabular data, then the $\chi^2$ statistic is:

$$\chi^2 = \sum_{i=1}^{k} \frac{(X_i - E[X_i])^2}{E[X_i]}$$

The $\chi^2$ statistic is always positive, and will be larger when the observed values $X_i$ are far from the expected values $E[X_i]$. In all cases we consider, the $\chi^2$ statistic will have approximately the $\chi^2$ distribution with $d$ degrees of freedom, for some $d < k$. The $p$-value for the test is the probability of a $\chi^2$ value as large or larger than the observed $\chi^2$. The R function `chisq.test` computes $\chi^2$ and the corresponding $p$-value.

The $\chi^2$ test is always a one-tailed test. For example, if we observe $\chi^2 = 10$ and have four degrees of freedom, the $p$-value corresponds to the shaded area in Figure 10.4.

The full theory behind the $\chi^2$ test is beyond the scope of this book, but in the remainder of this section we give some motivation for the formula for $\chi^2$ and the meaning of degrees of freedom. A reader less interested in theory could proceed to Section 10.3.1.

Consider the value in one particular cell of tabular data. For each of the $n$ observations in the sample, the observation either lies in the cell or it does not, hence the count in that one cell can be considered as a binomial rv $X_i$. Let $p_i$ be the probability a random observation is in that cell. Then $E[X_i] = np_i$ and $\sigma(X_i) = \sqrt{np_i(1 - p_i)}$. If $np_i$ is sufficiently large (at least 5, say) then $X_i$ is approximately normal and

$$\frac{X_i - np_i}{\sqrt{np_i(1 - p_i)}} \sim Z_i,$$

where $Z_i$ is a standard normal variable. Squaring both sides and multiplying by $(1 - p_i)$ we have

$$(1 - p_i)Z_i^2 \sim \frac{(X_i - np_i)^2}{np_i} = \frac{(X_i - E[X_i])^2}{E[X_i]}$$

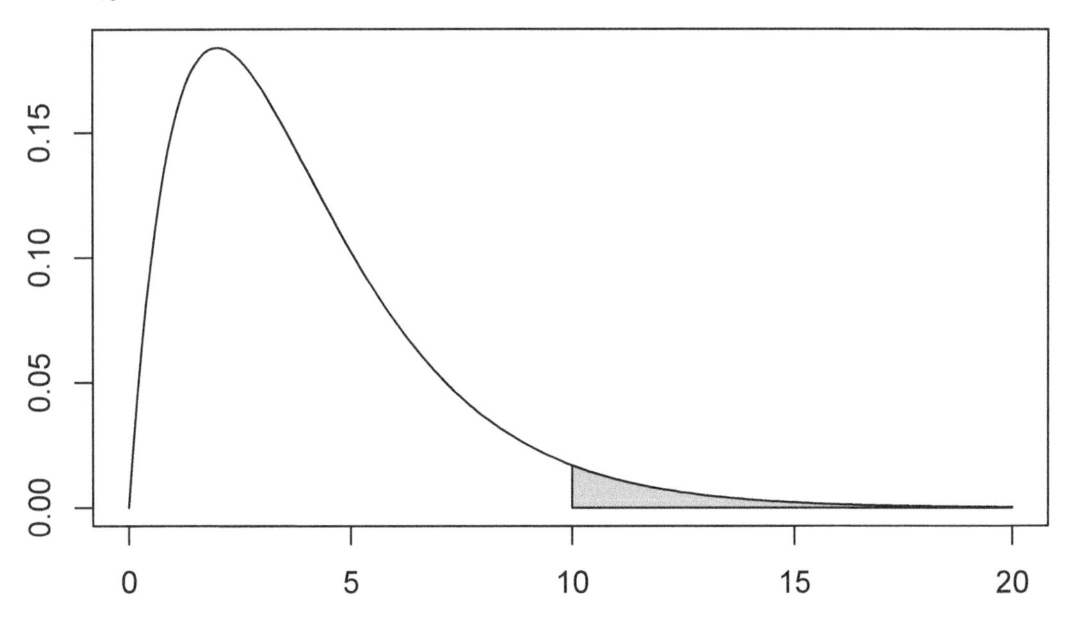


**FIGURE 10.4** $\chi^2$ distribution with $p$-value shaded.

As long as all cell counts are large enough, the $\chi^2$ statistic is approximately

$$\chi^2 = \sum_{i=1}^{k}(1 - p_i)Z_i^2$$

In this expression, the $Z_i$ are standard normal but *not independent* random variables. In many circumstances, one can rewrite these $k$ variables in terms of a smaller number $d$ of independent standard normal rvs and find that the $\chi^2$ statistic does have a $\chi^2$ distribution with $d$ degrees of freedom. The details of this process require some advanced linear algebra and making precise what we mean when we say $X_i$ are approximately normal. The details of the dependence are not hard to work out in the simplest case when the table has two cells.

**Example 10.4.** Consider a table with two cells, with $n$ observations, cell probabilities $p_1$, $p_2$, and cell counts given by the random variables $X_1$ and $X_2$:

| $X_1$ | $X_2$ |
|---|---|

This is simply a single binomial rv in disguise, with $X_1$ the count of successes and $X_2$ the count of failures. In particular, $p_1 + p_2 = 1$ and $X_1 + X_2 = n$. Notice that

$$\frac{X_1 - np_1}{\sqrt{np_1(1-p_1)}} + \frac{X_2 - np_2}{\sqrt{np_2(1-p_2)}} = \frac{X_1 + X_2 - n(p_1 + p_2)}{\sqrt{np_1p_2}} = \frac{n - n(1)}{\sqrt{np_1p_2}} = 0.$$

So the two variables $Z_1$ and $Z_2$ are not independent, and satisfy the equation $Z_1 + Z_2 = 0$. Then both can be written in terms of a single rv $Z$ with $Z_1 = Z$ and $Z_2 = -Z$. As long as $X_1$ and $X_2$ are both large, $Z_i$ will be approximately standard normal, and

$$\chi^2 = (1 - p_1)Z_1^2 + (1 - p_2)Z_2^2 = (1 - p_1 + 1 - p_2)Z^2 = Z^2.$$

We see that $\chi^2$ has the $\chi^2$ distribution with one df.

The table in this example has two entries, giving two possible counts $X_1$ and $X_2$. The constraint that these counts must sum to $n$ leaves only the single degree of freedom to choose $X_1$.

### 10.3.1 $\chi^2$ test for given probabilities

In this section, we consider data coming from a single categorical variable, typically displayed in a $1 \times k$ table:

| $X_1$ | $X_2$ | $\cdots$ | $X_k$ |
|---|---|---|---|

For our null hypothesis, we take some vector of probabilities $p_1, \ldots, p_k$ as given. Because there are $k$ cells to fill and a single constraint (the cells sum to $n$), the $\chi^2$ statistic will have $k - 1$ df.

The most common case is when we assume all cells are equally likely, in which case this approach is called the $\chi^2$ *test for uniformity*.

**Example 10.5.** Doyle, Bottomley, and Angell[5] investigated the "Relative Age Effect," the disadvantage of being the youngest in a cohort relative to being the oldest. The difference in outcomes can persist for years beyond any difference in actual ability relative to age difference.

In this study, the authors found the number of boys enrolled in the British elite soccer academies for under 18 years of age. They binned the boys into three age groups: oldest, middle, and youngest, with approximately 1/3 of all British children in each group. The number of boys in the elite soccer academies was:

| Oldest | Middle | Youngest |
|---|---|---|
| 631 | 321 | 155 |

The null hypothesis is that the boys should be equally distributed among the three groups, or equivalently $H_0 : p_i = \frac{1}{3}$. There are a total of 1107 boys in this study. Under the null hypothesis, we expect $369 = 1107/3$ in each group. Then

$$\chi^2 = \frac{(631 - 369)^2}{369} + \frac{(321 - 369)^2}{369} + \frac{(155 - 369)^2}{369} \approx 316.38.$$

The test statistic $\chi^2$ has the $\chi^2$ distribution with $2 = 3 - 1$ degrees of freedom, and a quick glance at that distribution shows that our observed 316.38 is impossibly unlikely to occur by chance. The $p$-value is essentially 0, and we reject the null hypothesis. Boys' ages in elite British soccer academies are not uniformly distributed across the three age bands for a given year.

In R, the computation is done with `chisq.test`:

[5] John R Doyle, Paul A Bottomley, and Rob Angell, "Tails of the Travelling Gaussian Model and the Relative Age Effect: Tales of Age Discrimination and Wasted Talent," *PLOS One* 12, no. 4 (April 2017): 1–22, https://doi.org/10.1371/journal.pone.0176206.

```
boys <- c(631, 321, 155)
chisq.test(boys, p = c(1 / 3, 1 / 3, 1 / 3))
```

```
##
##  Chi-squared test for given probabilities
##
## data:  boys
## X-squared = 316.38, df = 2, p-value < 2.2e-16
```

**Example 10.6.** Benford's Law is used in forensic accounting to detect falsified or manufactured data. When data, such as financial or economic data, occurs over several orders of magnitude, the first digits of the values follow the distribution

$$P(\text{first digit is } d) = \log_{10}(1 + 1/d)$$

```
benford <- log10(1 + 1 / (1:9))
round(benford, 2)
```

```
## [1] 0.30 0.18 0.12 0.10 0.08 0.07 0.06 0.05 0.05
```

The data `fosdata::rio_instagram` has the number of Instagram followers for gold medal winners at the 2016 Olympics. First, we extract the first digits of each athlete's number of followers:

```
rio <- fosdata::rio_instagram %>%
  mutate(first_digit = stringr::str_extract(n_follower, "[0-9]")) %>%
  filter(!is.na(first_digit))
```

Let's visually compare the counts of observed first digits (as bars) to the expected counts from Benford's Law (red dots):

```
rio %>% ggplot(aes(x = first_digit)) +
  geom_bar() +
  geom_point(
    data = data.frame(x = 1:9, y = benford * nrow(rio)),
    aes(x, y), color = "red", size = 5
  )
```

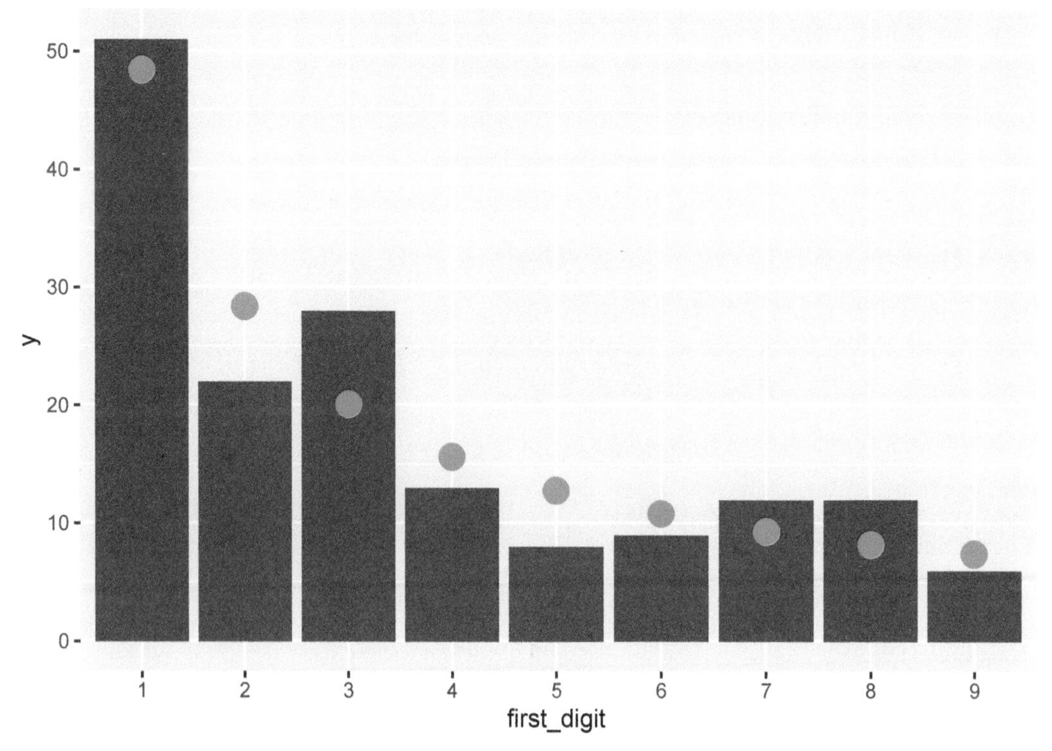


Is the observed data consistent with Benford's Law?

```
chisq.test(table(rio$first_digit), p = benford)
```

```
##
##  Chi-squared test for given probabilities
##
## data:  table(rio$first_digit)
## X-squared = 9.876, df = 8, p-value = 0.2738
```

The observed value of $\chi^2$ is 9.876, from a $\chi^2$ distribution with $8 = 9 - 1$ degrees of freedom. This is not extraordinary. The $p$-value is 0.2738 and we fail to reject $H_0$. The data is consistent with Benford's Law.

## 10.4 $\chi^2$ goodness of fit

In this section, we consider tabular data that is hypothesized to follow a parametric model. When the parameters of the model are estimated from the observed data, the model fits the data better than it should. Each estimated parameter reduces the degrees of freedom in the $\chi^2$ distribution by one.

When testing goodness of fit, the $\chi^2$ statistic is approximately $\chi^2$ with degrees of freedom given by the following:

$$\text{degrees of freedom} = \text{bins} - 1 - \text{parameters estimated from the data}.$$

We will explore this claim through simulation in Section 10.4.1.

**Example 10.7.** Goals in a soccer game arrive at random moments and could be reasonably modeled by a Poisson process. If so, the total number of goals scored in a soccer game should be a Poisson rv.

The data set `world_cup` from `fosdata` contains the results of the 2014 and 2015 FIFA World Cup soccer finals. Let's get the number of goals scored by each team in each game of the 2015 finals:

```
goals <- fosdata::world_cup %>%
  filter(competition == "2015 FIFA Women's World Cup") %>%
  tidyr::pivot_longer(cols = contains("score"), values_to = "score") %>%
  pull(score) # pull extracts the "score" column as a vector
table(goals)
```

```
## goals
##  0  1  2  3  4  5  6 10
## 30 40 20  6  3  2  1  2
```

We want to perform a hypothesis test to determine whether a Poisson model is a good fit for the distribution of goals scored. The Poisson distribution has one parameter, the rate $\lambda$. The expected value of a Poisson rv is $\lambda$, so we estimate $\lambda$ from the data:

```
lambda <- mean(goals)
```

Here $\lambda \approx 1.4$, meaning 1.4 goals were scored per game, on average. Figure 10.5 displays the observed counts of goals with the expected counts from the Poisson model Pois($\lambda$) in red.

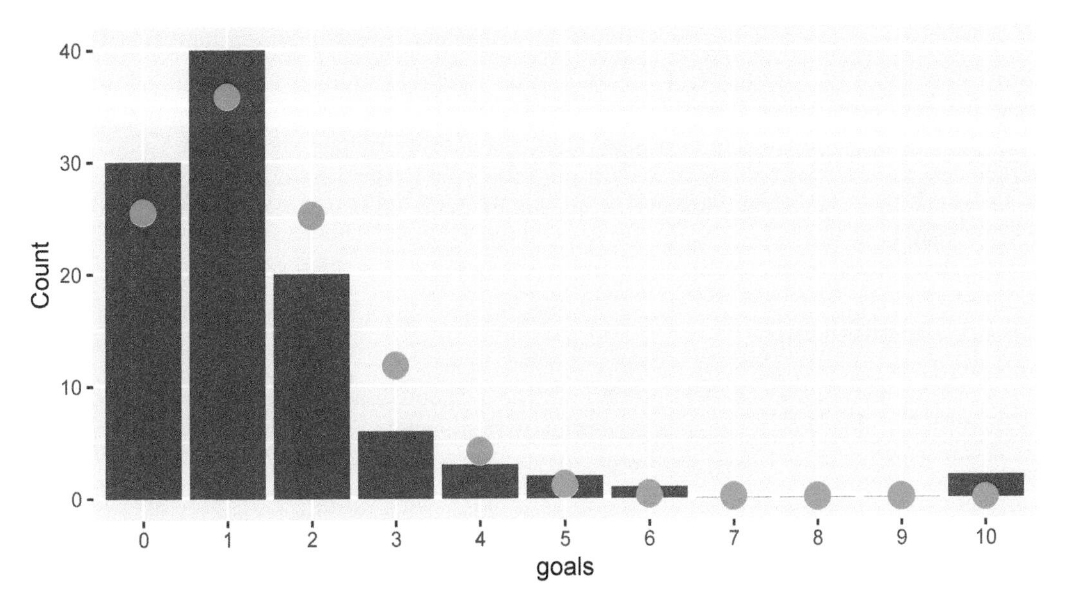


**FIGURE 10.5** Goals scored by each team in each game of the 2015 World Cup. Poisson model shown with red dots.

Since the $\chi^2$ test relies on the Central Limit Theorem, each cell in the table should have a large expected value to be approximately normal. Traditionally, the threshold is that a cell's expected count should be at least five. Here, all cells with 4 or more goals are too small.

The solution is to bin these small counts into one category, giving five total categories: zero goals, one goal, two goals, three goals, or 4+ goals. The observed and expected counts for the five categories are:

```
expected_goals <- 104 * c(
  dpois(0:3, lambda),
  ppois(3, lambda, lower.tail = FALSE)
)
observed_goals <- c(30, 40, 20, 6, 8)
```

The $\chi^2$ test statistic will have $3 = 5 - 1 - 1$ df, since:

- There are 5 bins.
- The bins sum to 104, losing one df.
- The model's one parameter $\lambda$ was estimated from the data, losing one df.

We compute the $\chi^2$ test statistic and $p$-value manually, because the `chisq.test` function is unaware that our expected values were modeled from the data, and would use the incorrect df.

```
chi_2 <- sum((observed_goals - expected_goals)^2 / expected_goals)
chi_2

## [1] 6.147538

pchisq(chi_2, df = 3, lower.tail = FALSE)

## [1] 0.1046487
```

The observed value of $\chi^2$ is 6.15. The $p$-value of this test is 0.105, and we would not reject $H_0$ at the $\alpha = .05$ level. This test does not give evidence against goal scoring being Poisson.

Note that there is one aspect of this data that is highly unlikely under the assumption that the data comes from a Poisson random variable: ten goals were scored on two different occasions. The $\chi^2$ test did not consider that, because we binned those large values into a single category. If you believe that data might not be Poisson because you suspect it will have unusually large values (rather than unusually many large values), then the $\chi^2$ test will not be very powerful.

### 10.4.1 Simulations

This section investigates the test statistic in the $\chi^2$ goodness of fit test via simulation. We observe that it does follow the $\chi^2$ distribution with df equal to bins minus one minus number of parameters estimated from the data.

Suppose that data comes from a Poisson variable $X$ with mean 2 and there are $N = 200$ data points.

```
test_data <- rpois(200, 2)
table(test_data)
```

| Goals | 0 | 1 | 2 | 3 | 4+ |
|---|---|---|---|---|---|
| Observed | 30 | 40 | 20 | 6 | 8 |
| Expected | 25.5 | 35.9 | 25.2 | 11.8 | 5.6 |

```
## test_data
##  0  1  2  3  4  5  6  7
## 24 53 54 35 19 12  2  1
```

The expected count in bin 5 is `200 * dpois(5,2)` which is 7.2, large enough to use. The expected count in bin 6 is only 2.4, so we combine all bins 5 and higher. In a real experiment, the sample data could affect the number of bins chosen, but we ignore that technicality.

```
test_data[test_data > 5] <- 5
table(test_data)

## test_data
##  0  1  2  3  4  5
## 24 53 54 35 19 15
```

Next, compute the expected counts for each bin using the rate $\lambda$ estimated from the data. Bins 0-4 can use `dpois` but bin 5 needs the entire tail of the Poisson distribution.

```
lambda <- mean(test_data)
p_fit <- c(dpois(0:4, lambda), ppois(4, lambda, lower.tail = FALSE))
expected <- 200 * p_fit
```

Finally, we produce one value of the test statistic:

```
observed <- table(test_data)
sum((observed - expected)^2 / expected)

## [1] 0.9263996
```

Naively using `chisq.test` with the data and the fit probabilities gives the same value of $\chi^2 = 0.9264$, but produces a $p$-value using 5 df, which is wrong. The function does not know that we used one df to estimate a parameter.

```
# wrong df produces incorrect p-value
chisq.test(observed, p = p_fit)

##
##  Chi-squared test for given probabilities
##
## data:  observed
## X-squared = 0.9264, df = 5, p-value = 0.9683
```

We now replicate to produce a sample of values of the test statistic to verify that 4 is the correct df for this test:

```
sim_data <- replicate(10000, {
  test_data <- rpois(200, 2) # produce data
  test_data[test_data > 5] <- 5 # re-bin to six bins
  observed <- table(test_data)

  lambda <- mean(test_data)
  expected <- 200 * c(
    dpois(0:4, lambda),
    ppois(4, lambda, lower.tail = FALSE)
  )
```

```
  sum((observed - expected)^2 / expected)
})

plot(density(sim_data),
  main = "Test statistic and chi-squared distributions"
)
curve(dchisq(x, 4), add = TRUE, col = "blue")
curve(dchisq(x, 5), add = TRUE, col = "red")
```

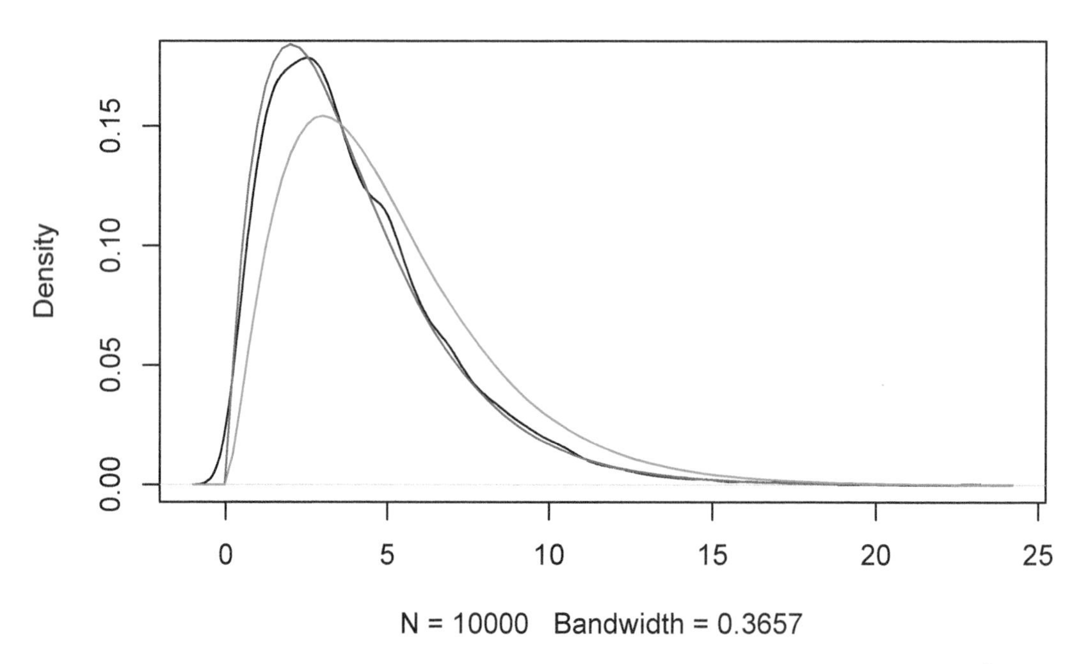


The black curve is the probability density from our simulated data. The blue curve is $\chi^2$ with 4 degrees of freedom, equal to (bins - parameters - 1). The red curve is $\chi^2$ with 5 degrees of freedom and does not match the observations. This seems to be pretty compelling.

## 10.5 $\chi^2$ tests on cross tables

Given two categorical variables $A$ and $B$, we can form a cross table with one cell for each pair of values $(A_i, B_j)$. That cell's count is a random variable $X_{ij}$:

| | $B_1$ | $B_2$ | $\cdots$ | $B_n$ |
|---|---|---|---|---|
| $A_1$ | $X_{11}$ | $X_{12}$ | $\cdots$ | $X_{1n}$ |
| $A_2$ | $X_{21}$ | $X_{22}$ | $\cdots$ | $X_{2n}$ |
| $\vdots$ | $\vdots$ | $\vdots$ | $\ddots$ | $\vdots$ |
| $A_m$ | $X_{m1}$ | $X_{m2}$ | $\cdots$ | $X_{mn}$ |

As in all $\chi^2$ tests, the null hypothesis leads to an expected value for each cell. In this setting, we require a probability $p_{ij}$ that an observation lies in cell $(i, j)$, $p_{ij} = P(A = A_i \cap B = B_j)$. These probabilities are called the *joint probability distribution* of $A$ and $B$.

The hypothesized joint probability distribution needs to come from somewhere. It could come from historical or population data, or by fitting a parametric model, in which case the methods of the previous two sections apply.

We assume that $B$ is random (and perhaps $A$ as well, but not necessarily) and we consider the null hypothesis that the probability distribution of $B$ is independent of the levels of $A$. Let $N$ be the total number of observations. If we let $a_i = \frac{1}{N}\sum_j X_{ij}$ denote the proportion of observations for which $A = A_i$ and $b_j = \frac{1}{N}\sum_i X_{ij}$ denote the proportion of responses for which $B = B_j$, then under the assumption of $H_0$ we would hypothesize that

$$p_{ij} = a_i b_j.$$

It follows that $E[X_{ij}] = N a_i b_j$.

The test statistic is

$$\chi^2 = \sum_{i,j} \frac{(X_{ij} - E[X_{ij}])^2}{E[X_{ij}]}.$$

When the expected cell counts $E[X_{ij}]$ are all large enough, the test statistic has approximately a $\chi^2$ distribution with (columns$-1$)(rows$-1$) degrees of freedom. There are two explanations for why this is the correct degrees of freedom, depending on the details of the experimental design. The mechanics of the test itself, however, do not depend on the experimental design. Sections 10.5.1 and 10.5.2 discuss the details.

### 10.5.1 $\chi^2$ test of independence

In the $\chi^2$ test of independence, the levels of $A$ and $B$ are both random. In this case, we are testing

$$H_0 : A \text{ and } B \text{ are independent random variables}$$

versus the alternative that they are not independent. The values of $p_{ij} = a_i b_j$ have a natural interpretation as $p_{ij} = P(A = A_i \cap B = B_j) = P(A = A_i)P(B = B_j)$.

To understand the degrees of freedom in the test for independence, the experimental design matters. We fix $N$ the total number of observations, and for each subject the two categorical variables $A$ and $B$ are measured (see Example 10.8). The row and column marginal sums of the cross table are random. Then:

- There are $mn$ cells.
- There are $m + n$ marginal probabilities $a_i$ and $b_j$ estimated from the data, and $\sum a_i = \sum b_i = 1$, so we lose $m + n - 2$ df.
- All cell counts must add to $N$, losing one df.
- $mn - (m + n - 2) - 1 = (m - 1)(n - 1)$

**Example 10.8.** Are grove snail color and banding patterns related? Figure 10.1 suggests that brown snails are more likely to be unbanded than the other colors.

In R, the $\chi^2$ test for independence is simple: we pass the cross table to `chisq.test`.

```
snailtable <- xtabs(Count ~ Color + Banding, data = fosdata::snails)
snailtable
```

```
##         Banding
## Color    X00000 X00300 X12345 Others
##   Brown     339     48     16     23
##   Pink      433    421    395    373
##   Yellow    126    222    352    156
```

```
chisq.test(snailtable)
```

```
##
##  Pearson's Chi-squared test
##
## data:  snailtable
## X-squared = 652.77, df = 6, p-value < 2.2e-16
```

The cross table is $3 \times 4$, so the $\chi^2$ statistic has $(3-1)(4-1) = 6$ df. The $p$-value is very small, so we reject $H_0$. Snail color and banding are not independent.

Let's reproduce the results of `chisq.test`. First, compute marginal probabilities.

```
a_color <- snailtable %>%
  marginSums(margin = 1) %>%
  proportions()
a_color
b_band <- snailtable %>%
  marginSums(margin = 2) %>%
  proportions()
b_band
```

```
## Color
##     Brown      Pink    Yellow
## 0.1466942 0.5585399 0.2947658
```

```
## Banding
##    X00000    X00300    X12345    Others
## 0.3092287 0.2379477 0.2627410 0.1900826
```

Next, compute the joint distribution $p_{ij} = a_i b_j$. This uses the matrix multiplication operator `%*%` and the matrix transpose `t` to compute all 12 entries at once. The result is multiplied by $N = 2904$ to get expected cell counts:

```
p_joint <- a_color %*% t(b_band)
expected <- p_joint * 2904
round(expected, 1)
```

```
##         Banding
## Color    X00000 X00300 X12345 Others
##   Brown   131.7  101.4  111.9   81.0
##   Pink    501.6  386.0  426.2  308.3
##   Yellow  264.7  203.7  224.9  162.7
```

Finally, compute the $\chi^2$ test statistic and the $p$-value, which match the results of `chisq.test`.

```
chi2 <- sum((snailtable - expected)^2 / expected)
chi2
```

```
## [1] 652.7671
```

```
pchisq(chi2, df = 6, lower.tail = FALSE)
```

```
## [1] 9.605329e-138
```

It is instructive to view each cell's contribution to $\chi^2$ graphically as a "heatmap" to provide a sense of which cells were most responsible for the dependence between color and banding.

```
((snailtable - expected)^2 / expected) %>%
  as.data.frame() %>%
  ggplot(aes(x = Banding, y = Color, fill = Freq)) +
  geom_tile() +
  scale_fill_gradient(low = "white", high = "red")
```

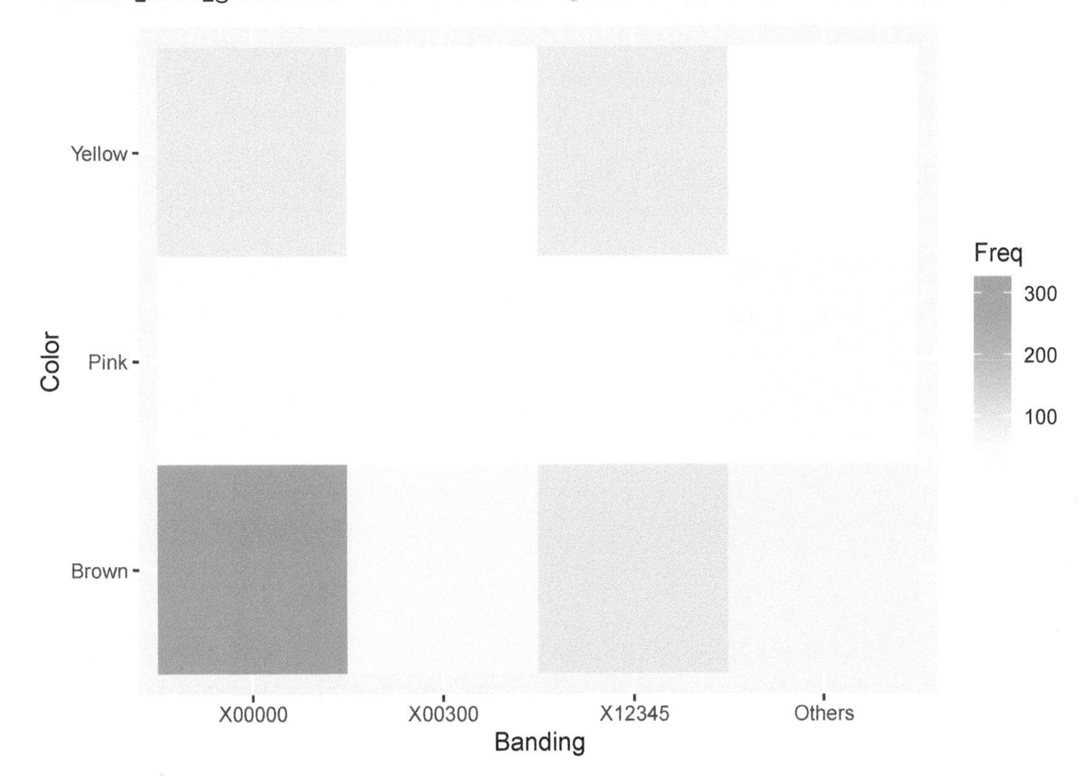


Clearly, most of the interaction between Banding and Color comes from the overabundance of unbanded (X00000) Brown snails. The authors of the original study were interested in environmental effects on color and bandedness of snails. It is possible, though a more thorough analysis would be required, that an environment that favors the survival of brown snails also favors unbanded snails.

**Example 10.9.** To what extent do animals display conformity? That is, will they forgo personal information in order to follow the majority? Researchers[6] studied conformity among

[6] Markus Germar et al., "Dogs (Canis Familiaris) Stick to What They Have Learned Rather Than Conform to Their Conspecifics' Behavior," *PLOS One* 13, no. 3 (March 2018): 1–16, https://doi.org/10.1371/journal.pone.0194808.

dogs. They trained a subject dog to walk around a wall in one direction in order to receive a treat. After training, the subject dog then watched other dogs walk around the wall in the opposite direction. If the subject dog changes its behavior to match the dogs it observed, it is evidence of conforming behavior.

The data from this experiment is available as the `dogs` data frame in the `fosdata` package.

```
dogs <- fosdata::dogs
```

This data set has quite a bit going on. In particular, each dog repeated the experiment three times, which means that it would be unwise to assume independence across trials. So, we will restrict to the first trial only. We also restrict to dogs that did not drop out of the experiment.

```
dogs_trial1 <- filter(dogs, trial == 1, dropout == 0)
```

Subject dogs participated under three conditions. The control group (condition = 0) observed no other dogs, and was simply asked to repeat what they were trained to do. Another group (condition = 1) saw one dog that went the "wrong" way around the wall three times. Another group (condition = 3) saw three different dogs that each went the wrong way around the wall one time.

We summarize the results of the experiment with a table showing the three experimental conditions in the three rows and whether the subject dog conformed or not in the two columns.

```
xtabs(~ condition + conform, data = dogs_trial1)
```

```
##          conform
## condition  0  1
##         0 12  3
##         1 20  9
##         3 17  9
```

The null hypothesis is that `conform` and `condition` are independent variables, so that the three groups of dogs would have the same conforming behavior. We store the cross table and apply the $\chi^2$ test for independence:

```
dogtable <- xtabs(~ condition + conform, data = dogs_trial1)
chisq.test(dogtable)
```

```
## Warning in chisq.test(dogtable): Chi-squared approximation may be
## incorrect
```

```
##
##  Pearson's Chi-squared test
##
## data:  dogtable
## X-squared = 0.9928, df = 2, p-value = 0.6087
```

The $p$-value is 0.61, so there is not significant evidence that the conform and condition variables are dependent. Dogs do not disobey training to conform, at least according to this simple analysis.

The $\chi^2$ test reports 2 df because we have 3 rows and 2 columns, and $(3-1)(2-1) = 2$. The test also produces a warning, because the expected cell count for conforming dogs under condition 0 is low. With a high $p$-value and good cell counts elsewhere, the lack of accuracy is not a concern.

> **Try It Yourself.**
> A link to the paper associated with the `dogs` data is given in `?fosdata::dogs`. Find the place in the paper where they perform the above $\chi^2$ test, and read the authors' explanation related to it.

### 10.5.2 $\chi^2$ test of homogeneity

In a $\chi^2$ test of homogeneity, one of the variables $A$ and $B$ is not random. For example, if an experimenter decides to collect data on cats by finding 100 American shorthair cats, 100 highlander cats, and 100 munchkin cats and measuring eye color for each of the 300 cats, then the number of cats of each breed is not a random variable. A test of this type is called a $\chi^2$ test of homogeneity, or a $\chi^2$ test with one fixed margin. However, we are still interested in whether the distribution of eye color depends on the breed of the cat, and we proceed exactly in the same manner as before, with a slightly reworded null hypothesis and a different justification of the degrees of freedom. We denote $B$ as the variable that is random. Our null hypothesis is:

$$H_0 : \text{ the distribution of } B \text{ does not depend on the level of } A$$

and the alternative hypothesis is that the distribution of $B$ does depend on the level of $A$. We compute degrees of freedom as follows:

- There are $mn$ cells.
- There are $n$ marginal probabilities $b_1, \ldots, b_n$. Since these must sum to 1, we lose $n - 1$ degrees of freedom.
- Each row sums to a fixed number, so we lose $m$ degrees of freedom.
- We do not lose any degrees of freedom for all bins summing to $N$, since that is implied by the column condition.
- Total degrees of freedom are $mn - (n - 1) - m = (m - 1)(n - 1)$, as in the case of the $\chi^2$ test of independence.

The mechanics of a $\chi^2$ test of homogeneity are the same as a $\chi^2$ test of independence.

**Example 10.10.** Consider the `sharks` data set[7] in the `fosdata` package. Participants were paid 25 cents to listen to either silence, ominous music, or uplifting music while possibly watching a video on sharks. An equal number were recruited for each type of music. They were then asked to give their rating from 1-7 on their willingness to help conserve endangered sharks. We are interested in whether the distribution of the participants' willingness to conserve sharks depends on the type of music they listened to.

We start by computing the cross table of the data.

```
sharks <- fosdata::sharks
shark_tabs <- xtabs(~ music + conserve, data = sharks)
shark_tabs %>%
  addmargins()
```

[7] Andrew P Nosal et al., "The Effect of Background Music in Shark Documentaries on Viewers' Perceptions of Sharks." *PLOS One* 11, no. 8 (2016): e0159279, https://doi.org/10.1371/journal.pone.0159279.

```
##               conserve
## music          1   2   3   4   5   6   7 Sum
##     ominous   10   6  13  28  43  49  54 203
##     silence    5   6  12  28  57  51  48 207
##     uplifting  6   6   7  24  59  44  60 206
##     Sum       21  18  32  80 159 144 162 616
```

The rows do not add up to exactly the same number because some participants dropped out of the study. We ignore this problem and continue.

```
chisq.test(shark_tabs)
```

```
##
##  Pearson's Chi-squared test
##
## data:  shark_tabs
## X-squared = 9.0547, df = 12, p-value = 0.6982
```

We see that there is not sufficient evidence to conclude that the distribution of willingness to help conserve endangered sharks depends on the type of music heard ($p = .6982$).

### 10.5.3 Two sample test for equality of proportions

An important special case of the $\chi^2$ test for independence is the two sample test for equality of proportions.

Suppose that $n_1$ trials are made from population 1 with $x_1$ successes, and that $n_2$ trials are made from population 2 with $x_2$ successes. We wish to test $H_0 : p_1 = p_2$ versus $H_a : p_1 \neq p_2$, where $p_i$ is the true probability of success from population $i$. We create a $2 \times 2$ table of values as follows:

| | Pop. 1 | Pop. 2 |
|---|---|---|
| Successes | $x_1$ | $x_2$ |
| Failures | $n_1 - x_1$ | $n_2 - x_2$ |

The null hypothesis says that $p_1 = p_2$. We estimate this common probability using all the data:

$$\hat{p} = \frac{\text{Successes}}{\text{Trials}} = \frac{x_1 + x_2}{n_1 + n_2}$$

The expected number of successes under $H_0$ is calculated from $n_1$, $n_2$, and $\hat{p}$:

| | Pop. 1 | Pop. 2 |
|---|---|---|
| Exp. Successes | $n_1\hat{p}$ | $n_2\hat{p}$ |
| Exp. Failures | $n_1(1 - \hat{p})$ | $n_2(1 - \hat{p})$ |

We then compute the $\chi^2$ test statistic. This has 1 df, since there were 4 cells, two constraints that the columns sum to $n_1$, $n_2$, and one parameter estimated from the data.

The test statistic and $p$-value can be computed with `chisq.test`. The `prop.test` function performs the same computation, and allows for finer control over the test in this specific setting.

**Example 10.11.** Researchers randomly assigned patients with wrist fractures to receive a cast in one of two positions, the VFUDC position and the functional position. The assignment of cast position should be independent of which wrist (left or right) was fractured. We produce a cross table from the data in `fosdata::wrist` and run the $\chi^2$ test for independence:

```
wrist_table <- xtabs(~ cast_position + fracture_side, data = fosdata::wrist)
wrist_table
```

```
##              fracture_side
## cast_position  1  2
##             1 18 32
##             2 27 28
```

```
chisq.test(wrist_table)
```

```
##
##  Pearson's Chi-squared test with Yates' continuity correction
##
## data:  wrist_table
## X-squared = 1.3372, df = 1, p-value = 0.2475
```

For `prop.test` we need to know the group sizes, $n_1 = 45$ with right-side fractures and $n_2 = 60$ with left-side fractures. We also need the number of successes, which we arbitrarily select as cast position 1.

```
prop.test(c(18, 32), c(45, 60))
```

```
##
##  2-sample test for equality of proportions with continuity
##  correction
##
## data:  c(18, 32) out of c(45, 60)
## X-squared = 1.3372, df = 1, p-value = 0.2475
## alternative hypothesis: two.sided
## 95 percent confidence interval:
##  -0.34362515  0.07695849
## sample estimates:
##    prop 1    prop 2
## 0.4000000 0.5333333
```

The `prop.test` function applies a continuity correction by default. `chisq.test` only applies continuity correction in this $2 \times 2$ case. There seems to be some disagreement on whether or not continuity correction is desirable. From the point of view of this text, we would choose the version that has observed type I error rate closest to the assigned rate of $\alpha$. Let's run some simulations, using $n_1 = 45$, $n_2 = 60$, and success probability $p = 50/105$ to match the wrist example.

```
p <- 50 / 105
n1 <- 45
n2 <- 60
data_ccorrected <- replicate(10000, {
  x1 <- rbinom(1, n1, p)
  x2 <- rbinom(1, n2, p)
  prop.test(c(x1, x2), c(n1, n2), correct = TRUE)$p.value
})
```

```
data_not_ccorrected <- replicate(10000, {
  x1 <- rbinom(1, n1, p)
  x2 <- rbinom(1, n2, p)
  prop.test(c(x1, x2), c(n1, n2), correct = FALSE)$p.value
})
```

```
mean(data_ccorrected < .05)
```

```
## [1] 0.0284
```

```
mean(data_not_ccorrected < .05)
```

```
## [1] 0.0529
```

We see that for this sample size and common probability of success, `correct = FALSE` comes closer to the desired type I error rate of 0.05, but is a bit too high. This holds across a wide range of $p$, $n_1$ and $n_2$. Using continuity correction tends to have effective error rates lower than the designed type I error rates, while `correct = FALSE` has type I error rates closer to the designed type I error rates.

**Example 10.12.** Consider the `babynames` data set in the `babynames` package. Is there a statistically significant difference in the proportion of girls named "Bella"[8] in 2007 and the proportion of girls named "Bella" in 2009?

We will need to do some data wrangling on this data and define a binomial variable `bella`:

```
bella_table <- babynames::babynames %>%
  filter(sex == "F", year %in% c(2007, 2009)) %>%
  mutate(name = ifelse(name == "Bella", "Bella", "Not Bella")) %>%
  xtabs(n ~ year + name, data = .)
bella_table
```

```
##       name
## year    Bella Not Bella
##   2007   2253   1918366
##   2009   4532   1830067
```

We see that the number of girls named "Bella" nearly doubled from 2007 to 2009. The two sample proportions test shows that this was highly significant.

```
prop.test(bella_table)$p.value
```

```
## [1] 2.982331e-192
```

[8]"Bella" was the name of the character played by Kristen Stewart in the movie *Twilight*, released in 2008. Fun fact, one of the authors has a family member who appeared in *The Twilight Saga: Breaking Dawn - Part 2.*

## 10.6 Exact and Monte Carlo methods

The $\chi^2$ methods of the previous sections all approximate discrete variables with continuous (normal) variables. Exact and Monte Carlo methods are very general approaches to testing tabular data, and neither method requires assumptions of normality.

*Exact methods* produce exact $p$-values by examining all possible ways the $N$ outcomes could fill the table. The first step of an exact method is to compute the test statistic associated to the observed data, often $\chi^2$. Then for each possible table, compute the test statistic and the probability of that table occurring, assuming the null hypothesis. The $p$-value is the sum of the probabilities of the tables whose associated test statistics are as extreme or more extreme than the observed test statistic. This $p$-value is exact because (assuming the null hypothesis) it is *exactly* the probability of obtaining a test statistic as or more extreme than the one coming from the data.

Unfortunately, the number of ways to fill out a table grows exponentially with the number of cells in the table (or more precisely, exponentially in the degrees of freedom). This makes exact methods unreasonably slow when $N$ is large or the table has many cells. *Monte Carlo methods* present a compromise that avoids assumptions but stays computationally tractable. Rather than investigate every possible way to fill the table, we randomly create many tables according to the null hypothesis. For each, the $\chi^2$ statistic is computed. The $p$-value is taken to be the proportion of generated tables that have a larger $\chi^2$ statistic than the observed data. Though we compute the $\chi^2$ statistic for the observed and simulated tables, we do not rely on assumptions about its distribution – it may not have a $\chi^2$ distribution at all.

**Example 10.13.** Return to the data on age cohorts in soccer, introduced in Example 10.5. There were three relative age groups in each cohort year: old, middle, and young. Our null hypothesis is that each age group should be equally likely for an elite soccer player in a given cohort. The data has $N = 1107$ boys, with 631, 321, and 155 in the old, middle, and young groups.

To apply Monte Carlo methods, we need to generate simulated $3 \times 1$ tables. We use the R function `rmultinom`, which generates multinomially distributed random number vectors. As with all random variable generation functions in R, the first argument to `rmultinom` is the number of simulations we want. Then there are two required parameters, the number of observations $N$ in each table and the null hypothesis probability distribution. Here are ten tables that might result from the experiment, one in each column.

```
sim_boys <- rmultinom(10, size = 1107, prob = c(1 / 3, 1 / 3, 1 / 3))
sim_boys
```

```
##      [,1] [,2] [,3] [,4] [,5] [,6] [,7] [,8] [,9] [,10]
## [1,]  355  337  373  357  412  387  370  364  378   410
## [2,]  394  367  383  368  363  366  365  375  388   349
## [3,]  358  403  351  382  332  354  372  368  341   348
```

From the first column, one possible outcome of the soccer study would be to find 355, 394, and 358 boys in the old, medium, and young groups. The next nine columns are also possible outcomes, each with $N = 1107$ observations. It is apparent that the observed value of 631 boys in the "old" group is exceptionally large under $H_0$.

To get a $p$-value, we first compute the $\chi^2$ statistic for the observed data:

```
observed_boys <- c(631, 321, 155)
expected_boys <- c(1 / 3, 1 / 3, 1 / 3) * sum(observed_boys)
sum((observed_boys - expected_boys)^2 / expected_boys)
```

```
## [1] 316.3794
```

The $\chi^2$ statistic is a measure of how far our observed group sizes are from the expected group sizes. For the observed boys, $\chi^2$ is 316.3794. Next compute the $\chi^2$ statistic for each set of simulated group sizes:

```
colSums((sim_boys - expected_boys)^2 / expected_boys)
```

```
##  [1] 2.5528455 5.9186992 1.4525745 0.8509485 8.8184282 1.5121951
##  [7] 0.0704607 0.1680217 3.3224932 6.8346883
```

Again, it is clear that the observed data is quite different than the data that was simulated under $H_0$. We should use more than 10 simulations, of course, but for this particular data you will never see a value as large as 316 in the simulations. The true $p$-value for this experiment is essentially zero.

R can carry out the Monte Carlo method within the `chisq.test` function:

```
chisq.test(observed_boys,
  p = c(1 / 3, 1 / 3, 1 / 3),
  simulate.p.value = TRUE
)
```

```
##
##  Chi-squared test for given probabilities with simulated
##  p-value (based on 2000 replicates)
##
## data:  observed_boys
## X-squared = 316.38, df = NA, p-value = 0.0004998
```

The function performed 2000 simulations and none of them had a higher $\chi^2$ value than the observed data. The $p$-value was reported as $1/2001$, because R always includes the actual data set in addition to the 2000 simulated values. This is a common technique that makes a relatively small absolute difference in estimates.

**Example 10.14.** Continuing with the boys elite soccer age data, we show how to apply the *exact multinomial test*.

The idea of the exact test is to sum the probabilities of all tables that lead to test statistics that are as extreme or more extreme than the observed test statistic. The table of boys is $3 \times 1$, and we need the three values in the table to sum to 1107. In Exercise 10.29, you are asked to show that there are 614386 possible ways to fill a $3 \times 1$ table with numbers that sum to 1107.

The `multinomial.test` function in the `EMT` package carries out this process.

```
EMT::multinomial.test(observed_boys, prob = c(1 / 3, 1 / 3, 1 / 3))
```

```
##
##  Exact Multinomial Test, distance measure: p
##
##     Events    pObs    p.value
##     614386       0          0
```

As before, the $p$-value is 0. The `EMT::multinomial.test` function can also run Monte Carlo tests using the parameter `MonteCarlo = TRUE`.

## Vignette: Tables

Tables are an often overlooked part of data visualization and presentation. They can also be **difficult** to do well! In this vignette, we introduce the `knitr::kable` function, which produces tables compatible with .pdf, .docx and .html output inside of your R Markdown documents.

To make a table using `knitr::kable`, create a data frame and apply kable to it.

```
knitr::kable(xtabs(Count ~ Banding + Color, data = snails))
```

| | Brown | Pink | Yellow |
|---|---|---|---|
| X00000 | 339 | 433 | 126 |
| X00300 | 48 | 421 | 222 |
| X12345 | 16 | 395 | 352 |
| Others | 23 | 373 | 156 |

**Example 10.15.** Suppose you are studying the `palmerpenguins::penguins` data set, and you want to report the mean, standard deviation, range, and number of samples of bill length in each species type. The `dplyr` package helps to produce the data frame, and we use kable options to create a caption and better column headings. The table is displayed as Table 10.1.

```
penguins <- palmerpenguins::penguins
penguin_table <- penguins %>%
  filter(!is.na(bill_length_mm)) %>%
  group_by(species) %>%
  summarize(
    mean(bill_length_mm),
    sd(bill_length_mm),
    paste(min(bill_length_mm), max(bill_length_mm), sep = " - "),
    n()
  )
knitr::kable(penguin_table,
  caption = "Bill lengths (mm) for penguins.",
  col.names = c("Species", "Mean", "SD", "Range", "# Birds"),
  digits = 2
)
```

TABLE 10.1: Bill lengths (mm) for penguins.

| Species | Mean | SD | Range | # Birds |
|---|---|---|---|---|
| Adelie | 38.79 | 2.66 | 32.1 – 46 | 151 |
| Chinstrap | 48.83 | 3.34 | 40.9 – 58 | 68 |
| Gentoo | 47.50 | 3.08 | 40.9 – 59.6 | 123 |

The kable package provides only basic table styles. To adjust the width and other features of table style, use the `kableExtra` package.

Another interesting use of tables is in combination with `broom::tidy`, which converts the outputs of many common statistical tests into data frames. Let's see how it works with `t.test`.

**Example 10.16.** Display the results of a $t$-test of the body temperature data from `fosdata::normtemp` in a table.

```
t.test(fosdata::normtemp$temp, mu = 98.6) %>%
  broom::tidy() %>%
  select(1:6) %>%
  knitr::kable(digits = 3)
```

| estimate | statistic | p.value | parameter | conf.low | conf.high |
|---:|---:|---:|---:|---:|---:|
| 98.249 | -5.455 | 0 | 129 | 98.122 | 98.376 |

We selected only the first six variables so that the table would better fit the page.

**Example 10.17.** As a final example, let's test groups of cars from `mtcars` to see if their mean mpg is different from 25. The groups we want are the four possible combinations of transmission (`am`) and engine (`vs`). This requires four $t$-tests, and could be a huge hassle! But, check this out:

```
mtcars %>%
  group_by(am, vs) %>%
  do(broom::tidy(t.test(.$mpg, mu = 25))) %>%
  select(1:8) %>%
  knitr::kable(
    align = "r", digits = 4,
    caption = "Is mean mpg 25 for combinations of trans and engine?
              A two-sided one sample $t$-test."
  )
```

TABLE 10.2: Is mean mpg 25 for combinations of trans and engine? A two-sided one sample $t$-test.

| am | vs | estimate | statistic | p.value | parameter | conf.low | conf.high |
|---:|---:|---:|---:|---:|---:|---:|---:|
| 0 | 0 | 15.0500 | -12.4235 | 0.0000 | 11 | 13.2872 | 16.8128 |
| 0 | 1 | 20.7429 | -4.5581 | 0.0039 | 6 | 18.4575 | 23.0282 |
| 1 | 0 | 19.7500 | -3.2078 | 0.0238 | 5 | 15.5430 | 23.9570 |
| 1 | 1 | 28.3714 | 1.8748 | 0.1099 | 6 | 23.9713 | 32.7716 |

## Exercises

Exercises 10.1 – 10.2 require material through Section 10.1.

**10.1.** Kahle, Sharon, and Baram-Tsabari[9] examined the reaction to posts to social media by The European Organization for Nuclear Research, better known as CERN. The data is available in the `cern` data frame in the `fosdata` package. Recreate the content of the table that appeared in their paper using `xtabs` and `addmargins`. To get the order of the levels the same, you will need to change the levels of the factor in the data set.

| Content Type | Platform | | | | | Total |
|---|---|---|---|---|---|---|
| | *Facebook* | *Twitter* English | *Twitter* French | *Google+* | *Instagram* | |
| News | 24 | 23 | 17 | 22 | 8 | **94** |
| GWII | 8 | 8 | 8 | 8 | 8 | **40** |
| TBT | 8 | 8 | 8 | 8 | 8 | **40** |
| Wow | 8 | 8 | 8 | 8 | 8 | **40** |
| **Total** | **48** | **47** | **41** | **46** | **32** | **214** |

doi:10.1371/journal.pone.0156409.t003

*Image credit: Kahle et al..*

**10.2.** Consider the `cern` data set in the `fosdata` package. Create a figure similar to Figure 10.1 which illustrates the total number of likes for each type of post, colored by the platform. French Twitter may not show up because it has so few likes.

Exercises 10.3 – 10.8 require material through Section 10.2.

**10.3.** Suppose you are testing $H_0 : p = 0.4$ versus $H_a : p \neq 0.4$. You collect 20 pieces of data and observe 12 successes. Use `dbinom` to compute the $p$-value associated with the exact binomial test, and check using `binom.test`.

**10.4.** Suppose you are testing $H_0 : p = 0.4$ versus $H_a : p \neq 0.4$. You collect 100 pieces of data and observe 33 successes. Use the normal approximation to the binomial to find an approximate $p$-value associated with the hypothesis test.

**10.5.** Shaquille O'Neal (Shaq) was an NBA basketball player from 1992–2011. He was a notoriously bad free throw shooter[10]. Shaq always claimed, however, that the true probability of him making a free throw was greater than 50%. Throughout his career, Shaq made 5,935 out of 11,252 free throws attempted. Is there sufficient evidence to conclude that Shaq indeed had a better than 50/50 chance of making a free throw?

**10.6.** Diaconis, Holmes and Montgomery[11] claim that vigorously flipped coins tend to come up the same way they started. In a real coin tossing experiment[12], two UC Berkeley students

[9]Kate Kahle, Aviv J Sharon, and Ayelet Baram-Tsabari, "Footprints of Fascination: Digital Traces of Public Engagement with Particle Physics on CERN's Social Media Platforms." *PLOS One* 11, no. 5 (2016): e0156409.

[10]Shaq is reported to have said, "Me shooting 40 percent at the foul line is just God's way of saying that nobody's perfect. If I shot 90 percent from the line, it just wouldn't be right."

[11]Persi Diaconis, Susan Holmes, and Richard Montgomery, "Dynamical Bias in the Coin Toss," *SIAM Review* 49, no. 2 (2007): 211–35.

[12]Priscilla Ku and Janet Larwood, "40,000 Coin Tosses Yield Ambiguous Evidence for Dynamical Bias," 2009, https://www.stat.berkeley.edu/~aldous/Real-World/coin_tosses.html.

tossed coins a total of 40 thousand times in order to assess whether this is true. Out of the 40,000 tosses, 20,245 landed on the same side as they were tossed from.

a. Find a (two-sided) 99% confidence interval for $p$, the true proportion of times a coin will land on the same side it is tossed from.
b. Clearly state the null and alternative hypotheses, defining any parameters that you use.
c. Is there sufficient evidence to reject the null hypothesis at the $\alpha = .05$ level based on this experiment? What is the $p$-value?

**10.7.** This exercise requires material from Section 6.7 or knowledge of loops. The curious case of the dishonest statistician – suppose a statistician wants to "prove" that a coin is not a fair coin. They decide to start flipping the coin, and after 10 tosses they will run a hypothesis test on $H_0 : p = 1/2$ versus $H_a : p \neq 1/2$. If they reject at the $\alpha = .05$ level, they stop. Otherwise, they toss the coin one more time and run the test again. They repeatedly toss and run the test until either they reject $H_0$ or they toss the coin 100 times (hey, they're dishonest **and** lazy). Estimate using simulation the probability that the dishonest statistician will reject $H_0$.

**10.8 (Hard).** Suppose you wish to test whether a die truly comes up "6" 1/6 of the time. You decide to roll the die until you observe 100 sixes. You do this, and it takes 560 rolls to observe 100 sixes.

a. State the appropriate null and alternative hypotheses.
b. Explain why `prop.test` and `binom.test` are not formally valid to do a hypothesis test.
c. Use reasoning similar to that in the explanation of `binom.test` above and the function `dnbinom` to compute a $p$-value.
d. Should you accept or reject the null hypothesis?

---

Exercises 10.9 – 10.12 require material through Section 10.3.

**10.9.** Suppose you are collecting categorical data that comes in three levels. You wish to test whether the levels are equally likely using a $\chi^2$ test. You collect 150 items and obtain a test statistic of 4.32. What is the $p$-value associated with this experiment?

**10.10.** Recall that the colors of M&M's supposedly follow this distribution:

| *Yellow* | *Red* | *Orange* | *Brown* | *Green* | *Blue* |
|---|---|---|---|---|---|
| 0.14 | 0.13 | 0.20 | 0.12 | 0.20 | 0.21 |

Imagine you bought 10,000 M&M's and got the following color counts:

| *Yellow* | *Red* | *Orange* | *Brown* | *Green* | *Blue* |
|---|---|---|---|---|---|
| 1357 | 1321 | 1946 | 1182 | 2052 | 2142 |

Does your sample appear to follow the known color distribution? Perform the appropriate $\chi^2$ test at the $\alpha = .05$ level and interpret.

**10.11.** The data set `fosdata::bechdel` has information on budget and earnings for many popular movies.

a. Is the `budget` data consistent with Benford's Law?
b. Is the `intgross` data consistent with Benford's Law?
c. Is the `domgross` data consistent with Benford's Law? (Hint: one movie had no domestic gross. Bonus: which one was it?)

**10.12.** The United States Census Bureau produces estimates of population for all cities and towns in the U.S. On the census website http://www.census.gov, find population estimates for all incorporated places (cities and towns) for any one state. Import that data into R. Do the values for city and town population numbers follow Benford's Law? Report your results with a plot and a $p$-value as in Example 10.6.

---

Exercises 10.13 – 10.17 require material through Section 10.4.

**10.13.** Did the goals scored by each team in each game of the 2014 FIFA Men's World Cup soccer final follow a Poisson distribution? Perform a $\chi^2$ goodness of fit test at the $\alpha = 0.05$ level, binning values 4 and above. Data is in `fosdata::world_cup`.

**10.14.** Consider the `austen` data set in the `fosdata` package. In this exercise, we are testing to see whether the number of times that words are repeated after their first occurrence is Poisson. Restrict to the first chapter of *Pride and Prejudice*, and count the number of times that each word is repeated, and see that we obtain the following table:

```
## .
##   0   1   2   3   4   5   6   7   8   9  10  11  12  13  14  16  17  20
## 201  50  16  13  12   2   5   5   2   1   4   2   1   2   1   2   1   1
##  21  28  30
##   1   1   1
```

Use a $\chi^2$ goodness of fit test with $\alpha = .05$ to test whether the distribution of repetitions of words is consistent with a Poisson distribution.

**10.15.** Powerball is a lottery game in which players try to guess the numbers on six balls drawn randomly. The first five are white balls and the sixth is a special red ball called the powerball. The results of all Powerball drawings from February 2010 to July 2020 are available in `fosdata::powerball`.

a. Plot the numbers drawn over time. Use color to distinguish the six balls. What do you observe? You will need `pivot_longer` to tidy the data.
b. Use a $\chi^2$ test of uniformity to check if all numbers ever drawn fit a uniform distribution.
c. Restrict to draws after October 4, 2015, and only consider the white balls drawn, `Ball1-Ball5`. Do they fit a uniform distribution?
d. Restrict to draws after October 4, 2015, and only consider Ball1. Check that it is not uniform. Explain why not.

**10.16.** In this exercise, we explore doing $\chi^2$ goodness of fit tests for continuous variables. Consider the `hdl` variable in the `adipose` data set in `fosdata`. We wish to test whether the data is normal using a $\chi^2$ goodness of fit test and 7 bins.

a. Estimate the mean $\mu$ and the standard deviation $\sigma$ of the HDL.
b. Use `qnorm(seq(0, 1, length.out = 8), mu, sigma)` to create the dividing points (`breaks`) between 7 equally likely regions. The first region is $(-\infty, 0.8988)$.
c. Use `table(cut(aa, breaks = breaks))` to obtain the observed distribution of values in bins. The expected number in each bin is the number of data points over 7, since each bin is equally likely.
d. Compute the $\chi^2$ test statistic as the difference between observed and expected squared, divided by the expected.
e. Compute the probability of getting this test-statistic or larger using `pchisq`. The degrees of freedom is the number of bins minus 3, one because the sum has to be 71 and the other because you are estimating two parameters from the data.

f. Is there evidence to conclude that HDL is not normally distributed?

**10.17.** Consider the `fosdata::normtemp` data set. Use a goodness of fit test with 10 bins, all with equal probabilities, to test the normality of the temperature data set. Note that in this case, you will need to estimate two parameters, so the degrees of freedom will need to be adjusted appropriately.

---

Exercises 10.18 – 10.28 require material through Section 10.5.

**10.18.** Clark and Westerberg[13] investigated whether people can learn to toss heads more often than tails. The participants were told to start with a heads up coin, toss the coin from the same height every time, and catch it at the same height, while trying to get the number of revolutions to work out so as to obtain heads. After training, the first participant got 162 heads and 138 tails.

a. Find a 95% confidence interval for $p$, the proportion of times this participant will get heads.
b. Clearly state the null and alternative hypotheses, defining any parameters.
c. Is there sufficient evidence to reject the null hypothesis at the $\alpha = .01$ level based on this experiment? What is the $p$-value?
d. The second participant got 175 heads and 125 tails. Is there sufficient evidence to conclude that the probability of getting heads is different for the two participants at the $\alpha = .05$ level?

**10.19.** Left digit bias is when people attribute a difference to two numbers based on the first digit of the number, when there is not really a large difference between the numbers. In an article[14], researchers studied left digit bias in the context of treatment choices for patients who were just over or just under 80 years old.

Researchers found that 265 of 5036 patients admitted with acute myocardial infarction who were admitted in the two weeks **after** their 80th birthday underwent Coronary-Artery Bypass Graft (CABG) surgery, while 308 out of 4426 patients with the same diagnosis admitted in the two weeks **before** their 80th birthday underwent CABG. There is no recommendation in clinical guidelines to reduce CABG use at the age of 80. Is there a statistically significant difference in the percentage of patients receiving CABG in the two groups?

---

Exercises 10.20 and 10.21 consider the psychology of randomness, as studied in Bar-Hillel et al.[15]

**10.20.** The researchers considered whether people are good at creating random sequences of heads and tails in a unique way. The researchers recruited 175 people and asked them to create a random sequence of 10 heads and tails, though the researchers were only interested in the first guess. Of the 175 people, 143 predicted heads on the first toss. Let $p$ be the probability that a randomly selected person will predict heads on the first toss. Perform a hypothesis test of $p = 0.5$ versus $p \neq 0.5$ at the $\alpha = 0.05$ level.

---

[13] Matthew P A Clark and Brian D Westerberg, "Holiday Review. How Random Is the Toss of a Coin?" *Canadian Medical Association Journal* 181, no. 12 (December 2009): E306–8.

[14] Andrew R Olenski et al., "Behavioral Heuristics in Coronary-Artery Bypass Graft Surgery." *N Engl J Med* 382, no. 8 (February 2020): 778–79.

[15] M Bar-Hillel, E Peer, and A Acquisti, "'Heads or Tails?' – a Reachability Bias in Binary Choice," *Journal of Experimental Psychology: Learning, Memory, and Cognition* 40, no. 6 (2014): 1656--1663, https://doi.org/10.1037/xlm0000005.

**10.21.** The researchers also considered whether the linguistic convention of naming heads before tails impacts participants' choice for their first imaginary coin toss. The authors recruited 54 people and told them to create a sample of size 10 by entering H for heads and T for tails. They recruited 51 people and told them to create a sample of size 10 by entering T for tails and H for heads. A total of 47 of the 54 people in Group 1 chose heads first, while 16 of the 51 people in Group 2 chose heads first. Perform a hypothesis test of $p_1 = p_2$ versus $p_1 \neq p_2$ at the $\alpha = .05$ level, where $p_i$ is the percentage of heads that people given instructions in Group $i$ would create as their first guess.

---

**10.22.** If someone offered you either one really great marble and three mediocre ones, or four mediocre marbles, which would you choose?

Third-grade children in Rijen, the Netherlands, were split into two groups.[16] In group 1, 43 out of 48 children preferred a blue and white striped marble to a solid red marble. In group 2, 12 out of 44 children preferred four solid red marbles to three solid red marbles and one blue and white striped marble. Let $p_1$ be the proportion of children who would prefer a blue and white marble to a red marble, and let $p_2$ be the proportion of children who would prefer three red marbles and one blue and white striped marble to four red marbles. Perform a hypothesis test of $p_1 = p_2$ versus $p_1 \neq p_2$ at the $\alpha = .05$ level.

**10.23.** A 2017 study[17] considered the care of patients with burns. A patient who stayed in the hospital for seven or more days past the last surgery for a burn is considered an extended postoperative stay. The researchers examined records and found that for patients with scalds, 30 did not have extended stays while 16 did have extended stays. For patients with flame burns, 51 did not have extended stays while 78 did have extended stays. Test whether the proportion of extended stays is the same for scald patients as for flame burn patients at the $\alpha = .05$ level.

**10.24.** Ronald Reagan became president of the United States in 1980. The `babynames::babynames` data set contains information on babies named "Reagan" born in the United States. Is there a statistically significant difference in the percentage of babies (of either sex) named "Reagan" in the United States in 1982 and in 1978? If so, which direction was the change?

**10.25.** Consider the `dogs` data set in the `fosdata` package. For dogs in trial 1 that were shown a single dog going around the wall in the "wrong" direction three times, is there a statistically significant difference in the proportion that stay and the proportion that switch depending on their start direction?

**10.26.** Consider the `sharks` data set in the `fosdata` package. Participants were assigned to listen to either silence, ominous music, or uplifting music while watching a video about sharks. They then ranked sharks on various scales.

a. Create a cross table of the type of music listened to and the response to `dangerous`; "how well does dangerous describe sharks."
b. Perform a $\chi^2$ test of homogeneity to test whether the ranking of how well "dangerous" describes sharks has the same distribution across the type of music heard.

---

[16] Ellen R K Evers, Yoel Inbar, and Marcel Zeelenberg, "Set-Fit Effects in Choice." *J Exp Psychol Gen* 143, no. 2 (April 2014): 504–9.

[17] Islam Abdelrahman et al., "Division of Overall Duration of Stay into Operative Stay and Postoperative Stay Improves the Overall Estimate as a Measure of Quality of Outcome in Burn Care," *PLOS One* 12, no. 3 (March 2017): e0174579–79.

**10.27.** Police sergeants in the Boston Police Department take an exam for promotion to lieutenant. In 2008, 91 sergeants took the lieutenant promotion test. Of them, 65 were white and 26 were Black or Hispanic.[18] The passing rate for white officers was 94%, while the passing rate for minorities was 69%. Was there a significant difference in the passing rates for whites and for minority test takers?

**FIGURE 10.6** Bicycle signage. (Image credit: Hess and Peterson.)

**10.28.** Hess and Peterson[19] studied whether bicycle signage can affect an automobile driver's perception of bicycle rights and safety. Load the `fosdata::bicycle_signage` data, and see the help page for descriptions of the variables.

a. Create a contingency table of the variables `bike_move_right2` and `treatment`.
b. Calculate the proportion of participants who agreed and disagreed for each type of sign treatment. Which sign was most likely to lead participants to disagree?
c. Perform a $\chi^2$ test of independence on the variables `bike_move_right2` and `treatment` at the $\alpha = .05$ level. Interpret your answer.

---

Exercise 10.29 requires material through Section 10.6.

**10.29.** In Example 10.14, we stated that the number of possible ways to fill a $3 \times 1$ table with non-negative integers that sum to 1107 is 614,386. Explain why this is the case. (Hint: if you know the first two values, then the third one is determined.)

---

[18]Zack Huffman, "Boston Police Promotion Exam Deemed Biased" (Courthouse News Service, November 18, 2015), https://www.courthousenews.com/boston-police-promotion-exam-deemed-biased/.

[19]George Hess and M Nils Peterson, ""Bicycles May Use Full Lane" Signage Communicates U.S. Roadway Rules and Increases Perception of Safety," *PLOS One* 10, no. 8 (August 2015): e0136973.

# 11

# *Simple Linear Regression*

Consider the data `Formaldehyde`, which is built into R:

```
Formaldehyde
```

```
##   carb optden
## 1  0.1  0.086
## 2  0.3  0.269
## 3  0.5  0.446
## 4  0.6  0.538
## 5  0.7  0.626
## 6  0.9  0.782
```

*Image credit: Rainis Venta.*

In this experiment, a container of the carbohydrate formaldehyde of known concentration `carb` was tested for its optical density `optden` (a measure of how much light passes through the liquid). The variable `carb` is called the *explanatory* variable, denoted in formulas by $x$. Explanatory variables are also sometimes called *predictor* variables or *independent* variables. The `optden` variable is the *response* variable $y$. In this data, the experimenters selected the values $x_1 = 0.1$, $x_2 = 0.3$, $x_3 = 0.5$, $x_4 = 0.6$, $x_5 = 0.7$, and $x_6 = 0.9$ and created solutions at those concentrations. They then measured the values $y_1 = 0.086, \ldots, y_6 = 0.782$.

A plot of this data (Figure 11.1) shows a nearly linear relationship between the concentration of formaldehyde and the optical density of the solution:

```
Formaldehyde %>%
  ggplot(aes(x = carb, y = optden)) +
  geom_point()
```

The goal of simple linear regression is to find a line that fits the data. The resulting line is called the *regression line* or the *best fit line*. R calculates this line with the `lm` command, which stands for *linear model*:

DOI: 10.1201/9781003004899-11

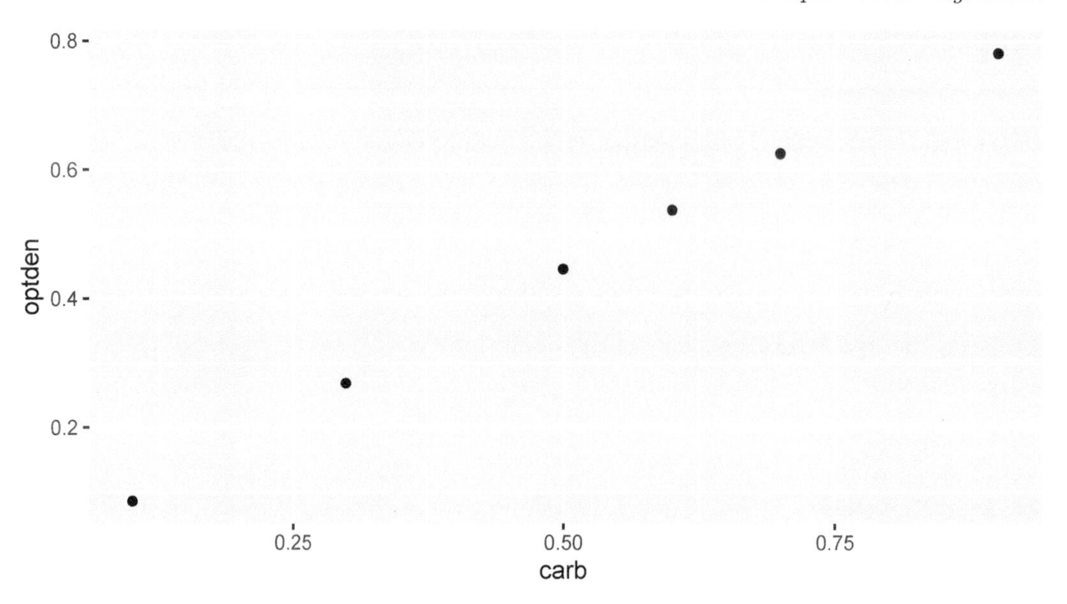


**FIGURE 11.1** The relationship between optical density and formaldehyde concentration is nearly linear.

```
lm(optden ~ carb, data = Formaldehyde)

##
## Call:
## lm(formula = optden ~ carb, data = Formaldehyde)
##
## Coefficients:
## (Intercept)         carb
##    0.005086     0.876286
```

In the `lm` command, the first argument is a *formula*. Read the tilde character as "explained by," so that the formula says that `optden` is explained by `carb`. The output gives the coefficients of a linear equation

$$\widehat{\text{optden}} = 0.0050806 + 0.876286 \cdot \text{carb}.$$

We put a hat on `optden` above to indicate that if we plug in a value for `carb`, we get an *estimated* value for `optden`. The ggplot geom `geom_abline` can plot a line given an intercept and slope. Figure 11.2 shows the data again with the regression line.

```
Formaldehyde %>% ggplot(aes(x = carb, y = optden)) +
  geom_point() +
  geom_abline(intercept = 0.005086, slope = 0.876286)
```

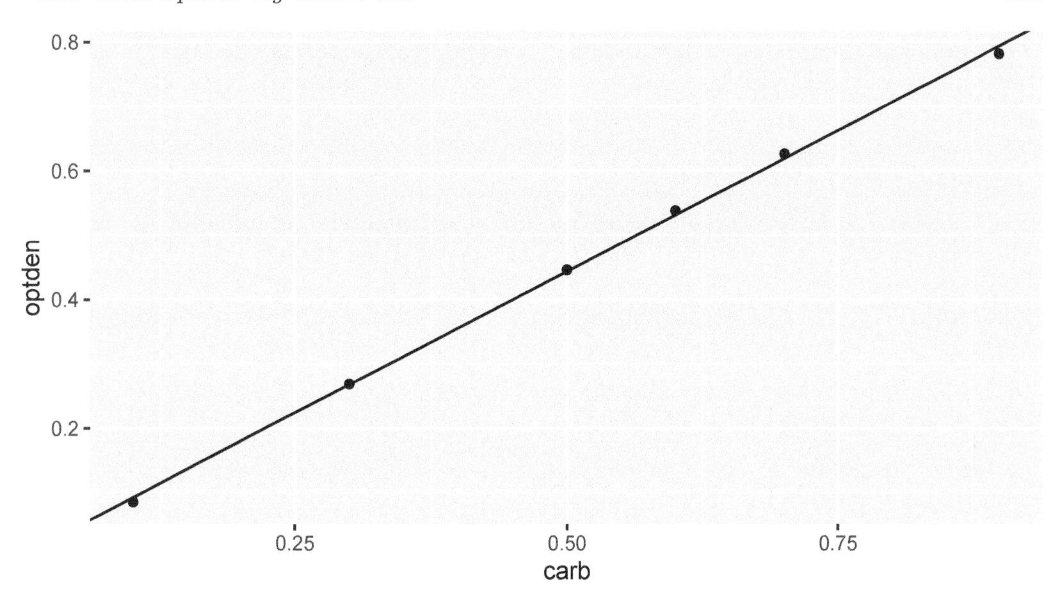


**FIGURE 11.2** Formaldehyde data and the regression line.

## 11.1 Least squares regression line

Assume data comes in the form of ordered pairs $(x_1, y_1), \ldots, (x_n, y_n)$. Values of the slope $b_1$ and the intercept $b_0$ describe a line $b_0 + b_1 x$. Among all choices of $b_0$ and $b_1$, there should be one which comes closest to following the ordered pairs $(x_1, y_1), \ldots, (x_n, y_n)$. In this section, we investigate what that means and we find the optimal choice of slope and intercept for two example data sets. Note that we are **not** assuming that the underlying data follows a line, or that it is a line up to some error term. We are simply finding the line which best fits the data points.

For each value of $x_i$, the line with intercept $b_0$ and slope $b_1$ goes through the point $b_0 + b_1 x_i$. The error associated with $b_0$ and $b_1$ at $x_i$ is the difference between the observed value $y_i$ and the value of the line at $x_i$, namely $y_i - (b_0 + b_1 x_i)$. This error is the vertical distance between the actual value of the data $y_i$ and the value estimated by our line.

The regression line chooses $b_0$ and $b_1$ to make the error terms as small as possible in the following sense.

**Definition 11.1.** The *regression line* for points $(x_1, y_1), \ldots, (x_n, y_n)$ is the line which minimizes the sum of squared errors (SSE)

$$SSE(b_0, b_1) = \sum_{i=1}^{n} (y_i - (b_0 + b_1 x_i))^2 .$$

We denote the values of $b_0$ and $b_1$ that minimize the SSE by $\hat{\beta}_0$ and $\hat{\beta}_1$, so the regression line is given by

$$\hat{\beta}_0 + \hat{\beta}_1 x.$$

The sum of squared errors is not the only possible way to measure the quality of fit, so sometimes the line is called the *least squares regression line.* Reasons for minimizing the SSE are that it is reasonably simple, it involves all data points, and it is a smooth function. Using SSE also has a natural geometric interpretation: minimizing SSE is equivalent to minimizing the length of the vector of errors.

The `lm` command in R returns a data structure which contains the values of $\hat{\beta}_0$ and $\hat{\beta}_1$. It also contains the errors $y_i - (\hat{\beta}_0 + \hat{\beta}_1 x_i)$, which are called *residuals.*

**Example 11.1.** Find $\hat{\beta}_i$ for the Formaldehyde data, calculate the residuals, and calculate the SSE.

In the notation developed above, the estimated intercept is $\hat{\beta}_0 = 0.005086$ and the estimated slope is $\hat{\beta}_1 = 0.876286$.

```
Formaldehyde_model <- lm(optden ~ carb, data = Formaldehyde)
Formaldehyde_model$residuals
```

```
##            1            2            3            4            5
## -0.006714286  0.001028571  0.002771429  0.007142857  0.007514286
##            6
## -0.011742857
```

```
sum(Formaldehyde_model$residuals^2)
```

```
## [1] 0.0002992
```

In this code, the result of the `lm` function is stored in a variable we chose to call `Formaldehyde_model`. Observe that the first and sixth residuals are negative, since the data points at $x_1 = 0.1$ and $x_6 = 0.9$ are both below the regression line. The other residuals are positive since those data points lie above the line. The SSE for this line is 0.0002992.

If we fit any other line, the SSE will be larger than 0.0002992. For example, using the line $0 + 0.9x$ results in a SSE of 0.0008, so it is not the best fit:

```
optden_hat <- 0.9 * Formaldehyde$carb # fit values
optden_hat - Formaldehyde$optden # compute errors
```

```
## [1] 0.004 0.001 0.004 0.002 0.004 0.028
```

```
sum((optden_hat - Formaldehyde$optden)^2) # compute SSE
```

```
## [1] 0.000837
```

**Example 11.2.** The **penguins** data set from the **palmerpenguins** package will appear in several places in this chapter. This has measurements of three species of penguins found in Antarctica, and was originally collected by Gorman et al. in 2007-2009.[1]

[1]Kristen B Gorman, Tony D Williams, and William R Fraser, "Ecological Sexual Dimorphism and Environmental Variability Within a Community of Antarctic Penguins (Genus Pygoscelis)." *PLOS One* 9, no. 3 (2014): e90081.

*Image credit: Allison Horst.*

For now, we focus on the chinstrap species, and investigate the relationship between flipper length and body mass. We begin with a visualization, using flipper length as the explanatory variable and body mass as the response. The geometry `geom_smooth` can fit various lines and curves to a plot. The argument `method = "lm"` indicates that we want to use a linear model, the regression line.

```
penguins <- palmerpenguins::penguins
chinstrap <- filter(palmerpenguins::penguins, species == "Chinstrap")
chinstrap %>%
  ggplot(aes(x = flipper_length_mm, y = body_mass_g)) +
  geom_point() +
  geom_smooth(method = "lm")
```

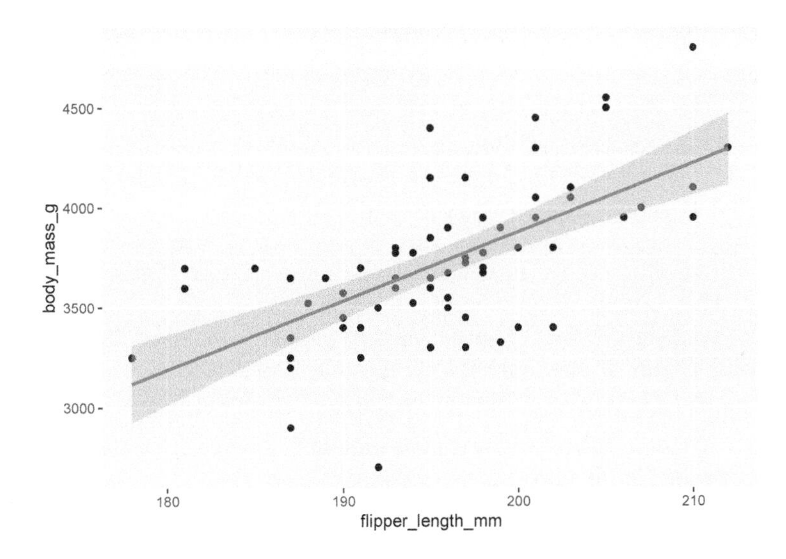


**FIGURE 11.3** Chinstrap penguin body mass as explained by flipper length.

The resulting plot, in Figure 11.3, shows each observation as a point, and the least squares regression line in blue. The upward slope of the line suggests that chinstrap penguins with longer flippers will have greater body mass, on average. Here are the intercept and slope of the regression line:

```
chinstrap_model <- lm(body_mass_g ~ flipper_length_mm, data = chinstrap)
chinstrap_model$coefficients
```

```
##       (Intercept) flipper_length_mm
##       -3037.19577          34.57339
```

The equation of the regression line is

$$\widehat{\text{mass}} = -3037.19577 + 34.57339 \cdot \text{flipper length}.$$

The estimated slope $\hat{\beta}_1 \approx 34.6$ means that for every additional millimeter of flipper length, we expect an additional 34.6 g of body mass. The estimated intercept $\hat{\beta}_0 \approx -3037$ has little meaning, because it describes the supposed mass of a penguin with flipper length zero.

The slope of the regression line often has a useful interpretation, while the intercept often does not.

The regression line associates a body mass of $-3037.19577 + 34.57339 \cdot 200 = 3877$ to a chinstrap penguin with flipper length 200 mm. There is some uncertainty in this estimate, which is expressed in ggplot by a gray band around the regression line. This is the 95% confidence band for the regression line. The regression line shown in the figure depends on the data. Repeating the study would produce new measurements and a new line. To quantify this, we will need to expand our assumptions, see Section 11.5 for details.

The R function `predict` computes the response associated with the given value of the predictor and the regression line. We will be using it here with two arguments: the `object` which is the model as returned by `lm`, and `newdata` which is a data frame of values for which we want to be making predictions. To find the body mass associated with a penguin with flipper length 200 mm, we would use:

```
predict(chinstrap_model, newdata = data.frame(flipper_length_mm = 200))
```

```
##        1
## 3877.483
```

**Example 11.3.** The data set `child_tasks` is available in the `fosdata` package:

```
child_tasks <- fosdata::child_tasks
```

It contains results from a study[2] in which 68 children were tested at a variety of executive function tasks. The variable `stt_cv_trail_b_secs` measures the speed at which the child can complete a "connect the dots" task called the "Shape Trail Test (STT)," which you may try yourself using Figure 11.4.

How does a child's age affect their speed at completing the Shape Trail Test? We choose `age_in_months` as the explanatory variable and `stt_cv_trail_b_secs` as the response variable, as shown in Figure 11.5.

```
child_tasks <- fosdata::child_tasks
child_tasks %>%
  ggplot(aes(x = age_in_months, y = stt_cv_trail_b_secs)) +
  geom_point() +
  geom_smooth(method = "lm")
```

[2] Chan and Morgan, "Assessing Children's Cognitive Flexibility with the Shape Trail Test."

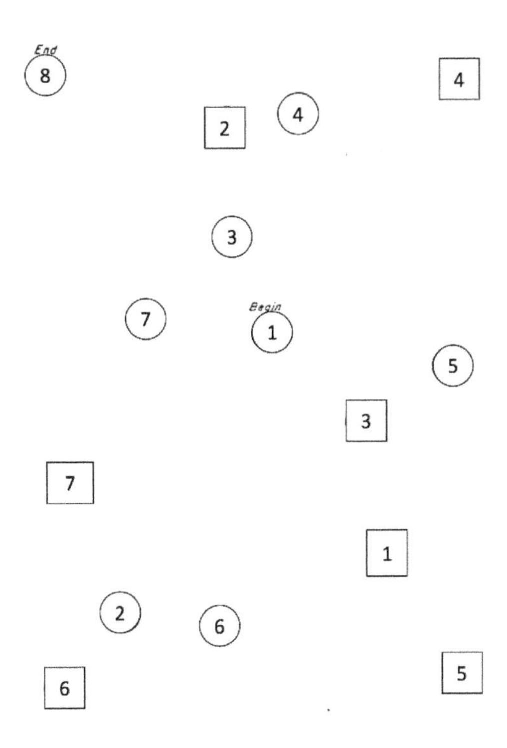


**FIGURE 11.4** STT trail B. Connect the circles in order.

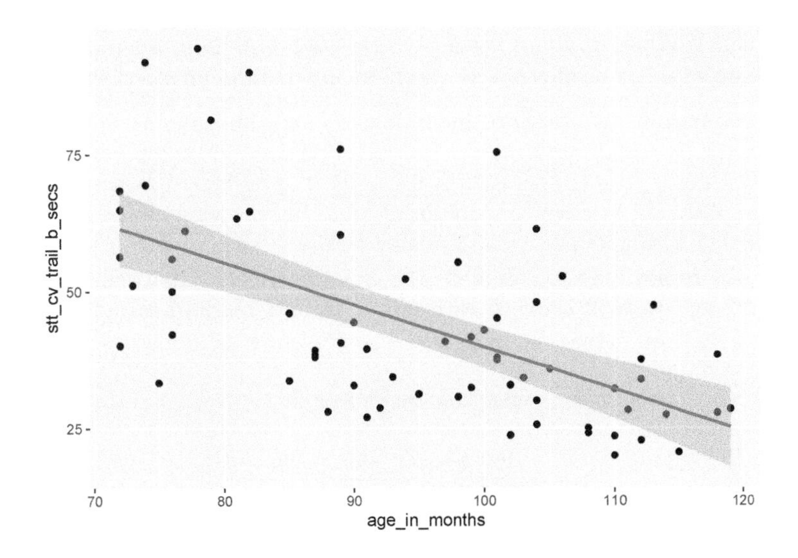


**FIGURE 11.5** Speed of children on Shape Trail Test as explained by age.

```
stt_model <- lm(stt_cv_trail_b_secs ~ age_in_months, data = child_tasks)
stt_model$coefficients
```

```
##   (Intercept) age_in_months
##   116.8865146    -0.7706084
```

The regression line is given by the equation

$$\widehat{\text{test time}} = 116.9 - 0.77 \cdot \text{age}.$$

The estimated slope $\hat{\beta}_1 = -0.77$ indicates that each additional month of age reduces the mean time for children to complete the STT by 0.77 seconds on average. The estimated intercept $\hat{\beta}_0 = 116.9$ is meaningless since it corresponds to a child of age 0, and a newborn child cannot connect dots.

If we wish to find the time to complete the puzzle associated with a child that is 90 months old, we can compute $116.9 - 0.77 \cdot 90 = 47.6$ seconds, or we can use the `predict` function as follows:

```
predict(stt_model, newdata = data.frame(age_in_months = 90))
```

```
##        1
## 47.53176
```

## 11.2 Correlation

The *correlation* $\rho$ of random variables $X$ and $Y$ is a number between -1 and 1 that measures the strength of the linear relationship between $X$ and $Y$. It is positive when $X$ and $Y$ tend to be large together and small together, and it is negative when large $X$ values tend to accompany small $Y$ values and vice versa. A correlation of 0 indicates no linear relationship, and a correlation of $\pm 1$ is achieved only when $X$ and $Y$ have an exact linear relationship.

If $X$ and $Y$ have means and standard deviations $\mu_X$, $\sigma_X$ and $\mu_Y$, $\sigma_Y$ respectively, then

$$\rho_{X,Y} = \frac{E\left[(X - \mu_X)(Y - \mu_Y)\right]}{\sigma_X \sigma_Y} = \frac{\text{Cov}(X, Y)}{\sigma_X \sigma_Y}.$$

Many times, we are not able to calculate the exact correlation between two random variables $X$ and $Y$, and we will want to estimate it from a random sample. Given a sample $(x_1, y_1), \ldots, (x_n, y_n)$, we define the *correlation coefficient* $r$ as follows:

**Definition 11.2.** The *sample correlation coefficient* is

$$r = \frac{1}{n-1} \sum_{i=1}^{n} \left(\frac{x_i - \bar{x}}{s_x}\right)\left(\frac{y_i - \bar{y}}{s_y}\right)$$

The $i^{\text{th}}$ term in the sum for $r$ will be positive whenever:

- Both $x_i$ and $y_i$ are larger than their means $\bar{x}$ and $\bar{y}$.
- Both $x_i$ and $y_i$ are smaller than their means $\bar{x}$ and $\bar{y}$.

It will be negative whenever

- $x_i$ is larger than $\bar{x}$ while $y_i$ is smaller than $\bar{y}$.
- $x_i$ is smaller than $\bar{x}$ while $y_i$ is larger than $\bar{y}$.

Since $r$ is a sum of these terms, $r$ will tend to be positive when $x_i$ and $y_i$ are large and small together, and $r$ will tend to be negative when large values of $x_i$ accompany small values of $y_i$ and vice versa.

For the rest of this chapter, when we refer to the *correlation* or the *sample correlation* between two random variables, we will mean the *sample correlation coefficient.*

The sample correlation coefficient is symmetric in $x$ and $y$, and is not dependent on the assignment of explanatory and response to the variables.

**Example 11.4.** The correlation between `carb` and `optden` in the `Formaldehyde` data set is $r = 0.9995232$, which is quite close to 1. The plot showed these data points were almost perfectly on a line.

```
cor(Formaldehyde$carb, Formaldehyde$optden)
```

```
## [1] 0.9995232
```

**Example 11.5.** Figure 11.6 shows the relationship between flipper length and body mass for all three penguin species.

```
penguins %>%
  ggplot(aes(x = flipper_length_mm, y = body_mass_g, color = species)) +
  geom_point(size = 0.1) +
  facet_wrap(vars(species))
```

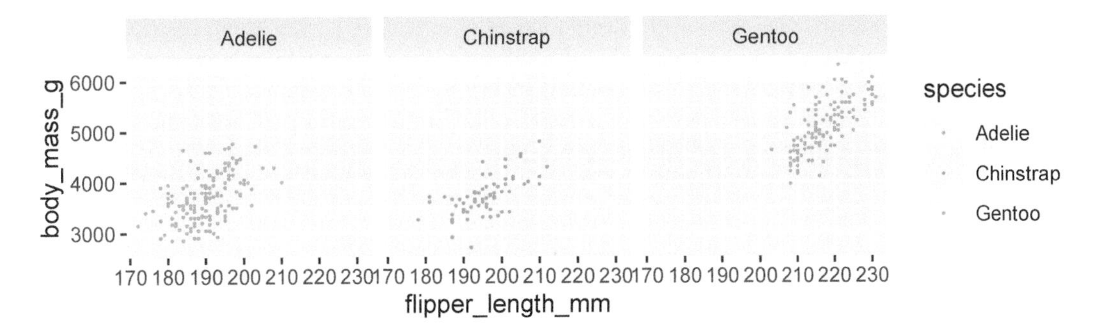


**FIGURE 11.6** Body mass and flipper length for three penguin species.

We compute the sample correlation coefficient $r$ for each species of penguin.

```
penguins %>%
  group_by(species) %>%
  summarize(r = cor(flipper_length_mm, body_mass_g, use = "complete.obs"))
```

```
## # A tibble: 3 x 2
##   species       r
##   <fct>     <dbl>
## 1 Adelie    0.468
```

```
## 2 Chinstrap 0.642
## 3 Gentoo    0.703
```

Gentoo penguins have the strongest linear relationship between flipper length and body mass, with $r = 0.703$. The Adelie penguins have the weakest, with $r = 0.468$. The difference is visible in the plots, where the points for Adelie penguins have a looser clustering. All three scatterplots do exhibit a clear linear pattern.

**Example 11.6.** In the child tasks study, the sample correlation between child's age and time on the STT trail B is $r = -0.593$.

```
cor(child_tasks$age_in_months, child_tasks$stt_cv_trail_b_secs)
```

```
## [1] -0.5928128
```

The negative correlation indicates that older children post faster times on the test. This is visible in the scatterplot as a downward trend as you read the plot from left to right.

Note that correlation is a unitless quantity. The term $(x_i - \bar{x})/\sigma_x$ has the same units (for $x$) in the numerator and denominator, so they cancel, and the $y$ term is similar. This means that a linear change of units will not affect the correlation coefficient:

```
child_tasks$age_in_years <- child_tasks$age_in_months / 12
cor(child_tasks$age_in_years, child_tasks$stt_cv_trail_b_secs)
```

```
## [1] -0.5928128
```

It it clear (from experience, not from a statistical point of view) that there is a causal relationship between a child's age and their ability to connect dots quickly. As children age, they get better at most things. However, **correlation is not causation**. There are many reasons why two variables might be correlated, and $x$ causes $y$ is only one of them.

As a simple example, the size of children's shoes is correlated with their reading ability. However, you cannot buy a child bigger shoes and expect that to make them a better reader. The correlation between shoe size and reading ability is due to a common cause: age. In this example, age is a *lurking variable*, important to our understanding of both shoe size and reading ability, but not included in the correlation.

## 11.3 Geometry of regression

In this section, we establish two geometric facts about the least squares regression line.

We assume data is given as $(x_1, y_1), \ldots, (x_n, y_n)$, and that the sample means and standard deviations of the data are $\bar{x}$, $\bar{y}$, $s_x$, $s_y$.

**Theorem 11.1.** *The least squares regression line:*

1. *Passes through the point* $(\bar{x}, \bar{y})$.
2. *Has slope* $\hat{\beta}_1 = r\frac{s_y}{s_x}$.

Before turning to the proof, we illustrate with penguins. Figure 11.7 shows body mass as explained by flipper length for chinstrap penguins. There are vertical dashed lines at $\bar{x}$ and $\bar{x} \pm s_x$. Similarly, horizontal dashed lines are at $\bar{y}$ and $\bar{y} \pm s_y$. The regression line is thick and blue. The central dashed lines intersect at the point $(\bar{x}, \bar{y})$, and this confirms Theorem 11.1, part 1: the regression line passes through the "center of mass" of the data.

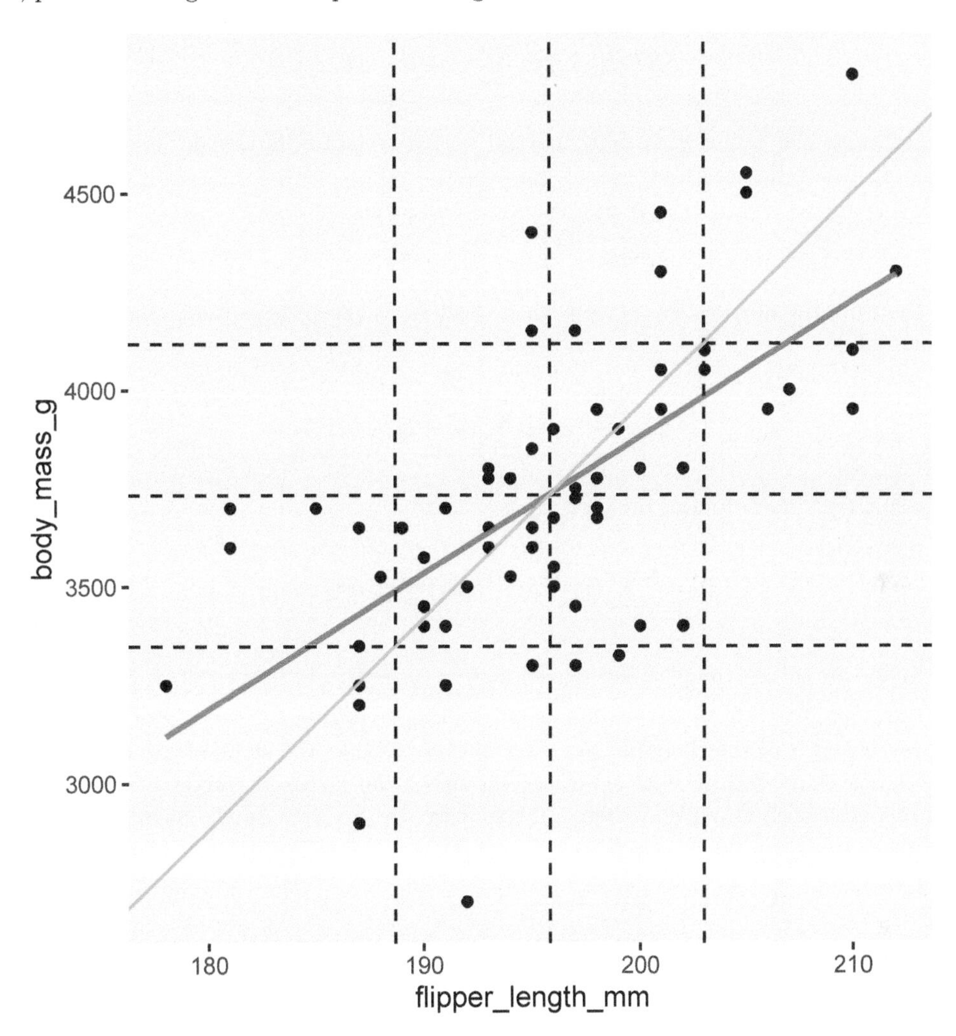


**FIGURE 11.7** Regression line (blue) and standard deviation line (orange) for body mass as described by flipper length for chinstrap penguins.

To better understand Theorem 11.1, part 2, the scale of Figure 11.7 has been adjusted so that one standard deviation is the same distance on both the $x$-axis and $y$-axis. The diagonal orange line in Figure 11.7 is called the *standard deviation line,* and it has slope $\frac{s_y}{s_x}$. Since one SD appears the same length both horizontally and vertically in the plot, the SD line appears at a 45° angle, with apparent slope 1. When looking at scatterplots, the SD line is the visual axis of symmetry of the data. But the SD line is *not* the regression line. The regression line is always flatter than the SD line. Since the regression line has slope $r\frac{s_y}{s_x}$, it appears in our scaled figure to have slope $r$, in this case 0.64. In other words, the correlation coefficient $r$ describes how much flatter the regression line will be than the diagonal axis of the data.

*Proof (of Theorem 11.1).* The regression line is given by $\hat{\beta}_0 + \hat{\beta}_1 x$. The error for each data point is $y_i - \hat{\beta}_0 - \hat{\beta}_1 x_i$. The coefficients $\hat{\beta}_0$ and $\hat{\beta}_1$ are chosen to minimize the two variable function

$$SSE(b_0, b_1) = \sum_i (y_i - b_0 - b_1 x_i)^2$$

This minimum will occur when both partial derivatives of SSE vanish. That is, the least squares regression line satisfies

$$\frac{\partial SSE}{\partial b_0} = 0 \quad \text{and} \quad \frac{\partial SSE}{\partial b_1} = 0.$$

First, compute the partial derivative with respect to $b_0$ (keeping in mind that all $x_i$ and $y_i$ are constants):

$$\frac{\partial SSE}{\partial b_0} = -2 \sum_i (y_i - b_0 - b_1 x_i)$$

Setting this equal to zero and dividing by $n$,

$$0 = \frac{1}{n} \sum_i (y_i - b_0 - b_1 x_i) = \bar{y} - b_0 - b_1 \bar{x}$$

Since $\hat{\beta}_0$ and $\hat{\beta}_1$ satisfy the previous equation, $\bar{y} = \hat{\beta}_0 + \hat{\beta}_1 \bar{x}$. In other words, the point $(\bar{x}, \bar{y})$ lies on the regression line.

This proves part 1 of the theorem. For part 2, observe that the slope of the regression line won't change if we shift all data points along either the $x$-axis or $y$-axis. So from here on, we assume that both $\bar{x}$ and $\bar{y}$ are zero. In that case, we manipulate the formulas for $s_x$ and the correlation coefficient $r$ as follows:

$$s_x = \sqrt{\frac{\sum (x_i - \bar{x})^2}{n-1}} \implies \sum_i x_i^2 = (n-1) s_x^2$$

$$r = \frac{1}{n-1} \sum \left( \frac{x_i - \bar{x}}{s_x} \right) \left( \frac{y_i - \bar{y}}{s_y} \right) \implies \sum_i x_i y_i = (n-1) s_x s_y r$$

Now compute the partial derivative of SSE with respect to $b_1$:

$$\frac{\partial SSE}{\partial b_1} = -2 \sum_i (y_i - b_0 - b_1 x_i) x_i$$

Setting this to zero, we find:

$$0 = \sum_i x_i y_i - b_0 \sum_i x_i - b_1 \sum_i x_i^2$$

Since $\bar{x} = 0$ the middle sum vanishes, and we have

$$0 = (n-1) s_x s_y r - b_1 (n-1) s_x^2$$

Finally, solve for $b_1$ to get $\hat{\beta}_1 = r \frac{s_y}{s_x}$, which is part 2 of the theorem. ∎

## 11.4 Residual analysis

Up until this point, we have not assumed anything about the nature of the data $(x_1, y_1), \ldots, (x_n, y_n)$. However, in order to quantify the uncertainty in our estimates for the slope and intercept of the least squares regression line, as well as to perform inference in other ways, we will need to make some additional assumptions.

**Assumptions 11.1 (Simple Linear Regression Assumptions).**
*In a simple linear regression model, there are two random variables $X$ and $Y$ and two values $\beta_0$ and $\beta_1$ such that the following hold.*

1. *Given that $X = x$, assume $Y = \beta_0 + \beta_1 x + \epsilon$, where $\epsilon$ is a random variable.*
2. *The random variable $\epsilon$ is a normal rv with mean 0 and standard deviation $\sigma$.*
3. *The mean and standard deviation of $\epsilon$ do not depend on $x$.*
4. *$\epsilon$ is independent across observations.*

Names for $\beta_0$ and $\beta_1$ include *parameters*, *constants*, and the *true values* of the intercept and slope. The notation $\hat{\beta}_0$ and $\hat{\beta}_1$ refers to estimates of the true intercept and slope.

We will write as a shorthand either $Y = \beta_0 + \beta_1 X + \epsilon$ or $y_i = \beta_0 + \beta_1 x_i + \epsilon_i$, depending on whether we are thinking of the process of obtaining responses from predictors in general (the first equation) or the process of obtaining responses from particular predictors (the second case). However, we consider the two formulations to be equivalent, and shorthand for the full assumptions given in Assumptions 11.1.

When performing inference using a linear model, it is essential to investigate whether or not the model actually makes sense, and is a reasonable description of the generative process. In this section, we will focus on five common problems that can occur with linear models:

- The relationship between the explanatory and response variables is not linear.
- The variance of $\epsilon$ is not constant across the $x$ values.
- $\epsilon$ is not normally distributed.
- There are outliers that adversely affect the model.
- $\epsilon$ is not independent across observations.

For each of the five common problems, we provide at least one data set which exhibits the listed problem.

The most straightforward way to assess problems with a model is by visual inspection of the residuals. This is more of an art than a science, and in the end it is up to the investigator to decide whether a model is accurate enough to act on the results.

Recall that the residuals are the vertical distances between the regression line and the data points. Figure 11.8 shows the residuals as vertical line segments. Black points are the penguin data, and the fitted points $(x_i, \hat{y}_i)$ are shown as open circles. The residuals are the lengths of the segments between the actual and fitted points.

The simplest way to visualize the residuals for a linear model is to apply the base R function `plot` to the linear model. The `plot` function produces four diagnostic plots, and when used interactively it shows the four graphs one at a time, prompting the user to hit return between graphs. In RStudio, the Plots panel has arrow icons that allow the user to view older plots, which is often helpful after displaying all four.

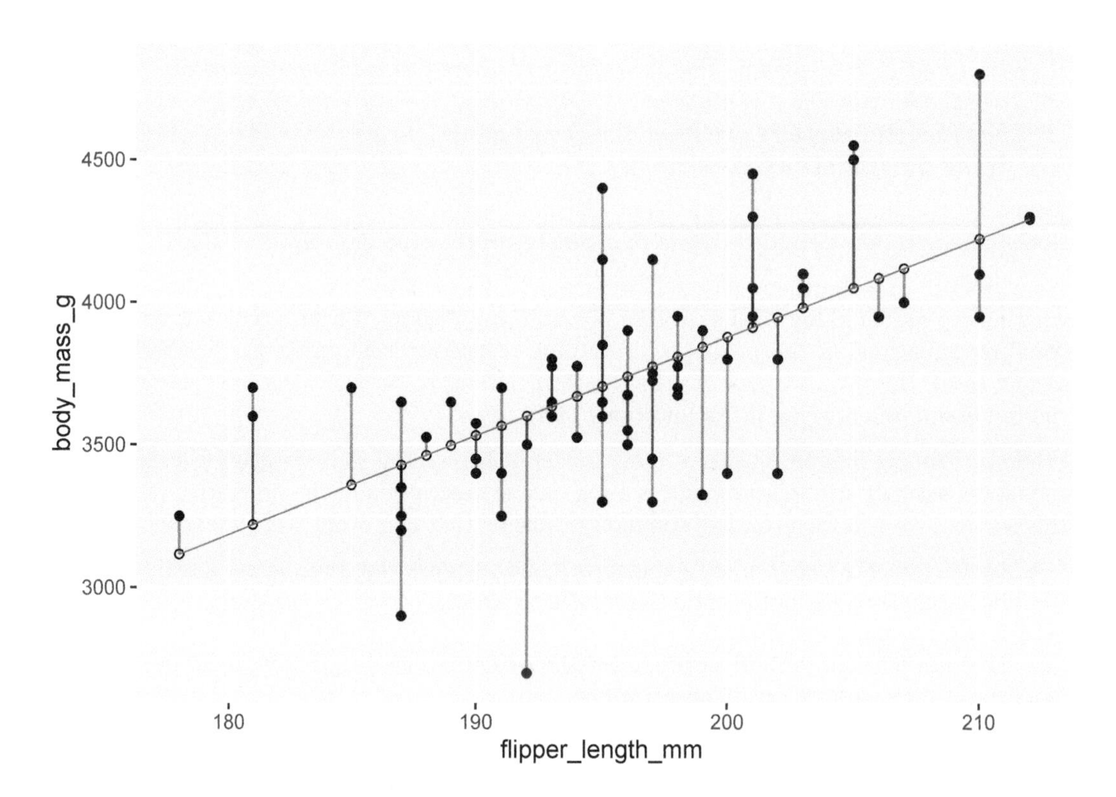


**FIGURE 11.8** Residuals shown as vertical segments for body mass as described by flipper length in chinstrap pengions.

As an example, we plot the residuals for the chinstrap penguin data. All four resulting plots are shown in Figure 11.9.

```
plot(chinstrap_model)
```

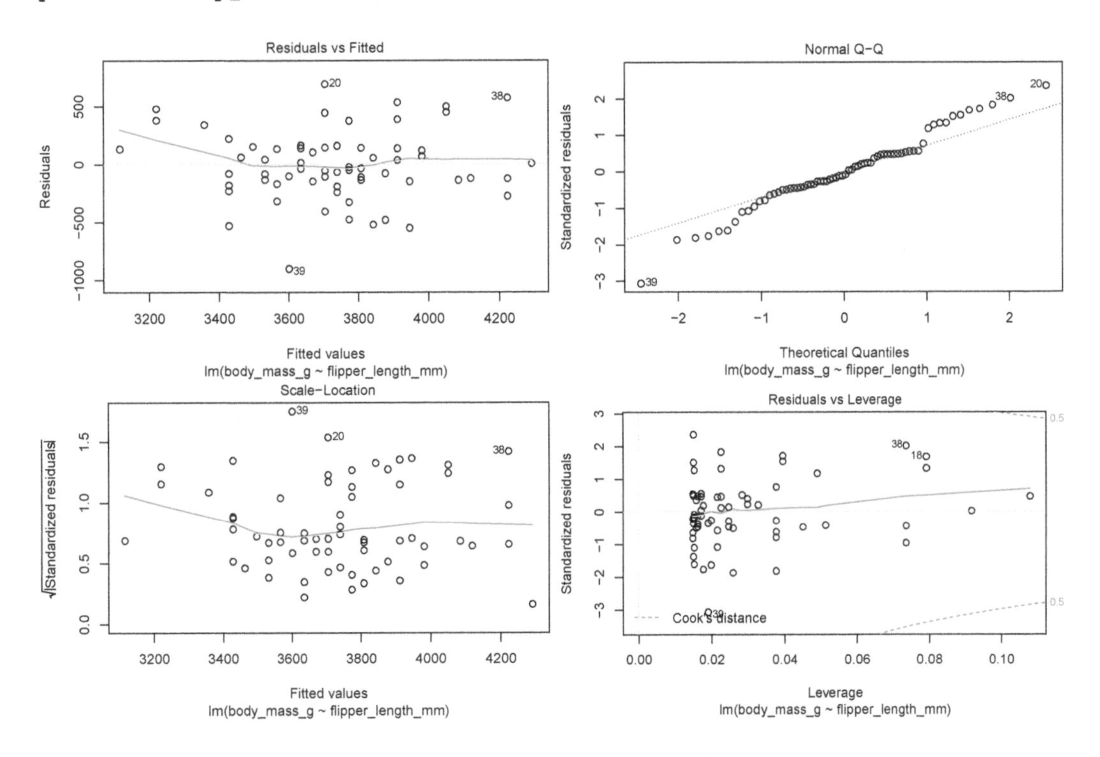


**FIGURE 11.9** Residual plots for chinstrap penguins model of body weight on flipper length.

The "Residuals vs Fitted" plot shows the residuals versus the fitted values $\hat{y}_i$. It might seem more natural to plot residuals against the explanatory variables $x_i$, and in fact for simple linear regression it does make sense to do that. However, $\hat{y}_i$ is a linear function of $x_i$, so the plot produced by `plot` looks the same as a plot of residuals against $x_i$, just with a different scale on the $x$-axis. The advantage of using fitted values comes with multiple regression, where it allows for a two-dimensional plot.

The regression line is chosen to minimize the SSE of the residuals. This implies that the residuals themselves have mean zero, since a non-zero mean would allow a better fit by raising or lowering the fit line vertically. The "Residuals vs Fitted" plot shows the mean of zero as a horizontal dashed line. In the "Residuals vs Fitted" plot, ideal residuals are spread uniformly across this line.

A pattern such as a U-shape is evidence of a *lurking variable* that we have not taken into account. A lurking variable could be information related to the data that we did not collect, or it could be that our model should include a quadratic term. The red curve is fitted to the residuals to make patterns in the residuals more visible.

The second plot is a normal qq plot of the residuals, as discussed in Section 7.2.5. Ideally, the points would fall more or less along the diagonal dotted line shown in the plot.

The "Scale Location" plot shows standardized residuals versus the fitted values. It is perhaps counterintuitive that the variance of residuals associated with predictors that are close to the mean of all predictors is generally **larger** than the variance of residuals associated with predictors far from the mean of all predictors, see Exercise 11.35. Standardizing the residuals takes this difference into account, so that by making them positive and taking the square root, the visual height of the dots should be equal on average when the variance of the errors is constant, which is a crucial assumption for inference. Ideally the vertical spread of dots will be constant across the plot. The red fitted curve helps to visualize the variance, and we are looking to see that the red line is more or less flat.

The last plot is an analysis of outliers and leverage. Outliers are points that fit the model worse than the rest of the data. Outliers with $x$-coordinates in the middle of the data tend to have less of an impact on the final model than outliers toward the edge of the $x$-coordinates. Data that falls outside the red dashed lines are high-leverage outliers, meaning that they (may) have a large effect on the final model.

The linear model for the chinstrap penguin data is a good fit, satisfies the assumptions of regression, and does not have high leverage outliers.

### 11.4.1 Linearity

In order to apply a linear model to data, the relationship between the explanatory and response variable should be linear. If the relationship is not linear, sometimes transforming the data can create new variables that do have a linear relationship. Alternatively, a more complicated non-linear model or multiple regression may be required.

**Example 11.7.** As part of a study to map developmental skull geometry, Li et al.[3] recorded the age and skull circumference of 56 young children. The data is in the `skull_geometry` data set in the `fosdata` package. The correlation between age and skull circumference is 0.8, suggesting a strong relationship, but is it linear? Here is a plot:

```
skull_geometry <- fosdata::skull_geometry
ggplot(skull_geometry, aes(x = age_mos, y = circumference_cm)) +
  geom_point() +
  geom_smooth(method = "lm") +
  labs(
    title = "Age vs. skull circumference in young children",
    subtitle = "The linear model is a poor fit"
  )
```

[3]Zhigang Li et al., "A Statistical Skull Geometry Model for Children 0-3 Years Old." *PLOS One* 10, no. 5 (2015): e0127322.

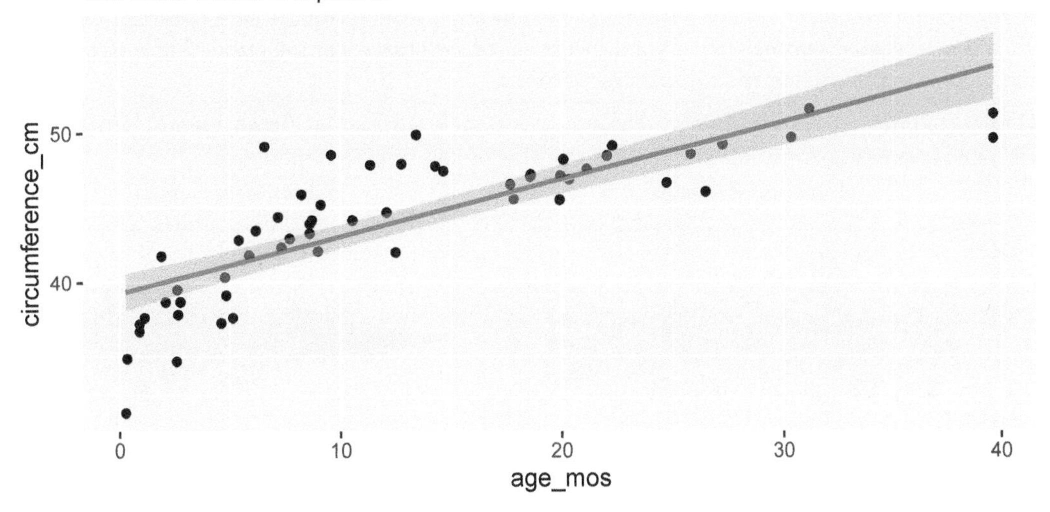


Observe that most of the younger and older children have skulls smaller than the line predicts, while children from 5 to 20 months have larger skulls. This U-shaped pattern suggests that the line is not a good fit for this data. The residual plot (Figure 11.10) makes the U-shape even more visible.

```
skull_model <- lm(circumference_cm ~ age_mos, data = skull_geometry)
plot(skull_model, which = 1)
# which=1 selects the first of the four plots for display
```

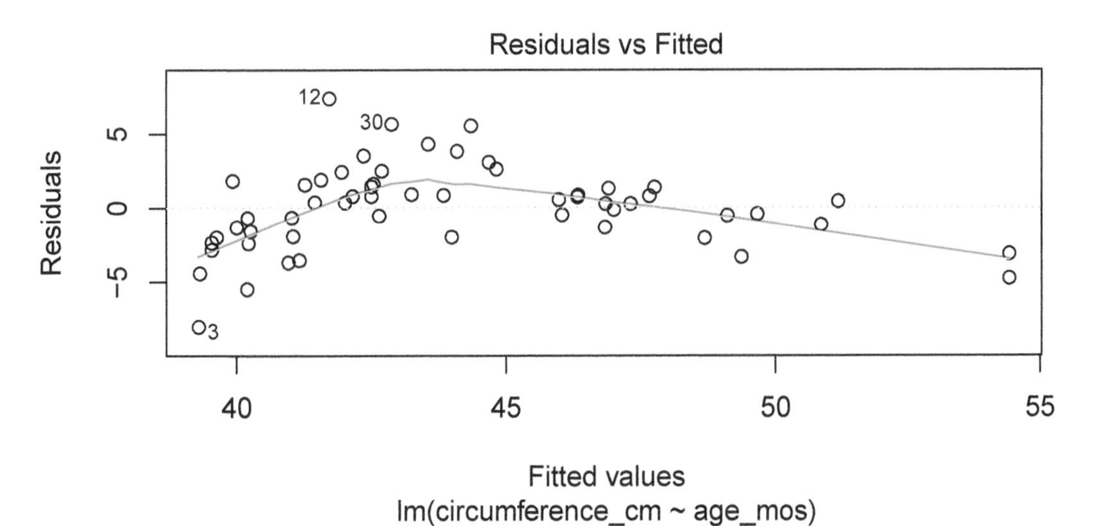


**FIGURE 11.10** Visible U-shape for residuals of the skull geometry model.

A good growth chart for skull circumference would fit a curve to this data, for example by using `geom_smooth(method = "loess")`.

### 11.4.2 Heteroscedasticity

A key assumption for inference with linear models is that the residuals have constant variance. This is called *homoscedasticity*. Its failure is called *heteroscedasticity*, and this is one of the most common problems with linear regression.

Heteroscedastic residuals display changing variance across the fitted values, often with variance increasing from left to right as the size of the response increases:

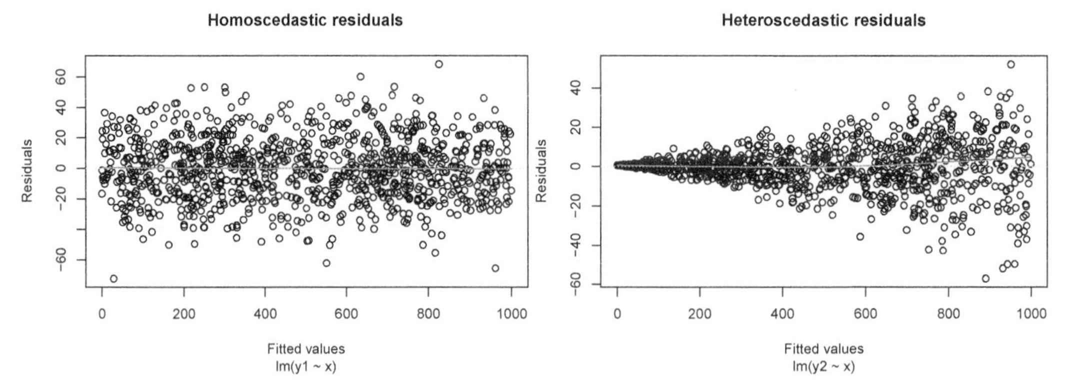


Heteroscedasticity frequently occurs because the true errors are a percentage of the data values, rather than an additive error. For example, many expenditures related to household income are heteroscedastic, since households spend a percentage of their income rather than a fixed amount.

**Example 11.8.** Consider the `leg_strength` data set in the `fosdata` package. Loss of leg strength is a predictor of falls in elderly adults, but measuring leg strength requires expensive laboratory equipment. Researchers[4] built a simple setup using a Nintendo Wii Balance Board and compared measurements using the Wii to measurements using the laboratory stationary isometric muscle apparatus (SID).

```
leg_strength <- fosdata::leg_strength
leg_strength %>%
  ggplot(aes(x = mean_wii, y = mean_sid)) +
  geom_point() +
  geom_smooth(method = "lm", se = FALSE) +
  labs(
    x = "SID", y = "WII",
    title = "Mean leg strength as measured by WII Balance Board vs SID"
  )
```

[4] A W Blomkvist et al., "Unilateral Lower Limb Strength Assessed Using the Nintendo Wii Balance Board: A Simple and Reliable Method." *Aging Clinical and Experimental Research* 29, no. 5 (October 2017): 1013–20, https://doi.org/10.1007/s40520-016-0692-5.

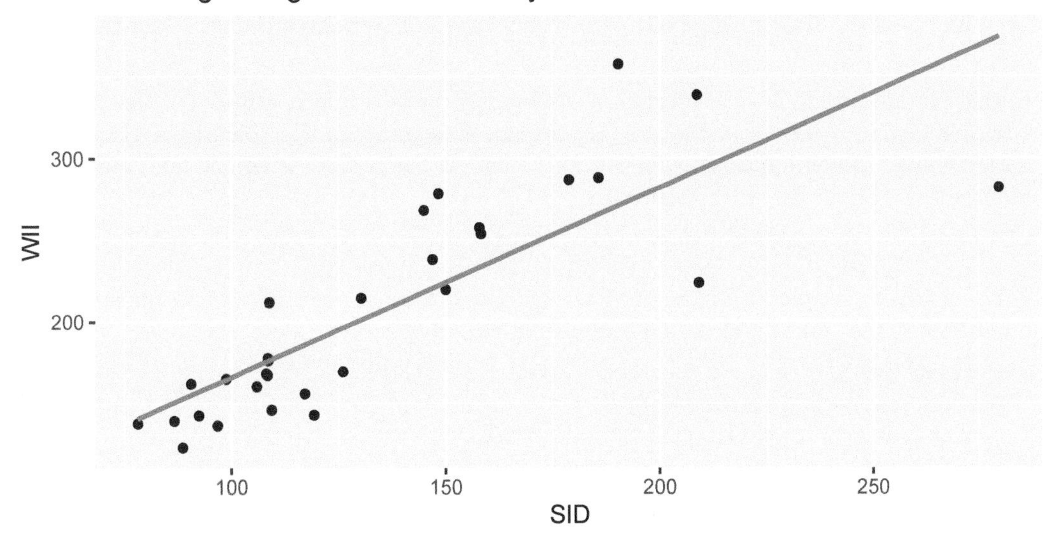


The goal of the experiment is to replace the accurate SID measurement with the inexpensive Wii measurement. The graph makes it clear that the Wii works less well for stronger legs. The "Residuals vs. Fitted" plot shows this heteroscedasticity as well, but the best view comes from the third of the diagnostic plots, shown in Figure 11.11.

```
wii_model <- lm(mean_sid ~ mean_wii, data = leg_strength)
plot(wii_model, which = 3)
```

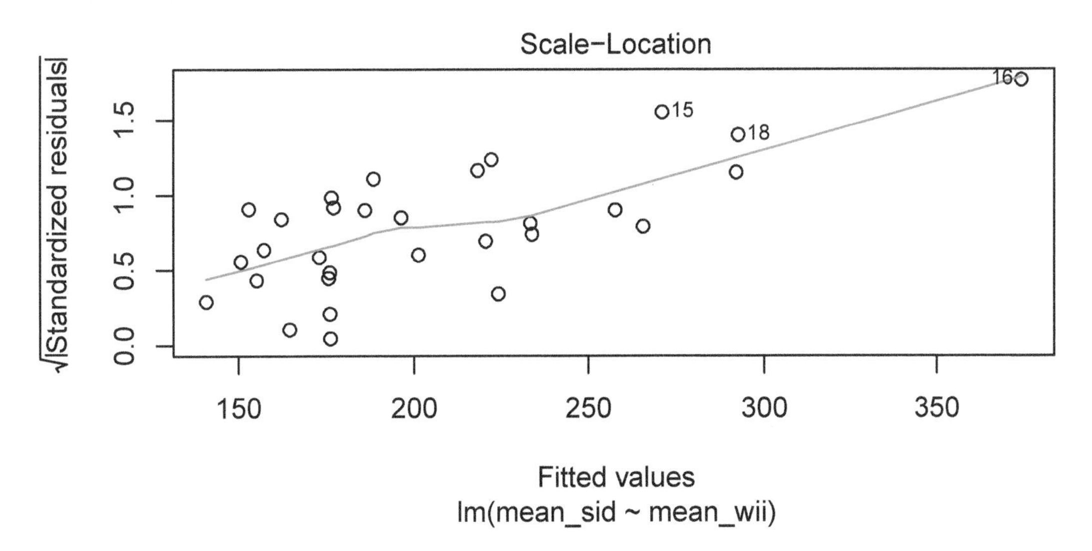


**FIGURE 11.11** Heteroscedasticity of residuals in the leg strength model.

There is a clear upward slant to the points in Figure 11.11, emphasized by the red line. This data is heteroscedastic. Using the linear model for inference would be questionable. You could still use the regression line to convert Wii balance readings to SID values, but you would need to be aware that higher Wii readings would have higher variance.

**Example 11.9.** Right-skewed variables are common. When variables are right-skewed, it means they contain a large number of small values and a small number of large values.

For such variables, we are more interested in the order of magnitude of the values: is the value in the 10's, 1000's, millions? Working with right-skewed variables often results in heteroscedasticity.

In this example, we look at Instagram posts by gold medal winning athletes in the 2016 Rio Olympic Games, which is in the `rio_instagram` data frame in the `fosdata` package. This data contains some duplicated values for athletes who won more than one gold medal. We remove those duplicates with `distinct`.

```
rio_clean <- distinct(fosdata::rio_instagram, name, .keep_all = TRUE)
```

For each athlete, we plot in Figure 11.12 the number of Instagram followers and the maximum number of "likes" for any of their posts during the Olympics.

```
rio_clean %>%
  filter(!is.na(n_follower)) %>%
  ggplot(aes(x = n_follower, y = max_like)) +
  geom_point()
```

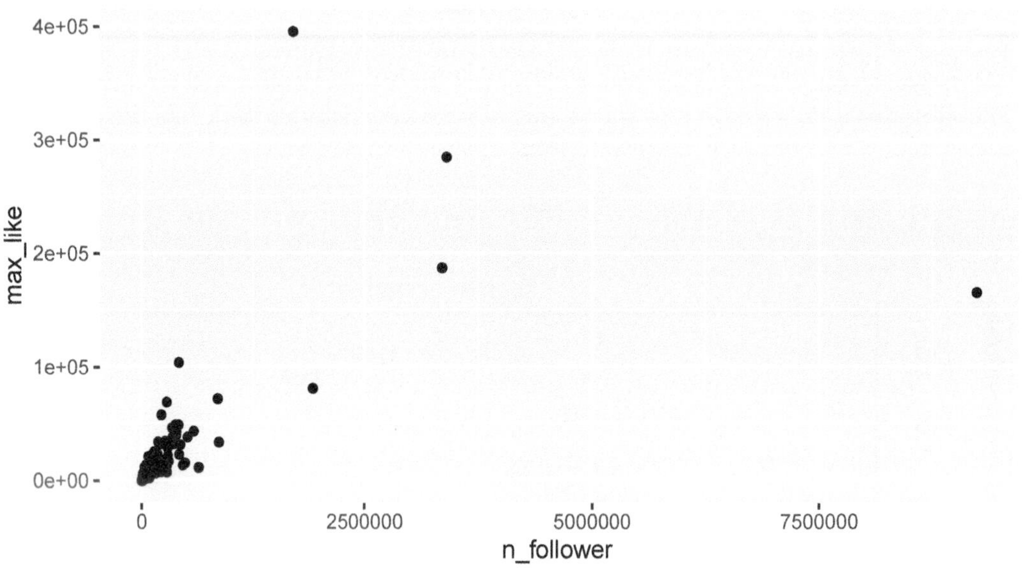

**FIGURE 11.12** Right-skewed variables in the Rio Instagram data set.

The vast majority of the athletes had small (less than 10000) followings, while a few (Usain Bolt, Simone Biles, and Michael Phelps) had over 2.5 million. Even if we filter out the large followings, the graph (Figure 11.13) is still badly heteroscedastic.

```
rio_clean %>%
  filter(!is.na(n_follower)) %>%
  filter(n_follower < 1000000) %>%
  ggplot(aes(x = n_follower, y = max_like)) +
  geom_point()
```

The solution is to transform the data with a *logarithmic transformation.* In R, the function `log` computes natural logarithms (base $e$). It is not important whether we use base 10 or natural logarithms, since they are linearly related (i.e., $\log_e(x) = 2.303 \log_{10}(x)$).

Figure 11.14 shows the Rio Instagram data after applying a log transformation.

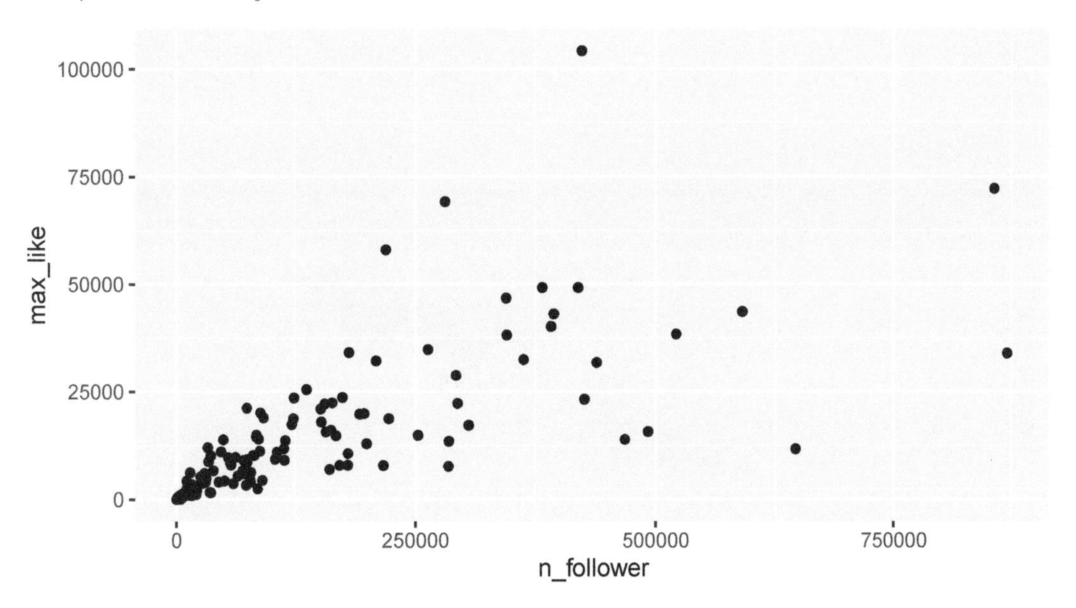


**FIGURE 11.13** Heteroscedasticity in the Rio Instagram data set.

```
rio_clean %>%
  filter(!is.na(n_follower)) %>%
  ggplot(aes(x = log(n_follower), y = log(max_like))) +
  geom_point() +
  geom_smooth(method = "lm")
```

The relationship between log followers and log likes is now visibly linear. Figure 11.15 shows the residual Scale-Location plot, and the residuals are homoscedastic.

```
rio_model <- lm(log(max_like) ~ log(n_follower), data = rio_clean)
plot(rio_model, which = 3)
```

Our linear model now involves the log variables. We compute the coefficients:

```
rio_model$coefficients
```

```
##      (Intercept) log(n_follower)
##       -0.5322455       0.8489013
```

The regression line is

$$\log(\widehat{\text{Max.Likes}}) = -0.532 + 0.849 \cdot \log(\text{N.follower})$$

We can use this model to make predictions and estimates, after converting to and from log scales. Or, we can exponentiate the regression equation to get the following relationship:

$$\widehat{\text{Max.Likes}} = 0.587 \cdot \text{N.follower}^{0.849}$$

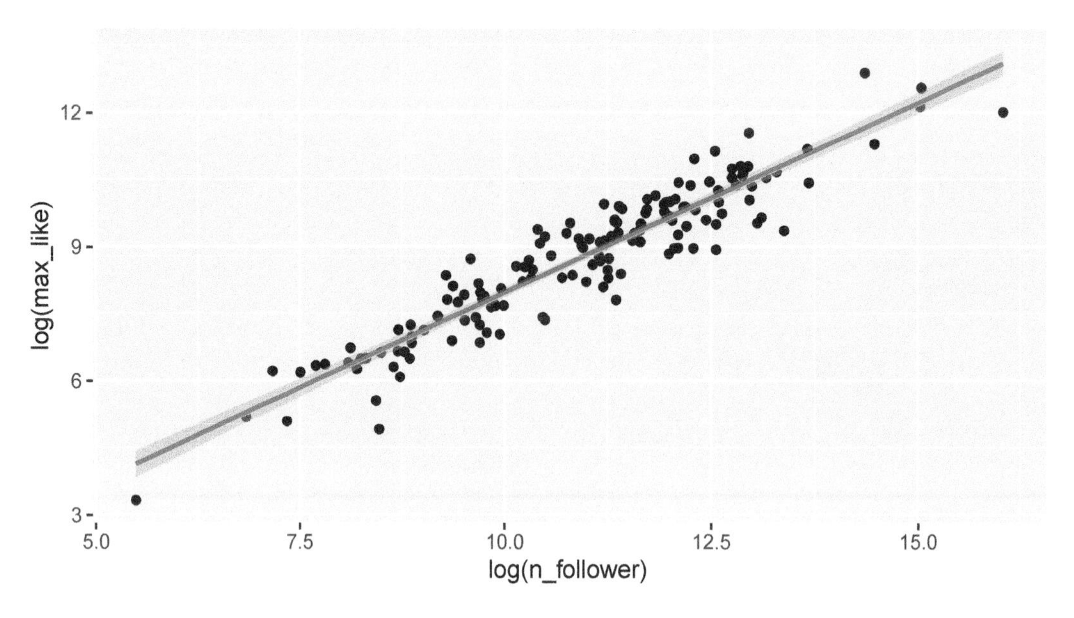


**FIGURE 11.14** Rio Instagram data after a log transformation.

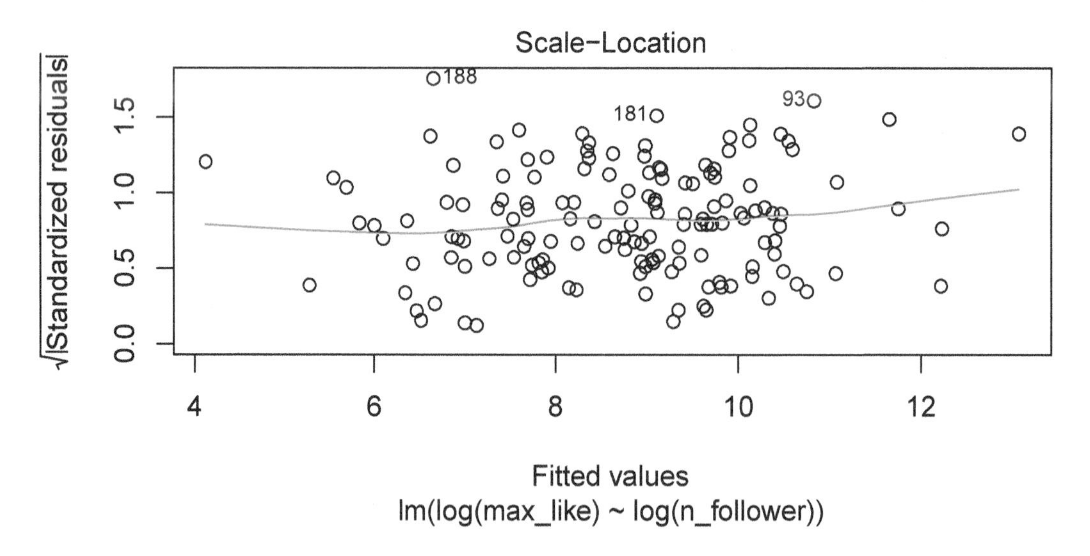


**FIGURE 11.15** Rio Instagram residuals after the log transformation are homoscedastic.

### 11.4.3 Normality

Assumption 2 in our model is that the error terms are **normally distributed**. We saw in Section 7.2 how to use qq plots to visualize whether a sample likely comes from a specified distribution. In this section, we see how to examine qq plots of residuals.

**Example 11.10.** The `barnacles` data set in the `fosdata` package describes the number of barnacles found on coral reefs. We model the density of barnacles (the number of barnacles per square meter) on the percentage of coral cover on the reef.

```
barnacles <- fosdata::barnacles
ggplot(barnacles, aes(x = coral_cover, y = barnacle_density)) +
  geom_point() +
  geom_smooth(method = "lm")
```

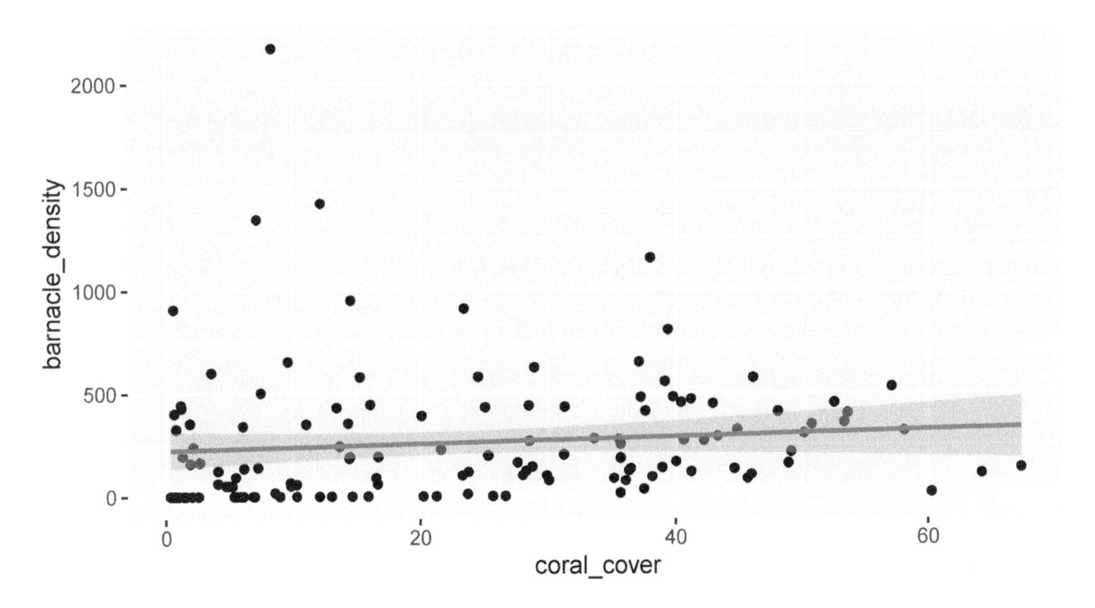


**FIGURE 11.16** Barnacle data with non-normal (right-skewed) residuals.

We see from Figure 11.16 that there doesn't appear to be a strong linear relationship between the predictor and response. If you look carefully, you will see that the errors appear to be right-skewed. Let's look at the normal qq plot of the residuals.

```
barn_mod <- lm(barnacle_density ~ coral_cover, data = barnacles)
plot(barn_mod, which = 2)
```

Figure 11.17 has a U-shape, which is an indicator of skewness. This data does not follow our model.

> **Try It Yourself.**
> Filter out reefs with no barnacles (a `count` of zero) and then model the log of barnacle density on the coral cover. Do the residuals appear approximately normal?

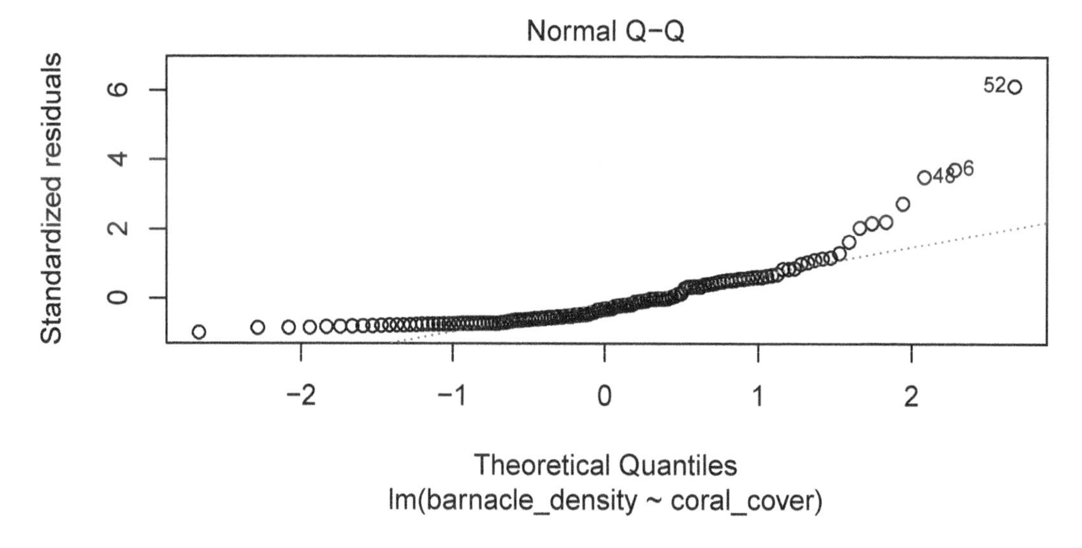


**FIGURE 11.17** Right-skewed residuals in the diagnostic qq plot.

### 11.4.4 Outliers and leverage

Linear regression minimizes the SSE of the residuals. If one data point is far from the others, the residual for that point could have a major effect on the SSE and therefore the regression line. A data point which is distant from the rest of the data is called an *outlier*.

Outliers can pull the regression line toward themselves. The regression line always goes through the center of mass of the data, the point $(\bar{x}, \bar{y})$. If the outlier's $x$-coordinate is far from the other $x$-coordinates in the data, then a small pivot of the line about the center of mass will make a large difference to the residual for the outlier. Because the outlier contributes the largest residual, it can have an outsized effect on the slope of the line. This phenomenon is called *leverage*.

Figure 11.18 shows a cloud of uncorrelated points, with two labeled $A$ and $B$. Point $A$ has an $x$-coordinate close to the mean $x$-coordinate of all points, so changing its $y$ value has a small effect on the regression line. This is shown in the second picture. Point $B$ has an $x$-coordinate that is extreme, and the third picture shows that changing its $y$ value makes a large difference to the slope of the line. Point $A$ has low leverage, point $B$ has high leverage.

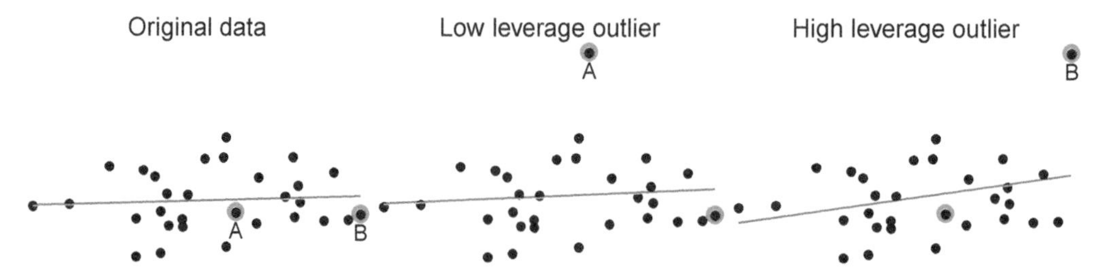


**FIGURE 11.18** High leverage point B has a relatively large effect on the regression line.

Dealing with high leverage outliers is a challenge. Usually it is a good idea to run the analysis both with and without the outliers, to see how large of an effect they have on the conclusions. A more sophisticated approach is to use *robust* methods. Robust methods for regression are analogous to rank based tests in that both are resistant to outliers. For more information on robust regression, see Venables and Ripley.[5]

**Example 11.11.** The `pulitzer`[6] data set contains data on the circulation of fifty major U.S. newspapers along with counts of the number of Pulitzer prizes won by each. Do Pulitzer prizes help newspapers maintain circulation?

In Figure 11.19 we plot the number of Pulitzer prizes won as the explanatory variable, and the percent change in circulation from 2004 to 2013 as the response. The `pulitzer` data is available in the `fivethirtyeight` package.

```
pulitzer <- fivethirtyeight::pulitzer
pulitzer <- rename(pulitzer, pulitzers = num_finals2004_2014)
ggplot(pulitzer, aes(x = pulitzers, y = pctchg_circ)) +
  geom_point() +
  geom_smooth(method = "lm")
```

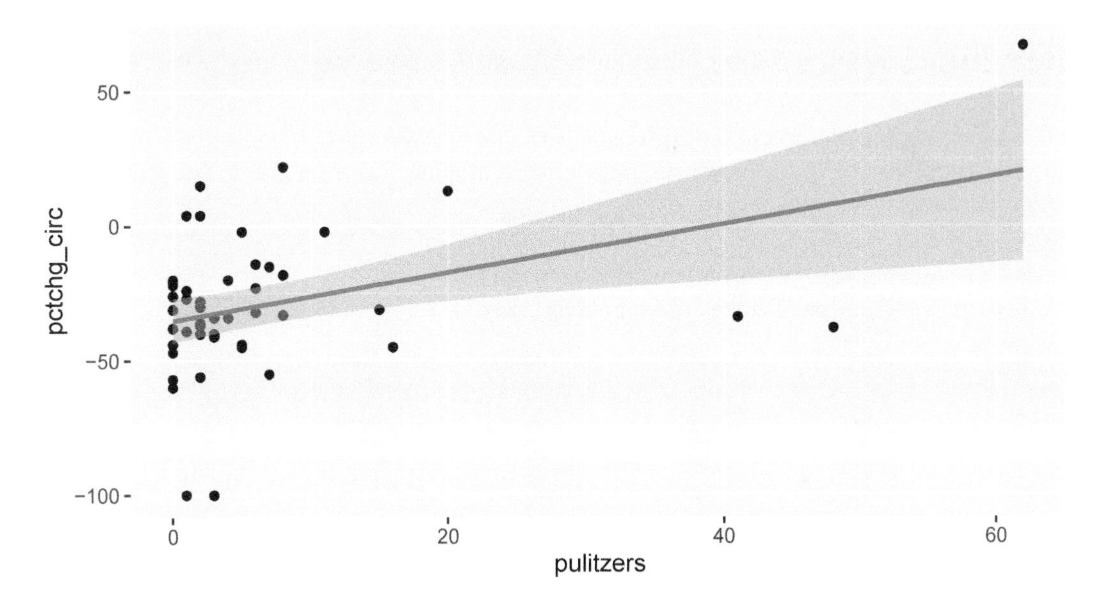


**FIGURE 11.19** The Pulitzer Prize data has high leverage outliers.

The regression line has positive slope, suggesting that more Pulitzer prizes correspond with an increase (or smaller decrease) in circulation. There are some high leverage outliers with a large number of Pulitzers compared to the other newspapers. The high leverage outliers show up in the "Residuals vs Leverage" diagnostic plot, see Figure 11.20.

```
pulitzer_full_model <- lm(pctchg_circ ~ pulitzers, data = pulitzer)
plot(pulitzer_full_model, which = 5)
```

Newspaper 3 is well outside the red "Cook's distance" lines. What paper is it?

```
pulitzer$newspaper[3]

## [1] "New York Times"
```

[5]W N Venables and B D Ripley, *Modern Applied Statistics with S*, 4th ed. (Springer, New York, 2002).

[6]Nate Silver, "Do Pulitzers Help Newspapers Keep Readers?" (FiveThirtyEight, April 15, 2014), https://fivethirtyeight.com/features/do-pulitzers-help-newspapers-keep-readers.

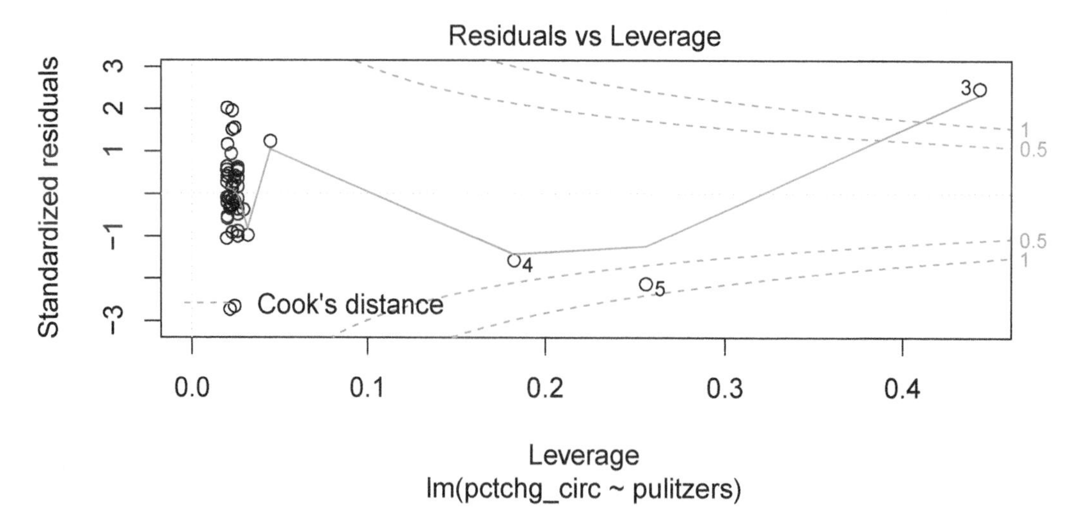


**FIGURE 11.20** Newspaper 3 is high leverage, outside the dashed Cook's distance lines.

It is good practice to perform the regression with and without the outliers, and see if it makes a difference. Figure 11.21 shows the data after removing *The New York Times,* together with the line of best fit.

```
pulitzer %>%
  filter(newspaper != "New York Times") %>%
  ggplot(aes(x = pulitzers, y = pctchg_circ)) +
  geom_point() +
  geom_smooth(method = "lm")
```

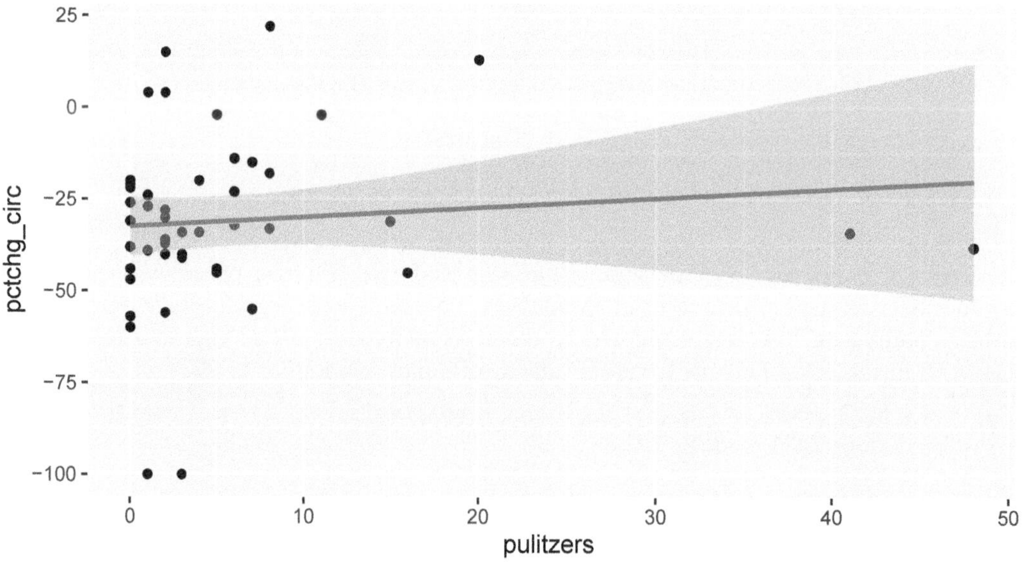

**FIGURE 11.21** Removing The New York Times removes most of the relationship between circulation and prizes.

We see that the line of best fit has changed dramatically with the removal of a single point, and that the apparent correlation in the first plot between Pulitzer Prizes and circulation is almost entirely due to *The New York Times.* Because most newspapers are struggling and *The New York Times* is one of the few that are not, it is hard to say if quality (as measured by prizes) does influence circulation. As always, correlation is not causation, and here even the correlation itself is questionable.

### 11.4.5 Independence

The model for linear regression assumes that the errors are independent. While there are some techniques that we can use to check for independence by looking at the residuals, you will also need to think carefully about the way in which the data was collected. We consider two examples that illustrate different ways that residuals may not be independent.

In this section, we will need to access the residuals of the model directly rather than through `plot(mod)`.

**Example 11.12.** Consider the `seoulweather` data set in the `fosdata` package. This data set gives the predicted high temperature (`ldaps_tmax_lapse`) and the observed high temperature (`next_tmax`) for summer days at 25 stations in Seoul, along with a lot of other data. Suppose we wish to model the **error** in the prediction on the current day high temperature (`present_tmax`). We need to modify our data a bit.

```
seoul <- fosdata::seoulweather
seoul <- mutate(seoul, temp_error = ldaps_tmax_lapse - next_tmax)
temp_mod <- lm(temp_error ~ present_tmax, data = seoul)
```

**Try It Yourself.**
Plot the residuals of the model and verify that they do not look too bad, though they are clearly not normal.

We are concerned with correlation among the residuals. It seems **very** likely that the residuals from one station are correlated with the residuals from the same day for a nearby station. This would be saying that if it is hotter than predicted somewhere, then it is likely also hotter than predicted a half-mile away.

Let's append the residuals to the original data frame for ease of use.

```
seoul <- seoul %>%
  mutate(resid = temp_error - predict(temp_mod, newdata = seoul))
```

Now, let's see if the residuals associated with station 1 are correlated with the residuals associated with station 2. We filter those two stations and pivot the data so we have two columns of residuals, one for each station.

```
wide_resid <- seoul %>%
  filter(station %in% c(1, 2)) %>%
  select(station, resid, date) %>%
  tidyr::pivot_wider(
    names_from = "station",
    names_prefix = "station_",
    values_from = resid
  )
```

Compute the sample correlation coefficient (avoiding NA values):

```
summarize(wide_resid, cor = cor(station_1, station_2, use = "complete.obs"))
```

```
## # A tibble: 1 x 1
##     cor
##   <dbl>
## 1 0.892
```

Figure 11.22 shows a scatterplot of the residuals in station 2 versus the residuals in station 1 on the same day. The correlation coefficient of 0.892 and the linear trend in the plot provide convincing evidence that the residuals are correlated.

```
ggplot(wide_resid, aes(x = station_1, y = station_2)) +
  geom_point()
```

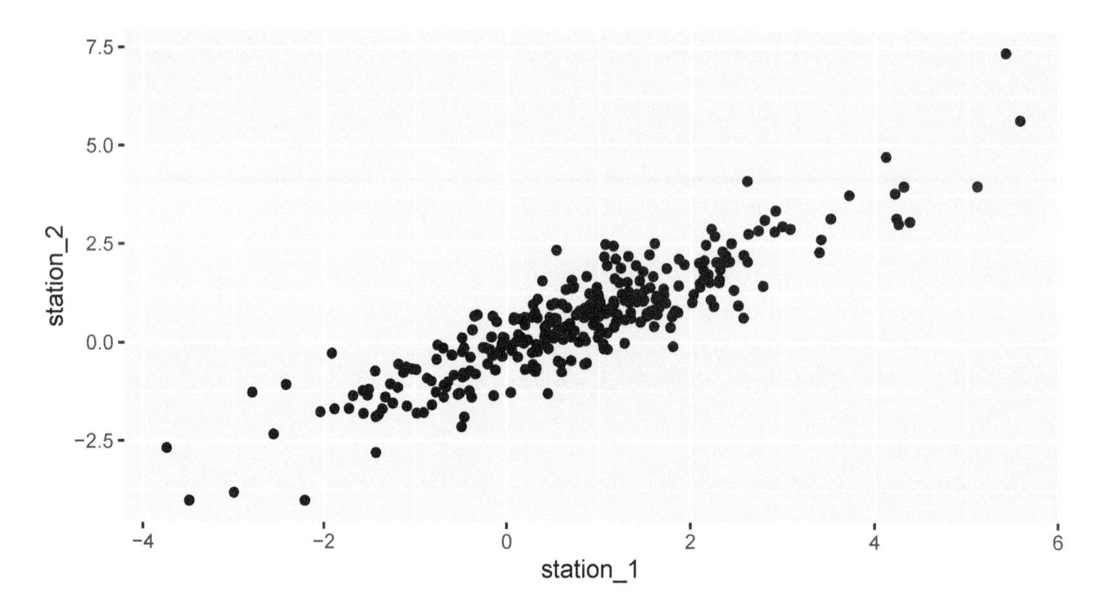


**FIGURE 11.22** Correlation between the residuals of two weather stations in Seoul.

In the next example, we examine another common source of dependence in data sets, which is dependence on time. This is known as *serial correlation.*

**Example 11.13.** Consider the daily high temperatures in Seoul from June 30 through August 3 at station 1 in 2016. We wish to model the high temperature on time using a linear model. The `lubridate` package helps clean up the dates and calculates the number of days since June 30, which we choose as our explanatory variable.

```
seoul_2016 <- seoul %>%
  mutate(date = lubridate::ymd(date)) %>%
  mutate(days_since_june30 = date - lubridate::ymd("2016/06/30")) %>%
  filter(station == 1, lubridate::year(date) == 2016)
```

Figure 11.23 shows the time series of high temperatures over the summer of 2016.

```
ggplot(seoul_2016, aes(x = date, y = present_tmax)) +
  geom_point() +
  geom_smooth(method = "lm")
```

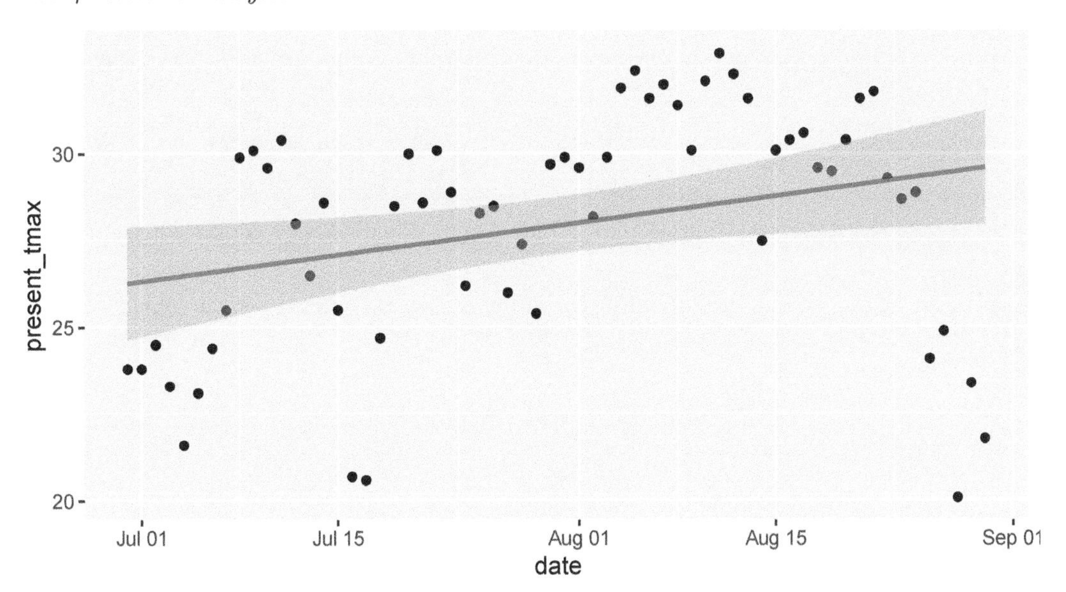


**FIGURE 11.23** Serial correlation of high temperatures in Seoul in 2016.

The upward slope of the line would seem to indicate that the daily high temperature increases throughout the late summer. However, we suspect serial correlation in the residuals: if the temperature is above the mean one day, then it will likely be above the mean the next day. Drawing conclusions based on this model is suspect.

To confirm the serial correlation, we construct the linear model, calculate the residuals, and then plot the residuals of one day versus the residuals of the next day. The function `lag` shifts the residual vector by one so the $y$-axis in Figure 11.24 gives the "next day" residual.

```
seoul_mod <- lm(present_tmax ~ days_since_june30, data = seoul_2016)

seoul_2016 <- seoul_2016 %>%
  mutate(resid = present_tmax - predict(seoul_mod, seoul_2016))

seoul_2016 %>% ggplot(aes(x = resid, y = lag(resid))) +
  geom_point() +
  labs(x = "Residual one day", y = "Residual the next day")
```

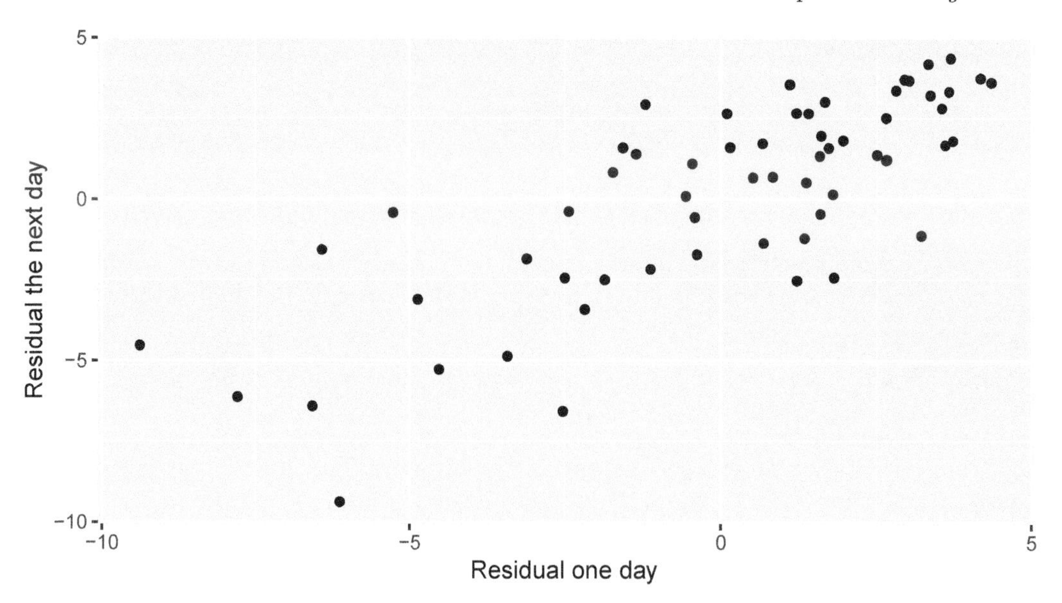


**FIGURE 11.24** Residuals from one day are correlated with those from the next day.

## 11.5 Inference

How can we detect a relationship between two numeric variables? Given a sample $(x_1, y_1), \ldots, (x_n, y_n)$, the correlation coefficient $r$ and the regression line $\hat{y} = \hat{\beta}_0 + \hat{\beta}_1 x$ both describe a linear relationship between $X$ and $Y$. However, the correlation and the regression line both depend on the sample. A different sample would lead to different statistics. How do we decide if the one sample we have available gives evidence of a linear relationship between the two variables? The answer is to use hypothesis testing.

The natural null hypothesis is usually that there is no linear relationship between $X$ and $Y$. To make this precise mathematically, observe that under this null hypothesis, the slope of the regression line would be zero. That is, the line would be flat. The hypothesis test then becomes:

$$H_0 : \beta_1 = 0, \quad H_a : \beta_1 \neq 0$$

Rejecting $H_0$ means that there is a linear relationship between $X$ and $Y$, or equivalently that $X$ and $Y$ are correlated.

To run this test, assume $H_0$, then observe a sample. Estimate the slope of the regression line from the sample. Then compute the probability of observing such a slope (or an even less likely slope) under the assumption that the true slope is actually 0. This probability is the $p$-value for the test.

In R, these computations are done by the `summary` command applied to the linear model constructed by `lm`.

**Example 11.14.** Is there a linear relationship between chinstrap penguin flipper length and body mass? If $\beta_1$ is the true slope of the regression line, we test the hypotheses

$$H_0 : \beta_1 = 0, \quad H_a : \beta_1 \neq 0$$

We previously examined residual plots (Figure 11.9) and determined that the assumptions of a linear model are met, so we proceed to inference.

```
summary(chinstrap_model)
```

```
##
## Call:
## lm(formula = body_mass_g ~ flipper_length_mm, data = chinstrap)
##
## Residuals:
##     Min      1Q  Median      3Q     Max
## -900.90 -137.45  -28.55  142.59  695.38
##
## Coefficients:
##                   Estimate Std. Error t value Pr(>|t|)
## (Intercept)      -3037.196    997.054  -3.046  0.00333 **
## flipper_length_mm   34.573      5.088   6.795 3.75e-09 ***
## ---
## Signif. codes:  0 '***' 0.001 '**' 0.01 '*' 0.05 '.' 0.1 ' ' 1
##
## Residual standard error: 297 on 66 degrees of freedom
## Multiple R-squared:  0.4116, Adjusted R-squared:  0.4027
## F-statistic: 46.17 on 1 and 66 DF,  p-value: 3.748e-09
```

The `summary` command produces quite a bit of output. Notice the `Call` is a helpful reminder of how the model was built – it was quite a ways back in this chapter!

The $p$-value we need for this hypothesis test is given in the `Coefficients` section, at the end of the `flipper_length_mm` line. It is labeled `Pr(>|t|)`. For this hypothesis test, $p = 3.75 \times 10^{-9}$ and it is significant. The `***` following the value means that the results are significant at the $\alpha = 0.001$ level.

We reject $H_0$ and conclude that there is a significant linear relationship between flipper length and body mass for chinstrap penguins.

**Example 11.15.** Lifting weights can build muscle strength. Scientific studies require detailed information about the body during weight lifting, and one important measure is the time under tension (TUT) during muscle contraction. However, TUT is difficult to measure. The best measurement comes from careful human observation of a video recording, which is impractical for a large study. In their 2020 paper[7], Viecelli et al. used a smartphone's accelerometer to estimate the time under tension for various common machine workouts. They also used a video recording of the workout to do the same thing. Could the much simpler and cheaper smartphone method replace the video recording?

The data is available as `accelerometer` in the `fosdata` package.

```
accelerometer <- fosdata::accelerometer
```

[7] Claudio Viecelli et al., "Using Smartphone Accelerometer Data to Obtain Scientific Mechanical-Biological Descriptors of Resistance Exercise Training," *PLOS One* 15, no. 7 (July 2020): e0235156.

In this example, we focus on the abductor machine during the Rep mode of the contraction. The data set contains observations tagged as outliers, which we remove.

```
abductor <- accelerometer %>%
  filter(machine == "ABDUCTOR") %>%
  filter(stringr::str_detect(contraction_mode, "Rep")) %>%
  filter(!rater_difference_outlier &
    !smartphone_difference_outlier &
    !video_smartphone_difference_outlier)
```

Let's start by looking at a plot. Following the authors of the paper, we will create a Bland-Altman plot. A Bland-Altman plot is used when there are two measurements of the same quantity, especially when one is considered the "gold standard." We plot the *difference* of the two measurements versus the *mean* of the two measurements in Figure 11.25.

```
abtimediff <- abductor %>%
  mutate(
    difference = smartphones_mean_ms - video_rater_mean_ms,
    mean_time = (video_rater_mean_ms + smartphones_mean_ms) / 2
  )
abtimediff %>%
  ggplot(aes(x = mean_time / 1000, y = difference / 1000)) +
  geom_point() +
  labs(
    x = "Mean time estimate (sec)",
    y = "Difference in Estimates (sec)"
  )
```

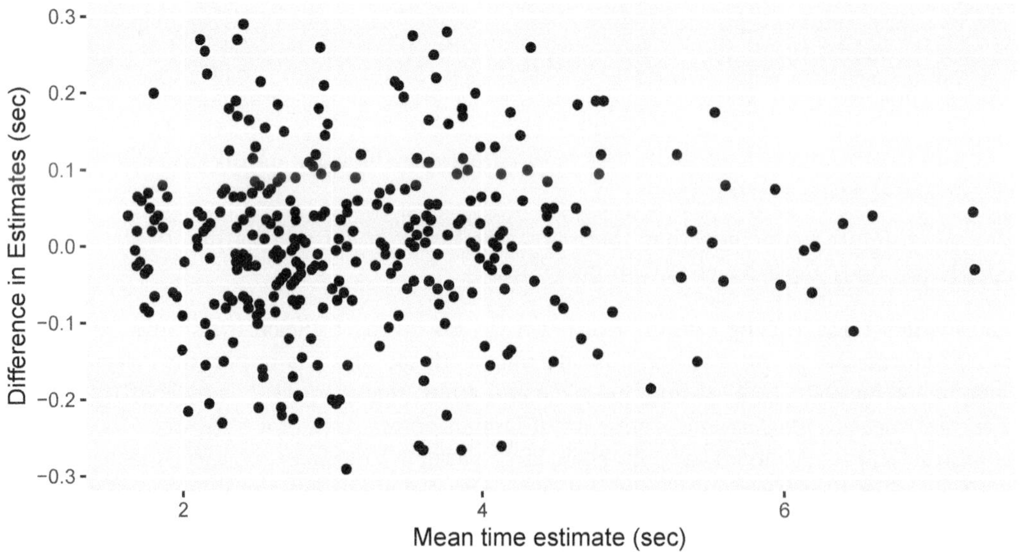

**FIGURE 11.25** Bland-Altman plot of video and smartphone measurements.

If the phone measurement is a good substitute for the video measurement, the difference of the measurements should have small standard deviation, have mean zero, and not depend on the mean of the measurements. We can use linear regression to assess the last two objectives.

The regression line through these points should have a slope and an intercept of 0. Indeed, if the intercept is not zero (but the slope is zero), then that would indicate a bias in the smartphone measurements. If the slope is not zero, then that would indicate that the smartphone estimate is not working consistently through all of the time intervals. So, let's perform a linear regression and see what we get.

```
accel_mod <- lm(difference ~ mean_time, data = abtimediff)
summary(accel_mod)$coefficients
```

```
##                 Estimate  Std. Error    t value  Pr(>|t|)
## (Intercept) -6.226185392 20.37164776 -0.3056299 0.7600870
## mean_time    0.003519751  0.00600059  0.5865675 0.5579117
```

**Try It Yourself.**
Check the residual plots of the model `accel_mod` and confirm that they are acceptable.

We cannot reject the null hypothesis that the slope is zero ($p = 0.56$), nor the null hypothesis that the intercept is zero ($p = 0.76$). It appears that the smartphone accelerometer measurement is a good substitute for video measurement of abductor time under tension, as long as the standard deviation is acceptable.

### 11.5.1 The summary command

Let's go through the rest of the output that R produces in the summary of a linear model. The full output of the command `summary(chinstrap_model)` is shown in the previous section. Here, we take it apart.

```
## Call:
## lm(formula = body_mass_g ~ flipper_length_mm, data = chinstrap)
```

The `Call` portion of the output reproduces the original function call that produced the linear model. Since that call happened much earlier in this chapter, it is helpful to have it here for review.

```
## Coefficients:
##                    Estimate Std. Error t value Pr(>|t|)
## (Intercept)       -3037.196    997.054  -3.046  0.00333 **
## flipper_length_mm    34.573      5.088   6.795 3.75e-09 ***
```

The `Coefficients` portion describes the coefficients of the regression model in detail. The `Estimate` column gives the intercept value and the slope of the regression line, which is $\widehat{\text{mass}} = -3037.196 + 34.573 \cdot \text{flipper length}$ as we have seen before.

The standard error column `Std. Error` is a measure of how accurately we can estimate the coefficient. We will not describe how this is computed. The value is always positive, and should be compared to the size of the estimate itself by dividing the estimate by the standard error. This division produces the $t$ value, i.e., $t = 34.573/5.088 = 6.795$ for flipper length.

Under the assumption that the errors are iid Norm(0, $\sigma$), the test statistic $t$ has a $t$ distribution, with $n - 2$ degrees of freedom. Here, $n = 68$ is the number of penguins in the chinstrap data. We subtract 2 because we used the data to estimate two parameters (the slope and the intercept) leaving only 66 degrees of freedom. The $p$-value for the test is the probability that a random $t$ is more extreme than the observed value of $t$, which explains the notation `Pr(>|t|)` in the last column.

To reproduce the $p$-values associated with `flipper_length_mm`, use

```
2 * (1 - pt(6.795, df = 66))
```

```
## [1] 3.743945e-09
```

where we have doubled the value because the test is two-tailed.

The final `Coefficients` column contains the significance codes, which are not universally loved. Basically, the more stars, the lower the $p$-value in the hypothesis test of the coefficients as described above. One problem with these codes is that the `(Intercept)` coefficient is often highly significant, even though it has a dubious interpretation. Nonetheless, it receives three stars in the output, which bothers some people.

```
## Residual standard error: 297 on 66 degrees of freedom
## Multiple R-squared:  0.4116, Adjusted R-squared:  0.4027
## F-statistic: 46.17 on 1 and 66 DF,  p-value: 3.748e-09
```

The residual standard error is an estimate of the standard deviation of the error term in the linear model. Recall that we are working with the model $y_i = \beta_0 + \beta_1 x_i + \epsilon_i$, where $\epsilon_i$ are assumed to be iid normal random variables with mean 0 and unknown standard deviation $\sigma$. The residual standard error is an estimate for $\sigma$.

Multiple R-squared is the square of the correlation coefficient $r$ for the two variables in the model.

The rest of the output is not interesting for simple linear regression, i.e., only one explanatory variable, but becomes interesting when we treat multiple explanatory variables in Chapter 13. For now, notice that the $p$-value on the last line is exactly the same as the $p$-value for the `flipper_length_mm` coefficient, which was the $p$-value needed to test for a relationship between flipper length and body weight.

### 11.5.2 Confidence intervals for parameters

The most likely type of inference on the slope we will make is $H_0 : \beta_1 = 0$ versus $H_a : \beta_1 \neq 0$. The $p$-value for this test is given in the summary of the linear model. However, there are times when we want a confidence interval for the slope, or to test the slope against a value other than 0. We illustrate with the following example.

**Example 11.16.** Consider the data set `Davis` in the package `carData`. The data consists of 200 observations of the variables `sex`, `weight`, `height`, `repwt`, and `repht`. Figure 11.26 shows a plot of patients' weight versus their reported weight.

```
Davis <- carData::Davis
ggplot(Davis, aes(x = repwt, y = weight)) +
  geom_point()
```

As we can see, there is a very strong linear trend in the data. There is also one outlier, who has reported weight of 56 and observed weight of 166. Let's remove that data point:

```
Davis <- filter(Davis, weight != 166)
```

Create a linear model for `weight` as explained by `repwt` and then extract the coefficients:

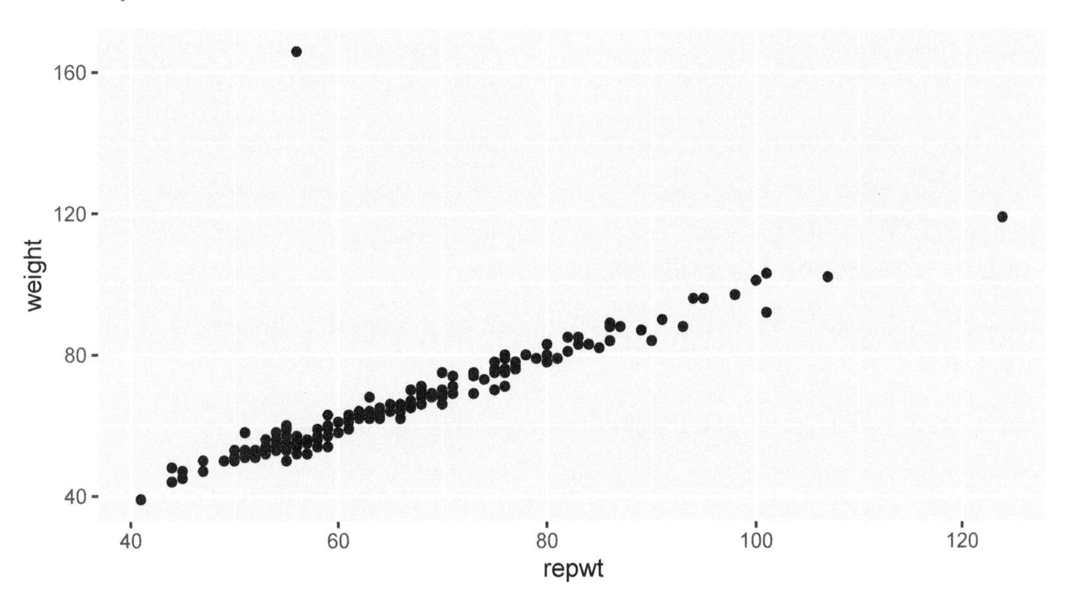


**FIGURE 11.26** Weight versus reported weight for 200 patients.

```
modDavis <- lm(weight ~ repwt, data = Davis)
summary(modDavis)$coefficients

##                Estimate Std. Error   t value      Pr(>|t|)
## (Intercept) 2.7338020 0.81479433  3.355205  9.672344e-04
## repwt       0.9583743 0.01214273 78.925763 1.385270e-141
```

**Try It Yourself.**
Check the residual plots of `modDavis` and confirm that they are acceptable.

In this example, we are not interested in testing the slope of the regression line against zero, since the linear relationship is obvious.

The statement we would like to test is whether people report their weights accurately. Perhaps as people vary further from what is considered a healthy weight, they report their weights toward the healthy side. Underweight people might overstate their weight and overweight people might understate their weight. So, an interesting question here is whether the slope of the regression line is 1.

Let's estimate the 95% confidence interval for the slope of the regression line. This interval is given by the estimate for the slope plus or minus approximately twice its standard error. From the regression coefficients data, the estimate is 0.95837 and the standard error is 0.01214, giving an interval of roughly $[.93, .98]$.

To get more precise information, we can use the R function `confint`

```
confint(modDavis, level = .95)

##                 2.5 %    97.5 %
## (Intercept) 1.1260247 4.3415792
## repwt       0.9344139 0.9823347
```

We are 95% confident that the true slope is in the interval $[0.934, 0.982]$.

The corresponding hypothesis test is $H_0 : \beta_1 = 1$ against $H_a : \beta_1 \neq 1$. Since the 95% confidence interval for the slope does not contain the value 1, we can reject $H_0$ in favor of $H_a$ at the $\alpha = .05$ level. We conclude that the slope is significantly different from 1, and that the population covered by this study does tend to report weights with a bias toward the mean weight.

### 11.5.3 Prediction intervals for response

Suppose we have found the line of best fit given the data, and it is represented by $\hat{y} = \hat{\beta}_0 + \hat{\beta}_1 x$. We would like to be able to predict what happens if we collect new data, possibly at $x$ values that weren't in the original sample. Ideally, we would like to provide an expected value as well as some sort of interval in which subsequent data will likely be.

Notationally, we will call $x_\star$ the new $x$ value. The value $\beta_0 + \beta_1 x_*$ is the true mean value of $Y$ when $X = x_\star$, but of course we do not know $\beta_0$ and $\beta_1$. We will use our estimates $\hat{\beta}_0$ and $\hat{\beta}_1$, which introduces one source of uncertainty in our prediction of the response at $X = x_*$.

Our model $Y = \beta_0 + \beta_1 x + \epsilon$ assumes that $\epsilon$ is normal with mean zero, so that values of $Y$ for a given $x_\star$ will be normally distributed around the true mean value of $y$ at $x_\star$, namely $\beta_0 + \beta_1 x_\star$. This introduces a second source of uncertainty in our prediction of the response when $X = x_*$.

The following two examples show how to create a point estimate and a prediction interval for the predicted response when $X = x_*$.

**Example 11.17.** Suppose you want to predict the body mass of a chinstrap penguin whose flipper length is known to be 204 mm. From the regression line,

$$\widehat{\text{mass}} = -3037.196 + 34.574 \cdot 204 = 4015.9$$

which predicts a body mass of 4015.9 g.

Recall that the `predict` command in R can perform the same computation, given the model and a data frame containing the $x$ values you wish to use to make predictions. Needing to wrap the explanatory variable in a data frame is a nuisance for making simple predictions, but will prove essential when using more complicated regression models.

```
predict(chinstrap_model, data.frame(flipper_length_mm = 204))
```

```
##        1
## 4015.777
```

A key assumption in the linear model is that the error terms $\epsilon_i$ are iid, so that they have the same variance for every choice of $x_i$. This allows estimation of the common variance from the data, and from that an interval estimate for $Y$. The `predict` command can provide a prediction interval using the `interval` option.[8] By default, this option produces a prediction interval that includes 95% of all $Y$ values for the given $x$ (we have already confirmed that the residuals for this model are acceptable):

**Example 11.18.**

[8]To find this option with R's built-in help, use `?predict` and then click on `predict.lm`, because we are applying predict to an object created by lm.

```
predict(chinstrap_model,
  data.frame(flipper_length_mm = 204),
  interval = "predict"
)
```

```
##        fit      lwr      upr
## 1 4015.777 3412.629 4618.925
```

The `fit` column is the prediction, `lwr` and `upr` describe the prediction interval. We are 95% confident that new observations of chinstrap penguins with flipper length 204 mm will have body weights between 3413 and 4619 g.

**Remark.** The choice of explanatory and response variable determines the regression line, and that choice is *not* reversible. We saw that a 204 mm flipper predicts a 4016 g body mass. Now switch the explanatory and response variables so that we can use body mass to predict flipper length:

```
reverse_model <- lm(flipper_length_mm ~ body_mass_g, data = chinstrap)
predict(reverse_model, data.frame(body_mass_g = 4015.777))
```

```
##       1
## 199.189
```

A 4016 g body mass predicts a 199 mm flipper length, shorter than the 204 mm you might have expected. In turn, a 199 mm flipper length predicts a body mass of 3849 g. Extreme values of flipper length predict extreme values of body mass, but the predicted body mass is relatively closer to the mean body mass. Similarly, extreme values of body mass predict extreme values of flipper length, but the predicted flipper length will be relatively closer to the mean flipper length. This phenomenon is referred to as *regression to the mean.*

**Example 11.19.** How quickly will 95% of seven-year-old children complete the Shape Trail Test trail B? We construct the model and confirm that the residual plots are acceptable (Figure 11.27).

```
stt_model <- lm(stt_cv_trail_b_secs ~ age_in_months, data = child_tasks)
plot(stt_model)
```

```
predict(stt_model, data.frame(age_in_months = 7 * 12), interval = "predict")
```

```
##        fit      lwr      upr
## 1 52.15541 22.72335 81.58747
```

The prediction interval is $[22.7, 81.6]$ so we are 95% confident that a seven-year-old child will complete trail B between 22.7 and 81.6 seconds.

**Remark.** Making predictions for $x$ values that are outside the observed range is called *extrapolation*, and is not recommended. In the child task study, the youngest children observed were six years old and the oldest were ten. Predictions about children outside of this age range would need to be treated with caution. For example, the regression line predicts that a 13-year-old child would complete trail B in negative three seconds, clearly impossible.

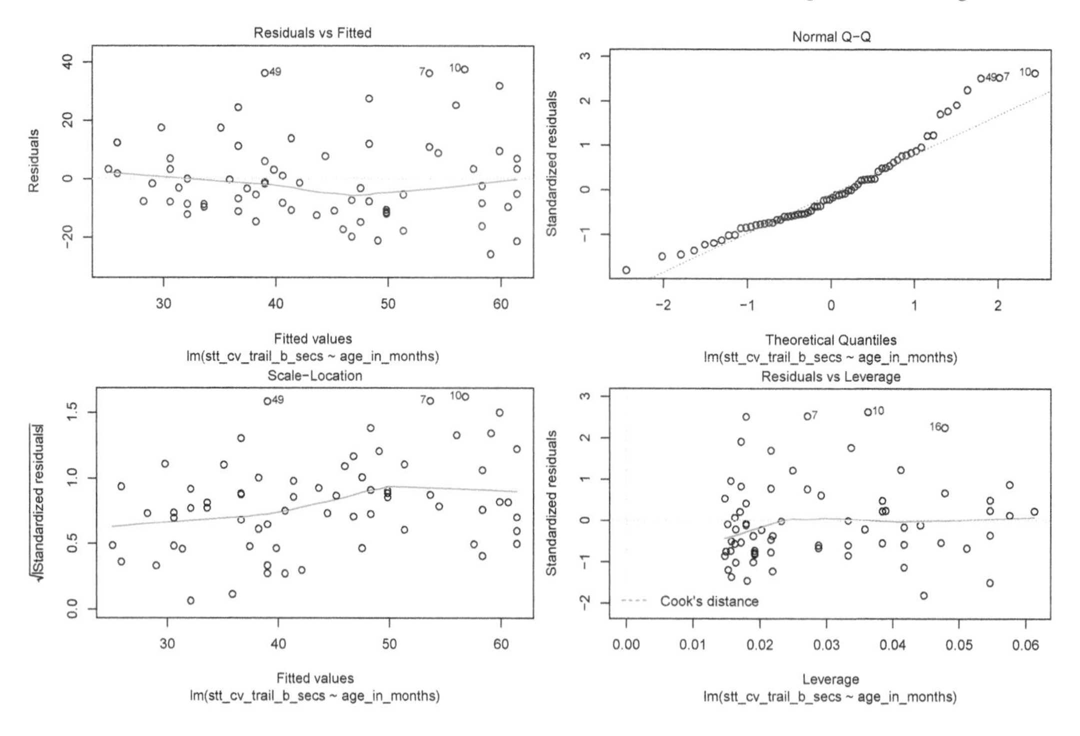


**FIGURE 11.27** Residual plots for the Shape Trail Test. The model assumptions are met.

### 11.5.4 Confidence intervals for response

The value $\hat{\beta}_0 + \hat{\beta}_1 x_*$ is the point estimate for the mean of the random variable $Y$, given $x_\star$. Confidence intervals for the response give an interval such that we are 95% (say) confident that the true mean of $Y$ when $X = x_*$ is within our interval. This is contrasted to the prediction interval, which gives an interval such that we are 95% sure that a subsequent observation of $Y$ when $X = x_*$ is within our interval. The confidence interval will always be contained within the prediction interval, because we do not need to take into account the random variation of a single observation; we are only concerned with estimating the mean.

For example, suppose you need to transport a chorus line of 80 chinstrap penguins, carefully selected to all have flipper length 204 mm. The average weight for such a penguin is 4016 g, so you expect your squad to weigh 321280 g or about 321 kg. How much will that total mass vary? The value 4016 g comes from the regression line at $x_\star = 204$. Refer back to the plot of this regression line in Figure 11.3, which shows the gray error band around the line. The estimated mean value 4016 g varies by the width of the gray band above $x_\star = 204$. The `predict` function can compute the range precisely (we have already checked that the residual plots are acceptable for this model):

```
predict(chinstrap_model,
  data.frame(flipper_length_mm = 204),
  interval = "confidence"
)
```

```
##        fit      lwr     upr
## 1 4015.777 3905.903 4125.65
```

This gives a 95% confidence interval of [3906, 4126] for the average mass of chinstrap penguins with flipper length 204 mm. Assuming that the mean of the 80 penguins is approximately the same as the true mean of all penguins of that weight, we can multiply by 80 and then be 95% confident that the 80 penguin dance troupe will weigh between 312.4 kg and 330 kg.

To summarize this section, let's plot the chinstrap penguin data again. This is a complicated plot, and it begins by building a new data frame called `bounds_data` which contains flipper length values from 178 to 212 and both confidence and prediction bounds for each of those $x$ values. Here we go.

```
flipper_range <- data.frame(flipper_length_mm = seq(178, 212, 2))

prediction_bounds <- predict(chinstrap_model,
  newdata = flipper_range, interval = "p"
)
colnames(prediction_bounds) <- c("p_fit", "p_lwr", "p_upr")
confidence_bounds <- predict(chinstrap_model,
  newdata = flipper_range, interval = "c"
)
colnames(confidence_bounds) <- c("c_fit", "c_lwr", "c_upr")

bounds_data <- cbind(flipper_range, confidence_bounds, prediction_bounds)

chin_plot <- chinstrap %>%
  ggplot(aes(x = flipper_length_mm, y = body_mass_g)) +
  geom_point() +
  geom_smooth(method = "lm") +
  labs(title = "Chinstrap penguin flipper length and body mass")

chin_plot <- chin_plot +
  geom_line(data = bounds_data, aes(y = c_lwr), linetype = "dashed") +
  geom_line(data = bounds_data, aes(y = c_upr), linetype = "dashed")

chin_plot <- chin_plot +
  geom_line(data = bounds_data, aes(y = p_lwr), color = "orange") +
  geom_line(data = bounds_data, aes(y = p_upr), color = "orange")

chin_plot
```

In Figure 11.28 the confidence interval about the regression line is shown in gray as usual and is also outlined with dashed lines. The prediction interval is shown in orange.

Some things to notice are as follows:

1. Roughly 95% of the data falls within the orange prediction interval. While prediction intervals are predictions of **future** samples, it is often the case that the original data also falls in the prediction interval roughly 95% of the time.
2. We are 95% confident that the true population regression line falls in the gray confidence band at any particular $x_*$.
3. The confidence band is narrowest near $(\overline{x}, \overline{y})$ and gets wider away from that central point. This is typical behavior.
4. We have confirmed the method that `geom_smooth` uses to create the confidence band.

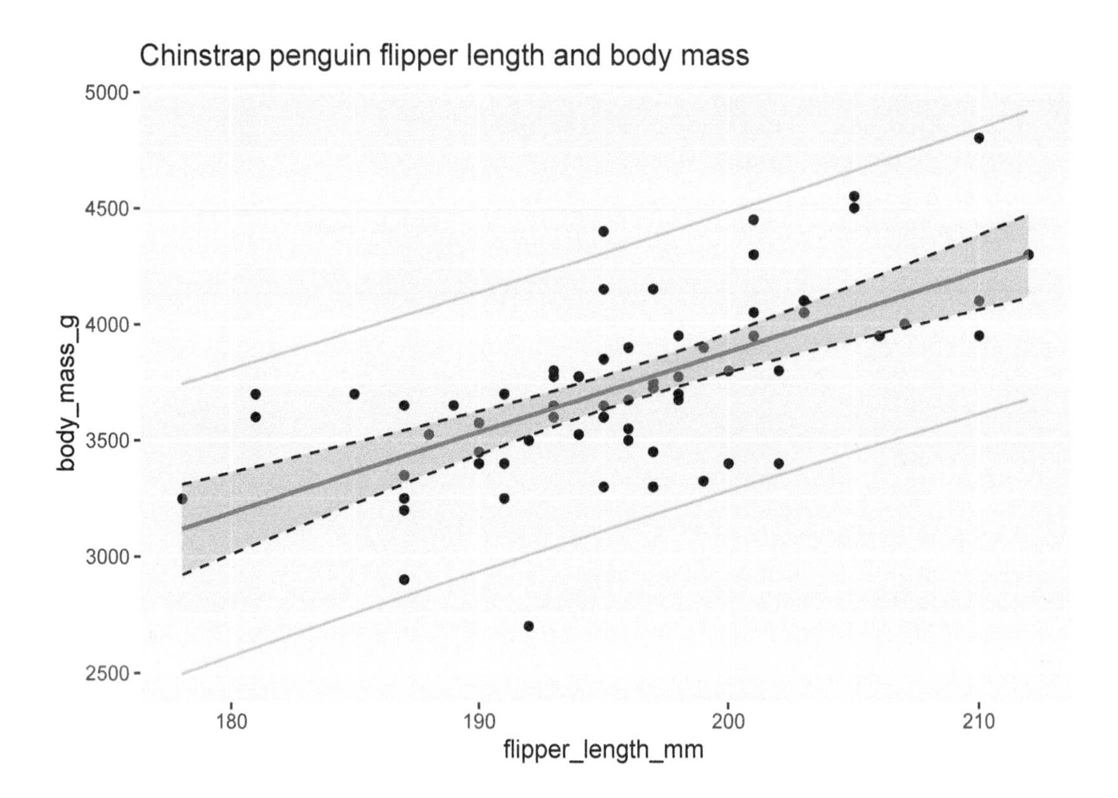


**FIGURE 11.28** Confidence bounds (dashed) and prediction bounds (orange).

## 11.6 Simulations for simple linear regression

In this section, we will use simulation for two purposes. The first is to create residual plots that we know follow the assumptions of our model in order to better assess residual plots of data. The second is to understand and confirm the inference of Section 11.5

### 11.6.1 Residuals

It can be tricky to learn what "good" residuals look like. One of the nice things about simulated data is that you can train your eye to spot departures from the assumptions. Let's look at two examples: one that satisfies the regression assumptions and one that does not. You should think about other examples you could come up with!

**Example 11.20.** Consider simulated data that follows the model $y_i = 2 + 3x_i + \epsilon_i$, where $\epsilon_i$ is iid normal with mean 0 and standard deviation 5. This model satisfies the assumptions required for linear regression, so it is an example of good behavior.

```
x <- runif(40, 10, 20) # 40 explanatory values between 10 and 20
epsilon <- rnorm(40, 0, 5) # 40 error terms, normally distributed
y <- 2 + 3 * x + epsilon
my_model <- lm(y ~ x)
my_model$coefficients
```

```
## (Intercept)           x
##     5.77620     2.77251
```

The estimate for the intercept is $\hat{\beta}_0 = 5.7762$ and for the slope it is $\hat{\beta}_1 = 2.77251$. The estimated slope is close to the true value of 3. The estimated intercept is less accurate because the intercept is far from the data, at $x = 0$, and so small changes to the line have a large effect on the intercept.

Printing the full summary of the model gives the residual standard error as 4.17229. The the true standard deviation of the error terms $\epsilon_i$ was chosen to be 5, so the model's estimate is good.

```
data.frame(x, y) %>% ggplot(aes(x = x, y = y)) +
  geom_point() +
  geom_smooth(method = "lm")
plot(my_model, which = 1)
```

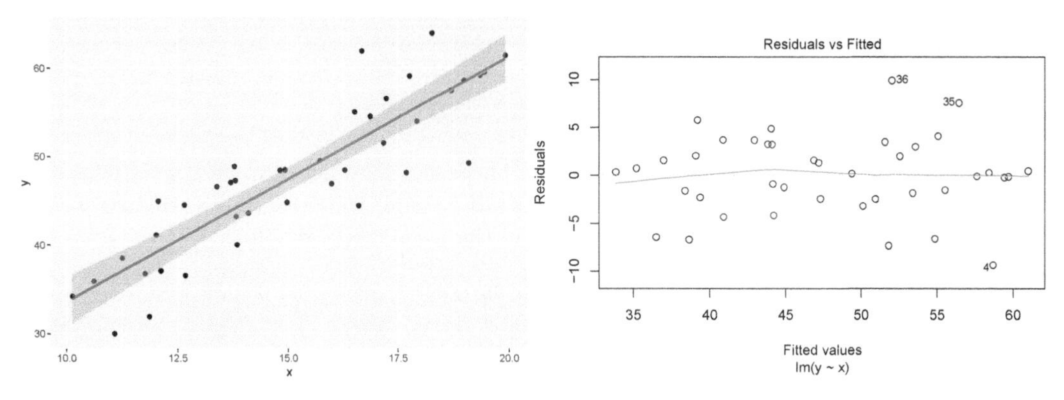


The regression line follows the line of the points and the residuals are more or less symmetric about 0.

**Example 11.21.** What happens if the errors in a regression model are non-normal? The data no longer satisfies the hypothesis needed for regression, but how does that affect the results?

This example shows the effect of skew residuals. As in Example 11.20, $y_i = 2 + 3x_i + \epsilon_i$ except that we now use iid exponential with rate $1/10$ for the error terms $\epsilon_i$, and subtract 10 so that the error terms still have mean 0.

```
x <- runif(40, 10, 20) # 40 explanatory values between 10 and 20
epsilon <- rexp(40, 1 / 10) - 10 # 40 error terms, mean 0 and very skew
y <- 2 + 3 * x + epsilon
my_model <- lm(y ~ x)

data.frame(x, y) %>% ggplot(aes(x = x, y = y)) +
  geom_point() +
  geom_smooth(method = "lm")
plot(my_model, which = 1)
```

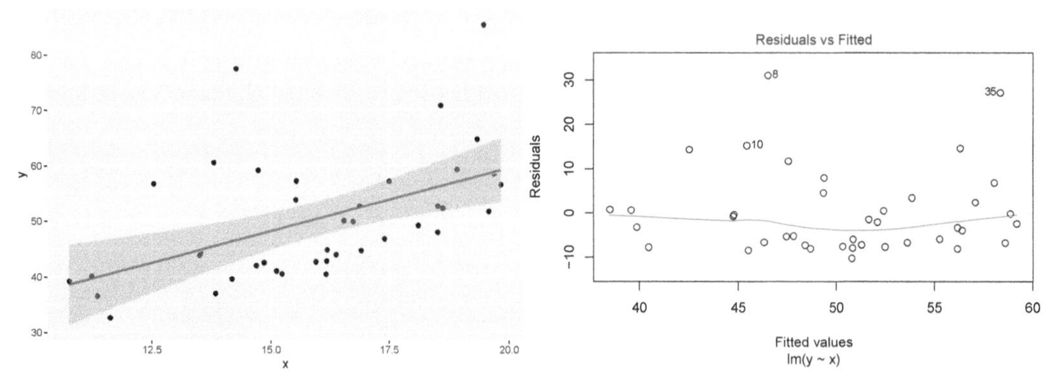


In the residuals plot, the residuals are larger above the zero line than below it, displaying the skewness of the residuals. The skewness is even more apparent in the Normal Q-Q plot as an upward curve:

```
plot(my_model, which = 2)
```

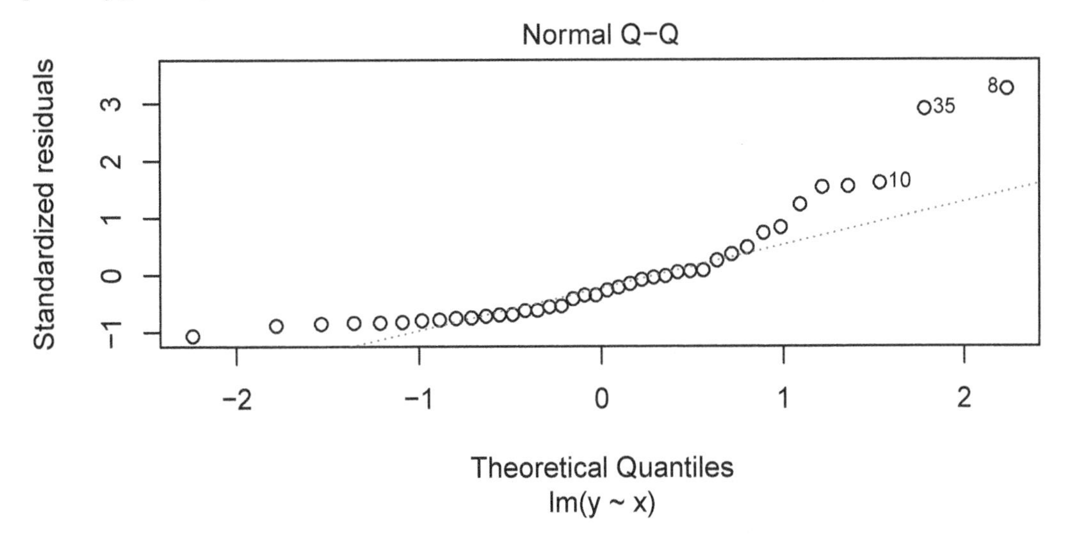


Now, we examine properties of the estimators for the slope and intercept when the errors are exponential rather than normal. Note that our model assumptions are not satisfied, so we are testing via simulation the robustness of our model to this type of departure from assumptions.

We first see whether the mean of the estimates for the slope and intercept are the correct values of 3 and 2; that is, we are seeing whether the estimates for slope and intercept are unbiased.

```
sim_data <- replicate(10000, {
  epsilon <- rexp(40, 1 / 10) - 10
  y <- 2 + 3 * x + epsilon
  my_model <- lm(y ~ x)
  my_model$coefficients
})
mean(sim_data[2, ]) # expected value of slope

## [1] 2.996852

mean(sim_data[1, ]) # expected value of intercept

## [1] 2.045073
```

This is pretty strong evidence that the estimates for the slope and intercept are still **unbiased** even when the errors are not normal.

Now, let's examine whether 95% confidence intervals for the slope contain the true value 95% of the time, as desired.

```
sim_data <- replicate(10000, {
  epsilon <- rexp(40, 1 / 10) - 10
  y <- 2 + 3 * x + epsilon
  my_model <- lm(y ~ x)
  confint(my_model)[2, 1] < 3 && confint(my_model)[2, 2] > 3
})
mean(sim_data)
```

```
## [1] 0.954
```

If you run the above code several times, you will see that the confidence interval for the slope contains the true value a bit more than 95% of the time. However, it is still working pretty close to advertised.

> **Try It Yourself.**
> About what percentage of time does the 95% confidence interval for the intercept contain the true value of the intercept when the error is exponential, as above?

### 11.6.2 Prediction intervals

In this section, we explore prediction intervals through simulation. We will find that a small error in modeling has minimal effect near the center of our data range, but can lead to large errors near the edge of observed data. Beyond the data's edge, making predictions is called *extrapolation* and can be wildly incorrect.

We begin by exploring prediction intervals a bit more. Suppose we have data that comes from the generative process $Y = 1 + 2X + \epsilon$, where $\epsilon \sim \text{Norm}(0, \sigma = 3)$. We collect data one time and create a prediction interval for when $x = 10$. What percentage of responses will fall in the 95% prediction interval? You would probably think 95%, but let's see what happens.

```
set.seed(14)
x <- runif(30, 0, 20)
y <- 1 + 2 * x + rnorm(30, 0, 3)
mod <- lm(y ~ x)
predict(mod, newdata = data.frame(x = 10), interval = "pre")
```

```
##        fit      lwr     upr
## 1 21.35936 15.76602 26.9527
```

We see that our prediction interval is from 15.76602 to 26.9527. Since $Y$ is normal with mean 21 and standard deviation 3 when $x = 10$, the probability that new values will be in the 95% prediction interval is about 93.6%, given by

```
pnorm(26.9527, 21, 3) - pnorm(15.76602, 21, 3)
```

```
## [1] 0.935863
```

What's going on? Why doesn't a 95% prediction interval contain 95% of future values? Ninety-five percent prediction intervals will contain 95% of future values assuming that the data is recollected and new prediction intervals are constructed each time before a new data point is collected. In terms of simulations, that means that we would have to recreate $y_i$, recreate our model and prediction interval, then draw one more sample and see whether it is in the prediction interval.

```
sim_data <- replicate(10000, {
  y <- 1 + 2 * x + rnorm(30, 0, 3)
  mod <- lm(y ~ x)
  new_data <- 1 + 2 * 10 + rnorm(1, 0, 3)
  pred_int <- predict(mod, newdata = data.frame(x = 10), interval = "pre")
  pred_int[2] < new_data && pred_int[3] > new_data
```

```
})
mean(sim_data)

## [1] 0.9521
```

Now we get close to the desired value of 95%, and larger simulations will yield closer and closer to 95%

Next, we compare the behavior of prediction intervals for two sets of simulated data:

- Data 1 (linear, assumptions correct): $y_i = 1 + 2x_i + \epsilon_i$
- Data 2 (non-linear, assumptions incorrect): $y = 1 + 2x_i + x^2/20 + \epsilon_i$

In both sets, the error terms $\epsilon_i \sim \text{Norm}(0, 1)$ and there are 20 $x$ values uniformly randomly distributed on the interval $(0, 10)$.

```
simdata <- data.frame(x = runif(20, 0, 10), epsilon = rnorm(20))
simdata <- mutate(simdata,
  y1 = 1 + 2 * x + epsilon,
  y2 = 1 + 2 * x + x^2 / 20 + epsilon
)
```

Data 1 meets the assumptions for linear regression, because the relationship between explanatory and response variables is linear and the error terms are iid normal. Data 2 does not meet the assumptions for linear regression, because the relationship between the variables is non-linear: there is a small quadratic term $x^2/20$.

Compare the two sets of simulated data visually:

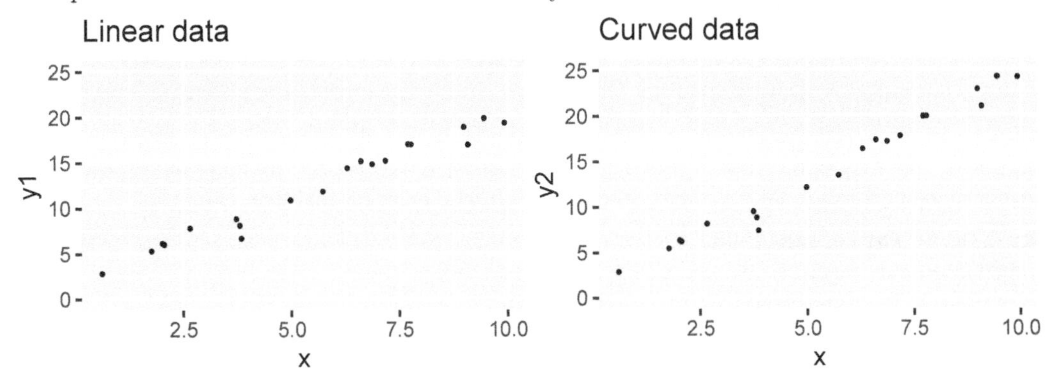


It is easy to believe that the quadratic term in the second set of data might go unnoticed.

Next, compute the 95% prediction intervals for both data sets at $x = 5$. We use a linear model in both cases, even though the linear model is not correct for the non-linear data.

```
model1 <- lm(y1 ~ x, data = simdata)
predict(model1, newdata = data.frame(x = 5), interval = "pre")

##        fit      lwr      upr
## 1 11.14539 9.158465 13.13232

model2 <- lm(y2 ~ x, data = simdata)
predict(model2, newdata = data.frame(x = 5), interval = "pre")

##        fit     lwr      upr
## 1 12.77723 10.7008 14.85366
```

The two prediction intervals are quite similar. In order to test whether the prediction interval is working correctly, we need to repeatedly create sample data, apply `lm`, create a new data point, and see whether the new data point falls in the prediction interval. We do this for the linear data first:

```
trials <- replicate(10000, {
  # generate data
  simdata <- data.frame(x = runif(20, 0, 10), epsilon = rnorm(20))
  simdata <- mutate(simdata, y1 = 1 + 2 * x + epsilon)

  # create a new data point at a random x location
  new_x <- runif(1, 0, 10)
  new_y <- 1 + 2 * new_x + rnorm(1)

  # build the prediction interval at x = new_x
  model1 <- lm(y1 ~ x, data = simdata)
  pred_int <- predict(model1,
    newdata = data.frame(x = new_x), interval = "pre"
  )

  # test if the new data point's y-value lies in the prediction interval
  pred_int[2] < new_y && new_y < pred_int[3]
})
mean(trials)
```

```
## [1] 0.9499
```

We find that the 95% prediction interval successfully contained the new point 94.99% of the time. That's pretty good! Repeating the simulation for the curved data, the 95% prediction interval is successful about 94.8% of the time, which is also quite good.

However, let's examine things more carefully. Instead of picking our new data point at a random $x$-coordinate, let's always choose $x = 10$, the edge of our data range.

```
trials <- replicate(10000, {
  # generate data
  simdata <- data.frame(x = runif(20, 0, 10), epsilon = rnorm(20))
  simdata <- mutate(simdata, y2 = 1 + 2 * x + x^2 / 20 + epsilon)

  # create a new data point at x=10
  new_x <- 10
  new_y <- 1 + 2 * new_x + new_x^2 / 20 + rnorm(1)

  # build the prediction interval at x = 10
  model2 <- lm(y2 ~ x, data = simdata)
  pred_int <- predict(model2,
    newdata = data.frame(x = 10), interval = "pre"
  )

  # test if the new data point's y-value lies in the prediction interval
  pred_int[2] < new_y && new_y < pred_int[3]
})
mean(trials)
```

```
## [1] 0.9029
```

The prediction interval is only correct 90.29% of the time. This isn't working well at all!

For the curved data, the regression assumptions are not met. The prediction intervals are behaving close to as designed for $x$ values near the mean, but fail twice as often as they should at $x = 10$.

When making predictions, be careful near the bounds of the observed data. Things get even worse as you move $x$ further from the known points. A statistician needs strong evidence that a relationship is truly linear to perform extrapolation.

## 11.7 Cross validation

In this chapter, we introduced descriptive modeling of data sets containing a single predictor and a single response variable. We chose to model the relationship via a straight line, and we minimized the SSE over all choices of slopes and intercepts. Our recurring example is regression line for body mass explained by flipper length for chinstrap penguins. You may be wondering why we didn't choose a more complicated model to describe the relationship. Indeed, quite a bit of error remained after we modeled the relationship with a line. Should we try to model that error? You may even be wondering why we don't use a function that goes through all of the data points, so that the SSE is 0. The purpose of this section is to address such concerns through the lens of *predictive modeling.* Predictive modeling is a big topic that we are only going to scratch the surface of. If you want to learn more after reading this section, we recommend the book by Kuhn and Johnson.[9]

We will consider two models, shown in Figure 11.29. Model one in blue is given by linear regression, the line of best fit. Model two is called a smooth spline, which is a piecewise polynomial, and it apparently follows the data quite well. In fact, the mean squared error for the smooth spline model is $4.5518 \times 10^4$, much smaller than the $8.5637 \times 10^4$ MSE for the linear model. However, your instincts should tell you that the smooth spline model is *overfit*, and will be less good at predicting new values of penguin measurements.

How might we assess the predictive value of the two models? One way would be to **leave out** some of the data, and build the models on the data that remains. We can then **predict** the values for the data that we left out, and see which model has a smaller MSE. This process is called *cross validation.* The data that we use to build the model is called the *training set*, and the data that we use to compute the MSE (or otherwise evaluate the model) is called the *testing set.*

It is very important to make sure that the model is built **only** using the training data. We do not want information from the test set leaking into the model, as that could make the model unrealistically accurate.

Let's start by making a train/test split of our data in such a way that there is exactly one observation in the test set. For the moment, we simply choose observation 5.

```
chinstrap <- palmerpenguins::penguins %>% filter(species == "Chinstrap")
test_obs <- 5
```

[9] Kuhn Max and Kjell Johnson, *Applied Predictive Modeling* (New York, NY: Springer, 2018).

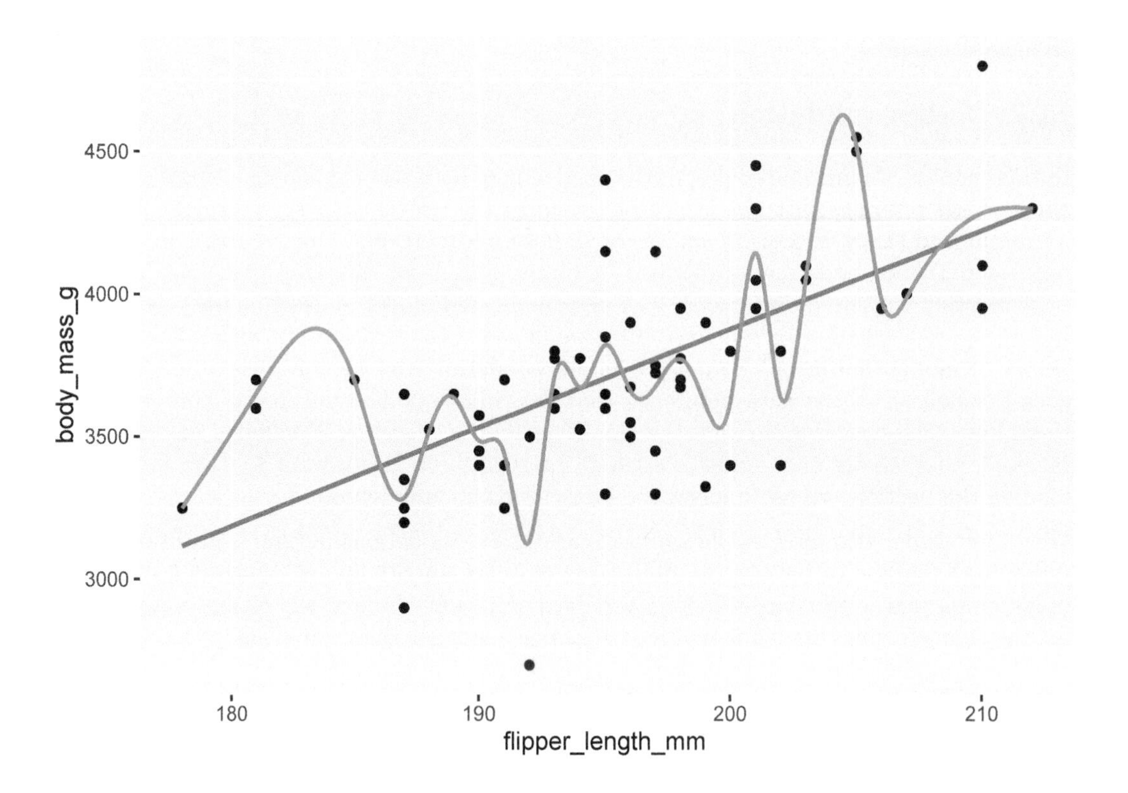


**FIGURE 11.29** Chinstrap penguin dimensions modeled by a linear function and a smooth spline. Which model has more predictive power?

```
test <- chinstrap[test_obs, ]
train <- chinstrap[-test_obs, ]
```

In this case, observation 5 is the test set, while the rest of the data is the train set. We now build our line of best fit, predict the body mass of the test penguin and compute the MSE.

```
mod1 <- lm(body_mass_g ~ flipper_length_mm, data = train)
bm1 <- predict(mod1, newdata = test)
err1 <- bm1 - test$body_mass_g
mean(err1^2)
```

```
## [1] 2451.342
```

Next, we do the same thing for the second model. The R function that computes the smooth spline is `smooth.spline` and it requires one parameter, the degrees of freedom. We choose 24 degrees of freedom since that is the largest possible value with 24 distinct flipper length measurements. The `smooth.spline` function returns a model which can be used to predict the body mass of the test penguin.

```
mod2 <- smooth.spline(
  x = train$flipper_length_mm,
  y = train$body_mass_g,
  df = 24
)
bm2 <- predict(mod2, test$flipper_length_mm)$y
err2 <- bm2 - test$body_mass_g
mean(err2^2)
```

```
## [1] 3776.056
```

For this test penguin, the linear model had a MSE of 2451 and the smooth spline had a MSE of 3776, so the linear model did a better job of predicting the test penguin's body mass. We should make this comparison for every penguin in the data set.

**Definition 11.3.** *Leave one out cross validation* (LOOCV) is a procedure in which we systematically remove each observation from the data, build the model on the remaining data, and compute the error in estimation of the test observation. Finally, compute the MSE over all tests.

Here we perform LOOCV on the chinstrap penguins data. Recall from Section 6.7 that `sapply` applies a function to all values in a vector. We define functions called `loo_linear` and `loo_spline` that compute the leave one out error for a single test penguin, using the linear and spline models, respectively. We then apply those functions to each of the 68 observations using `sapply`.

```
loo_linear <- function(test_obs) {
  test <- chinstrap[test_obs, ]
  train <- chinstrap[-test_obs, ]
  mod1 <- lm(body_mass_g ~ flipper_length_mm, data = train)
  bm1 <- predict(mod1, newdata = test)
  bm1 - test$body_mass_g
}
err_linear <- sapply(1:68, loo_linear)

loo_spline <- function(test_obs) {
```

```
  test <- chinstrap[test_obs, ]
  train <- chinstrap[-test_obs, ]
  mod2 <- smooth.spline(
    x = train$flipper_length_mm,
    y = train$body_mass_g,
    df = 24
  )
  bm2 <- predict(mod2, test$flipper_length_mm)$y
  bm2 - test$body_mass_g
}
err_spline <- sapply(1:68, loo_spline)
```

Compare the mean squared errors:

```
mean(err_linear^2)
```

```
## [1] 91245.43
```

```
mean(err_spline^2)
```

```
## [1] 101266
```

The linear model has a lower estimated MSE using LOOCV than the spline model, so it is doing a better job predicting the test penguin measurements.

Another popular approach to cross validation is $k$-fold cross validation. In this approach, the data is split into $k$ roughly equal size groups. Each group gets a turn as the test set. In repeated $k$-fold cross validation, this entire process is repeated multiple times.

## 11.8 Bias-variance tradeoff

Recall the bias-variance decomposition of an estimator $\hat{\theta}$ from Section 5.6.3:

$$\text{MSE}(\hat{\theta}) = E[(\hat{\theta} - \theta)^2] = \text{Bias}(\hat{\theta})^2 + \text{Var}(\hat{\theta})$$

We saw in that section that we could improve the mean squared error of the sample variance $S^2 = \frac{1}{n-1}\sum_{k=1}^{n}\left(X_i - \overline{X}\right)^2$ as an estimator for $\sigma^2$ by simultaneously **increasing** the bias and **decreasing** the variance of the estimator. A similar paradigm holds for model building. It is often the case that we can improve the MSE of a model by increasing the bias and decreasing the variance of the model.

To make things more precise, suppose data is generated by a process $y = f(x) + \epsilon$, where $\epsilon$ is normal with mean 0 and variance $\sigma^2$. Suppose also that we have a method for modeling the data which includes being able to predict values of $y$ for values of $x$. Then,

$$\text{MSE}(\text{model}) = \sigma^2 + \text{Bias}(\text{model})^2 + \text{Var}(\text{model}) \tag{11.1}$$

We examine this equality in the case that the model is only predicting values at one point. The MSE on the left-hand side refers to the MSE of the model for future values at that point. The bias of the model is the difference between the expected value of the predictions from the model and the true expected value $f(x)$. The variance of the model is the variance of the predictions at the one point.

**Example 11.22.** Let's suppose that data is generated by $Y = 1 + X + X^2 + \epsilon$, where $\epsilon \sim \text{Norm}(0, .25)$. Data is generated over the interval $[0, 1]$, and we model it with $Y = \beta_0 + \beta_1 X + \epsilon$, which is not the same as the generative process. We are interested in predictions at $x = 0.9$. The value of $\sigma^2$ in the formula above is 0.0625, which is normally not known when dealing with actual data. For example, we could have the following.

```
x <- seq(0, 1, length.out = 20)
y <- 1 + x + x^2 + rnorm(20, 0, .25)
plot(x, y)
```

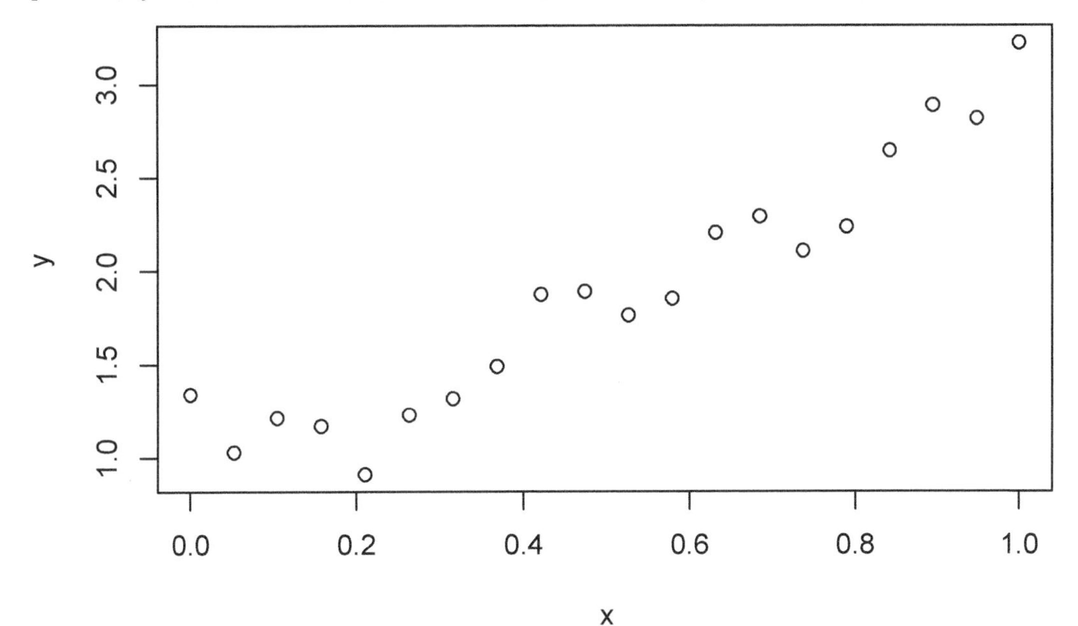


```
mod <- lm(y ~ x)
predict(mod, newdata = data.frame(x = 0.9))

##        1
## 2.691488
```

We first estimate the bias of our model by comparing the predicted values when $x = 0.9$ to the true value, which is $1 + 0.9 + 0.9^2 = 2.71$.

```
sim_data <- replicate(10000, {
  x <- seq(0, 1, length.out = 20)
  y <- 1 + x + x^2 + rnorm(20, 0, .25)
  mod <- lm(y ~ x)
  predict(mod, newdata = data.frame(x = 0.9))
})
mean(sim_data - 2.71)

## [1] -0.06892062
```

We see that we underestimate the true value by about 0.069.

Now, we estimate the variance of the model. We can do this by computing the variance of predictions from the model.

```
var(sim_data)
```

```
## [1] 0.008490911
```

The variance of the model is 0.008491. Finally, the MSE of the model is estimated via:

```
mean((sim_data - (1 + .9 + .9^2 + rnorm(10000, 0, .25)))^2)
```

```
## [1] 0.07635764
```

We compare this to the expression on the right-hand side of Equation (11.1):

```
0.25^2 + mean(sim_data - 2.71)^2 + var(sim_data)
```

```
## [1] 0.07574096
```

We see that we get good agreement between the two values.

We next turn to an example that may be counterintuitive. It is possible to decrease the MSE of the estimations coming from a model by **ignoring** information that is used in the generative process of the data. This will introduce bias, but the resulting reduction in variance can many times make up for the increase in bias.

**Example 11.23.** In this example, we use simulated data to illustrate the bias-variance tradeoff. Our simulated data follows the model $Y = 1 + 0.1X + \epsilon$, where $\epsilon \sim \text{Norm}(0, \sigma = 4)$.

```
n <- 101
x <- seq(0, 10, length.out = n)
y <- 1 + 0.1 * x + rnorm(n, 0, 4)
```

We compare two models for this data. The first model is constant: it assumes no dependence on x, and predicts every $y$ value as $\overline{y}$. The second model is a linear model with x as a predictor. The data and both models are shown in Figure 11.30.

The constant model is **biased**, because it will overestimate $y$ values associated with small $x$, and underestimate $y$ values associated with large $x$. On the other hand, it should have less variance than the linear model with slope term. Let's compute the LOOCV error for both models on this simulated data set. One new twist is that we now have functions that take two arguments (dat, the data frame, and test_obs, the observation to leave out). The sapply function will apply these functions to each test_obs while providing the data frame containing simulated data.

```
loo_const <- function(dat, test_obs) {
  test <- dat[test_obs, ]
  train <- dat[-test_obs, ]
  yhat <- mean(train$y)
  yhat - test$y
}
err_const <- sapply(1:n, loo_const, dat = data.frame(x, y))

loo_linear <- function(dat, test_obs) {
  test <- dat[test_obs, ]
  train <- dat[-test_obs, ]
  mod <- lm(y ~ x, data = train)
  yhat <- predict(mod, newdata = test)
```

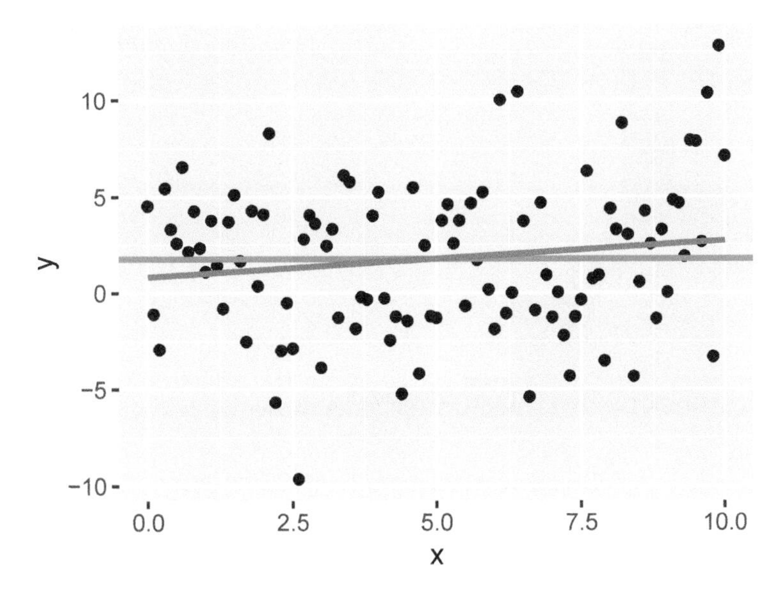


**FIGURE 11.30** Data with $y$ increasing slightly as a function of $x$. Linear model (blue) and constant model (red) are shown.

```
  yhat - test$y
}
err_linear <- sapply(1:n, loo_linear, dat = data.frame(x, y))

mean(err_const^2)

## [1] 15.96624

mean(err_linear^2)

## [1] 16.00987
```

We see in this example, the MSE for the constant model is slightly better, but they are so close that it seems like it could be by chance. Let's replicate the entire process, generating fresh data repeatedly. For each new set of $y$ values, we compute the LOOCV mean squared error for each model and then report the difference between them.

```
mse_diffs <- replicate(500, {
  y <- 1 + 0.1 * x + rnorm(n, 0, 4)
  err_const <- sapply(1:n, loo_const, dat = data.frame(x, y))
  err_linear <- sapply(1:n, loo_linear, dat = data.frame(x, y))
  mean(err_const^2) - mean(err_linear^2)
})
```

Figure 11.31 shows a histogram of the MSE differences, and it appears that in most trials the difference is negative, meaning that the constant model had smaller mean squared error than the linear model. To be sure, we run a Wilcoxon signed rank test to determine whether this difference is statistically significant.

```
wilcox.test(mse_diffs, mu = 0)

##
```

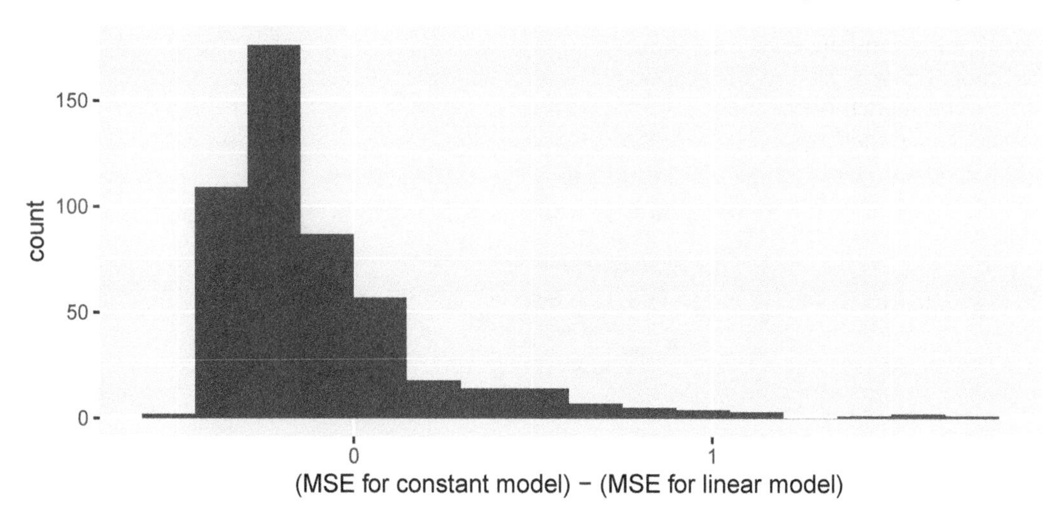


**FIGURE 11.31** Difference in MSE between constant and linear models.

```
##  Wilcoxon signed rank test with continuity correction
##
## data:  mse_diffs
## V = 30847, p-value < 2.2e-16
## alternative hypothesis: true location is not equal to 0
```

We see that, yes, the "incorrect" model which ignored the slope had a significantly **lower** LOOCV estimated MSE than the "correct" model, which included the slope. While the differences in this particular example are quite small, it can be the case that small differences are worth a lot. It can also be the case that choosing a biased model will lead to a much more noticeable reduction in MSE.

> **Try It Yourself.**
> With 500 samples, a `t.test` would also be appropriate on the MSE differences, even with this level of skewness. Perform a $t$-test on `mse_diffs` and confirm that it is also statistically significant.

---

## Vignette: Simple logistic regression

Regression is a very large topic, and one can do many, many more things than the simple linear models discussed in this chapter. One common task is to model the **probability** of an event occurring based on one or more predictor variables. For this, we can use logistic regression. In this vignette, we use the data set `malaria`[10] in the `fosdata` package to model the probability that a mouse will contract malaria based on the number of sporozoite parasites carried by the mosquito that bit the mouse.

[10] Thomas S Churcher et al., "Probability of Transmission of Malaria from Mosquito to Human Is Regulated by Mosquito Parasite Density in Naïve and Vaccinated Hosts." *PLOS Pathogens* 13, no. 1 (January 2017): e1006108.

Let $Y$ be a random variable that takes on the value 1 if the mouse contracts malaria, and 0 if not. Let $p = P(Y = 1)$ be the probability that the mouse contracts malaria. A linear regression model would be $p = \beta_0 + \beta_1 x$, where $x$ is the number of sporozoites. This type of model can sometimes be appropriate, but it can also lead to probabilities that are bigger than 1 or less than 0. In logistic regression, we use the model

$$\log \frac{p}{1-p} = \beta_0 + \beta_1 x,$$

or equivalently,

$$p = \frac{e^{\beta_0 + \beta_1 x}}{1 + e^{\beta_0 + \beta_1 x}}.$$

How to interpret this? Well, $\frac{p}{1-p}$ is the **odds** of the event occurring, so $\log \frac{p}{1-p}$ is the *log-odds* of it occurring. If the log-odds are a big positive number, then it is very likely for the event to occur. If they are zero, then the event is 50-50, and a big negative log-odds corresponds to a very unlikely event. We are modeling these log-odds linearly in terms of the predictor variable.

There are several ways of estimating $\beta_0$ and $\beta_1$. However, we will not go over the details except to say that they are implemented in the R function `glm`. We can find estimates for $\beta_0$ and $\beta_1$ via

```
malaria <- fosdata::malaria
malaria_mod <- glm(malaria ~ sporozoite,
  family = "binomial", data = malaria
)
broom::tidy(malaria_mod)
```

```
## # A tibble: 2 x 5
##   term        estimate std.error statistic  p.value
##   <chr>          <dbl>     <dbl>     <dbl>    <dbl>
## 1 (Intercept)   -2.48     0.428      -5.79 7.07e- 9
## 2 sporozoite     0.462    0.0645      7.17 7.74e-13
```

We see that the estimate for $\beta_0$ is -2.4761526 and the estimate for $\beta_1$ is 0.4622328. The value for the slope is significant at the $\alpha = .05$ level. Let's look at a plot.

```
ggplot(malaria, aes(x = sporozoite, y = malaria)) +
  geom_jitter(height = 0, width = 0.2) +
  geom_smooth(
    method = "glm",
    method.args = list(family = "binomial"),
    se = FALSE
  )
```

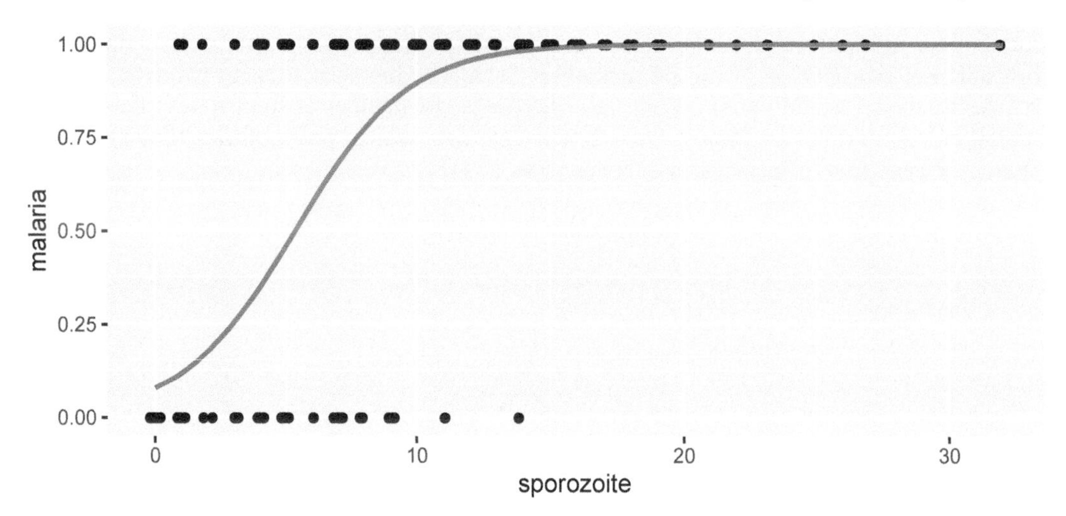


The blue curve is the estimate of the probability of contracting malaria based on the total number of sporozoites. We can see that when there are zero sporozoites, the model predicts a low probability of contracting malaria. When there are more than 20, then the probability is essentially 1. To estimate the probability of contracting malaria given a certain number of sporozoites more precisely, we can use the `predict` function. The following code estimates the **log odds** of contracting malaria when there are 8 sporozoites:

```
predict(malaria_mod, newdata = data.frame(sporozoite = 8))
```

```
##        1
## 1.221709
```

If we wish to find the **probability** of contracting malaria, we use:

```
predict(malaria_mod, newdata = data.frame(sporozoite = 8), type = "response")
```

```
##         1
## 0.7723642
```

There is much more to logistic regression than what we have shown in this vignette. We hope this taste of it will encourage you to further study!

## Exercises

Exercises 11.1 – 11.4 require material through Section 11.1.

**11.1.** The built-in data set `faithful` gives data on eruptions of the Old Faithful Geyser in Yellowstone National Park. Each observation has a length of time the eruption lasted (`eruptions`) and the length of time until the next eruption (`waiting`), both in minutes.

a. Find the equation of the regression line to explain waiting as a function of eruptions.
b. Plot the data and the regression line.
c. How long do you expect to wait after a 4.3 minute eruption?

**11.2.** The data set `barnacles` in the `fosdata` package has measurements of barnacle density on various coral reefs.

a. Find the equation of the regression line to explain barnacle density as a function of reef depth.
b. Predict the mean number barnacles per square meter for a 30-meter deep reef.

**11.3.** Consider the `juul` data set in `ISwR`.

a. Fit a linear model to explain igf1 level by age for people under age 20 who are in tanner puberty group 5.
b. Give the equation of the regression line and interpret the slope in terms of age and igf1 level.
c. Predict the mean igf1 levels for 16-year-olds who are in tanner puberty group 5.

**11.4.** Consider the `penguins` data set in the `palmerpenguins` package. Suppose we use the line we found for chinstrap penguins to model gentoo penguins instead.

a. Restrict to gentoo penguins and find the sum of squared error associated with the line

$$\text{body\_mass\_g} = -3037.2 + 34.6 \cdot \text{flipper\_length\_mm}.$$

b. Find the equation of the regression line for `body_mass_g` on `flipper_length_mm` for gentoo penguins.
c. Find the sum of squared error for the best fit line you found in part (b) and compare with your answer to part (a).

---

Exercises 11.5 – 11.8 require material through Section 11.2.

**11.5.** For each of the following four plots, indicate whether the sample correlation coefficient is strongly positive (greater than 0.3), weak (between -0.3 and 0.3), or strongly negative (less than -0.3).

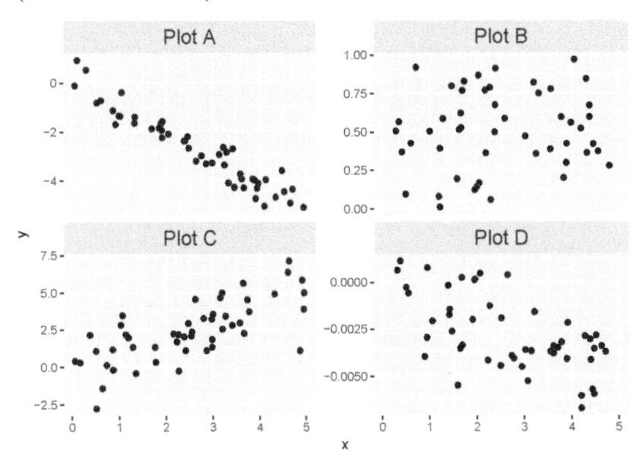


**11.6.** The data frame `datasaurus_dozen` is part of the `datasauRus` package. This data is actually 13 data sets, each one specified by a different value of the `dataset` variable. For each of the 13 data sets, compute $\bar{x}$,$\bar{y}$,$s_x$,$s_y$, and the correlation between $x$ and $y$. What do you observe? Plot a few of the individual data sets (including `dino`) and describe what you see.

**11.7.** In 2019, researchers[11] published a paper in which they found a correlation between

[11]Daisy Fancourt and Andrew Steptoe, "The Art of Life and Death: 14 Year Follow-up Analyses of Associations Between Arts Engagement and Mortality in the English Longitudinal Study of Ageing," *BMJ* 367 (2019).

arts engagement and mortality. *The New York Times* reported: "Another Benefit to Going to Museums? You May Live Longer."[12] The published paper accounted for many lurking variables by incorporating them in their statistical model, but still concluded with the statement that "A causal relationship cannot be assumed, and unmeasured confounding factors might be responsible for the association."

What are some lurking variables that might confound the correlation between museum attendance and mortality?

**11.8.** Let $X$ be a standard normal random variable.

a. The correlation of $X$ and $X^2$ is zero. Compute the sample correlation coefficient of $X$ and $X^2$ using 100,000 samples to confirm that this is at least approximately the case.
b. Are $X$ and $X^2$ independent?

---

Exercises 11.9 – 11.13 require material through Section 11.3.

**11.9.** Suppose you have 100 data points, and $\overline{x} = 3$, $s_x = 1$, $\overline{y} = 2$, $s_y = 2$, and the correlation coefficient is $r = 0.7$. Find the equation of the least squares regression line in slope-intercept form.

**11.10.** Theorem 11.1 gives a value for $\beta_1$ in terms of the data $(x_1, y_1), \ldots, (x_n, y_n)$. Use Theorem 11.1 to compute the value for $\beta_0$.

**11.11.** Consider the model $y_i = \beta_0 + x_i + \epsilon_i$. That is, assume that the slope is 1. Find the value of $\beta_0$ that minimizes the sum-squared error when fitting the line $y = \beta_0 + x$ to data given by $(x_1, y_1), \ldots, (x_n, y_n)$.

**11.12.** Consider the model $y_i = \beta_1 x_i + \epsilon_i$. That is, assume that the intercept is 0. Find the value of $\beta_1$ that minimizes the sum-squared error when fitting the line $y = \beta_1 x$ to data given by $(x_1, y_1), \ldots, (x_n, y_n)$.

**11.13.** Suppose that in Exercise 11.12, instead of minimizing the sums of squares of the *vertical* error $\sum(\beta_1 x_i - y_i)^2$, you decide to minimize the sums of squares of the *horizontal* error. This means that our model is that $x = \gamma y + \epsilon$ and we wish to minimize $\sum(x_i - \gamma y_i)^2$ over all choices of $\gamma$.

a. Find a formula for $\gamma$.
b. Find an example where the "best" line determined by minimizing vertical error is different from the "best" line determined by minimizing horizontal error. (Hint: almost any collection of points will do the trick.)

---

Exercises 11.14 – 11.19 require material through Section 11.4.

**11.14.** For each of the following eight residual plots, indicate whether the residual plot is evidence against the linear model being satisfied or not.

[12] Maria Cramer, "Another Benefit to Going to Museums? You May Live Longer" (The New York Times, December 22, 2019).

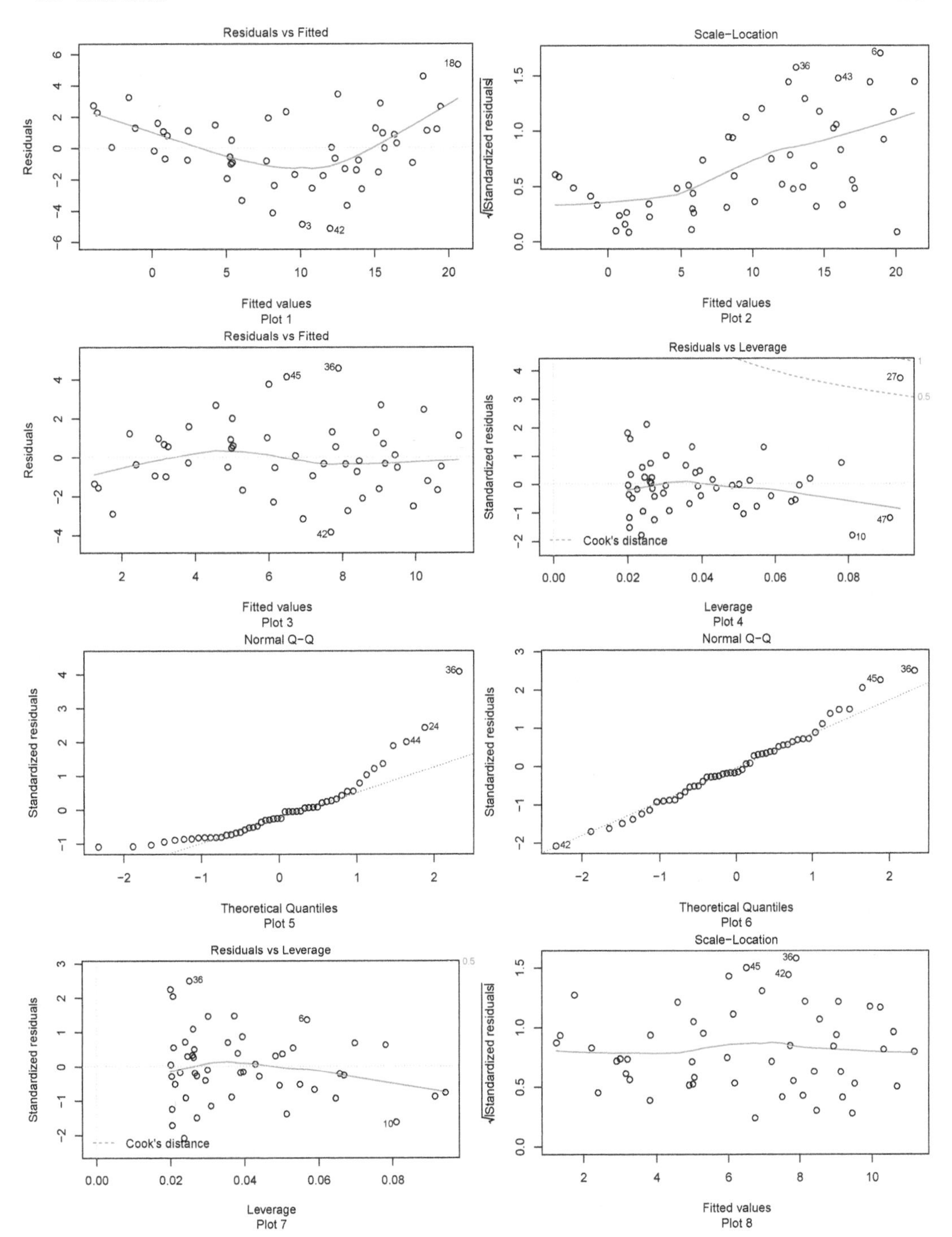


**11.15.** Consider the data set `Anscombe`, which is in the `carData` package. (Note: do not use the data set anscombe, which is preloaded in R.) The data contains, for each U.S. state, the per capita education expenditures (`education`) and the number of youth per 1000 people (`young`).

a. Would you expect number of youth and per capita education expenditures to be correlated? Why or why not?

b. What is the correlation coefficient between number of youth and per capita education expenditures?
c. Model expenditures as a function of number of youth. Interpret the slope of the regression line. Is it significant?
d. Examine your data and residuals carefully. Model the data minus the one outlier, and re-interpret your conclusions in light of this.

**11.16.** Consider the `cern` data set in the `fosdata` package. This data contains information on social media interactions of CERN. For the purposes of this problem, restrict to the `platform` Twitter.

a. Create a linear model of `likes` on `shares`, and examine the residuals.
b. Create a linear model of `log(likes)` on `log(shares)` and examine the residuals.
c. Which model seems to better match the assumptions of linear regression?

**11.17.** Consider the data set `crit_period` in the `fosdata` package. The critical period hypothesis posits that there is a critical period (young) for first exposure to a second language, and that first exposure past the critical period makes it much more difficult to attain high levels of second language proficiency.

a. Consider the North American subjects only, and plot `gjt` (scores on the grammaticality judgment test) versus `aoa` (age of onset of acquisition).
b. Find the coefficients of the regression line and plot it on top of the scatterplot.
c. What does the slope represent in terms of this data?
d. Examine the residuals. The critical period hypothesis predicts that there will be an elbow shape or L shape in the residuals versus fitted. Do you see that in your residuals?

**11.18.** Consider the `starwars` data set in the `dplyr` package.

a. Create a scatterplot of mass versus height of the *Star Wars* characters.
b. Create a linear model of mass as described by height for the *Star Wars* characters, and write the equation for the line of best fit.
c. Examine the residual plots and comment. Which character is the outlier?
d. Remove the large outlier and create a new linear model of mass as described by height for the *Star Wars* characters minus the outlier. Comment on the magnitude of the change of the slope and intercept.

**11.19.** The data set `engel` contains data on income and food expenditure for 235 working class Belgian households. To load this data, load the `quantreg` package and then enter the command `data(engel)`.

a. Make a plot of `foodexp` as explained by `income` and show the regression line on your plot.
b. Compute the linear model and plot the residuals.
c. The assumptions for linear regression are not satisfied. What is wrong with the residuals?

---

Exercises 11.20 – 11.30 require material through Section 11.5.

**11.20.** Consider the `juul` data set in `ISwR`.

a. Fit a linear model to explain igf1 level by age for people under age 20 who are in tanner puberty group 5, see Exercise 11.3.
b. Predict the igf1 level with 95 percent prediction bounds for a 16 year old who is in tanner puberty group 5.

**11.21.** Consider the `cern` data set in the `fosdata` package. Create a linear model of `log(likes)` on `log(shares)` for interactions in the Twitter platform (see Exercise 11.16). Find 95 percent confidence intervals for the slope and intercept for the model if the residuals are acceptable.

**11.22.** The first Vietnam draft lottery determined the order in which young men would be drafted in 1970 and sent to fight in Vietnam. Each of 366 possible birthdays were printed on paper, placed in plastic capsules, and drawn by hand from a glass jar (see Figure 11.32). The results of the lottery are in the data set `fosdata::draft`.

a. Make a scatterplot of `DraftNo` vs. `DayofYear`. Does it appear the draft numbers were evenly distributed across the year?
b. Test for a linear relationship between draft number and day of year. What do you conclude?

**FIGURE 11.32** Representative Alexander Pirnie (R-NY) drawing the first number in the Vietnam draft for 1970. (Source: Library of Congress)

**11.23.** Consider the `penguins` data set in the `palmerpenguins` package and restrict to `gentoo` penguins. Perform a hypothesis test of $H_0 : \beta_1 = 0$ versus $H_a : \beta_1 \neq 0$ for the model

$$\widehat{\text{body\_mass\_g}} = \beta_0 + \beta_1 \cdot \text{flipper\_length\_mm} + \varepsilon$$

at the $\alpha = .01$ level.

**11.24.** The data set `hot_dogs` is in the `fosdata` package. It contains information about various brands of hot dogs as listed in Consumer Reports, 1986.

a. Make a scatterplot showing sodium content as a function of calories. Color your points using the hot dog type.
b. Remove the Poultry hot dogs (why?) and then construct a linear model to explain sodium content as a function of calories. What is the equation of your regression line?
c. How much sodium does your model predict for a 140 calorie Beef or Meat hot dog?
d. Give the prediction interval that contains the sodium level for 95% of 140 calorie Beef or Meat hot dogs.

**11.25.** This exercise uses the `acorns` data from the `fosdata` package, and follows the analysis in Marcelo and Patterson, 1990[13]. The authors suggest that larger acorns allow for greater success in ecological niches, leading to a wider geographic range.

[13]Marcelo Aizen and William Patterson III, "Acorn Size and Geographical Range in the North American Oaks (Quercus L.)," *Journal of Biogeography* 17, no. 327 (1990), https://doi.org/10.2307/2845128.

a. Define new variables which are the log of `Acorn_size` and the log of `Range`. Check that `Acorn_size` and `Range` are very skewed but the logged variables are reasonably normal, except for one outlier. Remove the one outlier, *Quercus tomentella Engelm*, which is only found on a few small islands off the western coast.
b. Build a linear model of the log of Range on the log of acorn size, only for the Atlantic region oaks. Is there a significant relationship?
c. Build a linear model of the log of Range on the log of acorn size, only for the California region oaks. Is there a significant relationship?

**11.26.** Consider the `Batting` data from the `Lahman` data set. This problem uses home runs (HR) to explain total runs scored (R) in a baseball season.

a. Construct a data frame that has the total number of HR hit and the total number of R scored for each year in the data set.
b. Create a scatterplot of the data, with HR as the explanatory variable and R as the response. Use color to show the yearID.
c. Notice early years follow a different pattern than more modern years. Restrict the data to the years from 1970 to the present, and create a new scatterplot.
d. Fit a linear model to the recent data. How many additional R does the model predict for each additional HR? Is the slope significant?
e. Plot the best-fit line and the recent data on the same graph.
f. Create a prediction interval for the total number of runs scored in the league if a total of 4000 home runs are hit. Would this be a valid prediction interval if we found new data for 1870 that included HR but not R?

**11.27.** Consider the `ex0728` data set in `Sleuth2` (note: not in Sleuth3!). This gives the year of completion, height and number of stories for 60 randomly selected notable tall buildings in North America.

a. Create a scatterplot of Height (response) versus Stories (explanatory).
b. Which building(s) are unexpectedly tall for their number of stories?
c. Is there evidence to suggest that the height per story has changed over time?

**11.28.** Consider the data in `ex0823` in `Sleuth3`. This data set gives the wine consumption (liters per person per year) and heart disease mortality rates (deaths per 1000) in 18 countries.

a. Create a scatterplot of the data, with Wine as the explanatory variable. Does there appear to be a transformation of the data needed?
b. Does the data suggest that wine consumption is associated with heart disease?
c. Would this study be evidence that the odds of dying from heart disease would change for a person who increases their wine consumption to 75 liters per year?

**11.29.** Data in `fosdata::plastics` has measurements of plastic particles found in the snow in the Arctic and in Europe.

a. Check that `diameter` is approximately normally distributed but `length` is very skewed. Create a new variable which is the log of `length` and check that it is approximately normally distributed.
b. Make a scatterplot of `diameter` as explained by the log of `length` and add the regression line to the plot.
c. Create a linear model of `diameter` as explained by the log of `length`. Is there evidence of a relationship between `diameter` and the log of `length` at the $\alpha = 0.05$ level?

**11.30.** The `msleep` data set in the `ggplot2` package has information about 83 species of mammals.

a. Make a scatterplot of `sleep_total` on `brainwt`. Add the regression line to your plot.
b. What is the correlation between these two variables? Is the linear relationship significant at the 0.01 level?
c. Do the residuals support using the linear model for inference?

---

Exercises 11.31 – 11.35 are from Section 11.6.

**11.31.** In this exercise, regression assumptions are met perfectly. You will check how often the true mean of $y$ at a given $x$ value lies in the 95% confidence interval for that mean.

Follow these steps:

I. Create simulated data.

- Choose 30 $x$ values uniformly on the interval $[0, 10]$.
- Create 30 $y$ values following $y_i = 1 + 2x_i + \epsilon_i$, where $\epsilon_i \sim \text{Norm}(0, 1)$.

II. Use `predict` to compute the 95% confidence interval for the mean of $y$ at $x = x_1$, where $x_1$ is simply the first of your generated data values.
III. Check that the true mean, $1 + 2x_1$, is in the confidence interval you created.

Now replicate this process at least 1000 times to estimate the probability that the true mean is in the confidence interval.

**11.32.** In this exercise, regression assumptions are met perfectly. You will check how often new data at a given $x$ lies in the 95% prediction interval for the model at $x$.

Follow these steps:

I. Create simulated data.

- Choose 30 $x$ values uniformly on the interval $[0, 10]$.
- Create 30 $y$ values following $y_i = 1 + 2x_i + \epsilon_i$, where $\epsilon_i \sim \text{Norm}(0, 1)$.

II. Choose a random value $x_\star$ uniformly on the interval $[0, 10]$ and a new value $y_\star = 1 + 2x_\star + \epsilon_\star$.
III. Use `predict` to compute the 95% prediction interval at $x_\star$.
IV. Check that the new data $y_\star$ lies in the 95% prediction interval.

Now replicate this process at least 1000 times to estimate the probability that $y_\star$ is in the prediction interval.

**11.33.** In this exercise, regression assumptions are met perfectly. Follow the steps of Exercise 11.32 except only perform step I one time; do not recreate the data for each new value of $x_\star$.

Is your answer 95%? Run your simulation a few times to see whether your answer is consistent. Why would you not necessarily expect the answer to be 95% every time?

**11.34.** In this exercise, the data does not follow a linear model, violating a basic assumption of linear regression. You will check how often new data at a given $x$ lies in the 95% prediction interval for the model at $x$.

Follow the steps of Exercise 11.32 except use $y_i = x_i^2 + \epsilon_i$. Rather than choose $x_\star$ randomly, run the simulation with $x_\star = 0$. Then repeat when $x_\star = 5$.

a. What percentage of time is new data at $x_\star = 0$ in the 95% prediction interval at $x = 0$?
b. What percentage of time is new data at $x_\star = 5$ in the 95% prediction interval at $x = 5$?

**11.35.** In this exercise, we illustrate the fact mentioned in Section 11.4 that the standard deviation of a residual is larger near the mean of all predictors than near the extremes of all predictors.

a. Create data $x_1 = 1, x_2 = 2, \ldots, x_{19} = 19$ and $y_i = 1 + 2x_i + \epsilon_i$, where $\epsilon_i$ is normal with mean 0 and standard deviation 3.
b. The mean value of $x$ is 10. Model $y$ on $x$ and compute the residual when $x = 10$. Replicate this and verify that the standard deviation of the residuals when $x = 10$ is about 2.9.
c. Repeat part (b) for the same model and data, but compute the residual when $x = 1$ and verify that the standard deviation of the residuals when $x = 1$ is about 2.7.

---

Exercises 11.36 – 11.39 are from Section 11.7.

**11.36.** Consider the chinstrap penguins in the `palmerpenguins` data set. When flipper length is zero, one might reasonably expect the body mass to also be zero. Create a data frame `chinstrap` containing only the chinstrap penguins and a linear model **without intercept** of body mass on flipper length using `lm(body_mass_g ~ flipper_length_mm + 0, data = chinstrap)`. Estimate the MSE of the model using leave one out cross validation and compare to the MSE of 91245.43 found for the linear model in Section 11.7.

**11.37.** Section 11.7 computed the LOOCV MSE of a smooth spline model of chinstrap penguin body mass explained by flipper length. Find the LOOCV MSE for the smooth spline using 18, 10, and 3 degrees of freedom. Are any of these models better than the linear model?

**11.38.** Continuing Exercise 11.37, find the LOOCV MSE for the smooth spline using every value of df from 2 to 24. Make a plot showing the MSE as a function of df and add a horizontal line at the linear MSE of $9.1245 \times 10^4$. Interpret the results.

**11.39.** Example 11.3 used a regression line to model speed on the "shape trail B" test on age of children, with data from `fosdata::child_tasks`. Rather than a linear relationship, an inverse relationship of the form $\hat{y} = \beta_0 + \frac{\beta_1}{x}$ may be a better model for this data. You can build the inverse model using the formula `stt_cv_trail_b_secs ~ I(1/age_in_months)`.[14]

a. Compute the MSE for LOOCV of the linear model built in Example 11.3.
b. Compute the MSE for LOOCV of the inverse model.
c. Which model has more predictive ability as measured by LOOCV?
d. Make a plot showing the data, the linear model, and the inverse model. (Hint: use `predict` on the inverse model to create predicted $y$ values for every data point and then plot them with `geom_line`.)

[14] The `I` operator tells `lm` to treat its argument as a mathematical expression.

# 12

# *Analysis of Variance and Comparison of Multiple Groups*

In this chapter, we introduce one-way analysis of variance (ANOVA) through the analysis of a motivating example.

Viñals et al.[1] studied the effects of marijuana on mice. One group of ordinary ("wild type") mice were given a dose of tetrahydrocannabinol (THC), the active ingredient in marijuana. Another group of wild type mice were given a *vehicle*, which is a shot with the same inactive ingredients but no THC.

*Image credit: George Shuklin.*

The investigators measured the locomotor activity of the mice by placing each mouse in a box lined with photocells and then observing the total distance traveled by the mouse. The results were reported as the percentage movement relative to the untreated VEH group, so values over 100 represent mice that moved a lot, and values under 100 represent mice that moved less.

Data from the mouse study is in the `fosdata` package in the `mice_pot` data set.

```
mice_pot <- fosdata::mice_pot
```

The data consists of 42 observations of 2 variables. The variable `group` is the dosage of THC that each mouse received, and `percent_of_act` measures locomotion as the percent of baseline activity. Start by exploring the data with summary statistics in Table 12.1 and a boxplot.

```
mice_pot %>%
  group_by(group) %>%
  summarize(
    mean = mean(percent_of_act),
    sd = sd(percent_of_act),
    skew = e1071::skewness(percent_of_act),
```

[1] Xavier Viñals et al., "Cognitive Impairment Induced by Delta9-Tetrahydrocannabinol Occurs Through Heteromers Between Cannabinoid CB1 and Serotonin 5-HT2A Receptors," *PLOS Biology* 13, no. 7 (July 2015): e1002194.

DOI: 10.1201/9781003004899-12

```
    N = n()
  )

ggplot(mice_pot, aes(x = group, y = percent_of_act)) +
  geom_boxplot(outlier.size = -1) +
  geom_jitter(height = 0, width = .2)
```

The plot and summary statistics show that there is little difference between the first three groups, but once the dose gets up to 3 mg/kg, the mice appear to be moving less.

This chapter introduces a number of different hypothesis tests to test for a difference in the means of multiple groups. The first we consider is one-way *analysis of variance*, abbreviated ANOVA.

For the mice, one-way ANOVA will test whether the amount of movement is the same in all four groups. ANOVA requires values in each group to be normally distributed, equal variance across groups, and independence. From the boxplot, mouse locomotion appears to be normally distributed, and each group has approximately the same variance. We know from the experimental design that the measurements on each mouse are independent.

To run one-way ANOVA on the mouse experiment, first build a linear model of `percent_of_act` as explained by `group`, then run the `anova` command on the linear model:

**TABLE 12.1** Summary statistics for movement of mice

| group | mean | sd | skew | N |
|---|---|---|---|---|
| VEH | 100.000 | 25.324 | -0.190 | 15 |
| 0.3 | 97.323 | 31.457 | 0.133 | 9 |
| 1 | 99.052 | 26.258 | -0.106 | 12 |
| 3 | 70.668 | 20.723 | 0.034 | 10 |

```
mice_model <- lm(percent_of_act ~ group, data = mice_pot)
anova(mice_model)
```

```
## Analysis of Variance Table
##
## Response: percent_of_act
##           Df Sum Sq Mean Sq F value Pr(>F)
## group      3   6329 2109.65  3.1261 0.0357 *
## Residuals 42  28344  674.85
## ---
## Signif. codes:  0 '***' 0.001 '**' 0.01 '*' 0.05 '.' 0.1 ' ' 1
```

The next section will explain the complex output of the `anova` function fully. The $p$-value for the test is given by the `Pr(>F)` field, so $p = 0.0357$ for this data, and the `*` means significance at the $\alpha = 0.05$ level. We reject the hypothesis that the mean percent of activity is the same for all four groups. The association between TCH treatment and locomotion is unlikely to be due to chance.

## 12.1 ANOVA

One-way ANOVA is used when a single quantitative variable $Y$ is measured on multiple groups. We will only be interested in the case where there is one categorical variable (the grouping) to explain $Y$.

### 12.1.1 Groups and means

One-way ANOVA involves multiple observations over multiple groups, so the notation required is fairly complex:

- There are $N$ total observations.
- There are $k$ groups.
- There are $n_i$ observations in the $i$th group.
- $y_{ij}$ denotes the $j$th observation from the $i$th group.
- $\overline{y}_{i.}$ denotes the mean of the observations in the $i$th group.
- $\overline{y}_{.}$ denotes the mean of all observations.

**Example 12.1.** In `mice_pot` there are 4 groups so $k = 4$. The observed variable $Y$ is `percent_of_act`, percent of activity relative to baseline. From the following table, we see that $n_1 = 15$, $n_2 = 9$, $n_3 = 12$, and $n_4 = 10$.

```
table(mice_pot$group)
```

```
##
## VEH 0.3   1   3
##  15   9  12  10
```

There are 46 observations, so $N = 46$.

In group 1, the VEH group, we have $y_{11} = 54.2$, $y_{12} = 65.9$, and so on. This means the first mouse in group 1 has activity 54.2% of baseline, and the second had 65.9%. In group 2, the first mouse has activity $y_{21} = 98.8$, and in group three the first mouse has activity $y_{31} = 95.3$.

To compute the group means:

```
mice_pot %>%
  group_by(group) %>%
  summarize(round(mean(percent_of_act), 1))
```

```
## # A tibble: 4 x 2
##   group `round(mean(percent_of_act), 1)`
##   <fct>                            <dbl>
## 1 VEH                              100
## 2 0.3                               97.3
## 3 1                                 99.1
## 4 3                                 70.7
```

This tells us that $\overline{y}_{1\cdot} = 100$ and $\overline{y}_{4\cdot} = 70.7$. Group 1, the VEH group, is exactly 100 by definition, since the percentage of activity was based on what the VEH group did.

To get the overall mean, we use `mean(mice_pot$percent_of_act)`, which gives $\overline{y}_{\cdot\cdot} = 92.9$.

The first step to running an analysis of variance in R is to fit a linear model to the data with the `lm` command. The syntax is the same as when running simple linear regressions, a formula of the form `y ~ x`, where `x` is the explanatory variable, `y` is the response variable, and the `~` (tilde) character can be read as "explained by."

When the explanatory variable is categorical, `lm` chooses one of the $k$ groups defined by `x` to act as the base level, the first group alphabetically or in the factor order. It then introduces $k - 1$ variables taking their names from the names of the other groups defined by `x`. The linear model is presented as an "intercept," which is really the mean of the base level, and "coefficients" which are simply the differences between the base level group mean $\overline{y}_{1\cdot}$ and the other group means $\overline{y}_{i\cdot}$. Chapter 13 goes into more detail.

**Example 12.2.** Using `mice_pot`, form the linear model explaining percent of activity by treatment group.

```
lm(percent_of_act ~ group, data = mice_pot)
```

```
##
## Call:
## lm(formula = percent_of_act ~ group, data = mice_pot)
##
## Coefficients:
## (Intercept)     group0.3       group1       group3
##    100.0000      -2.6775      -0.9477     -29.3321
```

The intercept is 100, which is the mean percent of activity in the `VEH` group. The other treatment means are given by adding the appropriate coefficient to 100. For example, the mean percent of activity in the 3 mg/kg group is $100 - 29.3321 = 70.6679$.

Because the linear model presents the data as relative to the base level, the coefficients can only compare the groups to the base level, not to each other. Since the base level is often arbitrary, we do not perform inference directly on the model. Instead, we rely on analysis of variance.

### 12.1.2 The ANOVA model

For one-way ANOVA, the *cell means model* for observations is

$$y_{ij} = \mu_i + \epsilon_{ij},$$

Here, $y_{ij}$ represents the $j^{\text{th}}$ observation of the $i^{\text{th}}$treatment. The parameters $\mu_i$ represent the means of the treatment groups and will be estimated from data. The values of $\epsilon_{ij}$ account for the variation within the group. In the one-way ANOVA model, $\epsilon_{ij}$ are independent normal random variables with mean zero and common standard deviation $\sigma$.

Using estimates from the data each individual data point is described by

$$y_{ij} = \overline{y}_{\cdot} + (\overline{y}_{i\cdot} - \overline{y}_{\cdot}) + (y_{ij} - \overline{y}_{i\cdot})$$

The null hypothesis for ANOVA is that the group means are all equal. That is:

- $H_0$: $\mu_1 = \mu_2 = \cdots = \mu_k$,
- $H_a$: Not all $\mu_i$ are equal.

ANOVA is "analysis of variance" because we will compare the variation of data within individual groups with the variation between groups. We make assumptions (see Assumptions 12.1) to predict the behavior of this comparison when $H_0$ is true. If the observed behavior is extreme, then it gives evidence against the null hypothesis. The general idea of comparing variance within groups to the variance between groups is applicable in a wide variety of situations.

**Definition 12.1.** The *sum of squares within groups*[2] is the sum of squares of residuals:

$$SS_R = \sum_i \sum_j (y_{ij} - \overline{y}_{i\cdot})^2.$$

The *sum of squares between groups* is associated with the grouping factor as:

$$SS_B = \sum_i \sum_j (\overline{y}_{i\cdot} - \overline{y})^2.$$

The *sum of squares total* is

$$SS_{\text{tot}} = \sum_i \sum_j (\overline{y} - y_{ij})^2.$$

The key mathematical result is the following theorem.

**Theorem 12.1.** *The total variation is the sum of the variation between groups and the variation within the groups. Using the notation in Definition 12.1,*

$$SS_{tot} = SS_R + SS_B.$$

If the variation between groups is much larger than the variation within groups, then that is evidence that there is a difference in the means of the groups. The question then becomes: how large is large? In order to answer that, we will need to examine the $SS_B$ and $SS_R$ terms more carefully.

---

[2]This quantity is often denoted $SS_W$, but we choose $SS_R$ to match the R output, where it is denoted as the sum of squares of residuals.

**Assumptions 12.1 (for one-way ANOVA).**

1. *The population of each group is normally distributed.*
2. *The variances of the groups are equal.*
3. *The observations are independent.*

Assumption 1 is typical of parametric statistics, because it provides a way to identify the data with known distributions.

Assumption 2 is a big one, and is investigated thoroughly in Section 12.3. There is no theoretical reason why the groups would have equal variances, so this must be checked.

Assumption 3 is dependent on the experimental design. A common experimental design that violates Assumption 3 is to use the **same unit** in each group. Suppose the THC mouse experiment used the mice repeatedly, giving each of the four treatments to each mouse. Then the independence assumption would be false and one-way ANOVA would be inappropriate. This mirrors the distinction between paired and independent two sample tests.

**Theorem 12.2.** *The test statistic for one-way ANOVA is*

$$F = \frac{SS_B/(k-1)}{SS_R/(N-k)}.$$

*If the Assumptions (12.1) for one-way ANOVA are met, then $F$ has the $F$ distribution with $k-1$ and $N-k$ degrees of freedom.*

*Proof (sketch).* We have $k$ groups. There are $n_i$ observations in group $i$, and $N = \sum n_i$ total observations.

Notice that $SS_R$ is almost $\sum\sum(\overline{y}_{ij} - \mu_i)^2$, in which case $SS_R/\sigma^2$ would be a $\chi^2$ random variable with $N$ degrees of freedom. However, we are making $k$ replacements of means by their estimates, so we subtract $k$ to get that $SS_R/\sigma^2$ is $\chi^2$ with $N-k$ degrees of freedom. (This follows the heuristic we started in Section 10.4.)

As for $SS_B$, it is trickier to see, but $\sum_i\sum_j(y_{i\cdot} - \mu)^2/\sigma^2$ would be $\chi^2$ with $k$ degrees of freedom. Replacing $\mu$ by $\overline{y}$ makes $SS_B/\sigma^2$ a $\chi^2$ rv with $k-1$ degrees of freedom.

The test statistic is

$$F = \frac{SS_B/(k-1)}{SS_R/(N-k)} = \frac{(SS_B/\sigma^2)/(k-1))}{(SS_R/\sigma^2)/(N-k))}$$

This is the ratio of two $\chi^2$ rvs divided by their degrees of freedom; hence, it is $F$ with $k-1$ and $N-k$ degrees of freedom.

Though this derivation lacks details, it explains why Assumptions 12.1 are necessary. The assumption of equal variances is needed to cancel the unknown $\sigma^2$. The normality assumption is needed so that we get a known distribution for the test statistic. ■

### 12.1.3 Simulations

In this section, we confirm through simulation that the test statistic under the null hypothesis does have an $F$ distribution.

We must assume normality and equal variances, and we wish to simulate the test statistic under the null hypothesis that all means are equal. So, we assume three populations, all of which are normally distributed with mean 0 and standard deviation 1. We arbitrarily decide to use 10 samples from group 1, 20 from group 2, and 30 from group 3, giving $k = 3$ and $N = 60$.

```
three_groups <- data.frame(
  group = rep(1:3, times = c(10, 20, 30)),
  value = rnorm(60, 0, 1)
)
```

We could use built-in R functions to compute the test statistic, but it is worthwhile to understand all of the formulas so we do it step by step. We first make sure that `group` is a factor! We can also check that (in this case, anyway) $SS_R + SS_B = SS_{\text{tot}}$.

```
three_groups$group <- factor(three_groups$group)
sse <- three_groups %>%
  mutate(total_mean = mean(value)) %>%
  group_by(group) %>%
  mutate(
    mu = mean(value),
    diff_within = value - mu,
    diff_between = mu - total_mean,
    diff_tot = value - total_mean
  ) %>%
  ungroup() %>%
  summarize(
    sse_within = sum(diff_within^2),
    sse_between = sum(diff_between^2),
    sse_total = sum(diff_tot^2)
  )
sse
```

```
## # A tibble: 1 x 3
##   sse_within sse_between sse_total
##        <dbl>       <dbl>     <dbl>
## 1       45.2        1.46      46.7
```

Now that we have computed the three sums of squared errors, we verify that $SS_R + SS_B = SS_{\text{tot}}$.

```
sse$sse_within + sse$sse_between - sse$sse_total
```

```
## [1] 0
```

Now we can compute the test statistic.

```
k <- 3
N <- 60
(sse$sse_between / (k - 1)) / (sse$sse_within / (N - k))
```

```
## [1] 0.9198313
```

For simulation, it is faster to use the `anova` command to compute $F$ directly:

```
aov.mod <- anova(lm(value ~ group, data = three_groups))
aov.mod$`F value`[1]
```

```
## [1] 0.9198313
```

Now we are ready for the simulation.

```
sim_data <- replicate(1000, {
  three_groups <- data.frame(
    group = rep(1:3, times = c(10, 20, 30)),
    value = rnorm(60, 0, 1)
  )
  three_groups$group <- factor(three_groups$group)
  aov.mod <- anova(lm(value ~ group, data = three_groups))
  aov.mod$`F value`[1]
})
plot(density(sim_data),
  main = "F test statistic"
)
curve(df(x, 2, 57), add = T, col = 2)
```

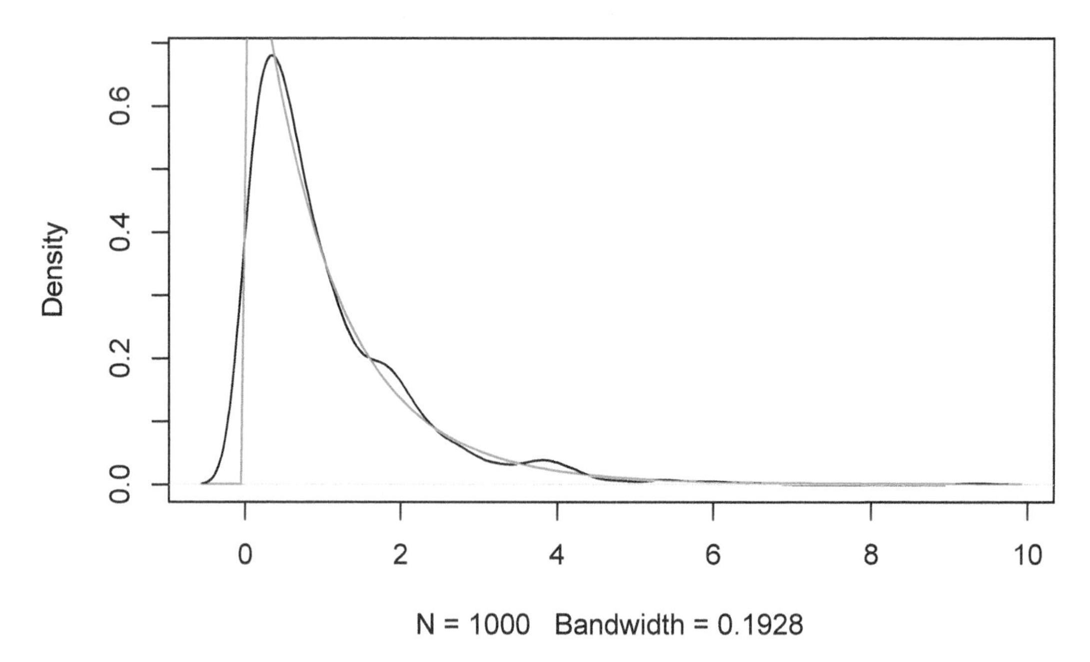


Indeed, the test statistic distribution matches the $F$ distribution with 2 and 57 degrees of freedom.

## 12.2 The ANOVA test

To perform analysis of variance, compute the test statistic $F$. Because large values of $F$ are more likely under $H_a$ than under $H_0$, while small values of $F$ are more likely under $H_0$ than under $H_A$, ANOVA is a one-sided test.

If $F$ is large enough, as determined by the appropriate $F$ distribution, then we reject $H_0 : \mu_1 = \mu_2 = \cdots = \mu_k$ in favor of the alternative hypothesis that not all of the $\mu_i$ are equal. More precisely, the $p$-value is the probability of observing a value as large as $F$ or larger from the $F(k-1, N-k)$ distribution.

To perform ANOVA in R, use `lm` to build a linear model of the numeric variable on the grouping variable. Then run `anova` on the model object.

When performing ANOVA in R, it is **essential** that the grouping variable is a factor. If the grouping variable is numeric, then R will fit a line to the data, which would be inappropriate. R will not warn you about this. For this reason, always check that the degrees of freedom matches the number of groups minus 1. If degrees of freedom is 1 and you have more than two groups, then you likely miscoded the grouping variable.

### 12.2.1 Example: THC mice

At the start of the chapter, we used `anova` to analyze the differences between four groups of mice given different doses of THC. In this section, we step through that computation carefully.

The data is in `fosdata::mice_pot`. We have already seen that the four distributions are approximately normal with approximately equal variance, and the measurements are independent. The null hypothesis is $H_0 : \mu_1 = \mu_2 = \mu_3 = \mu_4$, where $\mu_i$ is the true mean percent activity relative to baseline. The alternative hypothesis is that at least one of the means is different. We will test at the $\alpha = .05$ level.

```
mice_mod <- lm(percent_of_act ~ group, data = mice_pot)
anova(mice_mod)
```

```
## Analysis of Variance Table
##
## Response: percent_of_act
##           Df Sum Sq Mean Sq F value Pr(>F)
## group      3   6329 2109.65  3.1261 0.0357 *
## Residuals 42  28344  674.85
## ---
## Signif. codes:  0 '***' 0.001 '**' 0.01 '*' 0.05 '.' 0.1 ' ' 1
```

The results are significant ($p = .0357$), so we reject the null hypothesis that all of the means are the same.

We now explain the rest of the output from `anova`. The table contains all of the intermediate values used to compute $F$ as described in Section 12.1.2.

The two rows `group` and `Residuals` correspond to the between group and within group variation. The first column, `Df` gives the degrees of freedom in each case. Since $k = 4$, the between group variation has $k - 1 = 3$ degrees of freedom, and since $N = 46$, the within group variation (`Residuals`) has $N - k = 42$ degrees of freedom.

The `Sum Sq` column gives $SS_B$ and $SS_R$. The `Mean Sq` variable is the `Sum Sq` value divided by the degrees of freedom. These two numbers are the numerator and denominator of the test statistic, $F$. So here, $F = 2109.65/674.85 = 3.1261$.

To compute the $p$-value, we need the area under the tail of the $F$ distribution above $F = 3.1261$. Recall that this is a one-tailed test. Figure 12.1 shows the area corresponding to the $p$-value, and we may compute it with:

```
pf(3.1261, df1 = 3, df2 = 42, lower.tail = FALSE)
```

```
## [1] 0.03570167
```

This matches the `Pr(>F)` value used earlier to reject $H_0$.

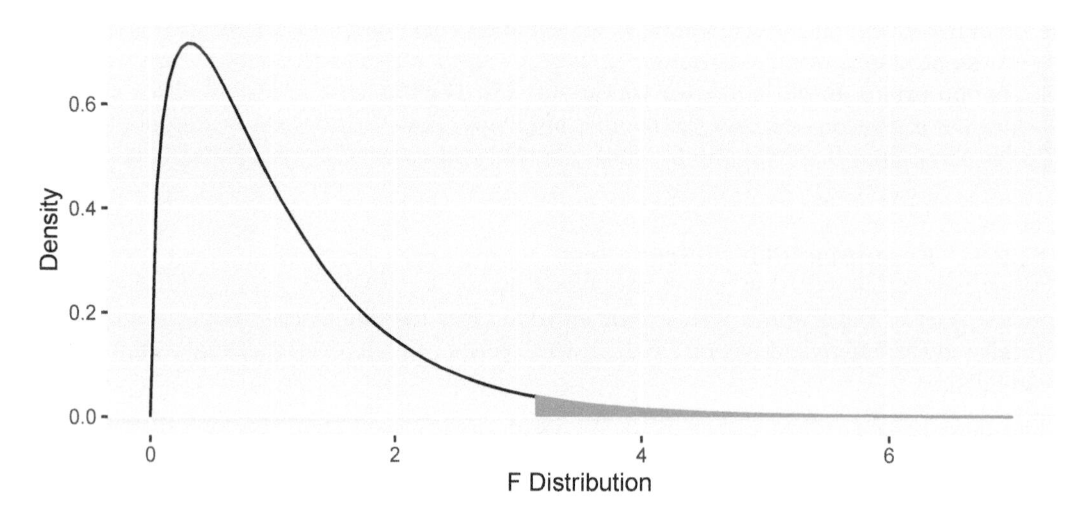


**FIGURE 12.1** ANOVA $p$-value is the one-tail area under the $F$ distribution.

After rejecting the null hypothesis that all of the means are the same, we are often interested in determining **which** means are different. See Exercise 12.21 for an approach for doing this using Tukey's Honest Significant Differences method.

### 12.2.2 Example: Humanization

Different social groups may hold strong opinions about each other. A 2018 study by Davies et al.[3] examined American opinions about Pakistanis, and whether those opinions changed when participants learned about humanitarian behavior by Pakistanis. American participants in the study were read a brief synopsis of the tragedy of Hurricane Katrina, and then:

1. Told nothing further (the control group).
2. Told about the aid sent by Pakistan, and told low numbers for the amount of that aid.
3. Told about the aid sent by Pakistan, and told high numbers for the amount of that aid.

Afterwards, the participants were asked how strongly they believed Pakistanis would have felt both primary and secondary emotions following the disaster, and the mean of their responses was taken. The secondary emotions in this study were grief, sorrow, mourning, anguish, guilt, remorse, and resentment.

We plan on analyzing whether there is a difference in the mean of the three groups' belief of how strongly Pakistanis would have felt secondary emotions. One issue with the technique that we are using is that we are *averaging* the ordinal scale responses to the seven questions on emotions, which is a common, but not always valid, thing to do.

[3]Thomas Davies et al., "From Humanitarian Aid to Humanization: When Outgroup, but Not Ingroup, Helping Increases Humanization," *PLOS One* 13, no. 11 (2018): e0207343.

We can load the data via the following.

```
humanization <- fosdata::humanization
```

We will first examine the data to see whether it appears to be approximately normal with equal variances across the groups. We are only concerned in this problem with the people who were told about Hurricane Katrina.

```
kat_humanization <- humanization %>%
  filter(stringr::str_detect(group, "Kat"))
ggplot(kat_humanization, aes(x = group, y = pak_sec)) +
  geom_boxplot() +
  geom_jitter(height = 0, width = 0.2)
```

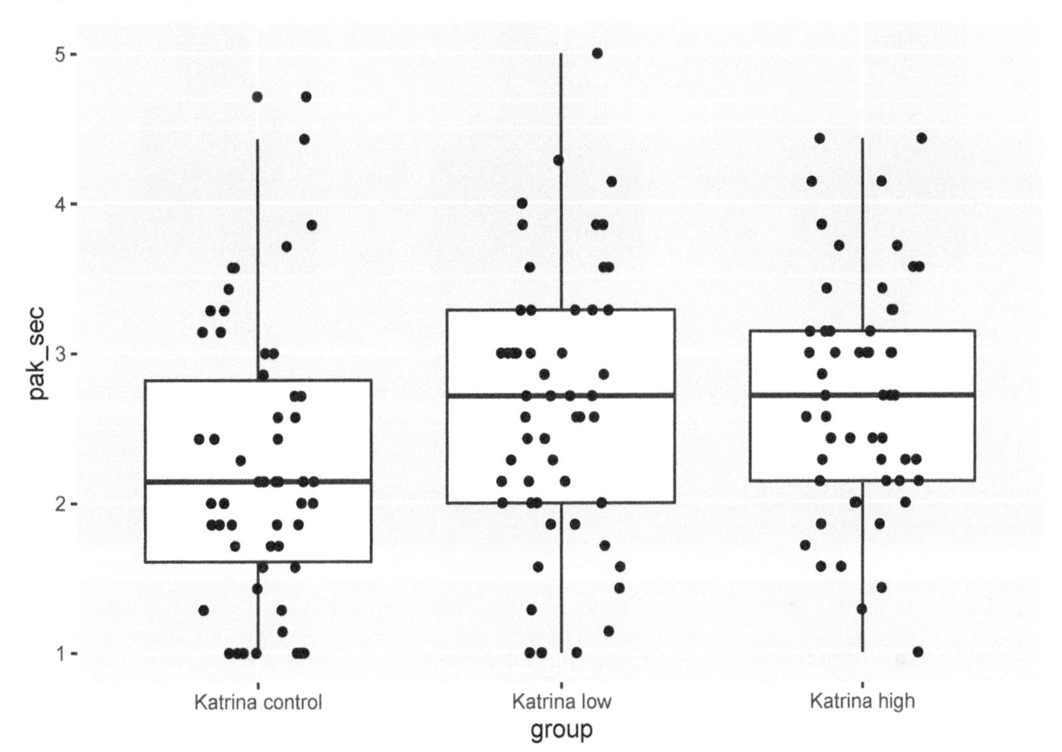


Those don't look too bad, except that the control group and the low group both have more values exactly equal to 1 than would be expected in a normal distribution. Let's look at the mean, standard deviation, and sample size in each group.

```
kat_humanization %>%
  group_by(group) %>%
  summarize(
    mean = mean(pak_sec),
    sd = sd(pak_sec),
    n = n()
  )
```

```
## # A tibble: 3 x 4
##   group             mean    sd     n
##   <fct>            <dbl> <dbl> <int>
## 1 Katrina control  2.24 0.927    54
```

```
## 2 Katrina low      2.65 0.911    53
## 3 Katrina high     2.72 0.794    54
```

It's also a good idea to do a histogram.

```
ggplot(kat_humanization, aes(x = pak_sec)) +
  geom_histogram(bins = 8) +
  facet_wrap(~group)
```

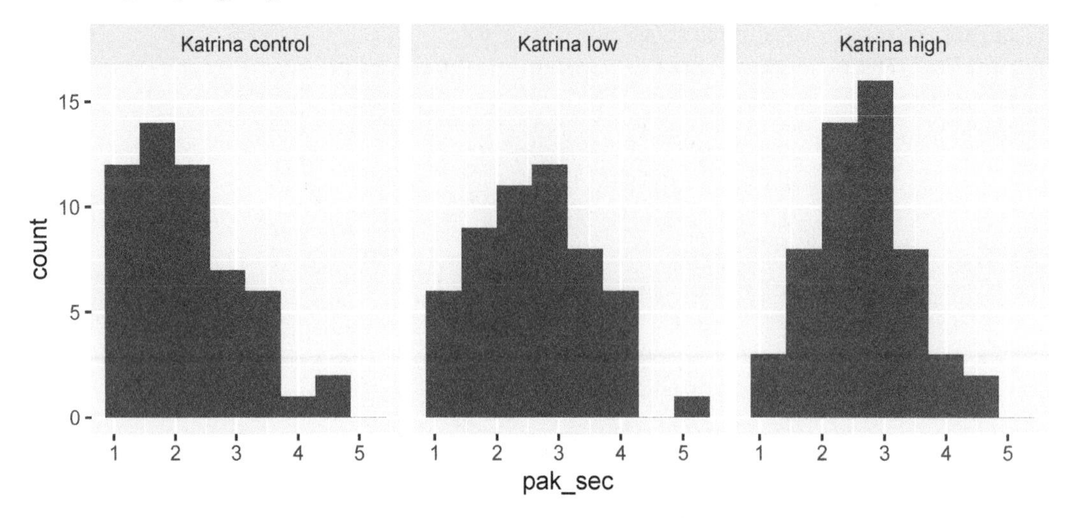


The control group may be moderately right-skewed, but we see nothing that would invalidate a one-way ANOVA analysis. To perform one-way ANOVA in R, first build a linear model for `pak_sec` as explained by `group`, and then apply the `anova` function to the model.

```
human_mod <- lm(pak_sec ~ group, data = kat_humanization)
anova(human_mod)
```

```
## Analysis of Variance Table
##
## Response: pak_sec
##            Df  Sum Sq Mean Sq F value   Pr(>F)
## group       2   7.456  3.7279  4.8269 0.009231 **
## Residuals 158 122.025  0.7723
## ---
## Signif. codes:  0 '***' 0.001 '**' 0.01 '*' 0.05 '.' 0.1 ' ' 1
```

The ANOVA table gives the $p$-value as `Pr(>F)`, here $p = 0.009231$. According to the $p$-value, we reject $H_0$ and conclude that not all three of the groups had the same mean. The test suggests that the story told had an effect on the opinions of the American participants.

## 12.3 Unequal variance

The equal variances assumption of ANOVA is often difficult to verify. In this section, we introduce a variant of one-way ANOVA that corrects for unequal variance. We then use simulation to explore the effect of unequal variance on the results of one-way ANOVA.

### 12.3.1 The oneway.test

The function `oneway.test` operates similarly to one-way ANOVA but performs an approximation procedure (called Welch's correction) to correct for unequal variance. We introduce `oneway.test` with an example of data involving the greying of chimpanzees.

The `fosdata::chimps` data comes from a 2020 study by Tapanes et al.[4] The investigators showed photographs of chimpanzees to human judges, who rated the greyness of chimpanzee facial hair on a scale of 1-6.

*Image credit: Ikiwaner.*

The goal of the study was to determine whether grey hair in chimpanzees increases with age, as it does with humans. The authors found that for chimpanzees aged 30 and older, there does not seem to be an age effect. We will use the data in a different way. The study included chimpanzees from three different locations: the Ngogo community in Uganda, the South community at Taï National Park, Ivory Coast, and the captive population of the New Iberia Research Center (NIRC). We wish to determine whether the mean grey hair is different for chimpanzees in the three groups.

Since the study found no age effect for older chimpanzees, we may ignore age in our analysis if we restrict to chimps aged 30 and older. One chimpanzee, Brownface, has two pictures in the data set. To maintain independence, we arbitrarily chose to remove his 2011 photograph. This choice did not affect the conclusions of our statistical analysis. We load and clean the data first:

---

[4]Elizabeth Tapanes et al., "Does Facial Hair Greying in Chimpanzees Provide a Salient Progressive Cue of Aging?" *PLOS One* 15, no. 7 (2020): e0235610.

```
old_chimps <- fosdata::chimps %>%
  filter(age >= 30) %>%
  filter(!(individual == "Brownface" & year == 2011))
```

Next, we make a boxplot to visually compare the three groups.

```
old_chimps %>%
  ggplot(aes(x = population, y = grey_score_avg)) +
  geom_boxplot()
```

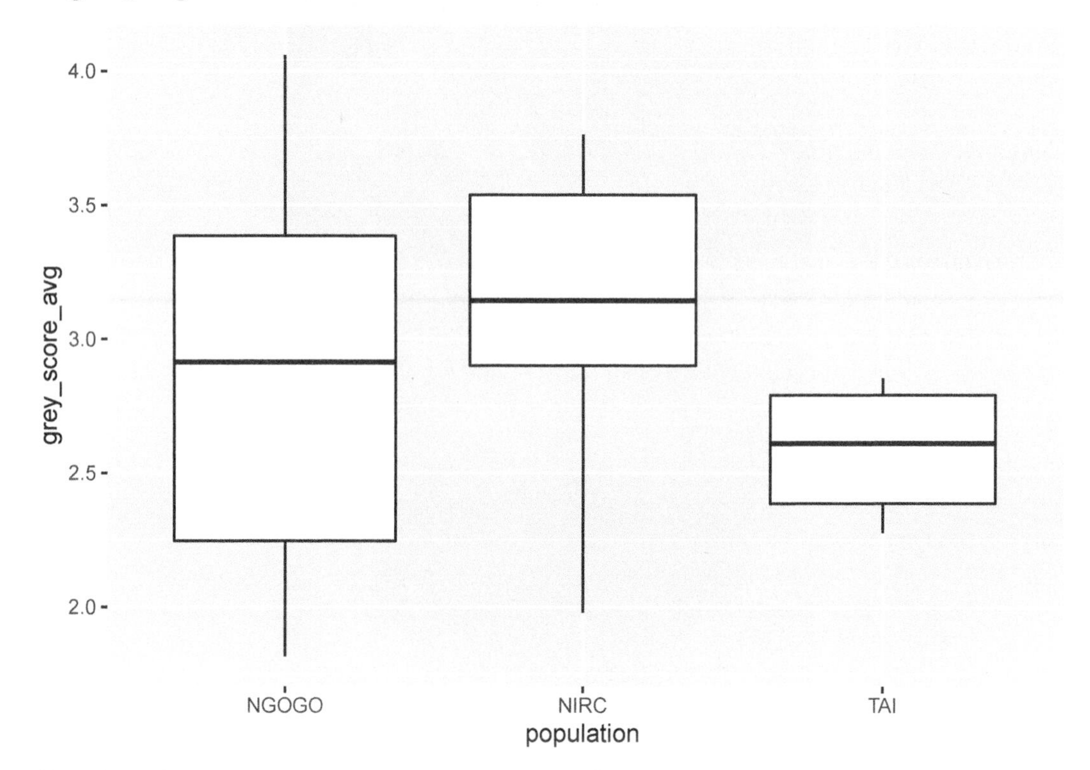


Based on this plot, it appears there may be a difference in mean grey score by population. However, the variances of the three groups are visibly different. The box and whiskers have more spread for NGOGO and less for TAI. Using one-way ANOVA on this data could give misleading results.

The one-way test has the same independence and normality assumptions as one-way ANOVA, and the same null and alternative hypotheses. For the chimps,

$$H_0 : \mu_{\text{NGOGO}} = \mu_{\text{NIRC}} = \mu_{\text{TAI}}, \quad H_a : \text{not all of the } \mu_i \text{ are equal.}$$

We test `grey_score_avg` as explained by the factor variable `population`:

```
oneway.test(grey_score_avg ~ population, data = old_chimps)
```

```
##
##  One-way analysis of means (not assuming equal variances)
##
## data:  grey_score_avg and population
## F = 4.2285, num df = 2.000, denom df = 18.927, p-value = 0.03033
```

The test results in $p$-value of .03033, so we reject $H_0$ at the $\alpha = 0.05$ level. We conclude that the three groups do not have the same greyness.

Incorrectly using `anova` in spite of the unequal variance would result in the less reliable value $p = 0.14$, which fails to detect a difference among the three groups.

### 12.3.2 Error simulations

For the chimpanzee data, unequal variance led to a striking difference between the results of one-way ANOVA and the results of `oneway.test`. In this section, simulations show how much unequal variances might affect the accuracy of ANOVA.

The simulations in this section assume the underlying populations are normal and independent, but have unequal variances. We will measure the effective type I error rates when $\alpha = .05$. The results will depend on whether or not our groups have equal or unequal sample sizes – generally the behavior of one-way ANOVA worsens when sample sizes are unequal.

**Equal sample sizes**

We create three groups of data, each size $n_i = 30$, all normally distributed with mean 0. Data in the first two groups has standard deviation 1, and in the third group we use standard deviation $\sigma$. The code below creates a function to run the simulation for various values of $\sigma$. For each $\sigma$, it calculates the probability of type I error, where we have $p < .05$ and incorrectly reject $H_0$.

```
sdevs <- c(0.01, 0.05, 0.1, 0.5, 5, 10, 50, 100)

sapply(sdevs, function(sigma) {
  pvals <- replicate(10000, {
    three_groups <- data.frame(
      group = factor(rep(1:3, each = 30)),
      values = rnorm(90, 0, rep(c(1, 1, sigma), each = 30))
    )
    anova(lm(values ~ group, data = three_groups))$`Pr(>F)`[1]
  })
  mean(pvals < 0.05)
})
```

The simulation takes a considerable time to run for 10,000 trials, and results in the following:

| $\sigma$ | 0.01 | 0.05 | 0.1 | 0.5 | 5 | 10 | 50 | 100 |
|---|---|---|---|---|---|---|---|---|
| $P(p < .05)$ | 0.0649 | 0.0639 | 0.0608 | 0.0572 | 0.0779 | 0.0866 | 0.0902 | 0.0911 |

The effective type I error is consistently too high. A correctly functioning test would have type I error of 0.05 in this setting. Things are worse when the third group has large variance, not quite as bad when it has small variance.

**Unequal sample sizes**

In this simulation, the three groups have sample sizes $n_1 = 30$, $n_2 = 30$, and $n_3$ which will vary from 10 to 500. For each value of $n_3$, we create data with three groups that are normal, mean 0, and have standard deviation 1, 1, and 5 respectively. In each replication, the $p$-value is checked against $\alpha = 0.05$.

```
sample_sizes <- c(10, 20, 50, 100, 500)
sapply(sample_sizes, function(n3) {
  pvals <- replicate(10000, {
    three_groups <- data.frame(
      group = factor(rep(1:3, times = c(30, 30, n3))),
      values = rnorm(60 + n3, 0, rep(c(1, 1, 5), times = c(30, 30, n3)))
    )
    anova(lm(values ~ group, data = three_groups))$`Pr(>F)`[1]
  })
  mean(pvals < .05)
})
```

| $n_3$ | 10 | 20 | 50 | 100 | 500 |
|---|---|---|---|---|---|
| $P(p < .05)$ | 0.3031 | 0.1483 | 0.0263 | 0.0028 | 0.0000 |

Unequal sample sizes exacerbate the problem! These type I error rates would be 5% if the test were working. When the different group's sample size is small relative to the other two, the error rate is as high as 30%. This means that 30% of the time, the test detects a difference in means where none exists. This is very bad, because it can lead investigators to find significance when none exists.

When the sample size is large, the rate falls to below 0.01%, indicating the test is failing to work as advertised. This is also bad because it will adversely affect the power of the test in instances where there is a difference in means.

The `oneway.test` does not depend on equal variances, so it should work properly under the same conditions:

```
sample_sizes <- c(10, 20, 50, 100, 500)
sapply(sample_sizes, function(n3) {
  pvals <- replicate(10000, {
    three_groups <- data.frame(
      group = factor(rep(1:3, times = c(30, 30, n3))),
      values = rnorm(60 + n3, 0, rep(c(1, 1, 5), times = c(30, 30, n3)))
    )
    oneway.test(values ~ group, data = three_groups)$p.value
  })
  mean(pvals < .05)
})
```

| $n_3$ | 10 | 20 | 50 | 100 | 500 |
|---|---|---|---|---|---|
| $P(p < .05)$ | 0.0482 | 0.0490 | 0.0502 | 0.0513 | 0.0489 |

As desired, the effective type I error rate when $\alpha = .05$ is approximately .05 in all cases.

If you suspect your populations may not follow the assumptions for ANOVA, simulations give a measure of how big a problem it will cause. For example, if you have two samples of size 30 and one of size 10, then you want to be pretty strict about the equal variance assumption. If your sample sizes are all 30, then you don't need to be as strict.

## 12.4 Pairwise $t$-tests

Suppose we have $k$ groups of observations and wish to detect a difference in means. It is tempting to perform $t$-tests on some or all of the groups, especially if two groups seem quite different. This sort of analysis is problematic because the selection of two groups biases the results. Given enough groups, even if all come from the same population, it is quite likely that two of them will appear to be different.

One approach that avoids the issue of selection is to test all possible pairs. For $k$ groups, this results in $\binom{k}{2}$ different $t$-tests. The resulting $p$-values must be adjusted upward to account for the multiple tests. This method is called *pairwise t-tests.*

Pairwise $t$-tests have the advantage of showing directly which pairs of groups have significant differences in means. They also apply in experiments where subjects are reused across all groups, when one-way ANOVA is not appropriate due to dependence (see Exercise 12.18).

On the other hand, pairwise $t$-tests do not directly test the hypothesis of interest, $H_0 : \mu_1 = \mu_2 = \cdots = \mu_k$ versus $H_a$: at least one of the means is different. Generally ANOVA will have more power (see Exercise 12.14).

The R command for doing pairwise $t$-tests is `pairwise.t.test`. This function requires two values: `x` the measured value or response vector and `g` the grouping variable.

**Example 12.3.** Perform pairwise $t$-tests to determine differences between groups for the `mice_pot` data set.

```
pairwise.t.test(mice_pot$percent_of_act, mice_pot$group)
```

```
##
##  Pairwise comparisons using t tests with pooled SD
##
## data:  mice_pot$percent_of_act and mice_pot$group
##
##     VEH   0.3   1
## 0.3 1.000 -     -
## 1   1.000 1.000 -
## 3   0.050 0.124 0.072
##
## P value adjustment method: holm
```

The only pair approaching significance is the test between the control (VEH) group and the treatment group that received the highest THC dose, 3 mg/kg. With this result, it would be hard to claim a relationship between THC and locomotive impairment. At the start of the chapter, we used ANOVA to calculate a single $p$-value of 0.0357, somewhat stronger evidence of a difference between the four groups of mice.

**Example 12.4.** Perform pairwise $t$-tests to determine differences between groups for the `humanization` data set from Section 12.2.2.

```
pairwise.t.test(kat_humanization$pak_sec, kat_humanization$group)
```

```
##
##  Pairwise comparisons using t tests with pooled SD
```

```
##
## data:  kat_humanization$pak_sec and kat_humanization$group
##
##               Katrina control Katrina low
## Katrina low   0.033           -
## Katrina high  0.013           0.647
##
## P value adjustment method: holm
```

There is a significant difference at $\alpha = .05$ between the control and both the low ($p = 0.033$) and high ($p = 0.013$) levels of aid stories, but no significant difference between the high and low level of aid stories ($p = 0.647$). Compare these values with the ANOVA $p$-value of 0.009 from Section 12.2.2, which showed an overall difference among the groups.

The `pairwise.t.test` function has additional options to control its behavior.

**p.adjust.method**
The method for adjusting the $p$-values, which can be "none," "holm," and "bonferroni." These options are explored in the next section.

**pool.sd**
Whether we want to pool the standard deviation across all of the groups to get an estimate of the common standard deviation, or whether we estimate the standard deviation of each group separately. Set to `TRUE` if we know that the variances of all the groups are equal, or possibly if some of the groups are small. Otherwise, `FALSE`.

**paired**
Set to `TRUE` if the observations are paired between each group. Default is `FALSE`.

**alternative**
Used for one-sided tests. The alternate hypothesis for each pair will be that the lower numbered group has a mean strictly less than (or greater than) the higher numbered group. This means that the ordering of groups is important.

### 12.4.1 FWER and $p$-value adjustments

This section investigates the methods for adjusting $p$-values when performing multiple tests. First, we define a measure for the failure of multiple tests.

**Definition 12.2.** Suppose that multiple hypothesis tests are conducted. The *Family Wide Error Rate* (FWER) is the probability of making at least one type I error in any of the tests.

**Example 12.5.** Estimate the FWER for an uncorrected pairwise $t$-test on six iid groups, each normal with mean 0 and sd 1. Use a sample size of $n_i = 10$ from each group.

As a first step, generate one set of data and observe the results. In the pairwise $t$-test, set `p.adjust.method` to "none" so no $p$-value correction is performed.

```
six_groups <- data.frame(
  values = rnorm(60),
  groups = factor(rep(1:6, each = 10))
)
pvals <- pairwise.t.test(six_groups$values,
  six_groups$groups,
  p.adjust.method = "none"
```

```
)$p.value
pvals
```

```
##           1         2         3         4         5
## 2 0.7694607        NA        NA        NA        NA
## 3 0.5048147 0.3383610        NA        NA        NA
## 4 0.9769938 0.7475434 0.5233001        NA        NA
## 5 0.9961214 0.7731745 0.5017338 0.9731171        NA
## 6 0.9774319 0.7911463 0.4870307 0.9544448 0.9813092
```

By looking at the results, we can tell that none of the 15 $p$-values are less than 0.05. To simulate, we use the `any` function to detect whether *any* of the $p$-values are less than 0.05, indicating a family-wide error.

```
sim_data <- replicate(10000, {
  six_groups <- data.frame(
    values = rnorm(60),
    groups = factor(rep(1:6, each = 10))
  )
  pvals <- pairwise.t.test(six_groups$values,
    six_groups$groups,
    p.adjust.method = "none"
  )$p.value
  any(pvals < .05, na.rm = TRUE)
})
```

```
mean(sim_data)
```

```
## [1] 0.3582
```

The simulation gives a FWER of 0.3582, so an uncorrected pairwise $t$-test would result in a type I error about 36% of the time. A correction to the $p$-values for multiple tests is certainly required!

The *Bonferroni correction* to the pairwise $t$-test multiplies each $p$-value by the total number of tests, capping $p$ at 1 if necessary. The Bonferroni correction is a conservative approach, and will lead to a FWER of less than the specified $\alpha$ level.

Repeating the simulation in Example 12.5 using `p.adjust.method = "bonferroni"` results in a FWER of about 0.04, which is indeed less than $\alpha = .05$.

A more sophisticated technique is the *Holm correction*. The FWER of Holm is the same as that of Bonferroni, but in general Holm is more powerful. For this reason, Holm is the preferred method and the default used by `pairwise.t.test`.

In Exercise 12.17, you are asked to show that the *power* of Holm is higher than the power of Bonferroni in a specific instance.

## Vignette: Reproducibility

Up to this point in the book, we have emphasized writing *reproducible code*. Reproducible code means that anyone can take the R script or R Markdown file from a statistical analysis and run it **as is** to get exactly the results that were reported. This is important because it can often be challenging to understand which exact details were used for a statistical test when reading the summary of an experiment. Providing reproducible code eliminates any confusion about what techniques were used.

A second notion of reproducibility is that of *reproducible science*. A crucial aspect of the scientific method is that experiments should be reproducible. There is currently a reproducibility crisis in several scientific areas, in the sense that it is believed that many published results would not be reproduced if the experiment was repeated. Here are some reasons for that.

**Significance**
: Let's say that a $p$-value less than .05 means that we can reject $H_0$. Then we will reject $H_0$ five percent of the time even when it is true. Now, consider how many thousands of statistical tests are run each day. Many (in fact, 5%) of those tests will result in $p < .05$ simply due to chance. This leads to a problem reproducing significance.

**Power and effect size**
: Some experiments do not have a sample size that is large enough to attain a reasonable power. When a test is underpowered and yet $p < .05$, frequently the **effect size** gets overestimated. That is, if we are interested not only in whether $H_0$ is false but also how big of a difference from $H_0$ there is, we will overestimate that difference in an underpowered test. This leads to a problem reproducing effect sizes. We recommend doing a power analysis whenever possible before collecting data.

**Data dredging**
: Some experimenters will take measurements on hundreds or thousands of variables, and look for interesting patterns. If the experimenters do enough tests and check on enough things, then eventually they will find something that is statistically significant. This is fine, as long as the experimenters report all of the tests that they conducted or considered in addition to the ones that were significant. Unethical researchers might not report on the data that wasn't significant, which is called $p$-hacking. Even if researchers report all of the tests that were run, people can be misled by the statistically significant results, and assign them more importance than deserved. Both $p$-hacking and misinterpretation lead to reproducibility issues. We recommend reporting all tests that were run (or considered, but not run based on the data). We also recommend making all data and scripts public whenever possible.

**Researcher degrees of freedom**
: There are many decisions that go into collecting and analyzing data. Consider outliers. If we think that data is an outlier or may have been miscoded, it is common practice to check with the person who collected the data. How do we decide what appears to be an outlier? If there are too many, then maybe we use a Wilcoxon rank sum test or other method that is robust to outliers rather than a $t$-test. If we are using a $t$-test, how do we decide whether `var.equal` is `TRUE` or `FALSE`? Do you perform a log transformation of the data? Is the alternative hypothesis one-sided or two-sided? There are any number of things that we may or may not do when analyzing a data set, and each one has an effect on the outcome that is hard to quantify. As a solution, we recommend *pre-registration*. Pre-registration means the experimenters state the details of the data collection and

analysis procedure before starting to collect data. Any deviance from the pre-registered procedure should be noted and justified.

**Replication incentives**

Replication studies are not seen as exciting and result in little social or financial reward. They are more difficult to publish, since they are not original results. Not surprisingly, replication studies are conducted far less often than original research. The dearth of replication studies has allowed research with statistical design flaws to linger and gain credence, leading to larger crises when the results cannot be reproduced.

## Exercises

---

Exercises 12.1 – 12.3 require material through Section 12.1.

**12.1.** Consider the `chimps` data set from the `fosdata` package. This data set contains 169 observations of (among other things) the average grey hair score `grey_score_avg` of chimps together with a `population` that the chimps come from. Suppose that the order of the populations is NIRC, NGOGO, TAI. In the notation of Section 12.1, find $n_1, n_2, n_3$ as well as $\overline{x}_{i\cdot}$ for $i = 1, 2, 3$.

**12.2.** If you have 30 independent observations of 3 groups, each of which is normal with the same mean and standard deviation, what is the distribution of the one-way ANOVA test statistic $F$? Include degrees of freedom.

**12.3.** Create a simulation that shows that the one-way ANOVA test statistic follows an $F$ distribution under the null hypothesis. Assume that you have 4 groups, each with 25 subjects, and each with mean 1 and standard deviation 2. Show via simulation that the ANOVA test statistic $F$ is an $F$ random variable with $k - 1$ and $N - k$ degrees of freedom.

---

Exercises 12.4 – 12.8 require material through Section 12.2.

**12.4.** Consider the `weight_estimate` data in the `fosdata` package. Children and adults were asked to estimate the weight of an object that an actor picks up. Conduct a one-way ANOVA test at the $\alpha = .05$ level to test whether the means of the weight estimates for the 300 g object are the same across the age groups in the study.

**12.5.** Consider the `case0502` data from `Sleuth3`. Dr. Benjamin Spock was tried in Boston for encouraging young men not to register for the draft. It was conjectured that the judge in Spock's trial did not have appropriate representation of women. The jurors were supposed to be selected by taking a random sample of 30 people (called venires), from which the jurors would be chosen. In the data `case0502`, the percent of women in 7 judges' venires are given.

a. Create a boxplot of the percent women for each of the 7 judges. Comment on whether you believe that Spock's lawyers might have a point.
b. Determine whether there is a significant difference in the percent of women included in the 6 judges' venires *who aren't Spock's* judge.
c. Determine whether there is a significant difference in the percent of women induced in

Spock's venires versus the percent included in the other judges' venires combined. (Your answer to part (b) should justify doing this.)

**12.6.** Consider the data in `ex0524` in `Sleuth3`. This gives 2584 data points, where each subject's IQ was measured in 1979 and their income was measured in 2005. Create a boxplot of income for each of the IQ quartiles. Is there evidence to suggest that there is a difference in mean income among the 4 groups of IQ? Check the assumptions necessary to do one-way ANOVA. Does this cause you to doubt your result?

**12.7.** The built-in data set `chickwts` measures weights of chicks after six weeks eating various types of feed. Test if the mean weight is the same for all feed types. Are the assumptions for one-way ANOVA reasonable?

**12.8.** The data set `MASS::immer` from the `MASS` package has data on a trial of barley varieties performed by Immer, Hayes, and Powers in the early 1930's. Test if the total yield (sum of 1931 and 1932) depends on the variety of barley. Are the assumptions for ANOVA reasonable?

---

Exercises 12.9 – 12.13 require material through Section 12.3.

**12.9.** The built-in data set `InsectSprays` records the counts of living insects in agricultural experimental units treated with different insecticides.

a. Plot the data. Three of the sprays seem to be more effective (less insects collected) than the other three sprays. Which three are the more effective sprays?
b. Use one-way ANOVA to test if the three effective sprays have the same mean. What do you conclude?
c. Use one-way ANOVA to test if the three less effective sprays have the same mean. What do you conclude?
d. Would it be appropriate to use one-way ANOVA on the entire data set? Why or why not?
e. Use `oneway.test` to test the null hypothesis that all of the mean insect counts are the same for the various sprays versus the alternative that they are not, at the $\alpha = .01$ level.

**12.10.** The built-in data set `morley` gives speed of light measurements for five experiments done by Michelson in 1879.

a. Create a boxplot with one box for each of the five experiments.
b. Compute the sample standard deviation and sample size for each of the five experiments.
c. Use ANOVA to test if the five experiments have the same mean, and report your findings. Careful: the `Expt` variable is coded as an integer.
d. Use `oneway.test` to test if the five experiments have the same mean, and report your findings.

**12.11.** The `msleep` data set in the `ggplot2` package has information about 83 species of mammals.

a. How many mammal species are in each `vore` group?
b. Make a boxplot to show how `sleep_total` depends on `vore`.
c. The group variances are apparently unequal, so use `oneway.test` to test if the mean sleep total depends on vore.
d. Run the ANOVA anyway. What $p$ value does it report?

Notice that the unequal variance combined with unequal group sizes made a huge difference in the $p$-value.

**12.12.** Estimate the effective type I error rate when performing one-way ANOVA on 4 groups at the $\alpha = .05$ level. Assume groups 1-3 are size 20 and group 4 is size 100. Assume all groups have mean 0, groups 1-3 have standard deviation 1, and group 4 has standard deviation 2.

**12.13.** Repeat Exercise 12.12, but assume all groups have size 20. Is the effective type I error closer to the $\alpha = .05$ level, or further from it?

---

Exercises 12.14 – 12.22 require material through Section 12.4.

**12.14.** Suppose you have 6 groups of 15 observations each. Groups 1-3 are normal with mean 0 and standard deviation 3, while groups 4-6 are normal with mean 1 and standard deviation 3. Compare via simulation the power of the following tests of $H_0 : \mu_1 = \cdots = \mu_6$ versus the alternative that at least one pair is different at the $\alpha = .05$ level.

a. One-way ANOVA. (Hint: `anova(lm(formula, data))[1,5]` pulls out the $p$-value of ANOVA.)
b. `pairwise.t.test`, where we reject the null hypothesis if any of the paired $p$-values are less than 0.05.

The next two exercises use the data set `flint` from the `fosdata` package. The data set `fosdata::flint` contains data on lead levels in water samples taken from Flint, Michigan, during the "Flint water crisis" of 2015. Three lead measurements were taken at each house, Pb1, Pb2, and Pb3, on first draw, after 45 seconds, and after 2 minutes of water running. As observed in 9.16, the lead levels are highly skew.

**12.15.** a. Make a boxplot showing the distribution of the log levels of Pb1 for each ward. Notice that ward 0 contains only a single sample. In fact, there is no ward 0 in Flint. Remove that data point before continuing.
b. Do the log Pb1 levels for each ward appear normally distributed? Are the variances approximately equal across wards?
c. Use ANOVA to determine if there is a difference in log Pb1 levels across the wards of Flint. Report a $p$-value with your answer.

**12.16.** a. Make a boxplot showing the distribution of the log levels of Pb1, Pb2, and Pb3. Hint: use `pivot_longer` to tidy the data.
b. Do the three log levels appear normally distributed? Are their variances approximately equal?
c. Explain why it is inappropriate to use one-way ANOVA to test for a difference among the means of the three log lead levels.
d. Use a pairwise $t$-test to determine if there is a difference in log lead level between the first draw, 45 second, and 2 minute samples.

**12.17.** Suppose you have three populations, all of which are normal with standard deviation 1, and with means 0, 0.5, and 1. You take samples of size 30 and perform a pairwise $t$-test.

a. Estimate the percentage of times that at least one of the null hypotheses being tested is rejected when using the "bonferroni" $p$-value adjustment.
b. Estimate the percentage of times that at least one of the null hypotheses being tested is rejected when using the "holm" $p$-value adjustment.

**12.18.** Consider the data set `cows_small` in the `fosdata` package. In this data set, cows in Davis, CA were sprayed with water on hot days to try to cool them down. Each cow was measured with no spray (control), nozzle TK-0.75, and nozzle TK-12.

a. Is one-way ANOVA appropriate to use on this data set? If so, do it. If not, explain why not.
b. Perform pairwise $t$-tests on the three groups with `holm` $p$-value adjustment and explain.

**12.19.** Suppose you perform pairwise $t$-tests on 6 iid groups.

a. How many tests are required?
b. At the $\alpha = 0.05$ level, what is the probability of type I error for a single one of these tests?
c. Assume the pairwise tests are independent. Compute the exact probability that a type I error occurs on at least one of these tests.
d. Compare your answer from part (c) to results of the simulation in Example 12.5. Are the pairwise tests independent?

**12.20.** Consider the `weight_estimate` data in the `fosdata` package. Children and adults were asked to estimate the weight of an object that an actor picks up. In Exercise 12.4, you conducted an ANOVA test at the $\alpha = .05$ level to test whether the means of the weight estimates for the 300 g object are the same across the age groups in the study. Use `pairwise.t.test` to test all pairs of means at the $\alpha = .05$ level.

**12.21.** Consider the `frogs` data set in the `fosdata` package. This data set contains measurements from 6 species of frogs, including one new species, *dakha*, that the authors of the paper found. When establishing that a group of animals is a new species, it can be important to find physiological differences between the group of animals and animals that are nearest genetically. Read more about the data set by typing `?fosdata::frogs`.

a. Create a boxplot of the distance from front of eyes to the nostril versus species. Does it appear that the data satisfies the assumptions for one-way ANOVA?
b. Test $H_0$ (all of the mean distances of the species are the same) versus $H_a$ (not all of the mean distances of the species are the same) at the $\alpha = .05$ level.
c. If you reject $H_0$, look up `TukeyHSD` in base R, and use it to perform a post-hoc analysis to determine which pairs of mean bill depths are the same, and which are different at the $\alpha = .05$ level. The function `TukeyHSD` requires ANOVA to be done with the following command: `mod_aov <- aov(en ~ species, data = frogs)`. Write up your result in an organized fashion with explanations.

**12.22.** Consider the `penguins` data set in the `palmerpenguins` package that was discussed in Chapter 11.

a. Create a boxplot of the variable `bill_depth_mm` by `species`. Does it appear that the data satisfies the assumptions for one-way ANOVA?
b. Test $H_0$ (all of the mean bill depths of the species are the same) versus $H_a$ (not all of the mean bill depths are the same) at the $\alpha = .01$ level.
c. If you reject $H_0$, perform a post-hoc analysis as in Exercise 12.21 to determine which pairs of mean bill depths are the same, and which are different at the $\alpha = .01$ level.

# 13

## *Multiple Regression*

This chapter deals with multiple regression, where a single numeric response variable depends on multiple explanatory variables. There is a natural tension between explaining the response variable well and keeping the model simple. Our main purpose will be to understand this tension and find a balance between the two competing goals.

### 13.1 Two explanatory variables

We consider housing sales data from King County, WA, which includes the city of Seattle. The data set **houses** in the **fosdata** package contains a record of every house sold in King County from May 2014 through May 2015. For each house, the data includes the sale price along with many variables describing the house and its location.

Our goal is to model the sale price on the other variables. There are many variables that we could use to do this, but in this section we will only consider two of them; namely, the square footage in the house (`sqft_living`) and the size of the lot (`sqft_lot`). To keep the size of the problem small, we restrict to the urban ZIP code 98115.

```
houses_98115 <- fosdata::houses %>%
  filter(zipcode == "98115") %>%
  select(price, sqft_living, sqft_lot)
```

There are now 583 observations. We start by doing some summary statistics and visualizations.

```
summary(houses_98115)
```

```
##      price          sqft_living       sqft_lot    
##  Min.   : 200000   Min.   : 620   Min.   :  864  
##  1st Qu.: 456750   1st Qu.:1270   1st Qu.: 4080  
##  Median : 567000   Median :1710   Median : 5300  
##  Mean   : 619900   Mean   :1835   Mean   : 5444  
##  3rd Qu.: 719000   3rd Qu.:2285   3rd Qu.: 6380  
##  Max.   :2300000   Max.   :5770   Max.   :30122
```

It appears that the variables may be right-skewed. Let's compute the skewness:

```
apply(houses_98115, 2, e1071::skewness)
```

```
##       price sqft_living    sqft_lot 
##   2.0482534   0.9114968   2.9309820
```

DOI: 10.1201/9781003004899-13

The variables price and sqft_lot are right-skewed and the variable sqft_living is moderately right-skewed. Let's look at histograms.

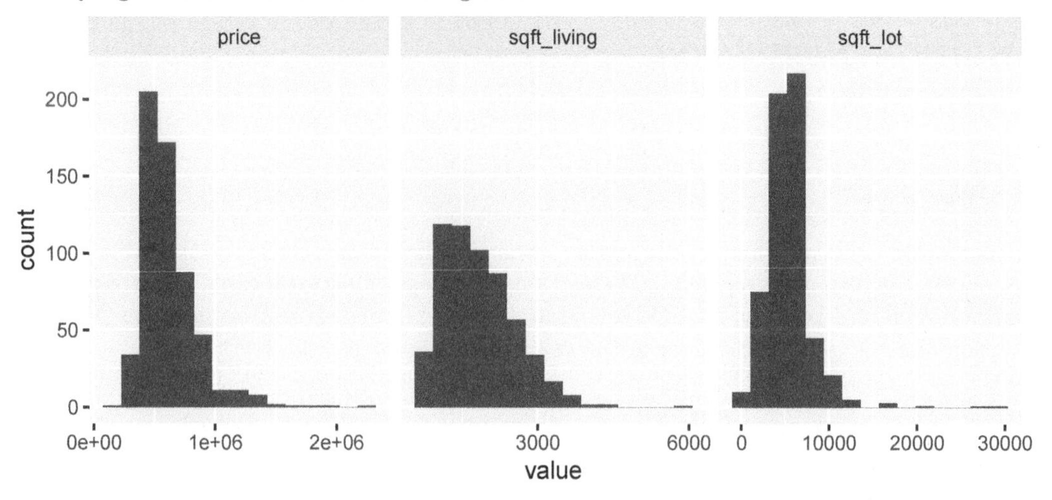


```
houses_98115 %>%
  tidyr::pivot_longer(cols = c(price, sqft_living, sqft_lot)) %>%
  ggplot(aes(value)) +
  geom_histogram(bins = 15) +
  facet_wrap(~name, scales = "free_x")
```

Now let's make two scatterplots.

```
ggplot(houses_98115, aes(x = sqft_lot, y = price)) +
  geom_point()
ggplot(houses_98115, aes(x = sqft_living, y = price)) +
  geom_point()
```

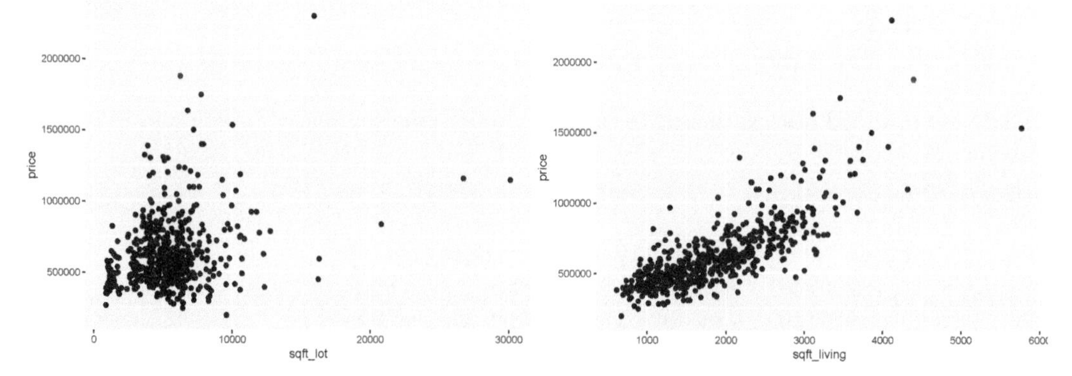


Looking at the two plots separately, it appears that the variance of price is not constant across the explanatory variables. Combining this with the fact that the variables are skewed, we take logs of all three variables:

```
log_houses <- fosdata::houses %>%
  filter(zipcode == "98115") %>%
  mutate(
    log_price = log(price),
    log_sqft_living = log(sqft_living),
    log_sqft_lot = log(sqft_lot)
```

```
  ) %>%
  select(matches("^log"))
```

The scatterplots are improved, though log of price versus log of lot size does not look great.

```
ggplot(log_houses, aes(x = log_sqft_lot, y = log_price)) +
  geom_point()
ggplot(log_houses, aes(x = log_sqft_living, y = log_price)) +
  geom_point()
```

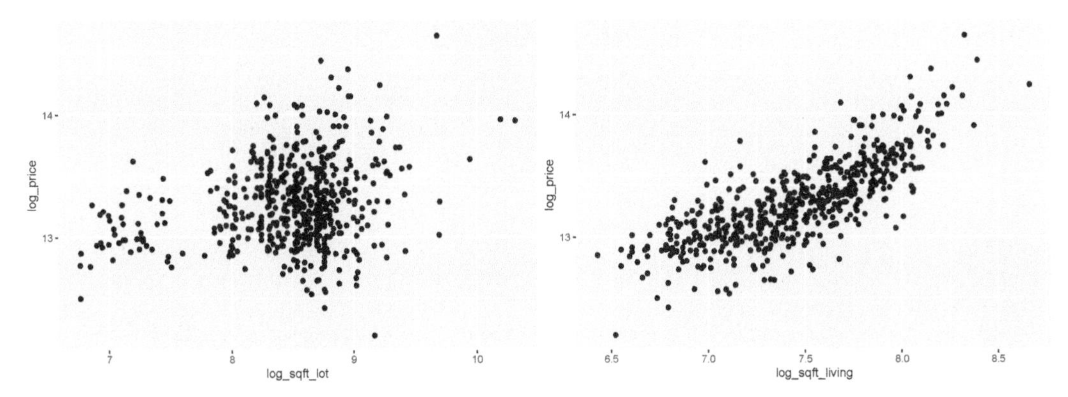


We continue by modeling the log of price on the log of the two explanatory variables. With only two explanatory variables, it is still possible to visualize the relationship. For scatterplots in three dimensions, we recommend `plotly`. We will not be utilizing all of the features of `plotly`, so we present the code that produces Figure 13.1 without fully explaining the syntax.

```
plotly::plot_ly(log_houses,
  x = ~log_sqft_living,
  y = ~log_sqft_lot,
  z = ~log_price,
  type = "scatter3d",
  marker = list(size = 2)
)
```

Plots made with `plotly` are interactive, so we recommend running the code yourself so that you can rotate and view the scatterplot. With a full view of the data, it seems reasonable that a **plane** could model the relationship between $x$, $y$, and $z$. How would we find the equation of the plane that is the best fit? We write the equation of the plane as $z = b_0 + b_1x + b_2y$ for some choices of $b_0, b_1$, and $b_2$. For each $b_0, b_1$, and $b_2$ we can predict the values of $z$ at the values of the explanatory variables by computing $\hat{z}_i = b_0 + b_1x_i + b_2y_i$. The associated error is $\hat{z}_i - z_i$, and we find the values of $b_0, b_1$, and $b_2$ that minimize the sum of squares of the error. We denote the values that minimize the SSE as $\hat{\beta}_0$, $\hat{\beta}_1$, and $\hat{\beta}_2$.

**Model**: Our model is

$$z_i = \beta_0 + \beta_1x_i + \beta_2y_i + \epsilon_i$$

where $\beta_0, \beta_1$, and $\beta_2$ are unknown parameters, and $\epsilon_i$ are independent, identically distributed normal random variables. The $i$th response is given by $z_i$, and the $i$th predictors are $x_i$ and $y_i$.

For example, if $b_0 = b_1 = b_2 = 1$, then we can compute the sum of squared error via

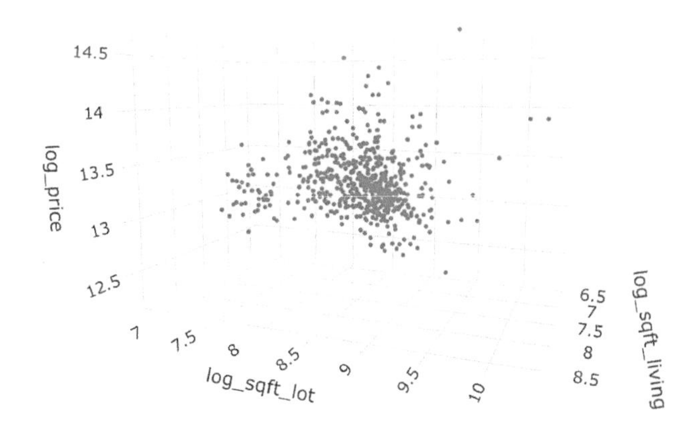


**FIGURE 13.1** Three-dimensional plot created with plotly.

```
b_0 <- 1
b_1 <- 1
b_2 <- 1
log_houses %>%
  mutate(
    z_hat = b_0 + b_1 * log_sqft_living + b_2 * log_sqft_lot,
    resid = log_price - z_hat
  ) %>%
  summarize(SSE = sum(resid^2))
```

```
##         SSE
## 1 7961.38
```

> **Try It Yourself.**
> By hand, adjust the values of $b_0, b_1$, and $b_2$ in the code above to try and minimize the SSE.

We can find the values of $b_0, b_1$, and $b_2$ that minimize the SSE using `lm`. The format is very similar to the single explanatory variable case.

```
price_mod <- lm(log_price ~ log_sqft_living + log_sqft_lot,
  data = log_houses
)
summary(price_mod)
```

```
##
## Call:
## lm(formula = log_price ~ log_sqft_living + log_sqft_lot, data = log_houses)
##
```

```
## Residuals:
##      Min       1Q   Median       3Q      Max
## -0.62790 -0.12085 -0.00189  0.12904  0.75567
##
## Coefficients:
##                 Estimate Std. Error t value Pr(>|t|)
## (Intercept)      8.14199    0.19922   40.87   <2e-16 ***
## log_sqft_living  0.65130    0.02303   28.29   <2e-16 ***
## log_sqft_lot     0.03422    0.01755    1.95   0.0517 .
## ---
## Signif. codes:  0 '***' 0.001 '**' 0.01 '*' 0.05 '.' 0.1 ' ' 1
##
## Residual standard error: 0.211 on 580 degrees of freedom
## Multiple R-squared:  0.6021, Adjusted R-squared:  0.6007
## F-statistic: 438.8 on 2 and 580 DF,  p-value: < 2.2e-16
```

We ignore most of the output for now and focus on the estimates of the coefficients. Namely, $\hat{\beta}_0 = 8.14199$, $\hat{\beta}_1 = 0.65130$, and $\hat{\beta}_2 = 0.03422$.

For comparison, these coefficients lead to a SSE of 25.8:

```
beta_0 <- 8.14199
beta_1 <- 0.65130
beta_2 <- 0.03422
log_houses %>%
  mutate(
    z_hat = beta_0 + beta_1 * log_sqft_living + beta_2 * log_sqft_lot,
    resid = log_price - z_hat
  ) %>%
  summarize(SSE = sum(resid^2))
```

```
##       SSE
## 1 25.8307
```

This value is **much** smaller than the SSE we got from our first guess of $b_0 = b_1 = b_2 = 1$!

> **Try It Yourself.**
> Compute the SSE with $b_0 = 8.15, b_1 = 0.65$, and $b_2 = 0.034$ and confirm that the SSE increases.

Let's use `plotly` to visualize the regression plane. We include the code for completeness, but do not explain the details of how to do this. Plotly produces interactive plots that do not render well on paper, so we recommend running this code in R to see the results.

```
# Compute a grid of x,y values for the plane
x <- seq(min(log_houses$log_sqft_living),
  max(log_houses$log_sqft_living),
  length.out = 10
)
y <- seq(min(log_houses$log_sqft_lot),
  max(log_houses$log_sqft_lot),
  length.out = 10
)
grid <- expand.grid(x, y) %>%
  rename(log_sqft_living = Var1, log_sqft_lot = Var2) %>%
```

```
  select(matches("^log")) %>%
  modelr::add_predictions(price_mod)
z <- reshape2::acast(grid, log_sqft_lot ~ log_sqft_living,
  value.var = "pred"
)

plotly::plot_ly(
  data = log_houses,
  x = ~log_sqft_living,
  y = ~log_sqft_lot,
  z = ~log_price,
  type = "scatter3d",
  marker = list(size = 2, color = "black")
) %>%
  plotly::add_trace(
    x = ~x,
    y = ~y,
    z = ~z,
    type = "surface"
  )
```

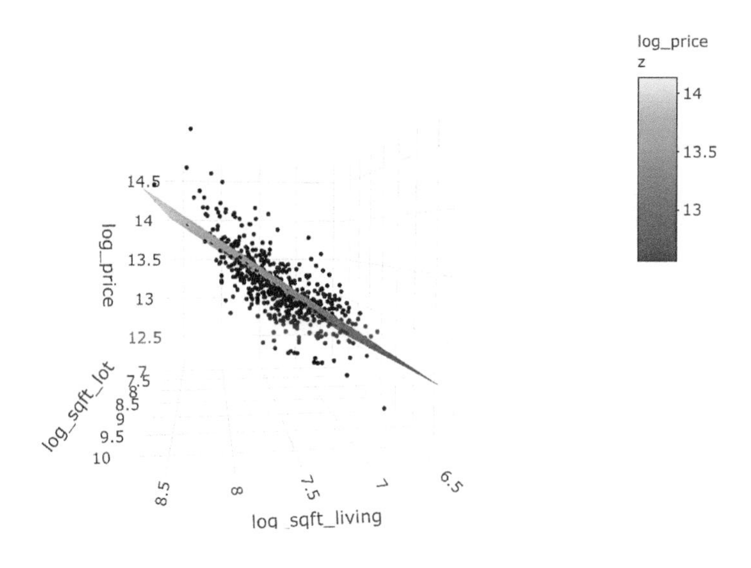


Let's now examine the summary of lm when there are two explanatory variables.

```
summary(price_mod)
```

```
##
## Call:
## lm(formula = log_price ~ log_sqft_living + log_sqft_lot, data = log_houses)
##
## Residuals:
##      Min       1Q   Median       3Q      Max
```

```
## -0.62790 -0.12085 -0.00189  0.12904  0.75567
##
## Coefficients:
##                 Estimate Std. Error t value Pr(>|t|)
## (Intercept)      8.14199    0.19922   40.87   <2e-16 ***
## log_sqft_living  0.65130    0.02303   28.29   <2e-16 ***
## log_sqft_lot     0.03422    0.01755    1.95   0.0517 .
## ---
## Signif. codes:  0 '***' 0.001 '**' 0.01 '*' 0.05 '.' 0.1 ' ' 1
##
## Residual standard error: 0.211 on 580 degrees of freedom
## Multiple R-squared:  0.6021, Adjusted R-squared:  0.6007
## F-statistic: 438.8 on 2 and 580 DF,  p-value: < 2.2e-16
```

We proceed through the summary line-by-line to explain what each value represents.

### Call

This gives the call to `lm` that was used to create the model, in this case

```
lm(formula = log_price ~ log_sqft_living + log_sqft_lot, data = log_houses)
```

Having the call reproduced can be useful when you are comparing multiple models, or when a function creates a model for you without your typing it in directly.

### Residuals

This gives the values of the estimated value of the response minus the actual value of the response. Our assumption is that $Z = \beta_0 + \beta_1 X + \beta_2 Y + \epsilon$, where $\epsilon$ is normal. In this case, we should expect that the residuals are symmetric about 0, so we would like to see that the median is approximately 0, the 1st and 3rd quartiles are about equal in absolute value, and the min and max are about equal in absolute value.

### Coefficients

This gives the estimate for the coefficients, together with the standard error, $t$-value for use in a $t$-test of significance, and $p$-value. The $p$-value in row $i$ is for the test of $H_0 : \beta_i = 0$ versus $H_a : \beta_i \neq 0$. In this instance, we reject the null hypotheses that the slope associated with `log_sqft_living` is zero at the $\alpha = .05$ level, and fail to reject that the slope of `log_sqft_lot` is zero at the $\alpha = .05$ level. The $\epsilon_i$ must be normal in order for these $p$-values to be accurate.

### Residual standard error

This can be computed from the residuals as follows:

```
N <- nrow(log_houses)
sqrt(sum(residuals(price_mod)^2) / (N - 3))
```

```
## [1] 0.2110348
```

Recall our rule of thumb that we divide by the sample size minus the number of estimated parameters when estimating $\sigma$. In this case, there are three estimated parameters. The degrees of freedom is just the number of samples minus the number of parameters estimated.

#### Multiple R-squared

This can be interpreted as the percent of variance in the response that is explained by the model. It can be computed as follows:

```
var_response <- var(log_houses$log_price)
var_residuals <- var(residuals(price_mod))
(var_response - var_residuals) / var_response
```

```
## [1] 0.6020751
```

The Adjusted $R^2$ uses the residual standard error (that has been adjusted for the number of parameters) rather than the raw variance of the residuals.

```
(var_response - sum(residuals(price_mod)^2) / (N - 3)) / (var_response)
```

```
## [1] 0.600703
```

The multiple $R^2$ value will increase with the addition of more variables, where the adjusted $R^2$ has a penalty for adding variables, so can either increase or decrease when a new variable is added.

#### F statistic

The test statistic which is used to test $H_0 : \beta_1 = \beta_2 = \cdots = 0$ versus not all of the slopes are 0. Note this does not include a test for the intercept. The $\epsilon_i$ must be normal for this to be accurate. We compute the value via

```
ssresid <- sum(residuals(price_mod)^2)
ssresponse <- sum((log_houses$log_price - mean(log_houses$log_price))^2)
(ssresponse - ssresid) / 2 / (ssresid / (N - 3))
```

```
## [1] 438.7808
```

## 13.2 Categorical variables

This section details how R incorporates categorical variables into linear models, expanding on the brief discussion in Section 12.1.1. Let's expand the ZIP codes in our model to include a ZIP code in the suburbs, which may have different characteristics. These are the same ZIP codes we used for data visualization in Section 7.3.1.

```
houses_2zips <- filter(fosdata::houses, zipcode %in% c("98038", "98115"))
```

We again take the log of `price`, `sqft_lot` and `sqft_living`.

```
houses_2zips <- mutate(houses_2zips,
  log_sqft_lot = log(sqft_lot),
  log_sqft_living = log(sqft_living),
  log_price = log(price)
)
houses_2zips$zipcode <- factor(houses_2zips$zipcode)
```

It would be a mistake to treat **zipcode** as a numeric variable! It is converted to a factor instead.

**FIGURE 13.2** House price as a function of living space and lot size in two Seattle area ZIP codes (all variables logged).

Figure 13.2 shows the dependence of price on living space, lot size, and zip code.

### 13.2.1 Equal slopes model

From Figure 13.2, it appears that 98038 and 98115 may have similar slopes for `log_price` when modeled by `log_sqft_living` and `log_sqft_lot`, although changing ZIP code causes a vertical shift. Based on that, it is reasonable to model the price by two parallel planes, one for each ZIP code. To do this with `lm`, we include all three explanatory variables on the right-hand side of the formula.

```
house_modE <- lm(log_price ~ log_sqft_living + log_sqft_lot + zipcode,
  data = houses_2zips
)
summary(house_modE)
```

```
##
## Call:
## lm(formula = log_price ~ log_sqft_living + log_sqft_lot + zipcode,
##     data = houses_2zips)
##
## Residuals:
##      Min       1Q   Median       3Q      Max
## -0.96278 -0.09510  0.00084  0.09298  0.68666
##
```

```
## Coefficients:
##                  Estimate Std. Error t value Pr(>|t|)
## (Intercept)       6.82806    0.12150   56.20   <2e-16 ***
## log_sqft_living  0.67827    0.01562   43.43   <2e-16 ***
## log_sqft_lot     0.08329    0.00659   12.64   <2e-16 ***
## zipcode98115     0.69684    0.01182   58.97   <2e-16 ***
## ---
## Signif. codes:  0 '***' 0.001 '**' 0.01 '*' 0.05 '.' 0.1 ' ' 1
##
## Residual standard error: 0.1857 on 1169 degrees of freedom
## Multiple R-squared:  0.7999, Adjusted R-squared:  0.7994
## F-statistic:  1558 on 3 and 1169 DF,  p-value: < 2.2e-16
```

The equation to predict the log of price comes from the coefficients of the model:

$$\widehat{\log(\text{price})} = 6.83 + 0.678 \cdot \log(\text{sqft\_living}) + 0.083 \cdot \log(\text{sqft\_lot}) + 0.697 \cdot \text{zipcode98115}.$$

The `zipcode98115` variable is a dummy variable that takes two values. It is 1 when the ZIP code is 98115 and otherwise 0. Both ZIP codes have the same coefficients on the living and lot variables, but there is an additive adjustment when the house is in ZIP code 98115 rather than in 98038. This number, 0.697, could be interpreted as the amount that log(price) increases between ZIP code 98038 and 98115. Exponentiation turns this additive factor into a multiplicative factor. Since $e^{0.697} \approx 2$, properties in 98115 are worth about twice as much as comparably sized properties in 98038.

The ZIP code 98038 is the "base" level of the `zipcode` variable. The base level is whichever level appears first when listing the levels of a factor, for example with `levels(houses_2zips$zipcode)`. If you want to change the base level to 98115, reorder the factor levels so 98115 comes first:

```
houses_2zips <- houses_2zips %>%
  mutate(zipcode = factor(zipcode, levels = c("98115", "98038")))
house_modE <- lm(log_price ~ log_sqft_living + log_sqft_lot + zipcode,
  data = houses_2zips
)
summary(house_modE)$coefficients
```

```
##                    Estimate Std. Error   t value      Pr(>|t|)
## (Intercept)      7.52489382 0.11779971  63.87871  0.000000e+00
## log_sqft_living  0.67827046 0.01561607  43.43414 3.778832e-246
## log_sqft_lot     0.08329213 0.00659035  12.63850  2.059559e-34
## zipcode98038    -0.69683562 0.01181683 -58.96974  0.000000e+00
```

### 13.2.2 Interaction terms

Although the slopes of the lines in Figure 13.2 appear similar, a more accurate model of house prices might include variable slopes for each ZIP code. Adding *interaction terms* to the model allows the coefficients to vary with ZIP code. Interaction terms involve two variables, separated by a colon (`:`) character. In the case of house prices, we add `log_sqft_living:zipcode` and `log_sqft_lot:zipcode` to the model.

```
house_modV <- lm(log_price ~ log_sqft_living + log_sqft_lot + zipcode +
  log_sqft_living:zipcode + log_sqft_lot:zipcode,
data = houses_2zips
)
summary(house_modV)
```

```
##
## Call:
## lm(formula = log_price ~ log_sqft_living + log_sqft_lot + zipcode + ...
## Residuals:
##      Min       1Q   Median       3Q      Max
## -0.94919 -0.09573  0.00371  0.09358  0.75567
##
## Coefficients:
##                             Estimate Std. Error t value Pr(>|t|)
## (Intercept)                  8.14199    0.17375  46.861  < 2e-16 ***
## log_sqft_living              0.65130    0.02008  32.433  < 2e-16 ***
## log_sqft_lot                 0.03422    0.01531   2.236 0.025558 *
## zipcode98038                -1.81575    0.25254  -7.190 1.16e-12 ***
## log_sqft_living:zipcode98038 0.08308    0.03175   2.617 0.008995 **
## log_sqft_lot:zipcode98038    0.05716    0.01695   3.373 0.000769 ***
## ---
## Signif. codes:  0 '***' 0.001 '**' 0.01 '*' 0.05 '.' 0.1 ' ' 1
##
## Residual standard error: 0.184 on 1167 degrees of freedom
## Multiple R-squared:  0.8038, Adjusted R-squared:  0.8029
## F-statistic:   956 on 5 and 1167 DF,  p-value: < 2.2e-16
```

From the summary[1], we see that there is a small but statistically significant difference between the linear relationship of log of price to the other variables in the two ZIP codes.

To estimate the log price of a house in ZIP code 98115 (recall we changed the base level), use

$$\widehat{\log(\text{price})} = 8.142 + 0.651 \cdot \log(\text{sqft\_living}) + 0.034 \cdot \log(\text{sqft\_lot}).$$

To estimate the expected value of the log of the price of a home in ZIP code 98038, we must adjust each coefficient by its interaction term. So

$$\widehat{\log(\text{price})} = 8.142 - 1.816 + (0.651 + 0.083) \cdot \log(\text{sqft\_living}) + (0.03422 + 0.05716) \cdot \log(\text{sqft\_lot}).$$

The `predict` function works on these models as well as on simple linear models. Let's check our results using `predict` for a house in ZIP code 98038 with `log_sqft_living = 8` and `log_sqft_lot = 9`.

```
predict(house_modV,
  newdata = data.frame(
    log_sqft_living = 8,
    log_sqft_lot = 9, zipcode = "98038"
  )
)
```

[1]Throughout this chapter, we truncate the output of `summary(mod)` in two ways. If the call is long, we only show the first line of the call. If the variable names are long enough to cause the Coefficients table to wrap, we truncate the variable names.

```
##        1
## 13.02381
```

```
log_sqft_living <- 8
log_sqft_lot <- 9
8.14199 - 1.81575 + (0.65130 + 0.08308) * log_sqft_living +
  (0.03422 + 0.05716) * log_sqft_lot # agrees with above
```

```
## [1] 13.0237
```

Finally, we note that `predict` has options for prediction intervals and confidence intervals which work for multiple linear regression as well. If we want a 95% prediction interval for the log price of a house with `log_sqft_living = 8` and `log_sqft_lot = 9`, we see that it is $[12.66, 13.39]$, which corresponds roughly to $[315000, 651000]$ in the non-log scale.

```
predict(house_modV,
  newdata = data.frame(
    log_sqft_living = 8,
    log_sqft_lot = 9, zipcode = "98038"
  ),
  interval = "prediction"
) %>% exp()
```

```
##        fit      lwr      upr
## 1 453072.1 315496.9 650638.2
```

### 13.2.3 Cross validation

We have seen two models of housing price on living space, lot size, and ZIP code. One model assumed equal slopes, the other included interaction terms. This choice of models is another example of the bias-variance tradeoff introduced in Section 11.8. Interaction terms decrease the bias of the model (by fitting the data better) but come at a cost of additional variance. In this section we apply leave one out cross validation to compare the predictive value of an equal slopes model to a model including interaction terms in a new data set.

**Example 13.1.** The data set `hot_dogs` is in the `fosdata` package.

```
hot_dogs <- fosdata::hot_dogs
```

This data comes from *Consumer Reports* and describes the sodium and calorie content various brands of hot dogs. The categorical variable `type` is one of Beef, Meat, or Poultry.

Figure 13.3 shows sodium content as explained by calories and type of hot dog.

The left plot in Figure 13.3 is an equal slopes model, generated by

```
hd_Eslope <- lm(sodium ~ calories + type, data = hot_dogs)
```

The right plot shows a variable slopes model, which includes the interaction term `calories:type`. A shortcut to include all interaction terms is to use `*` rather than `+` between the explanatory variables:

```
hd_Vslope <- lm(sodium ~ calories * type, data = hot_dogs)
summary(hd_Vslope)
```

```
##
```

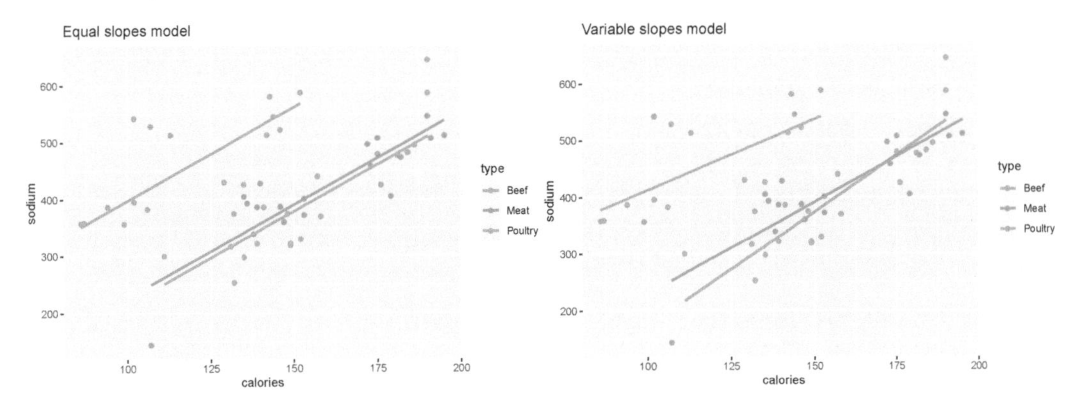


**FIGURE 13.3** Two models of hot dog sodium content.

```
## Call:
## lm(formula = sodium ~ calories * type, data = hot_dogs)
##
## Residuals:
##      Min       1Q   Median       3Q      Max
## -116.916  -28.180   -8.961   35.798  124.694
##
## Coefficients:
##                       Estimate Std. Error t value Pr(>|t|)
## (Intercept)          -228.3313    87.5770  -2.607  0.01213 *
## calories                4.0133     0.5529   7.259 2.96e-09 ***
## typeMeat              137.1460   123.3106   1.112  0.27159
## typePoultry           391.9615   114.0463   3.437  0.00122 **
## calories:typeMeat      -0.8016     0.7733  -1.037  0.30511
## calories:typePoultry   -1.5263     0.8195  -1.862  0.06868 .
## ---
## Signif. codes:  0 '***' 0.001 '**' 0.01 '*' 0.05 '.' 0.1 ' ' 1
##
## Residual standard error: 54.57 on 48 degrees of freedom
## Multiple R-squared:  0.7065, Adjusted R-squared:  0.6759
## F-statistic: 23.11 on 5 and 48 DF,  p-value: 9.698e-12
```

Observe that the `type` variable has base level Beef, and that there is a 0/1 valued variable for each of the other two levels, Meat and Poultry. From the summary, we see that the interaction terms are not significant at the 0.05 level.

Now we perform LOOCV on the two models. For the equal slopes model:

```
loo_Eslope <- function(k) {
  train <- hot_dogs[-k, ]
  test <- hot_dogs[k, ]
  hd_Eslope <- lm(sodium ~ calories + type, data = train)
  test$sodium - predict(hd_Eslope, test)
}
errs_E <- sapply(1:nrow(hot_dogs), loo_Eslope)
mean(errs_E^2)
```

```
## [1] 3345.211
```

For the variable slopes model:

```
loo_Vslope <- function(k) {
  train <- hot_dogs[-k, ]
  test <- hot_dogs[k, ]
  hd_Vslope <- lm(sodium ~ calories * type, data = train)
  test$sodium - predict(hd_Vslope, test)
}
errs_V <- sapply(1:nrow(hot_dogs), loo_Vslope)
mean(errs_V^2)
```

```
## [1] 3562.832
```

For this data set, LOOCV prefers the equal slopes model, whose SSE is 3345, about 217 smaller than the SSE for the variable slopes model.

## 13.3 Variable selection

One of the goals of linear regression is to be able to choose a *parsimonious* set of explanatory variables: a small set of variables that have great explanatory power. The problem of choosing a parsimonious set is one of *variable selection*, and there are many ways to approach this problem.

This section presents one possible approach to variable selection, *stepwise regression* using $p$-values. The method proceeds as follows:

1. Organize the variables that naturally belong together into groups, if possible.
2. Remove the non-significant variables in a group in decreasing order of $p$-value. Check after each removal to see whether the significance of other variables has changed. Keep categorical variables if any of the levels of the variable are significant.
3. Once all remaining variables in the group are significant, use `anova` to test whether the collection of removed variables was significant. If so, test one-by-one to try to find one to put back in the model.
4. Move to the next group of variables and continue until all variables in the model are significant.

The rest of this section illustrates the process with an extended example.

The data set `fosdata::conversation` is adapted from a psychological study by Manson et. al.[2] This study had 210 university students participate in a 10-minute conversation in small groups. After the conversation, the students were paired up to play a game known as the "prisoner's dilemma." The participants also provided various demographic information, and took a test to indicate whether they suffered from subclinical psychopathy. The goal of the study was to determine whether and/or how conversational behavior depends on demographic data.

To illustrate variable selection, we are going to model the `proportion_words` variable on

[2] Joseph H Manson et al., "Subclinical Primary Psychopathy, but Not Physical Formidability or Attractiveness, Predicts Conversational Dominance in a Zero-Acquaintance Situation." *PLOS One* 9, no. 11 (2014): e113135.

the other variables in the data set. The `proportion_words` variable gives the proportion of words spoken by that student among all words spoken by everyone during the conversation.

> **Try It Yourself.**
> Read the help page for `conversation`. Use the `summary` command to get an overview of all variables in the data set at once. Observe that most of the numeric variables have been standardized so as to have approximately mean 0 and standard deviation 1.

Many of the variables in this data refer to the prisoner's dilemma portion of the experiment. That part of the experiment happened **after** the conversations, so we will omit them from our model. Those variables include `camera` and eight more that all begin with the string `indiv`. The variable `oldest` has 72 missing values and is poorly documented, so we remove it as well. Also as part of this data cleanup, we convert `gender` to a factor.

```
converse <- fosdata::conversation %>%
  select(!camera) %>%
  select(!matches("^indiv")) %>%
  select(!oldest) %>%
  mutate(gender = factor(gender, levels = c(0, 1), labels = c("M", "F")))
```

This leaves 18 variables, consisting of 17 explanatory variables and our one response variable `proportion_words`. These 17 explanatory variables split naturally into two groups: demographic data and conversational data. The demographic data includes data on gender, appearance, psychopathy, the student's major, fighting ability (as estimated based on a picture), and more. The conversational data measures the number of interruptions, the number of times that a person started a new topic on a per word basis, and the proportion of times a person started a new topic.

Let's explore with some visualizations. We arbitrarily pick two of the explanatory variables to plot versus the response:

```
ggplot(converse, aes(x = f1_psychopathy, y = proportion_words)) +
  geom_point() +
  geom_smooth(method = "lm")
ggplot(converse, aes(x = interruptions_per_min, y = proportion_words)) +
  geom_point() +
  geom_smooth(method = "lm")
```

The results, in Figure 13.4, suggest that neither of the two variables have strong predictive ability for the response by themselves. Perhaps with all of the variables, things will look better:

```
mod <- lm(proportion_words ~ ., data = converse)
summary(mod)
```

```
##
## Call:
## lm(formula = proportion_words ~ ., data = converse)
##
## Residuals:
##       Min        1Q    Median        3Q       Max
## -0.183163 -0.040952  0.002491  0.044088  0.188120
##
## Coefficients:
```

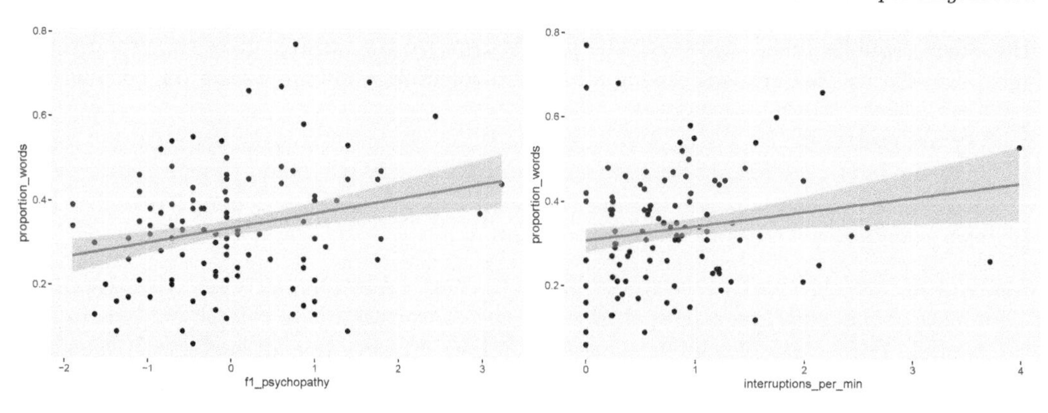


**FIGURE 13.4** Proportion of words spoken modeled on psychopathy (left) and interruptions per minute (right).

```
##                          Estimate Std. Error t value Pr(>|t|)
## (Intercept)             0.2959771  0.0305151   9.699  < 2e-16 ***
## genderF                 0.0051863  0.0118649   0.437  0.66252
## f1_psychopathy         -0.1579680  0.4577533  -0.345  0.73040
## f2_psychopathy         -0.0980854  0.2457922  -0.399  0.69029
## total_psychopathy       0.2237067  0.5963622   0.375  0.70799
## attractiveness         -0.0022099  0.0055057  -0.401  0.68859
## fighting_ability       -0.0286368  0.0086793  -3.299  0.00115 **
## strength                0.0186639  0.0085663   2.179  0.03057 *
## height                 -0.0057262  0.0055482  -1.032  0.30333
## median_income           0.0001271  0.0002090   0.608  0.54379
## highest_class_rank      0.0242887  0.0115059   2.111  0.03607 *
## major_presige          -0.0005630  0.0005683  -0.991  0.32313
## dyad_status_difference -0.0178477  0.0070620  -2.527  0.01230 *
## proportion_seque...rts  0.4855446  0.0409178  11.866  < 2e-16 ***
## interruptions_per_min   0.1076160  0.0144719   7.436 3.35e-12 ***
## sequence_starts_...100 -0.0493571  0.0061830  -7.983 1.28e-13 ***
## interruptions_pe...100 -0.0708835  0.0094431  -7.506 2.22e-12 ***
## affect_words           -0.0039201  0.0027227  -1.440  0.15155
## ---
## Signif. codes:  0 '***' 0.001 '**' 0.01 '*' 0.05 '.' 0.1 ' ' 1
##
## Residual standard error: 0.07303 on 192 degrees of freedom
## Multiple R-squared:  0.751,  Adjusted R-squared:  0.729
## F-statistic: 34.07 on 17 and 192 DF,  p-value: < 2.2e-16
```

A couple of things to notice. First, two of the other three conversation variables are significant when predicting the proportion of words spoken. Second, the multiple $R^2$ is .751, which means that about 75% of the variance in the proportion of words spoken is explained by the explanatory variables.

Let's start removing variables in groups. The three psychopathy variables have high $p$-values, so we begin with them, removing one at a time.

To remove variables from an existing model, use the `update` command. The `update` command takes the model as its first argument and the update we are making as its second. The second argument is given as a formula `~ . - variable_name`, which means we update by subtracting the named variable.

Rather than show the full model at each step, we show only selected variables and format them nicely (using `kable`).

```
mod2 <- update(mod, ~ . - f1_psychopathy)
```

| term | estimate | std.error | statistic | p.value |
|---|---|---|---|---|
| (Intercept) | 0.295 | 0.030 | 9.745 | 0.000 |
| f2_psychopathy | -0.013 | 0.008 | -1.656 | 0.099 |
| total_psychopathy | 0.018 | 0.009 | 2.102 | 0.037 |

```
summary(mod2)$r.squared
```

```
## [1] 0.7508613
```

```
summary(mod2)$adj.r.squared
```

```
## [1] 0.7302073
```

We see that after removing `f1_psychopathy`, the variables `total_psychopathy` and `f2_psychopathy` had much smaller $p$-values than before. It is common to see such large changes in significance when removing variables that are linked or correlated in some way. Also, note that the $R^2$ did not decrease much when we removed `f1_psychopathy`, and the adjusted $R^2$ increased from .729 to .730. These are good signs that it is OK to remove `f1_psychopathy`.

In the updated model, `total_psychopathy` has coefficient significantly different from 0 ($p =$ .037), while `f2_psychopathy` is not significant at the $\alpha = .05$ level. Since `f2_psychopathy` is weaker, we try removing it too:

```
mod3 <- update(mod2, ~ . - f2_psychopathy)
```

| term | estimate | std.error | statistic | p.value |
|---|---|---|---|---|
| (Intercept) | 0.296 | 0.030 | 9.731 | 0.000 |
| total_psychopathy | 0.007 | 0.006 | 1.293 | 0.198 |

```
summary(mod3)$r.squared
```

```
## [1] 0.7473232
```

```
summary(mod3)$adj.r.squared
```

```
## [1] 0.7277863
```

Here we get conflicting information about whether to remove `f2_psychopathy`. The variable was not significant at the $\alpha = .05$ level, but the adjusted $R^2$ decreased from .730 to .728 when we removed it.

Different variable selection techniques would make different choices at this point. The $p$-value technique determines that `total_psychopathy` is not significant ($p = 0.198$) so it should be removed.

```
mod4 <- update(mod3, ~ . - total_psychopathy)
```

Here is the full model summary after removing all three psychopathy variables.

```
summary(mod4)
```

```
##
## Call:
## lm(formula = proportion_words ~ gender + attractiveness + fighting_a...
## Residuals:
##       Min        1Q    Median        3Q       Max
## -0.180736 -0.038420 -0.000994  0.043020  0.202987
##
## Coefficients:
##                          Estimate Std. Error t value Pr(>|t|)
## (Intercept)             0.2933090  0.0303800   9.655  < 2e-16 ***
## genderF                -0.0037965  0.0108581  -0.350 0.726983
## attractiveness         -0.0028329  0.0055096  -0.514 0.607706
## fighting_ability       -0.0304228  0.0083761  -3.632 0.000359 ***
## strength                0.0211761  0.0083132   2.547 0.011628 *
## height                 -0.0082203  0.0053578  -1.534 0.126585
## median_income           0.0001295  0.0002088   0.620 0.535722
## highest_class_rank      0.0244283  0.0114397   2.135 0.033976 *
## major_presige          -0.0004255  0.0005441  -0.782 0.435162
## dyad_status_difference -0.0153037  0.0069836  -2.191 0.029610 *
## proportion_seque...rts  0.5013595  0.0399347  12.554  < 2e-16 ***
## interruptions_per_min   0.1138530  0.0141768   8.031 8.97e-14 ***
## sequence_starts_...100 -0.0481660  0.0061777  -7.797 3.73e-13 ***
## interruptions_pe...100 -0.0731355  0.0093695  -7.806 3.53e-13 ***
## affect_words           -0.0050216  0.0026643  -1.885 0.060946 .
## ---
## Signif. codes:  0 '***' 0.001 '**' 0.01 '*' 0.05 '.' 0.1 ' ' 1
##
## Residual standard error: 0.07331 on 195 degrees of freedom
## Multiple R-squared:  0.7451, Adjusted R-squared:  0.7268
## F-statistic: 40.72 on 14 and 195 DF,  p-value: < 2.2e-16
```

The adjusted $R^2$ fell again, this time to .727. Now that we have removed all of the variables related to psychopathy, we can check to see whether **as a group** they are significant. That is, we will test $H_0$ (the coefficients of the psychopathy variables are all 0) versus $H_a$ (not all of the coefficients are zero).

```
anova(mod4, mod)
```

```
## Analysis of Variance Table
##
## Model 1: proportion_words ~ gender + attractiveness + fighting_ability...
## Model 2: proportion_words ~ gender + f1_psychopathy + f2_psychopathy +...
##   Res.Df    RSS Df Sum of Sq      F Pr(>F)
## 1    195 1.0481
## 2    192 1.0240  3  0.024138 1.5087 0.2136
```

With a $p$-value of .2136, we do not reject the null hypothesis that all of the coefficients are zero, and so we leave the psychopathy variables out of the model. If we had rejected the null hypothesis, then we would have needed to add back in one or more of the variables to the model. It is important when doing stepwise variable selection that you periodically check to see whether you need to add back in one or more of the variables that you removed.

Next, we begin removing variables that describe the person, such as gender and attractiveness. We step in order of decreasing $p$-values within that group of variables.

```
mod5 <- update(mod4, ~ . - gender)
mod6 <- update(mod5, ~ . - attractiveness)
mod7 <- update(mod6, ~ . - median_income)
summary(mod7)
```

```
##
## Call:
## lm(formula = proportion_words ~ fighting_ability + strength +   ...
## Residuals:
##       Min        1Q    Median        3Q       Max
## -0.181215 -0.037676 -0.000914  0.042230  0.202280
##
## Coefficients:
##                          Estimate Std. Error t value Pr(>|t|)
## (Intercept)             0.3024526  0.0214860  14.077  < 2e-16 ***
## fighting_ability       -0.0300948  0.0082729  -3.638 0.000351 ***
## strength                0.0207458  0.0082203   2.524 0.012397 *
## height                 -0.0080111  0.0052692  -1.520 0.130012
## highest_class_rank      0.0243218  0.0112871   2.155 0.032382 *
## major_presige          -0.0004917  0.0005239  -0.939 0.349084
## dyad_status_difference -0.0144526  0.0067093  -2.154 0.032439 *
## proportion_seque...rts  0.5008202  0.0393863  12.716  < 2e-16 ***
## interruptions_per_min   0.1127051  0.0138752   8.123 4.81e-14 ***
## sequence_starts_...100 -0.0488678  0.0060006  -8.144 4.22e-14 ***
## interruptions_pe...100 -0.0725000  0.0092657  -7.825 2.99e-13 ***
## affect_words           -0.0048069  0.0025190  -1.908 0.057807 .
## ---
## Signif. codes:  0 '***' 0.001 '**' 0.01 '*' 0.05 '.' 0.1 ' ' 1
##
## Residual standard error: 0.07289 on 198 degrees of freedom
## Multiple R-squared:  0.7442, Adjusted R-squared:   0.73
## F-statistic: 52.37 on 11 and 198 DF,  p-value: < 2.2e-16
```

```
anova(mod7, mod4)
```

```
## Analysis of Variance Table
##
## Model 1: proportion_words ~ fighting_ability + strength + height + hig...
## Model 2: proportion_words ~ gender + attractiveness + fighting_ability...
##   Res.Df    RSS Df Sum of Sq      F Pr(>F)
## 1    198 1.0520
## 2    195 1.0481  3 0.0038477 0.2386 0.8693
```

After removing those three variables, the adjusted $R^2$ is back up to 0.73, and ANOVA indicates it is not necessary to put any of the variables back in. We continue.

```
mod8 <- update(mod7, ~ . - major_presige)
mod9 <- update(mod8, ~ . - height)
mod10 <- update(mod9, ~ . - dyad_status_difference)
anova(mod10, mod7)
```

```
## Analysis of Variance Table
##
## Model 1: proportion_words ~ fighting_ability + strength + highest_clas...
## Model 2: proportion_words ~ fighting_ability + strength + height + hig...
##   Res.Df    RSS Df Sum of Sq      F  Pr(>F)
## 1    201 1.0893
## 2    198 1.0520  3  0.037357 2.3438 0.07427 .
## ---
## Signif. codes:  0 '***' 0.001 '**' 0.01 '*' 0.05 '.' 0.1 ' ' 1
```

```
summary(mod10)
```

```
##
## Call:
## lm(formula = proportion_words ~ fighting_ability + strength +    ...
## Residuals:
##       Min        1Q    Median        3Q       Max
## -0.192950 -0.041033  0.003062  0.042783  0.211376
##
## Coefficients:
##                           Estimate Std. Error t value Pr(>|t|)
## (Intercept)               0.295980   0.020301  14.580  < 2e-16 ***
## fighting_ability         -0.033053   0.008021  -4.121 5.52e-05 ***
## strength                  0.021673   0.008193   2.645  0.00881 **
## highest_class_rank        0.023024   0.011185   2.058  0.04084 *
## proportion_seque...rts    0.502863   0.039371  12.772  < 2e-16 ***
## interruptions_per_min     0.113354   0.013994   8.100 5.22e-14 ***
## sequence_starts_...100   -0.048002   0.005906  -8.128 4.41e-14 ***
## interruptions_pe...100   -0.072946   0.009305  -7.840 2.59e-13 ***
## affect_words             -0.005876   0.002418  -2.430  0.01597 *
## ---
## Signif. codes:  0 '***' 0.001 '**' 0.01 '*' 0.05 '.' 0.1 ' ' 1
##
## Residual standard error: 0.07362 on 201 degrees of freedom
## Multiple R-squared:  0.7351, Adjusted R-squared:  0.7246
## F-statistic: 69.73 on 8 and 201 DF,  p-value: < 2.2e-16
```

The adjusted $R^2$ has fallen back down to .7246, but that is not a big price to pay for removing so many variables from the model. ANOVA does not reject $H_0$ that all the coefficients of the variables that we removed are zero at the $\alpha = .05$ level. We are left with 8 significant explanatory variables.

Let's check the diagnostic plots.

```
plot(mod10)
```

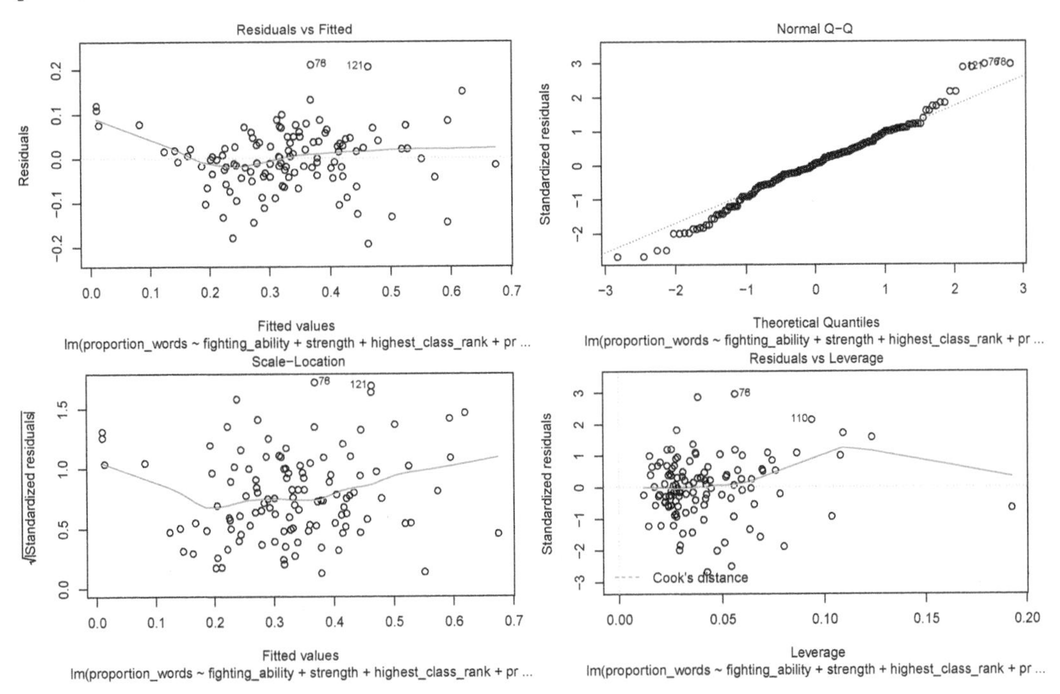


These look pretty good, except that there may be 3-4 values that don't seem to be following the rest of the trend. See for example the residuals versus fitted plot with the 4 values on the left.

Doing stepwise regression like this is not without its detractors. One issue is that the final model is very much dependent on the order in which we proceed. If we had removed `f2_psychopathy` first, how would that have affected things? Can we justify our decision to remove `f1_psychopathy` first? Certainly not. The resulting model that we obtained is simply **a** model for the response, not necessarily the best one. It is also very possible that the variables that we removed are also good predictors of the response. We should not make the conclusion that the other variables aren't related.

That being said, stepwise regression for variable selection is quite common in practice, which is why we include it in this book.

The technique we used above was based on $p$-values. One can also use other measures of goodness of fit of a model to perform stepwise regression. A common one is the Akaike Information Criterion (AIC). The AIC algorithm is nice because at each decision point, it tests what would happen if we add back in any of the variables that we have removed, as well as what would happen if we remove any of the variables that we currently have. We won't explain how AIC works, but we show how to use an implementation of it from the MASS package to find a parsimonious model

```
mod_aic <- MASS::stepAIC(mod, trace = FALSE)
summary(mod_aic)
```

```
##
## Call:
## lm(formula = proportion_words ~ f2_psychopathy + total_psychopathy +...
## Residuals:
```

```
##       Min        1Q    Median        3Q       Max
## -0.189974 -0.036895  0.000572  0.044182  0.190963
##
## Coefficients:
##                         Estimate Std. Error t value Pr(>|t|)
## (Intercept)             0.304026   0.020184  15.063  < 2e-16 ***
## f2_psychopathy         -0.013849   0.007641  -1.812 0.071434 .
## total_psychopathy       0.018827   0.007876   2.390 0.017773 *
## fighting_ability       -0.032058   0.008090  -3.963 0.000103 ***
## strength                0.020196   0.008139   2.481 0.013919 *
## highest_class_rank      0.024566   0.011126   2.208 0.028394 *
## dyad_status_difference -0.015754   0.006712  -2.347 0.019909 *
## proportion_seque...rts  0.479287   0.039657  12.086  < 2e-16 ***
## interruptions_per_min   0.106362   0.014115   7.536 1.70e-12 ***
## sequence_starts_...100 -0.047826   0.005855  -8.168 3.63e-14 ***
## interruptions_pe...100 -0.071390   0.009274  -7.698 6.42e-13 ***
## affect_words           -0.005436   0.002393  -2.272 0.024170 *
## ---
## Signif. codes:  0 '***' 0.001 '**' 0.01 '*' 0.05 '.' 0.1 ' ' 1
##
## Residual standard error: 0.07246 on 198 degrees of freedom
## Multiple R-squared:  0.7472, Adjusted R-squared:  0.7332
## F-statistic: 53.21 on 11 and 198 DF,  p-value: < 2.2e-16
```

In this case, the AIC derived model includes all of the variables that we included, plus `f2_psychopathy`, `total_psychopathy`, and `dyad_status_difference`. These variables do not all pass the $p$-value test, but notice that the adjusted $R^2$ of the AIC derived model is higher than the adjusted $R^2$ of the model that we chose using $p$-values. The two models have their pros and cons, and as is often the case, there is not just one final model that could make sense.

## Vignette: External data formats

By this point in the book, we hope that you are comfortable doing many tasks in R. We like to imagine that when someone gives you data in the future, your first thought will be to open it up in R. Unfortunately, not everyone is as civilized as we are, and they may have saved their data in a format that doesn't interact well with R. This vignette gives some guidance for what to do in that case.

As a general recommendation, never make modifications directly to the original data. That data should stay in the exact same state, and ideally, you would document every change and decision that you make while getting the data ready to be read into R. In particular, never write over the original data set.

### Text formats

Text data is data stored as a sequence of simple characters (ASCII or Unicode, usually). It is easy to inspect and manipulate with any text editor. On the other hand, it typically does not include formatting or metadata (information *about* the data), and may be larger in size than some compressed formats. Despite, or really because of, their simplicity, textual data formats have proven to be the best storage format for interoperability and resistance to loss over time. If you have important data to release to the world, do the world a favor and make it available in a text format.

The most popular textual format is "comma separated values," or CSV. CSV files have the `.csv` extension, and can be read and written with `read.csv` and `write.csv`:

```
# make some data
mydata <- data.frame(
  number = 1:10,
  letter = sample(c("A", "B", "C", "D"), 10, replace = TRUE)
)
# write it to a CSV
write.csv(mydata, file = "myfile.csv")
# read it back
mydata <- read.csv("myfile.csv")
```

Within a CSV file, each record is a row, and within that row data is separated into fields. The first row of a CSV file should contain the names of the fields, or variables. If it doesn't, specify that `header = FALSE` and R will assign generic variable names to your columns.

The default field separator is a comma, hence the C in CSV. You may specify a different separator to `read.csv`. For example, files with the `.tsv` extension are tab separated and need the `sep = "\t"` option.

Other text formats with records in rows and variables in columns can be read with `read.table`, and in fact `read.csv` is just a wrapper that eventually calls `read.table`. Another possible approach for reading text files is with `scan`, which reads typed data. The `readr` package, part of the tidyverse, is also a powerful tool.

JSON is a text based format that is growing in popularity. Unlike row and column formatted table data, JSON is hierarchical, allowing for complicated object structures and relationships. Since R works most naturally with rectangular data, dealing with JSON is inherently challenging. R does provide packages (`rjson`, `jsonlite`) that will read JSON formats, but some data manipulation will almost certainly be required after loading.

### Spreadsheets

Excel spreadsheet data is stored in files with `.xls` or `.xlsx` extensions. The `readxl` package provides `readxl::read_xls("your_file_name.xls", sheet = 1)` or `readxl::read_xlsx(your_file_name.xlsx, sheet = 1)` to import Excel data. Unfortunately, the `readxl` commands will only work well if the file was created by someone who understands the importance of rectangular data. Many, if not most, of the data that is stored in Excel format has data and metadata stored in the file itself. For example, cells might be merged for formatting, there may be 5 rows of headers describing the variables, or there might be cells performing computations – those aren't data!

Working with complex Excel spreadsheets inside of R is possible, but quite challenging. It is almost always easier to open the file in a spreadsheet program such as Google Sheets, Excel, or LibreOffice, and perform some pre-processing by hand. After cleaning the data in the spreadsheet, save it to a `.csv` file which can be read into R. Unfortunately, the by-hand nature of this process is not reproducible. At a minimum, keep a copy of the original Excel file.

### Binary formats

If the data file extension is .RData or .Rda, your file is likely an R Data file. This is the easiest case to deal with for an R user! The save command saves data into R Data files, and the load command loads them. Unlike most other file reading commands, you do not store the result of load into a variable. Instead, the load command directly adds one or more objects to your workspace. It returns a vector of the objects it created, so you know what load has changed in your environment.

```
save(mydata, file = "myfile.Rda")
what <- load("myfile.Rda")
what
```

```
## [1] "mydata"
```

Other statistical packages have their own specialized file formats. Unlike text data, without a system designed to interpret the format, you have little hope of getting at the data. Fortunately, R has packages that can read most specialized formats. The haven package is part of the tidyverse, and provides commands read_dat(), read_stata(), read_spss(), and read_sas(), among others. These read .dat (MATLAB), .dta (Stata), .sav (SPSS), and .sas7bdat (SAS) files. The haven package usually works well, though the format of the imported data can seem a bit odd.

For less common formats, either convert to a format that R can read, or look for a specialty package that can read your data type natively. Chances are, there is one out there.

## Exercises

Exercises 13.1 – 13.7 require material through Section 13.1.

**13.1.** Consider the adipose data set in the fosdata package. The goal is to estimate the visceral adipose tissue amount in patients, based on the other measurements. The authors of the study[3] did separate analyses for males and females; in this exercise, we consider only male subjects.

a. Model vat on waist_cm and stature_cm and examine the residuals.
b. Model log(vat) on waist_cm and stature_cm and examine the residuals.
c. Which model would you choose based on the residuals versus fitted plot?

**13.2.** Continuing from Exercise 13.1, in this exercise we consider only female patients.

a. The vat measurement of some females was compromised. Remove any female whose vat value is less than or equal to 5.
b. Model log(vat) on waist_cm and stature_cm, so log(vat) = beta0 + beta1 * waist_cm + beta2 * stature_cm.
c. What is the $p$-value for the test of $H_0 : \beta_1 = \beta_2 = 0$ versus the alternative that at least one is not zero?
d. What is the $p$-value for the test $H_0 : \beta_1 = 0$ versus $H_a : \beta_1 \neq 0$?
e. What is a 95% confidence interval for beta_2?

[3]Michelle G Swainson et al., "Prediction of Whole-Body Fat Percentage and Visceral Adipose Tissue Mass from Five Anthropometric Variables." *PLOS One* 12, no. 5 (2017): e0177175.

f. Find a 95% prediction interval for `vat` for a new patient who presents with `waist_cm = 70` and `stature_cm = 170`.

**13.3.** Exercise 11.25 showed a significant correlation between acorn size and geographic range for North American oak species. It is possible that tree size is a lurking variable that might explain both acorn size and geographic range. Data is in `fosdata::acorns`.

a. Only using Atlantic region species, perform a multiple regression modeling the log of geographic range on both tree height and the log of acorn size.
b. Which of the explanatory variables are significant?
c. If you remove tree size from the model, how much does $R^2$ change?

In their paper, Marcelo and Patterson[4] concluded that tree height is not a source of spurious correlation between acorn size and geographic range.

**13.4.** The data set `barnacles` in the `fosdata` package has measurements of barnacle density on coral reefs in two locations: the Flower Garden Beds in the Gulf of Mexico, and the U.S. Virgin Islands.

a. Use ggplot to plot barnacle density as a function of depth, color your points by location, and show the linear regression lines for the two locations.
b. The two lines appear parallel, so a parallel slope model is appropriate. Form a linear model of barnacle density on location and depth.
c. Predict the mean barnacle density at a depth of 30 meters separately for each location.
d. Explain the meaning of the `locationUSVI` coefficient in this model. What does $-270.92428$ mean?

**13.5.** The built-in data set `swiss` contains a standardized fertility measure and socio-economic indicators for each of 47 French-speaking provinces of Switzerland at about 1888.

a. Investigate the distribution of the `Catholic` variable. What do you observe? What does this tell you about Catholics and Protestants in 19th century Switzerland?
b. Produce a plot of Fertility as a function of Education. Use the `cut` function to divide the Catholic variable into three classes, and color your points with three colors for those three classes.
c. Form a linear model of Fertility on the other five variables. Which variables are significant at the 0.05 level?
d. Drop any variables that are not significant and make a new linear model. How does this change the adjusted $R^2$?

**13.6.** In this problem, we will use simulation to show that the $F$ statistic computed by `lm` follows the $F$ distribution.

a. Create a data frame where `x1` and `x2` are uniformly distributed on $[-2, 2]$, and $y = 2 + \epsilon$, where $\epsilon$ is a standard normal random variable.
b. Use `lm` to create a linear model of $y = \beta_0 + \beta_1 x_1 + \beta_2 x_2$.
c. Use `summary(mod)$fstatistic[1]` to pull out the $F$ statistic.
d. Put inside `replicate`, and create 2000 samples of the $F$ statistic (for different response variables).
e. Create a histogram of the $F$ statistic and compare to the pdf of an $F$ random variable with the correct degrees of freedom. (There should be 2 numerator degrees of freedom and $N$ - 3 denominator degrees of freedom.)

**13.7.** In this problem, we examine robustness to normality in `lm`.

[4] Aizen and Patterson III, "Acorn Size and Geographical Range in the North American Oaks (Quercus L.)."

a. Create a data frame of 20 observations where `x1` and `x2` are uniformly distributed on $[-2, 2]$, and $y = 1 + 2 * x_1 + \epsilon$, where $\epsilon$ is exponential with rate 1.
b. Use `lm` to create a linear model of $y = \beta_0 + \beta_1 x_1 + \beta_2 x_2$, and use `summary(mod)$coefficients[3,4]` to pull out the $p$-value associated with a test of $H_0 : \beta_2 = 0$ versus $H_a : \beta_2 \neq 0$.
c. Replicate the above code and estimate the proportion of times that the $p$-value is less than .05.
d. How far off is the value from what you would expect to get if the assumptions had been met?

---

Exercises 13.8 – 13.10 require material through Section 13.2.

**13.8.** The `penguins` data from the `palmerpenguins` package has body measurements of three species of penguin.

a. Make a scatterplot of `body_mass_g` as a function of `flipper_length_mm`, color by `species`, and show the regression lines from the variable slopes model (just use `geom_smooth` with `method=lm`).
b. Fit a linear model of `body_mass_g` on `flipper_length_mm`, `species`, and include the interaction between `flipper_length_mm` and `species`.
c. Which interaction terms are significant? Explain your answer based on the plot.

**13.9.** Continue using the `penguins` data. Use LOOCV to compare two models of penguin body mass on species and flipper length: the equal slopes model and the variable slopes model. Report the SSE for each, and decide which model has better predictive value.

**13.10.** Consider the data set `fosdata::fish`. This data set consists of the weight, species, and 5 other measurements of fish.

a. The `species` variable is coded as integer. Is this the appropriate coding for doing regression?
b. Find a linear model of `weight` on `species`, `length1` and the interaction between `species` and `length1`.
c. What is the expected weight of a fish that is species 3 and `length1` of 24.1?
d. What is the expected difference of weight between a fish of species 3 and `length1` of 23.1 and a fish of species 3 and `length1` of 22.1?

---

Exercises 13.11 – 13.17 require material through Section 13.3.

**13.11.** Consider again the `conversation` data set in the `fosdata` package. Suppose that we wanted to model the proportion of words spoken on variables that would be available **before** the conversation.

a. Find a parsimonious model of the proportion of words spoken on `gender`, `f1_psychopathy`, `f2_psychopathy`, `total_psychopathy`, `attractiveness`, `fighting_ability`, `strength`, `height`, `median_income`, `highest_class_rank`, `major_presige` and `dyad_status_difference`.
b. What is the $R^2$ value? Discuss.

**13.12.** Consider the `adipose` data set in the `fosdata` package.

a. Create a parsimonious model of the logarithm of `vat` on the other numeric variables for male patients. Which variables are kept?
b. What is the expected value of the logarithm of `vat` for a male patient who presents

with age = 20, ldl = 2, hdl = 1.5, trig = 0.6, glucose = 4.6, stature_cm = 170, waist_cm = 72, hips_cm = 76 and bmi = 19?

**13.13.** Consider the data set fosdata::cigs_small. The Federal Trade Commission tested 1200 brands of cigarettes in 2000 and reported the carbon monoxide, nicotine, and tar content along with other identifying variables. The data in cigs_small contains one randomly selected cigarette from each brand that made a "100" size cigarette in 2000. Find a parsimonious model of carbon monoxide content co on the variables nic, tar, pack and menthol.

**13.14.** Consider the data set fosdata::fish. This data set consists of the weight, species, and 5 other measurements of fish.

a. Find a parsimonious model of weight on the other variables. Note that there are many missing values for sex. Confirm that the residuals vs fitted plot shows a lot of curvature.
b. Since volume is length cubed, it might make sense to add the squares and cubes of the length variables. Add them and then find a parsimonious model for weight on the variables. If you include a length variable to a power, you should include the lower powers as well. Are the residuals better?

**13.15.** Consider the data set tlc in the ISwR package. Model the variable tlc inside the data set tlc on the other variables. Include plots and examine the residuals.

**13.16.** Consider the cystfibr data set in the ISwR package. Find a parsimonious linear model of the variable pemax on the other variables. Be sure to read about the variables so that you can guess which variables might be grouped together.

**13.17.** Consider the seoulweather data set in the fosdata package. We wish to model the **error** in the next day forecast max temperature on the variables that we knew on the present day. (That is, all variables except for next_tmax and next_tmin. Also remove date, latitude, longitude, slope, and dem from the model, as they are confounded with station.)

a. Why is the assumption of independence of the error unlikely to be true if we model the error on all of the other variables?
b. Restrict to **only** station 1, and explain why this would remove the most obvious source of dependence.
c. Find a parsimonious model of the error on the other variables.
d. What percentage of the error in the next day prediction is explained by your model?

# *Image Credits*

Section 2.2. [Candy] *Evan Amos.* Public domain https://commons.wikimedia.org/wiki/File:Plain-M&Ms-Pile.jpg . . . . . . . . . . . . . . . . . . . . 40

Section 3.1. [Lynx kitten] **Figure 3.1**. *Bernard Landgraf.* This file is licensed under the Creative Commons Attribution-Share Alike 3.0 Unported license. https://commons.wikimedia.org/wiki/File:Lynx_kitten.jpg 58

Section 3.6. [Cloud chamber] **Figure 3.5**. *Julien Simon.* This file is licensed under the Creative Commons Attribution-Share Alike 3.0 Unported license. https://commons.wikimedia.org/wiki/File:Radioactivity_of_a_Thorite_mineral_seen_in_a_cloud_chamber.jpg . . . . . . . . . . . 77

Section 7.2. [Bechdel Test] *Branson Reese.* Used by permission. https://www.branson-reese.com/comics . . . . . . . . . . . . . . . . . . . . . . . . . 208

Section 8.0. [Brake pedal setup] **Figure 8.1**. *Hasegawa et al.* This is an open access article distributed under the terms of the Creative Commons Attribution License. https://journals.plos.org/plosone/article?id=10.1371/journal.pone.0236053 . . . . . . . . . . . . . . . . . . . . . . . 253

Section 8.6. [Apples on tree] *George Chernilevsky.* Public domain. https://commons.wikimedia.org/wiki/File:Apples_on_tree_2011_G1.jpg . . . . . 276

Section 8.8. [Weight estimate] *Sciutti, et al.*. This is an open access article distributed under the terms of the Creative Commons Attribution License. https://journals.plos.org/plosone/article?id=10.1371/journal.pone.0224979 . . . . . . . . . . . . . . . . . . . . . . . . . . . . . . 291

Section 9.0. [Exhaled air sampler] **Figure 9.1**. *Milton et al.* This is an open-access article distributed under the terms of the Creative Commons Attribution License. https://www.ncbi.nlm.nih.gov/pmc/articles/PMC3591312/figure/ppat-1003205-g002/ . . . . . . . . . . . . . . . . . . 305

Section 10.6. [Social media table] *Kahle et al.*. This is an open access article distributed under the terms of the Creative Commons Attribution License. https://journals.plos.org/plosone/article?id=10.1371/journal.pone.0156409 . . . . . . . . . . . . . . . . . . . . . . . . . . . . . . 365

Section 10.6. [Road signage] **Figure 10.6**. *Hess and Peterson.* This is an open access article distributed under the terms of the Creative Commons Attribution License. https://journals.plos.org/plosone/article?id=10.1371/journal.pone.0136973 . . . . . . . . . . . . . . . . . . . . . . . 370

Section 11.0. [Cuvettes] *Rainis Venta.* This file is licensed under the Creative Commons Attribution-Share Alike 3.0 Unported license. https://commons.wikimedia.org/wiki/File:Kolorimeetriline_valkude_sisalduse_m~o~otmine_Bradford_meetodil..JPG . . . . . . . . . . . . . . 371

Section 11.1. [Penguins] *Allison Horst.* CC0-1.0 Licensed. https://github.com/allisonhorst/palmerpenguins . . . . . . . . . . . . . . . . . . . . . . . . 375

# Index

# *Index of Data Sets and Packages*